Sustainable Catalysis for Biorefineries

Green Chemistry Series

Titles in the series:

1: The Future of Glycerol: New Uses of a Versatile Raw Material
2: Alternative Solvents for Green Chemistry
3: Eco-Friendly Synthesis of Fine Chemicals
4: Sustainable Solutions for Modern Economies
5: Chemical Reactions and Processes under Flow Conditions
6: Radical Reactions in Aqueous Media
7: Aqueous Microwave Chemistry
8: The Future of Glycerol: 2nd Edition
9: Transportation Biofuels: Novel Pathways for the Production of Ethanol, Biogas and Biodiesel
10: Alternatives to Conventional Food Processing
11: Green Trends in Insect Control
12: A Handbook of Applied Biopolymer Technology: Synthesis, Degradation and Applications
13: Challenges in Green Analytical Chemistry
14: Advanced Oil Crop Biorefineries
15: Enantioselective Homogeneous Supported Catalysis
16: Natural Polymers Volume 1: Composites
17: Natural Polymers Volume 2: Nanocomposites
18: Integrated Forest Biorefineries
19: Sustainable Preparation of Metal Nanoparticles: Methods and Applications
20: Alternative Solvents for Green Chemistry: 2nd Edition
21: Natural Product Extraction: Principles and Applications
22: Element Recovery and Sustainability
23: Green Materials for Sustainable Water Remediation and Treatment
24: The Economic Utilisation of Food Co-Products
25: Biomass for Sustainable Applications: Pollution Remediation and Energy
26: From C-H to C-C Bonds: Cross-Dehydrogenative-Coupling
27: Renewable Resources for Biorefineries
28: Transition Metal Catalysis in Aerobic Alcohol Oxidation
29: Green Materials from Plant Oils
30: Polyhydroxyalkanoates (PHAs) Based Blends, Composites and Nanocomposites

31: Ball Milling Towards Green Synthesis: Applications, Projects, Challenges
32: Porous Carbon Materials from Sustainable Precursors
33: Heterogeneous Catalysis for Today's Challenges: Synthesis, Characterization and Applications
34: Chemical Biotechnology and Bioengineering
35: Microwave-Assisted Polymerization
36: Ionic Liquids in the Biorefinery Concept: Challenges and Perspectives
37: Starch-based Blends, Composites and Nanocomposites
38: Sustainable Catalysis: With Non-endangered Metals, Part 1
39: Sustainable Catalysis: With Non-endangered Metals, Part 2
40: Sustainable Catalysis: Without Metals or Other Endangered Elements, Part 1
41: Sustainable Catalysis: Without Metals or Other Endangered Elements, Part 2
42: Green Photo-active Nanomaterials
43: Commercializing Biobased Products: Opportunities, Challenges, Benefits, and Risks
44: Biomass Sugars for Non-Fuel Applications
45: White Biotechnology for Sustainable Chemistry
46: Green and Sustainable Medicinal Chemistry: Methods, Tools and Strategies for the 21st Century Pharmaceutical Industry
47: Alternative Energy Sources for Green Chemistry
48: High Pressure Technologies in Biomass Conversion
49: Sustainable Solvents: Perspectives from Research, Business and International Policy
50: Fast Pyrolysis of Biomass: Advances in Science and Technology
51: Catalyst-free Organic Synthesis
52: Hazardous Reagent Substitution: A Pharmaceutical Perspective
53: Alternatives to Conventional Food Processing: 2nd Edition
54: Sustainable Synthesis of Pharmaceuticals: Using Transition Metal Complexes as Catalysts
55: Intensification of Biobased Processes
56: Sustainable Catalysis for Biorefineries

How to obtain future titles on publication:
A standing order plan is available for this series. A standing order will bring delivery of each new volume immediately on publication.

For further information please contact:
Book Sales Department, Royal Society of Chemistry, Thomas Graham House, Science Park, Milton Road, Cambridge, CB4 0WF, UK
Telephone: +44 (0)1223 420066, Fax: +44 (0)1223 420247
Email: booksales@rsc.org
Visit our website at www.rsc.org/books

Sustainable Catalysis for Biorefineries

Edited by

Francesco Frusteri
Institute for Advanced Energy Technologies "Nicola Giordano", Italy
Email: francesco.frusteri@itae.cnr.it

Donato Aranda
Universidade Federal do Rio de Janeiro, Brazil
Email: donato@eq.ufrj.br

and

Giuseppe Bonura
Institute for Advanced Energy Technologies "Nicola Giordano", Italy
Email: giuseppe.bonura@itae.cnr.it

ROYAL SOCIETY
OF **CHEMISTRY**

Green Chemistry Series No. 56

Print ISBN: 978-1-78262-963-4
PDF ISBN: 978-1-78801-356-7
EPUB ISBN: 978-1-78801-478-6
ISSN: 1757-7039

A catalogue record for this book is available from the British Library

The Royal Society of Chemistry is a charity, registered in England and Wales, Number 207890, and a company incorporated in England by Royal Charter (Registered No. RC000524), registered office: Burlington House, Piccadilly, London W1J 0BA, UK, Telephone: +44 (0) 207 4378 6556.

For further information see our web site at www.rsc.org

Printed in the United Kingdom by CPI Group (UK) Ltd, Croydon, CR0 4YY, UK

Preface

The global scenario, combined with the crisis of the chemical industry and the growing level of environmental awareness, offers an opportunity to focus key resources and investments in the bio-based industry and advanced technologies using renewable sources, also finding system solutions to address specific environmental problems. In this perspective, the establishment of a bio-economy model, based on biorefineries for the creation of innovative products with high added value, such as biochemicals and bioplastics, will be possible to develop a "green chemistry" in synergy with traditional chemistry. Such success has a critical point to be addressed.

Chemical transformations necessarily involve chemical bonds being broken and rearranged. Catalytic tools and processes are recognized as the best driving system to promote high yields and high selectivity in those reactions, in other words, bringing competitiveness to the biorefineries. This will help to reduce the dependence on imports, to reward the environmental quality and to develop economically and environmentally sustainable production processes so to justify the huge investments and research and innovation efforts.

Biorefineries integrated along the territory, with dedicated logistics, can give proper priority to the use of biomass, respecting local biodiversity. Particular attention is therefore given to the development of new sustainable catalytic processes of biomass transformation, to enhance human resources and know-how, to reinvigorate local and regional competitiveness, and to attract domestic and international investments.

Green Chemistry Series No. 56
Sustainable Catalysis for Biorefineries
Edited by Francesco Frusteri, Donato Aranda and Giuseppe Bonura

Published by the Royal Society of Chemistry, www.rsc.org

This book represents an opportunity to highlight and address the worldwide transition to a more sustainable development model, aimed at promoting the most effective or promising catalytic processes for the conversion of bio-based components into high added value products as platform chemicals and intermediates.

Francesco Frusteri, Donato Aranda and Giuseppe Bonura

Contents

Green Chemistry Series No. 56
Sustainable Catalysis for Biorefineries
Edited by Francesco Frusteri, Donato Aranda and Giuseppe Bonura

Published by the Royal Society of Chemistry, www.rsc.org

CHAPTER 1

Catalysts for Co-processing Biomass in Oil Refining Industry

EDUARDO FALABELLA SOUSA-AGUIAR*[a], VITOR LOUREIRO XIMENES[b], JOÃO MONNERAT ARAUJO RIBEIRO DE ALMEIDA[a], PEDRO NOTHAFT ROMANO[a] AND YURI CARVALHO[a]

[a]Federal University of Rio de Janeiro (UFRJ), School of Chemistry, Department of Organic Processes, Centro de Tecnologia, Bloco E, Ilha do Fundão, Rio de Janeiro, Brazil; [b]Petrobras Research Centre (CENPES), Cidade Universitária, Ilha do Fundão, Q7, CEP 21949-900, Rio de Janeiro, Brazil
*E-mail: efalabella@eq.ufrj.br

1.1 Introduction

The refining industry has been confronted by challenges whose solution is actually non-trivial. A growing demand for cleaner fuels is generating more stringent environmental regulations all over the world. Additionally, it must be borne in mind that the quality of current crude oil is inferior to that of the crude oil produced several years ago. Indeed, the crude oil nowadays may be rather heavy, acidic and more impure; hence, it will require new steps in the refining process, such as, for instance, desulphurisation units. Such new units, however, are very energy consuming and will certainly reduce the

Green Chemistry Series No. 56
Sustainable Catalysis for Biorefineries
Edited by Francesco Frusteri, Donato Aranda and Giuseppe Bonura

Published by the Royal Society of Chemistry, www.rsc.org

overall thermal efficiency of the refinery, not to mention the profitability thereof. Moreover, conventional desulphurisation processes often require hydrogen, whose production *via* Shift reaction will also produce CO_2, a well-known greenhouse gas. Thus, society's demand to improve the air quality by using cleaner fuels may come at the expense of higher greenhouse-gas emissions during the production of such fuels. Globalisation and oscillating customer choice, as well as the growing pressure to reduce emissions, are parameters that ought to be taken into account when the overall profitability of the refinery is discussed. Undoubtedly, the aforementioned requirements are changing the traditional goals of petroleum refineries, frequently imposing a riddle whose answer is quite intricate. Clearly, alternative intelligent solutions are not straightforward and must be developed, requiring much effort and research. Nevertheless, the reduction of greenhouse-gas emissions by fostering the deployment of alternative raw materials such as biomass is an option that must not be disregarded. In fact, biomass utilisation is playing an important role in the concept entitled "the refinery of the future". At any rate, regardless of the new chemical route chosen, the role of catalysis is sovereign, since it is easier to change the catalyst, rather than changing the entire production systems to generate a new drop-in product. New catalysts must be developed, thereby avoiding rather expensive investment in the hardware of the refinery.

As already stated, the refining of the future will deal with the presence of biomass as an alternative feedstock to the traditional oil feedstock. Therefore, it will comprise the concept of Biorefineries, which, according to the 2008 Farm Act,[1–5] may be defined as "a facility (including equipment and processes) that converts renewable biomass into biofuels and bio-based products, and may produce electricity". More recently, another important concept described by the term *Integrated Biorefinery*[6] has been coined. An integrated biorefinery is capable of efficiently converting a broad range of biomass feedstocks into affordable biofuels, biopower, and other bioproducts. By definition, the integrated biorefinery must cope with the problem of residues. Hence, integrated biorefineries are similar to conventional refineries; however, integrated biorefineries still require much research. In other words, new processes must be developed in order to reduce production costs and improve competitiveness. Essentially, the potential of residues must be explored and innovative chemical routes have to be proposed.

The concept of an integrated biorefinery may be applied to several types of traditional refineries, using different raw materials.[7,8] Furthermore, aiming at the better utilisation of existing facilities, co-processing is often indicated. Co-processing is, by definition, the utilisation of blends in already existing units. In principle, vegetable oils can be rather easily co-processed in the installed facilities of refineries. In addition, different types of bio-oils, resulting from both catalytic and non-catalytic pyrolysis of lignocellulosic biomass, can also undergo processing in different units of the refinery.

Vegetable oils are more easily co-processed in existing refinery facilities. Indeed, co-processing of vegetable oils can be incorporated into a refiner's operating strategy with minimal detriment to catalyst stability or yields; however, the importance of the base feedstock and operational conditions must not be overlooked.

Coconut, sunflower, maize, olive, peanut and cottonseed oils are some of the potential oils proposed for studies in co-processing. Nevertheless, soybean, palm and rapeseed oils, which are readily available, are the most studied vegetable oils.[9]

The basic scheme of vegetable oil processing in refineries encompasses two units: the hydroprocessing unit and the fluid catalytic cracking (FCC) unit. As far as the hydroprocessing unit is concerned, the idea is rather simple. Blend certain amounts of vegetable oils with the regular feedstock of the unit and then allow the operation to proceed as usual. Of course, operational conditions will have to be adjusted to the new feedstock, since a new reaction scheme will take place. As a matter of fact, the following reactions will be carried out:

(a) In the first step of the reaction pathway, the unsaturated fatty acid chains will be rapidly converted into fully saturated *n*-paraffins;
(b) In the second step, the bonds between fatty acids and glycerol must be broken (cleavage of a carbon–oxygen bond), thereby ensuring that the products will have appropriate size for the diesel pool.

It is obvious that a convenient catalyst should be developed to promote both reactions.

Regarding FCC units, the same philosophy is used, that is to say, the policy of blending. However, unlike hydrotreating units where the catalyst cannot be changed without the unit being shutdown to reload the reactor, continuous replacement of catalyst in the FCC unit permits the refiner to change the inventory and use tailor-made catalyst formulations to optimize yields.

The problems of co-processing bio-oils are often related to the instability thereof. Bio-oils produced *via* fast pyrolysis present a low-viscosity, single-phase liquid. The deployment of such bio-oils requires that these initial properties be retained. Unfortunately, bio-oils may undergo several reactions, which will provoke an increase in viscosity with time. In fact, bio-oils resulting from fast pyrolysis are not a product of thermodynamic equilibrium, being produced *via* short contact times and rapid cooling or quenching.

The main reactions that may take place upon storage of bio-oils are the following:[9]

- Organic acids with alcohols to form esters and water;
- Organic acids with olefins to form esters;
- Aldehydes and water to form hydrates;
- Aldehydes and alcohols to form hemiacetals, or acetals and water;
- Aldehydes to form oligomers and resins;
- Aldehydes and phenolics to form resins and water;

- Aldehydes and proteins to form oligomers;
- Organic sulphur to form oligomers;
- Unsaturated compounds to form polyolefins;
- Air oxidation to form more acids and reactive peroxides that catalyze the polymerization of unsaturated compounds.

All those reactions will generate a product slate that will bring about an increase in the viscosity of bio-oils with time of storage. For that reason, a "stability parameter" has been proposed.[10] This parameter represents the slope of a plot of viscosity (cp) against time (h). This plot is frequently a straight line with a positive derivative, which represents the stability parameter. It is worth noticing that aging effects are a function of temperature, being accelerated as the temperature increases. An Arrhenius plot of the stability parameter gives straight lines, which is a clear indication that the increase in viscosity is a result of chemical reactions.[11]

It is clear that aging is a result of several reactions, most of them undesirable. Such reactions will increase the average molecular weight of the bio-oil. Hence, chemical properties of bio-oils will certainly change upon aging. For that reason, it is likely that the stability parameter will affect the performance of different bio-oils whenever co-processing is concerned.

Ramirez and co-workers[10,12,13] have issued several interesting patents in which bio-oils from different sources may undergo upgrading in order to be processed in the refineries. Indeed, different types of treatments may improve the upgrading of bio-oils. Such possibilities are depicted in Figure 1.1. It must be borne in mind, however, that the required upgrading process will depend on the type of bio-oil produced and the stability thereof. Most of the existing biomass conversion processes produce low-quality bio-oils, since they are highly unstable and present high oxygen contents. These bio-oils will often require extensive secondary upgrading and a special refining scheme, which may even include more than one upgrading step.

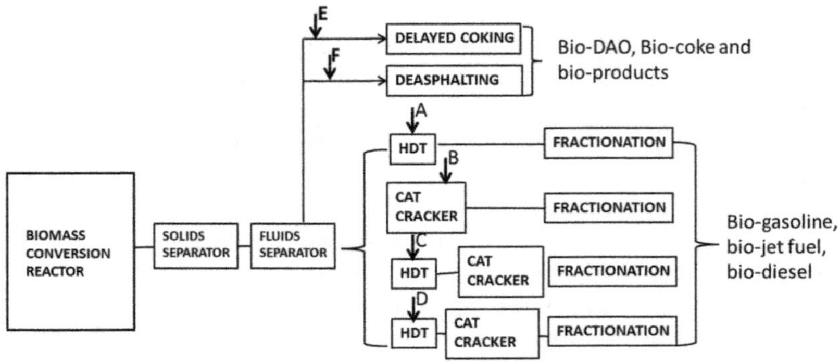

Figure 1.1 Potential use of bio-oil in a refinery. (Adapted from ref. 13).

Other traditional processes carried out in the refineries such as delayed coking and deasphalting may employ alternative feedstocks in which vegetable oils or bio-oils are blended with the regular feedstock. Hereinafter, such possibilities will be discussed along with hydroprocessing and FCC.

1.2 Bio-oil in FCC

Fluidized catalytic cracking (FCC) is a conversion process in petroleum refining schemes, which aims to reduce the molecular mass of molecules from the streams produced by petroleum fractionation. Usually, the FCC feed contains fractions of molecules produced in the atmospheric and vacuum distillation of petroleum, which are known as gasoils. It is also possible to complete the FCC feed with streams produced by other types of fractionation, such as the deasphalted oil obtained in the solvent deasphalting process. The reactions that occur in the FCC process convert these heavy fractions into more noble products, maximizing the yields in the distillation range of LPG, diesel and, mainly, gasoline. Among the processes found in refineries, FCC is the one that has been most studied to allow the insertion of bio-oil in the refining scheme through co-processing strategies. The acidic character of the FCC cracking catalyst allows the occurrence of reactions that promote the removal of oxygen from the bio-oil molecules, eliminating it in the form of CO, CO_2 and H_2O.

For a long time, the deoxygenation processes of molecules originating from biomass conversion in the presence of zeolitic catalysts have been studied. In 1986, Chen *et al.* evaluated the conversion of sugars catalyzed by ZSM-5 zeolite, reporting the production of hydrocarbons, coke, water and carbon oxides.[14] With bio-oil, initially, the most used approach, and the object of several publications, consisted of the improvement of this renewable stream through its conversion *via* acid catalysis. Adjaye *et al.* in 1995[15,16] were pioneers in the attempt to upgrade a bio-oil produced by Ensyn (a leader Company involved in the production of biocrudes from forest and agriculture residues) using a fixed bed zeolite catalyst. Production of coke, gases, water, and an organic phase containing several hydrocarbons was observed. It was shown that the yields of hydrocarbons were dependent on the type of catalyst, with HZSM-5 generating the highest selectivity to these products. In addition, it was found that the aromaticity of the organic phase was also higher with the use of HZSM-5, followed by H-mordenite, revealing a possibility to alter the product profile only by changing the type of catalyst.

At that time, the incentive to develop advanced biofuels was not so intense. Yet, several studies on the same theme were followed and the effectiveness of zeolitic catalysts in the conversion of bio-oil to hydrocarbons was already scientifically known. The reactions occurring in the presence of the acidic site of zeolites remove oxygen from the bio-oil molecules in the form of water and carbon oxides. Therefore, the mechanisms require the simultaneous rejection of carbon and/or hydrogen in order to reduce the oxygen content. With this knowledge, Chen *et al.*[14] proposed the use of a modified hydrogen–carbon ratio to evaluate the most suitable conditions of treatment of the bio-oil through catalytic cracking – the effective C:H ratio. This parameter differentiates the oxygen-bound hydrogen from the carbon-bound hydrogen, since they have different characteristics for the catalytic cracking process.

Another pioneering group in bio-oil catalytic cracking studies was Samolada *et al.*, and they evaluated the non-catalytic hydrogenation of bio-oils and their subsequent processing in a bench scale FCC unit (50–300 g h^{-1}) with commercial catalysts. An increase in coke yield was observed; nevertheless, it was shown that its co-processing with LCO in the proportion 15/85 was able to reduce this tendency. Besides, it was possible to produce a specified gasoline with high octane number, despite the reduction in the yield of this fraction.[17]

In the last decade, with the incentive to increase biofuel participation in energy matrices, especially biofuels from lignocellulosic material, the intensification of studies contemplating bio-oil improvement and its use as an automotive fuel was observed. In this context, and due to the well-known ability of zeolite catalysts to convert bio-oil-derived molecules to hydrocarbons under FCC conditions, research groups around the world have begun to study the co-processing of bio-oil with fossil streams in FCC. In 1997, a study was published in the report AIR2-CT93-1086, indicating the FCC feed as a potential point of insertion of bio-oil in the refining scheme.[18] Nevertheless, the lack of its feasibility due to bio-oil insolubility in fossil streams was noted.

In 2007, Corma *et al.* published the first scientific results of renewable feed cracking along with the fossil stream. The scenario was promising for advanced biofuels, with incentives through public policies, international cooperation agreements and high petroleum prices. It was the beginning of the dissemination studies on the development of a process that would serve as a basis for a gradual transition between the fossil and the renewable matrix. This new approach was also taking advantage of the existing units in the refining schemes and the knowledge acquired over time from catalytic cracking processes. They studied the co-processing of gasoils and oxygenated molecules in FCC bench units, publishing the results of the performance evaluation of several catalysts in the conversion of glycerol, sorbitol and mixtures of these two compounds with vacuum gasoil.[19] In 2008, Domine *et al.* published a co-processing study using model bio-oil and diesel molecules, finding higher deactivation rates in the presence of oxygenated compounds. They also observed CO production due to the removal of oxygenated functional groups, as well as an increase in the coke yield and a consequent reduction in hydrocarbon conversion due to the blockage of acidic sites.[20]

The first study beyond the bench scale was published by Lappas *et al.* in 2009, which co-processed 2.5% of hydrotreated bio-oil with 85% of vacuum gas oil in a FCC pilot unit, diluting it with 12.5% of light cycle oil (LCO) to facilitate the feed injection and dispersion. Despite the small differences, they observed an increase in the yields of coke and naphtha, with a reduction in the gas yield.[21]

Graça *et al.* concentrated their efforts on the understanding of the catalytic mechanisms and the proposal of more efficient catalysts for the cracking of feeds containing oxygenated compounds. In 2009, they published a study in which three model molecules of bio-oil (phenol, acetic acid and hydroxy-acetone) were mixed to a gasoil in order to evaluate the co-processing in a bench scale FCC unit.[22] In this study, two different commercial catalytic systems were used, one being 100% e-CAT and the other one containing 10% ZSM-5. Increased conversion was observed when the oxygenated molecules were co-processed with e-CAT. With the additive, it was observed that the presence of oxygenates reduced the fossil molecule conversion, which was attributed to the greater interaction between these molecules and the ZSM-5 sites.

Subsequently, Graça *et al.* published a series of studies with the objective of specifically evaluating the impact of phenolic compounds in catalytic cracking with different types of catalytic systems.[23–26] The tests used phenol and guaiacol as model molecules of bio-oil and *n*-heptane and methyl-cyclohexane as gasoil model molecules. It was found that in the co-processing of phenolic compounds with paraffinic feeds, the deleterious effects of oxygenated compounds on catalyst deactivation were not as significant as with naphthenic feeds. The size effect of the oxygenated molecule on the deactivation of the systems was also evaluated, concluding that guaiacol, due to its higher kinetic diameter, was preferentially adsorbed at the intersections of the channels near or at the entrance of the catalyst pores, facilitating their desorption when subjected to FCC temperatures. It was concluded, therefore, that phenol causes more damage than guaiacol in relation to FCC catalytic system deactivation when operating in co-processing mode.

Graça *et al.* published a new study evaluating the effect of co-processing guaiacol with *n*-heptane, but now also with actual gasoil, using several zeolitic catalysts.[27] The systems tested were HZSM-5, HY, commercial e-CAT and e-CAT with ZSM-5. Again, a negative effect of the presence of guaiacol in relation to the catalyst deactivation rate was observed, especially with HZSM-5 and with e-CAT and in the presence of the actual feed. With the other systems, the deactivation intensification in the presence of oxygenates was reduced. Regarding product quality, the presence of phenolic compounds in the gasoline range was observed, indicating that there are molecules of this chemical class that are inert to cracking and boil in the temperature range of naphtha.

FCC co-processing was one of the routes contemplated for the transformation of bio-oil into biofuels. However, other routes have also been evaluated and developed, such as, for example, the hydrogenation of bio-oil. In this process, discussed in more detail hereinafter, oxygen rejection occurs through hydrodeoxygenation reactions, and may even generate a hydrocarbon product without heteroatoms. However, the severity of the process, the need for catalysts with noble metals and the high demand for hydrogen indicate that the cost associated with the production of hydrocarbons by this route may hardly reach economic viability. Nonetheless, less severe hydrogenation processes could be used as a pre-treatment of bio-oil for FCC co-processing.

Mercarder *et al.* decided to evaluate the complementarity of the hydrogenation and cracking processes for the transformation of bio-oil into biofuel. They understood that this strategy could significantly reduce the cost of hydrogenation by reducing the desired degree of hydrodeoxygenation, as well as facilitating cracking by raising the bio-oil effective C/H ratio. They published two studies evaluating the effects of hydrodeoxygenation conditions of a bio-oil produced by VTT, with the aim of co-processing the hydrogenated product with atmospheric residue in the FCC.[28,29] It was shown that the higher severity of the hydrogenation was responsible for the increase of carbon recovery in the organic product, but it did not significantly affect the feasibility of bio-oil co-processing in a bench scale FCC unit. It was also observed that the conversion of the most reactive oxygenated functional groups, which would be responsible for the increase of the repolymerization reactions, already occurs under milder hydrogenation conditions. In the experiments without atmospheric residue, a significant increase in coke and gas yields was observed, revealing the importance of the presence of a non-hydrogen deficient fossil stream during catalytic co-cracking.

Using the same strategy, in 2011, Fogassy *et al.* evaluated the co-processing of 20% hydrotreated bio-oil with gasoil in a MAT unit using HZSM-5, HY and FCC equilibrium catalysts.[30] Steric hindrances and catalyst acidity effects were observed in the feed conversion, which resulted in a scheme of cracking mechanisms during co-processing. It was also shown that the higher pore size and the higher acidity of zeolite HY favoured the overcracking of molecules in the naphtha and LCO ranges, increasing the yield of gases, LPG and coke. Furthermore, HY zeolite allowed the occurrence of bimolecular reactions, increasing the concentration of paraffinic and aromatic compounds in the liquid fractions. It was also verified that the lignin oligomers did not enter into the pores of the catalytic systems, being cracked on the catalyst surface, with consequent coke generation and blocked access to the acid sites.

Since the hydrogenation of the bio-oil could allow its subsequent processing in FCC, other processes that had the capacity to reduce its oxygen content could act in the same way. In 2012, Agblevor *et al.* used a product of the catalytic pyrolysis of biomass, denominated in this study as bio-crude, as a renewable feed for FCC co-processing.[31] This bio-crude is more stable than bio-oil due to the more efficient removal of reactive functional groups by deoxygenation reactions promoted during pyrolysis in the presence of a catalyst. The experiment feeds consisted of mixtures of 15% $m\ m^{-1}$ bio-crude and 85% $m\ m^{-1}$ gasoil, and were evaluated in an ACE unit (Advanced Cracking Evaluation) using an equilibrium catalyst. Evaluating the results under isoconversion, the yields of naphtha, light cycle oil (LCO), heavy cycle oil

(HCO) and gas liquid processing (GLP) did not change when the bio-crude was introduced, but the liquid fractions showed an increased presence of aromatic carbon.

One of the relevant issues in bio-oil improvement studies is the ability of the process to retain renewable carbon in noble fractions, such as gasoline and diesel. Fogassy *et al.*, in 2012, presented a discussion on the distribution of renewable carbon during bio-oil co-processing experiments in FCC.[32] To do this, they conducted a series of experiments by co-processing 10% hydrogenated bio-oil with 90% vacuum gas oil in a bench scale FCC unit. In relation to the analytical technique used to determine the bio-carbon distribution, ^{14}C isotopic analysis was used as an indication of the renewable carbon content. It has been observed that most of the carbon present in bio-oil is transformed into gases (10.6% bio-carbon) and coke (15.8% bio-carbon), whereas gasoline has approximately only 7% bio-carbon. In this study, the increase in coke yield was explained by two mechanisms: repolymerization of oxygenated molecules, especially phenolic compounds; and hydrogen deficiency, due to the occurrence of deoxygenation reactions *via* dehydration.

Schuurman *et al.* discussed in a review published in 2013 the impacts of bio-oil co-processing on FCC units.[33] They noticed that most studies show that the co-processing of HDO with gasoil does not imply a loss in gasoline yield, but results in a reduction in the production of bottoms and GLP, as well as an increase of coke yield. In addition, the gasoline produced by the co-processing strategy has higher levels of aromatic and olefinic compounds, which in general lead to an increase in octane numbers. This characteristic – higher levels of unsaturation – is attributed to the increase in the consumption of hydrogen, which is produced through hydrogen transfer, by the deoxygenation of the molecules from the bio-oil. The decrease of the reduction character of the reaction media leads to lower rates of hydrogenation of unsaturated compounds. This phenomenon can also explain the increase in coke production. Another relevant fact is the inertia of the phenolic compounds to the FCC catalysts, so that they always appear in the products obtained by co-processing of bio-oil. In the review, the mechanisms involved in the co-processing are also discussed, indicating the preferential adsorption of the oxygenated compounds in the strongest external acid sites, mainly Lewis acid sites (EFAL), reducing their availability for the cracking of hydrocarbons, especially those of greater molecular mass, which do not access the pores. Because of this preferential adsorption, there is an increase in the yield of coke, which in turn ends up blocking the pores and reducing the conversion rates, as shown in several studies with model molecules.

A study comparing the effectiveness of processes to improve the quality of bio-oil before FCC co-processing was published in 2014. Thegarid *et al.* conducted a study comparing the co-processing of 10% hydrotreated bio-oil and catalytic bio-oil (CPO) with vacuum gas oil in a bench scale FCC unit.[34] With both bio-oils, an increase in gas production was observed, with CO_2 appearing in their compositions and hydrogen content decreasing. No significant changes were observed in the gasoline yield, although the aromatic content increased in the co-processing experiments, especially at high catalyst-to-oil (CTO) ratios. Phenol molecules were found in the co-processing products, being even more significant in cases with CPO. LCO was more converted in the presence of bio-oils and a great increase in coke yield was observed. Lastly, they concluded from the results that the bio-oil produced by catalytic pyrolysis is more appropriate to be co-processed in FCC, especially due to the greater retention of renewable carbon in liquid products.

Gueudré *et al.* published a study using the same data as Thegarid *et al.* (2014),[34] but now focusing on the effects related to coke.[35] It was verified that the coke yield increased in both co-processing cases, being higher in the presence of the hydrogenated bio-oil. In addition, it was found that only 40% of the coke generated during co-processing has a renewable origin, which can be explained by the occurrence of higher rates of hydrogen transfer from the fossil molecules and their consequent aromatization and transformation into coke, when the bio-oil is present. They also verified that the co-processing is responsible for the production of a coke with distinct chemical characteristics, making it more easily manageable due to the presence of oxygen and less graphitic structure. However, this coke generated by the bio-oil is responsible for accelerating the catalyst deactivation, changing its acidity and crystallinity more quickly.

Lindfors *et al.*, in 2015, compared some possible steps upstream of FCC co-processing.[36] They carried out a study contemplating the producing and co-cracking of bio-oil with low humidity, catalytic bio-oil or hydrotreated bio-oil with 80% vacuum gas oil in a FCC bench unit. An increase in gasoline yield was observed in the co-processing experiments, with catalytic bio-oil being the best case in this respect. It was explained that possibly some aromatic compounds present in the catalytic bio-oil are already in the gasoline range and they are not cracked during the process, explaining the higher naphtha yield. Regarding coke yield, all co-processing experiments generated a higher amount of solid fraction. However, it is possible to distinguish the result of the HDO sample, whose value was the closest to pure gasoil. Thus, it was concluded that the oxygen content could not be considered as the only criterion to predict the tendency to produce coke during co-cracking of bio-oil. It was also suggested that a better correlation can be obtained by using the micro carbon residue and the effective H/C ratio to estimate coke yield.

In 2015, Pinho *et al.* published the first results of bio-oil co-processing with gasoil on a demonstration scale unit, with a feed rate of 150 kg h^{-1}.[37] The bio-oil used was produced by BTG in a rotary cone unit using pine sawdust. Two blends were tested, one with 10% and the other with 20% bio-oil. Due to the immiscibility of the feeds and the chemical instability of the bio-oil, it was necessary to inject the streams in different axial positions, thus avoiding heating the bio-oil at a temperature above 50 °C. It was observed that the 10% bio-oil test did not promote significant changes in the yield profile of the products, including gasoline and coke. On the other hand, at 20%, there was a significant reduction in gasoline, in addition to increased coke production. In order to quantify the renewable carbon in the gasoline range, the products were sent for isotopic analysis of ^{14}C. The results showed that when 10% bio-oil is co-processed, only 2% of the gasoline carbon was from renewable origin, while when 20% bio-oil underwent co-processing, the content was between 3 and 5%. Through the mass balance, it was possible to determine that this value corresponds to 30% of the carbon present in the biomass, and this value is higher than that estimated by Fogassy *et al.* (2012).[32] This study also revealed a scale effect, mainly related to the segregated injection, which makes the results from demonstration units better than those obtained on bench scale, especially in relation to the coke yield. Due the early bio-oil injection and its small proportion, temperature and CTO tend to be much higher when the bed meets the renewable charge. The higher temperature promotes a thermal cracking of the renewable current, which facilitates its secondary conversion into the pores of the catalyst.

By 2015, Graça *et al.* were continuing their studies on catalyst synthesis and evaluation, focusing on the co-processing of oxygenated molecules in FCC. They published the results of a study evaluating the effect of FCC catalysts containing ZSM-5 zeolites with different textural properties in the co-processing of model diesel and bio-oil molecules.[38] The aim was to minimize the deactivation caused by the presence of phenolic compounds, as reported in previous publications. Several ZSM-5 catalysts were prepared by alkaline post-treatment of the original sample, with the main objective of altering their pore distribution. It was found that mesopore formation, keeping the existence of microporosity, reduces the rate of deactivation of the catalyst in the presence of phenol, without any loss of shape selectivity.

Wang *et al.*, in 2016, produced catalytic bio-oil and co-processed it with 90% gasoil in a FCC pilot unit.[39] No operational problems were observed up to 10% bio-oil, and the co-processing led to higher conversions than that with only the fossil feed in the same CTO. In addition, there was no change in the yield profile of most of the products. The only exception was the gaseous fraction, especially hydrogen, which presented lower yields due to its transfer between fossil and renewable molecules. Finally, it was possible to estimate through isotopic analysis of carbon 14 that 7% of the carbon present in the gasoline range had a renewable origin when 10% bio-oil was co-processed, which is significantly higher than that found during the thermal bio-oil co-processing.

In 2016, Ibarra *et al.* conducted experiments co-processing 20% bio-oil with gasoil in order to better understand the mechanisms of coke generation and its characteristics.[40] The conversions were greater when bio-oil was cracked, either pure or mixed with gasoil. In the latter case, a synergistic effect was also observed, which made the conversion of the mixture higher than those determined for the pure feeds. Although the presence of bio-oil is responsible for higher coke yields, it was noticed that the deposit generated during co-processing is less deleterious than with gasoil alone. The solid deposit originating from bio-oil has a larger fraction soluble in dichloromethane, which is lighter than the insoluble coke. It was observed that, during co-processing, there was an attenuation of the coke deposit, especially that from insoluble coke, explained here by the higher amount of steam when the bio-oil is also fed, which comes from the own renewable feed. Steam is responsible for the rectification of insoluble coke precursors from catalyst pores, and it competes with hydrocarbons for the acid sites, reducing the condensation reaction rates. It was also observed that the insoluble coke is usually deposited in the catalyst meso and macropores, while the coke generated during the bio-oil cracking is deposited in the micropores of the catalyst. Still, hydrogen transfer reactions between gasoil and bio-oil also act to reduce coke formation from oxygenates and the high temperatures help in the cracking of the coke from the bio-oil, contributing to reducing the deactivation route *via* oxygenates.

A new study on demonstration scale was published in 2017. Pinho *et al.* tested the co-processing of 5% $m\ m^{-1}$ and 10% $m\ m^{-1}$ bio-oil, confirming the positive effect of the axially segregated injection.[41] In the presence of bio-oil, a reduction in gas and LPG production was observed, with an increase in the olefinicity of the C3 and C4 molecules. In one of the series of experiments, gasoline overcracking was observed for all the feeds, maximizing yields between 68% and 69% conversion. At these points, gasoline yields were similar in the tests with and without bio-oil. LCO yield increased only in the 10% test and the yield of bottom products increased slightly for the two cases with bio-oil. Regarding the presence of oxygenated compounds in the cracked effluent, only phenolic compounds were detected, which, depending on the amount, may be a benefit to fuel stability. Finally, it was determined by carbon 14 isotopic analysis that, for the tests with 5% bio-oil, there is 1% renewable carbon in gasoline.

More recently, Gueudré *et al.* published a new study evaluating the effect of the bio-oil hydrogenation severity in subsequent co-processing in a FCC bench scale unit.[42] They observed that all tests with bio-oil showed increased naphtha yields, but no influence of the hydrogenation step on this variable was observed. On the other hand, the gasoline quality experienced an optimum hydrodeoxygenation severity point, determined by the MON and RON maximum values found in gasoline, with a H_2 consumption of 202 *N*L per kg of pyrolysis liquid, equivalent to an intermediate hydrogenation.

The intensification of studies and the significant number of publications and forums on co-processing of bio-oil in FCC have been improving the maturity of this route. Some difficulties, such as the increase in the deactivation rate of the catalytic systems, have already been addressed and have technological solutions. The increase in the number of studies has also provided intense analytical developments in the characterization of oxygen-containing matrices, and these techniques have improved the understanding of the phenomena involved in co-processing. Tests in a demonstration unit, conducted by Pinho *et al.*, demonstrated the advantages of scaling up and indicated the technical feasibility of the route for bio-oil insertion into the refining scheme, bringing the possibility of producing second-generation biofuels by co-processing closer.

1.3 Co-processing in Hydrotreating Units

Hydrotreating (HDT) processes have been employed for decades in the petroleum refining industries with different purposes, such as fuel contaminant removal or upgrading of heavier streams.[7] In the past 5 years, the demand for renewable jet fuel and green diesel, both commonly produced by hydrotreatment of vegetable oil, has spiked. This has occurred mostly because of new regulations regarding CO_2 emissions. These green fuels are usually blended with fossil fuels before reaching the final consumer. This represents a great disadvantage from the logistics point of view, if there is one plant dedicated to producing biofuels and the fossil fuels are produced in refineries. This logistic operation devoted to blending fossil fuels and green fuels comes at a cost not only from the economic point of view, but also from the environmental one. On the other hand, when vegetable oil is co-processed in hydrotreatment units in refineries with fossil streams to produce fuels, there is no need for blending. Moreover, the capital expenditure (CAPEX) needed for co-processing vegetable oil in HDT units is almost negligible if compared to the capital invested for the construction of a new plant devoted to hydroprocessed esters and fatty acids (HEFA process), especially if the HDT unit has been working in subcapacity mode.

The traditional catalysts employed in the HDT process are tungsten/molybdenum sulphide catalysts supported on alumina and promoted with Ni and Co. The reactor is a trickle-bed type and the process is carried out at high temperature and pressure. The most relevant reactions in HDT are hydrodesulphurisation (I, HDS), hydrodenitrogenation (II, HDN), hydrodemetallization (III, HDM) and hydrodeoxygenation (IV, HDO). In the case of co-processing vegetable oil, HDO and HDN are undoubtedly especially pertinent. These reactions involve the hydrogenolysis of the carbon–heteroatom bond. Depending on the source of the vegetable oil, other reactions such as the hydrogenation of olefins (HO) can be very relevant as well.[7,43,44]

In the traditional HDT process, generally, a hydrocarbon base or a mixture of hydrocarbons and vegetable oil containing no more than 20% renewable oil is mixed with a hydrogen stream, resulting in the feed for the reactor. This feed is pre-heated and injected into the first catalytic bed at temperatures between 220–350 °C. Most of the reactions are exothermal, as are most hydrogenations. To achieve a better temperature control inside the reactor, the amount of hydrogen mixed with the feed stream is not sufficient for full conversion. Because of that, another stream of hydrogen must be injected between the first and second bed of the reactor.

From the process engineering perspective, one of the challenges when co-processing vegetable oil in HDT units is the high exothermicity of the triglyceride conversion associated with the generation of water, CO and CO_2

gases. That is the reason why no more than 20% of renewable oil can be co-processed in traditional HDT units without major modifications in the process hardware. From the catalysis point of view, much work has been devoted to understanding the effect of co-processing vegetable oil on the activity and stability of the traditional HDT catalysts.

In this context, a lot of work has been done investigating the effect of additional water, CO and CO_2 on conventional HDT catalytic systems.[45–48] The CO_2 produced due to decarboxylation greatly affects the activity of the HDT catalysts. One reason for that is the competition for hydrogen between the desired reactions HDS/HDN/HDO and the undesired reaction triggered by CO_2 addition to the systems.[46] Among these undesired reactions, the more relevant is the RWGS reaction, consuming H_2 for the formation of CO. It is also interesting to observe that depending on the operational conditions, the effect of CO_2 can be more relevant. For the case of HDN, more severe conditions lead to greater activity loss. On the other hand, for HDS, higher temperatures result in less activity loss. Another important reaction taking place in the HDT reactor when co-processing renewable oil is decarbonylation, leading to CO and water formation. The effect of CO has also been extensively investigated in the literature,[45,49,50] and it has been very well established that the presence of CO in the reaction causes a loss of activity in HDN, HDO and especially HDS. This loss of activity can be attributed mainly to competitive adsorption of CO on the catalyst active sites. Finally, the water formed due to deoxygenation and decarbonylation also affects the activity of traditional HDT catalysts. This subject is a matter of controversy, with some studies attributing a slight decrease of activity to the presence of water,[51] while others argue that water does not significantly affect the catalytic activity during HDT.[52,53] There are even results which indicate an increase of activity as a result of low concentrations of H_2O in HDT streams.[52] More recently, Berzegianni, Dagonikou and Sklari[45] published a paper as an attempt to clarify that question. The results of this work support the idea that H_2O addition has no significant effect on catalytic activity. Moreover, their results demonstrate that even the loss of surface area of HDT catalysts during long times on streams cannot be attributed to the presence of water.

1.4 Co-processing Bio-oil in HDT Units

A simplified but representative composition of a regular bio-oil would be around 50 wt% carbon, 7 wt% hydrogen and 43 wt% oxygen.[54] Thus, the high oxygen content of bio-oils, which translates also into high water content, implies a much lower energy density (15–19 MJ kg^{-1}) when compared to petroleum oil (40 MJ kg^{-1}).[55] Moreover, due to the high instability of bio-oil, associated with problems of phase separation and immiscibility with regular fossil fuels, high natural acidity and viscosity, the use of bio-oil directly as a liquid fuel is problematic.[56,57] Therefore, proper bio-oil upgrading is essential to overcome bio-oil natural deficiencies and to use it efficiently as a liquid fuel. Hydrotreatment (HDT) is a promising method for bio-oil upgrading, in which oxygen levels can be significantly reduced, by means of water formation upon reaction with hydrogen. Traditionally, HDT of bio-oil is performed under high H_2 pressures (>70 bar) and multiple temperature stages are employed.[58,59]

One of the major problems related to bio-oil upgrading *via* hydrotreatment is the severe catalyst deactivation observed during reaction.[59,60] Phenolic compounds present in bio-oil are believed to strongly adsorb on the surface of the catalyst, forming species that are precursors of coking.[55] Accordingly, carbon deposits, or coke formation, are the main reason for catalyst deactivation during hydrotreatment of bio-oil and intense efforts have been devoted in order to design new catalysts less propitious to coke deposits. In more severe cases, even reactor plugging may occur due to intense carbon deposit formation.[61] The most studied catalysts for bio-oil HDT are noble metals, which are rather expensive, or transition metal sulphides, which are already employed as catalysts for petroleum oil HDT. Oxygen content in bio-oil can be reduced to levels as low as 0.2 wt% by means of hydrodeoxygenation.[62]

The main idea behind co-processing bio-oil with petroleum oil is to benefit from the vast and consolidated infrastructure of oil refineries. In this sense, not only capital, but also operational costs could be drastically reduced and become essential for the economic feasibility of bio-oil upgrading. In addition, by blending bio-oil and petroleum oil feeds or by just performing, essentially, the same unit operations with bio-oil that are already done with petroleum oil, a very similar product slate could be obtained.[58] Hence, drop-in fuels and chemicals could be produced with no need for drastic changes in the already consolidated supply chain.

Nevertheless, as already mentioned before, raw bio-oil has some unique properties that make the direct blending of bio-oil with petroleum oil, for instance, in a hydrotreatment unit, not trivial at all. Several authors have proposed that the direct blending of these two streams would be unfeasible and at least a mild pre-HDT of bio-oil, in order to lower its oxygen content and acidity, would be necessary.[58,62–65] A reduction in oxygen levels to at least 7 wt% is recommended before blending bio-oil with other refinery streams.[65] Differently from petroleum oil, bio-oil has low nitrogen and sulphur contents, so where bio-oil HDT is concerned, the main reaction occurring is hydrodeoxygenation (HDO), rather than hydrodenitrogenation (HDN) or hydrodesulphurisation (HDS). Moreover, the number of heteroatoms, mainly oxygen, present in bio-oils is an order of magnitude higher when compared to petroleum oil.[62,63] Obviously, there are several other reactions that may also occur such as decarbonylation, decarboxylation, cracking/hydrocracking, isomerisation, among others. The type of catalyst support used may have a strong influence on the extent of these reactions. Although removing oxygen *via* HDO has some drawbacks, such as high hydrogen consumption, removing oxygen *via* decarbonylation/decarboxylation, yielding CO and CO_2, reduces the overall carbon efficiency of the process.[62] In a nutshell, removing oxygen as water while avoiding saturation of the remaining hydrocarbons would be the best outcome.[58,66] In that way, not only would hydrogen consumption and carbon loss be minimized, but also unsaturated hydrocarbons, such as alkenes and aromatics, that have higher value, would be produced. However, it is known that removing oxygen *via* HDT is easier if the hydrocarbon is already saturated.[58] Therefore, this offers another difficulty to overcome where proper catalyst design is concerned. Ideally, the catalytic system should be very active to HDO, but not so much for the saturation of hydrocarbons, robust, in the sense of low coke formation and stability properties, and cheap. These characteristics introduce a barrier that should be surpassed in order to design appropriate catalysts for bio-oil HDT.

During the first studies regarding bio-oil HDT, it was found that efficiently performing HDT in a single stage, with high temperatures (>300 °C), was challenging due to severe coke and heavy tar formation that, eventually, resulted in reactor clogging and catalyst deactivation.[58,65,66] Therefore, a two-step approach was developed by Elliot *et. al.*,[67] in which the first stage involved a mild HDT step at lower temperatures (<300 °C) in order to stabilize the bio-oil and reduce coke like material formation.[66,68] After a second stage, at temperatures of ~350 °C, a significant reduction in the bio-oil oxygen content could be achieved. However, coke formation can still be an issue even with this strategy. The yield of upgraded product, in this case, was 30–50 wt% of the original bio-oil, the rest being gaseous products, water and coke.[65]

As Co-MoS_2/γ-Al_2O_3 and Ni-MoS_2/γ-Al_2O_3 are traditional HDT catalysts employed in the refinery, they were a natural starting choice for bio-oil HDO. When this type of catalyst is employed, Co and Ni actually play the role of promoters, donating electrons to the molybdenum atoms. As a result, the bond between molybdenum and sulphur is weakened, generating a sulphur vacancy site, which is, indeed, the active site for HDS and HDO reactions.[69] With the progress of the reaction, the catalyst may oxidize, which translates into a decrease in activity. Therefore, co-feeding H_2S, in order to regenerate the sulphide sites, might be necessary to keep the catalyst activity.[69] Transitional and noble metal based catalysts have also been applied to bio-oil HDO with a certain degree of success. However, the use of noble metals as catalysts is rather dependent on the content of sulphur in the bio-oil, as they are sensitive to high S levels and could be readily poisoned, not to mention their high prices.[63]

Where HDT catalysts are concerned, γ-Al_2O_3 is the most common support utilised. However, in the case of bio-oil HDO, the presence of water allied with high temperatures can promote the conversion of alumina into boehmite (AlO(OH)), which in turn can oxidize the Ni particles present in the catalyst. Moreover, the relatively high acidity of alumina is responsible for a great affinity with phenolic compounds that, ultimately, are responsible for coke formation.[62,63,69] Therefore, several other supports have been studied as HDO catalysts. Carbon and SiO_2 appear to be promising alternatives due to their neutral nature, able to avoid coke formation. However, the use of carbon as support introduced a new problem regarding catalyst regeneration, since the catalysts could not be regenerated through a simple calcination procedure.[64,65] Titania (TiO_2), ceria (CeO_2) and zirconia (ZrO_2), are also some interesting alternatives, because even though they present some intrinsic acidity, they have the ability to activate oxy-compounds, resulting in an increase in the activity.[69]

Another possible strategy to overcome the problems associated with carbon deposition during bio-oil HDT is to add a hydrogen donor, such as methanol, tetralin or decalin, to the bio-oil feed. The added hydrogen donors play several roles in this case. First of all, they improve the extent of cracking and hydrogenation reactions but also, by diluting the bio-oil feed, they decrease the probability of polymerization at high temperatures.[61]

Table 1.1 sum ups some relevant studies concerning bio-oil HDT from recent years. Baldauf *et al.*[70] reported high deoxygenation rates of 88–99.9% for bio-oil upgrading in a continuous bench scale unit. The authors employed commercial CoMo and NiMo HDT catalysts and oil yields of 32–36% were obtained at temperatures ranging from 350 to 370 °C. However, rapid catalyst deactivation, coking and plugging were observed. Elliot *et al.*[71] studied different approaches for bio-oil HDT. First, a Pd/C catalyst was employed to perform the HDT of different types of bio-oils at temperatures ranging from 310–360 °C. In this case, solid coke encrusted catalyst particles along with some degree of corrosion in the reactor wall were observed. In a next step, the authors used the previously hydrotreated bio-oil and performed a hydrocracking reaction with a commercial sulphided catalyst provided by UOP. The use of pre-stabilized oil allowed operation at higher temperatures (405 °C) and lower hydrogen pressures without coke formation in the catalyst bed. The last alternative studied was the use of a non-isothermal process where both Pd/C and the commercial catalyst were employed in the same reactor without a uniform temperature profile. The temperatures ranged from 250–410 °C. The results are shown in Table 1.1.

Wildschut *et al.*[72] studied several heterogeneous noble metal catalysts (Ru/C, Ru/TiO_2, Ru/Al_2O_3, Pt/C and Pd/C) and typical hydrotreatment catalysts (NiMo–S/Al_2O_3 and CoMo–S/Al_2O_3) for bio-oil deep and mild HDO. The authors concluded that the Pd/C catalyst was the most suitable for mild HDO (250 °C, 100 bar H_2). However, for deep HDO (350 °C, 200 bar H_2), Ru/C appears to be the best choice, even though Pd/C also performed well. The authors observed that in a general manner, TiO_2 and C supported catalysts performed much better than alumina supported ones. Venderbosch *et al.*[73] evaluated a configuration of four reactors in series working at different temperatures and employing a Ru/C catalyst. Even though the strategy was successful, the use of four different reactors may have a great impact on the costs of the process.

Ardiyanti *et al.*[74] evaluated a series of bimetallic NiCu catalysts supported on CeO_2–ZrO_2, ZrO_2, SiO_2, TiO_2, Sibunite and C for the hydrotreatment of fast pyrolysis oil in a batch reactor at 200 bar H_2 pressure. The reactions

Table 1.1 Relevant studies in bio-oil HDT.

Author	Catalyst	*T* (°C)	Pressure (bar)	Deoxygenation level (wt%)	Oil yield (wt%)	Time (h)[a]
Baldauf *et al.*[70]	$CoMo/Al_2O_3$	370	–	99.9	32.8	**4**
	$CoMo/Al_2O_3$	350	–	95.1	32.7	**4**
	$NiMo/Al_2O_3$	370	–	97.2	33.6	**6.67**
Elliott *et al.*[71]	Pd/C	340	138	71	62	**4**
	Pd/C + Commercial C	250–410	138	98.4	50	**6.67**
	Commercial	405	103	98.8	61	**5**
Wildschut *et al.*[72]	$CoMo/Al_2O_3$	250	100	41	27	4
	$NiMo/Al_2O_3$	250	100	36	30	4
	Ru/Al_2O_3	250	100	37	23	4
	Ru/TiO_2	250	100	50	37	4
	Ru/C	250	100	45	35	4
	Pd/C	250	100	56	43	4
	Pt/C	250	100	36	58	4
	$CoMo/Al_2O_3$	350	200	82	25	4
	$NiMo/Al_2O_3$	350	200	75	27	4
	Ru/Al_2O_3	350	200	79	35	4
	Ru/TiO_2	350	200	77	66	4
	Ru/C	350	200	86	53	4
	Pd/C	350	200	85	65	4
	Pt/C	350	200	76	27	4
Venderbosch *et al.*[73]	Ru/C	175	200	17	95	*0.1*
	Ru/C	225	200	35	50	*0.1*
	Ru/C	175/225/375/400	241	45	54	*0.18*
	Ru/C	350/375/400/400	230	64	52	*0.45*
Elliott *et al.*[77]	Ru–S/C + NiMo–S	170/390	138	99	45	**5.26**
	Ru–S/C + CoMo–S/C	170/390	138	95	43	**5.26**

Ardiyanti *et al.*[74]	NiCu/CeO_2–ZrO_2	150/350	200	67.1	40.6	1/3
	NiCu/ZrO_2	150/350	200	59.9	38.9	1/3
	NiCu/TiO_2	150/350	200	62.3	45.5	1/3
	NiCu/CRH	150/350	200	55.9	44.9	1/3
	NiCu/Sibunite	150/350	200	86.5	35.9	1/3
	NiCu/δ-Al_2O_3	150/350	200	57.4	42.2	1/3
Xu *et al.*[78]	Ru/C	250	20	36	52	3
	Ru/C	300	20	48	42.3	3
	Ru/C[b]	250	20	46	84.6	3
	Ru/C[b]	280	20	53	76.5	3
	Ru/C[b]	300	20	68	69.4	3
Ardiyanti *et al.*[79]	NiPd/SiO_2	150/350	120[c]	58.6	39.9	1/3
	NiCu/SiO_2	150/350	120[c]	58.1	44.3	1/3
	NiCu/SiO_2–ZrO_2–La_2O_3	150/350	120[c]	67.8	37.4	1/3

[a]Bold values are the inverse of LHSV. Italic values are the inverse of WHSV.
[b]1-butanol was used as solvent.
[c]Pressure at room temperature.

were performed at two temperatures stages, the first at 150 °C for a period of 1 h, and the second at 350 °C for 3 h, in order to avoid polymerization of the bio-oil and resulting coke formation. The authors observed that the highest HDO activity was from the NiCu/TiO_2 catalyst, however a moderate leaching of both Ni and Cu was observed. NiCu/δ-Al_2O_3 presented the lowest leaching levels. Moreover, the NiCu/TiO_2 catalyst was responsible for the production of an oil product with the highest quality among all catalysts.

Moreover, several other authors, instead of working directly with bio-oil, decided to work with model compounds such as guaiacol. For instance, Gutierrez *et al.*[75] evaluated the performance of mono and bimetallic noble metal catalysts supported on zirconia, such as, Rh/ZrO_2, Pt/ZrO_2, Pd/ZrO_2, RhPd/ZrO_2, RhPt/ZrO_2 and PdPt/ZrO_2 on guaiacol HDO at 300 °C. The authors also compared the performance of the noble metal catalysts with a traditional CoMo-S/Al_2O_3 catalyst and discovered that the Rh-containing catalysts were more active and less susceptible to coke formation than the conventional catalyst.

In conclusion, although bio-oil is a very promising feedstock, its natural properties require that at least one upgrading step is carried out before co-feeding it with petroleum oil in refineries. Therefore, one of the main objectives of researchers working with bio-oil co-processing is to set the minimum degree of pre-treatment needed to introduce a bio-oil feed in refineries without having a major impact on them. In addition, there are several studies being carried out aiming at developing new catalysts more suitable to bio-oil hydrotreatment particularities. The major hurdle to be surpassed is the tendency of coke formation over the catalyst surface. Therefore, the area of co-processing bio-oil in HDT units still needs to be studied and new catalytic and process alternatives must be presented in order to make the idea practicable.

1.5 Conclusions

The traditional refining industry has been challenged to comply with environmental pressure and market requirements. In this context, biomass utilisation is playing an important role and catalysis is the main tool for developing "the refinery of the future". Many studies have been carried out on new technologies with potential to produce hydrocarbon fuels from sustainably sourced feedstocks. It has been shown that co-processing vegetable oils and bio-oils in already existing refinery units might be an interesting alternative since the impact on the CAPEX would be almost negligible compared to the construction of new dedicated units. Indeed, co-processing utilises the original refinery infrastructure by blending bio-derived oils with the regular feedstock, lowering the costs of fuel production from renewables. However, challenges regarding feedstock storage and handling due to the instability of bio-oils and vegetable oils have to be surpassed. In addition, the presence of impurities may corrode equipment and poison the catalyst.

Generally speaking, one should consider some important key aspects[76] related to the perspective of co-processing renewable oils in existing refinery units:

- Guarantee of safety in operation and handling materials. Hazards and generation of new by-products must be taken into account.
- Reliability concerning product quality for complying with clients' property specifications.
- Predictability of the amount of products meeting the market demand and how feedstock oscillations will affect quality and productivity.
- Assured profitability, considering feedstock price and eventual additional costs related to maintenance, catalyst deactivation and inputs.

Regarding the latest research works on co-processing, the main concern still lies in the presence of oxygen compounds in the feedstock, leading to the formation of by-products, with possible catalyst deactivation by poisoning or undesired side-reactions. Moreover, the blending strategy is limited due to the hydrogen deficiency of the feedstock, which increases coke formation if no fossil feed is added to set the hydrogen carbon ratio. For that reason, new catalytic systems and process alternatives are still in development for making the technology practicable.

References

1. C. Briens, J. Piskorz and F. Berruti, *Int. J. Chem. React. Eng.*, 2018, **6**, DOI: 10.2202/1542-6580.1674.
2. G. W. Huber, S. Iborra and A. Corma, *Chem. Rev.*, 2006, **106**, 4044.
3. D. J. Hayes, http://www.carbolea.ul.ie/files/State_of_Play.pdf, 2007, last access May 2007.
4. J. C. Erickson, *Int. Sugar J.*, 2007, **109**, 163.
5. E. Sousa-Aguiar, N. Pereira, D. Alexandre, G. Aranda and A. de Souza Antunes, *Renewable Resour. Renewable Energy*, CRC Press, 2011, p. 391.
6. E. F. Sousa-Aguiar, L. G. Appel, P. C. Zonetti, A. d. C. Fraga, A. A. Bicudo and I. Fonseca, *Catal. Today*, 2014, **234**, 13.
7. E. F. Sousa-Aguiar, J. M. A. R. de Almeida, P. N. Romano and Y. Carvalho, *Industrial Biorenewables*, John Wiley & Sons, Inc., 2016, p. 451.
8. J. Roberto Gomes, S. Bezergianni, J. L. Zotin and E. F. Sousa-Aguiar, *Catalytic Hydrogenation for Biomass Valorization*, The Royal Society of Chemistry, 2015, p. 204.
9. J. P. Diebold, *A Review of the Chemical and Physical Mechanisms of the Storage Stability of Fast Pyrolysis Bio-oils*, National Renewable Energy Laboratory Golden, CO, 2000.
10. M. M. R. Corredores, X. Tong and J. Sorrells, Google Patents, 2016.

11. J. Cai, S. W. Banks, Y. Yang, S. Darbar and T. Bridgwater, *Energy Fuels*, 2016, **30**, 4999.
12. M. M. Ramirez-Corredores and C. Zhang, Google Patents, 2016.
13. M. M. R. Corredores and V. S. Iglesias, Google Patents, 2016.
14. N. Chen, T. Degnan and L. Koenig, *CHEMTECH*, 1986, **16**, 506.
15. J. D. Adjaye and N. N. Bakhshi, *Fuel Process. Technol.*, 1995, **45**, 161.
16. J. D. Adjaye and N. N. Bakhshi, *Fuel Process. Technol.*, 1995, **45**, 185.
17. M. Samolada, W. Baldauf and I. Vasalos, *Fuel*, 1998, **77**, 1667.
18. R. a. D. Maggi and B. Delmon, *Production, Treatment and Utilization of Bio-oils from Pyrolysis for Energy and Alternative Fuels and Chemicals*, AIR2-CT93-1086, UCL Contributions, 1994.
19. A. Corma, G. W. Huber, L. Sauvanaud and P. O'Connor, *J. Catal.*, 2007, **247**, 307.
20. M. E. Domine, A. C. van Veen, Y. Schuurman and C. Mirodatos, *ChemSusChem*, 2008, **1**, 179.
21. A. A. Lappas, S. Bezergianni and I. A. Vasalos, *Catal. Today*, 2009, **145**, 55.
22. I. Graça, F. R. Ribeiro, H. Cerqueira, Y. Lam and M. De Almeida, *Appl. Catal., B*, 2009, **90**, 556.
23. I. Graça, J.-D. Comparot, S. Laforge, P. Magnoux, J. Lopes, M. Ribeiro and F. R. Ribeiro, *Appl. Catal., A*, 2009, **353**, 123.
24. I. s. Graça, J.-D. Comparot, S. b. Laforge, P. Magnoux, J. M. Lopes, M. F. Ribeiro and F. Ramôa Ribeiro, *Energy Fuels*, 2009, **23**, 4224.
25. I. Graça, A. Fernandes, J. Lopes, M. Ribeiro, S. Laforge, P. Magnoux and F. R. Ribeiro, *Appl. Catal., A*, 2010, **385**, 178.
26. I. Graça, A. Fernandes, J. Lopes, M. Ribeiro, S. Laforge, P. Magnoux and F. R. Ribeiro, *Fuel*, 2011, **90**, 467.
27. I. Graça, J. Lopes, M. Ribeiro, F. R. Ribeiro, H. Cerqueira and M. De Almeida, *Appl. Catal., B*, 2011, **101**, 613.
28. F. de Miguel Mercader, M. Groeneveld, S. Kersten, N. Way, C. Schaverien and J. Hogendoorn, *Appl. Catal., B*, 2010, **96**, 57.
29. F. de Miguel Mercader, M. J. Groeneveld, S. R. Kersten, C. Geantet, G. Toussaint, N. W. Way, C. J. Schaverien and K. J. Hogendoorn, *Energy Environ. Sci.*, 2011, **4**, 985.
30. G. Fogassy, N. Thegarid, Y. Schuurman and C. Mirodatos, *Energy Environ. Sci.*, 2011, **4**, 5068.
31. F. A. Agblevor, O. Mante, R. McClung and S. Oyama, *Biomass Bioenergy*, 2012, **45**, 130.
32. G. Fogassy, N. Thegarid, Y. Schuurman and C. Mirodatos, *Green Chem.*, 2012, **14**, 1367.
33. Y. Schuurman, G. Fogassy and C. Mirodatos, in *The Role of Catalysis for the Sustainable Production of Bio-fuels and Bio-chemicals*, ed. A. A. Lappas and M. Stöcker, Elsevier, Amsterdam, 2013, p. 321.
34. N. Thegarid, G. Fogassy, Y. Schuurman, C. Mirodatos, S. Stefanidis, E. Iliopoulou, K. Kalogiannis and A. Lappas, *Appl. Catal., B*, 2014, **145**, 161.
35. L. Gueudré, N. Thegarid, L. Burel, B. Jouguet, F. Meunier, Y. Schuurman and C. Mirodatos, *Catal. Today*, 2015, **257**, 200.

36. C. Lindfors, V. Paasikallio, E. Kuoppala, M. Reinikainen, A. Oasmaa and Y. Solantausta, *Energy Fuels*, 2015, **29**, 3707.
37. A. de Rezende Pinho, M. B. de Almeida, F. L. Mendes, V. L. Ximenes and L. C. Casavechia, *Fuel Process. Technol.*, 2015, **131**, 159.
38. I. Graça, A. Carmo, J. Lopes and M. Ribeiro, *Fuel*, 2015, **140**, 484.
39. C. Wang, M. Li and Y. Fang, *Ind. Eng. Chem. Res.*, 2016, **55**, 3525.
40. Á. Ibarra, A. Veloso, J. Bilbao, J. M. Arandes and P. Castaño, *Appl. Catal., B*, 2016, **182**, 336.
41. A. de Rezende Pinho, M. B. de Almeida, F. L. Mendes, L. C. Casavechia, M. S. Talmadge, C. M. Kinchin and H. L. Chum, *Fuel*, 2017, **188**, 462.
42. L. Gueudré, F. Chapon, C. Mirodatos, Y. Schuurman, R. Venderbosch, E. Jordan, S. Wellach and R. M. Gutierrez, *Fuel*, 2017, **192**, 60.
43. L. A. Sousa, J. L. Zotin and V. Teixeira da Silva, *Appl. Catal., A*, 2012, **449**, 105.
44. A. A. Lappas, S. Bezergianni and I. A. Vasalos, *Catal. Today*, 2009, **145**, 55.
45. S. Bezergianni, V. Dagonikou and S. Sklari, *Fuel Process. Technol.*, 2016, **144**, 20.
46. S. Bezergianni and V. Dagonikou, *Can. J. Chem. Eng.*, 2015, **93**, 1017.
47. C. N. Satterfield and C. M. Smith, *Ind. Eng. Chem. Process Des. Dev.*, 1986, **25**, 942.
48. M. Badawi, J. F. Paul, S. Cristol, E. Payen, Y. Romero, F. Richard, S. Brunet, D. Lambert, X. Portier, A. Popov, E. Kondratieva, J. M. Goupil, J. El Fallah, J. P. Gilson, L. Mariey, A. Travert and F. Maugé, *J. Catal.*, 2011, **282**, 155.
49. A. Pinheiro, N. Dupassieux, D. Hudebine and C. Geantet, *Energy Fuels*, 2011, **25**, 804.
50. M. Philippe, F. Richard, D. Hudebine and S. Brunet, *Appl. Catal., B*, 2013, **132–133**, 493.
51. O. İ. Şenol, T. R. Viljava and A. O. I. Krause, *Catal. Today*, 2005, **106**, 186.
52. R. Z. Lee and F. T. T. Ng, *Catal. Today*, 2006, **116**, 505.
53. E. N. Vlasova, I. V. Deliy, A. L. Nuzhdin, P. V. Aleksandrov, E. Y. Gerasimov, G. I. Aleshina and G. A. Bukhtiyarova, *Kinet. Catal.*, 2014, **55**, 481.
54. I. s. Graça, J. M. Lopes, H. S. Cerqueira and M. F. Ribeiro, *Ind. Eng. Chem. Res.*, 2012, **52**, 275.
55. K. Routray, K. J. Barnett and G. W. Huber, *Energy Technol.*, 2017, **5**, 80.
56. J. L. Carrasco, S. Gunukula, A. A. Boateng, C. A. Mullen, W. J. DeSisto and M. C. Wheeler, *Fuel*, 2017, **193**, 477.
57. X. Zhang, Q. Zhang, T. Wang, B. Li, Y. Xu and L. Ma, *Fuel*, 2016, **179**, 312.
58. M. W. Nolte and B. H. Shanks, *Energy Technol.*, 2016, **5**, 7.
59. M. K. Poddar, A. Rai, M. R. Maurya and A. K. Sinha, *RSC Adv.*, 2016, **6**, 113720.
60. M. Gholizadeh, R. Gunawan, X. Hu, S. Kadarwati, R. Westerhof, W. Chaiwat, M. M. Hasan and C.-Z. Li, *Fuel Process. Technol.*, 2016, **150**, 132.
61. K. Jacobson, K. C. Maheria and A. K. Dalai, *Renewable Sustainable Energy Rev.*, 2013, **23**, 91.
62. A. H. Zacher, M. V. Olarte, D. M. Santosa, D. C. Elliott and S. B. Jones, *Green Chem.*, 2014, **16**, 491.

63. D. C. Elliott, *Curr. Opin. Chem. Eng.*, 2015, **9**, 59.
64. M. Gholizadeh, R. Gunawan, X. Hu, F. de Miguel Mercader, R. Westerhof, W. Chaitwat, M. M. Hasan, D. Mourant and C.-Z. Li, *Fuel Process. Technol.*, 2016, **148**, 175.
65. M. S. Talmadge, R. M. Baldwin, M. J. Biddy, R. L. McCormick, G. T. Beckham, G. A. Ferguson, S. Czernik, K. A. Magrini-Bair, T. D. Foust and P. D. Metelski, *Green Chem.*, 2014, **16**, 407.
66. D. C. Elliott, *Energy Fuels*, 2007, **21**, 1792.
67. D. Elliott, E. Baker, J. Piskorz, D. Scott and Y. Solantausta, *Energy Fuels*, 1988, **2**, 234.
68. M. Ozagac, C. Bertino-Ghera, D. Uzio, M. Rivallan, D. Laurenti and C. Geantet, *Biomass Bioenergy*, 2016, **95**, 194.
69. P. M. Mortensen, J.-D. Grunwaldt, P. A. Jensen, K. Knudsen and A. D. Jensen, *Appl. Catal., A*, 2011, **407**, 1.
70. W. Baldauf, U. Balfanz and M. Rupp, *Biomass Bioenergy*, 1994, **7**, 237.
71. D. C. Elliott, T. R. Hart, G. G. Neuenschwander, L. J. Rotness and A. H. Zacher, *Environ. Prog. Sustainable Energy*, 2009, **28**, 441.
72. J. Wildschut, F. H. Mahfud, R. H. Venderbosch and H. J. Heeres, *Ind. Eng. Chem. Res.*, 2009, **48**, 10324.
73. R. Venderbosch, A. Ardiyanti, J. Wildschut, A. Oasmaa and H. Heeres, *J. Chem. Technol. Biotechnol.*, 2010, **85**, 674.
74. A. Ardiyanti, S. Khromova, R. Venderbosch, V. Yakovlev, I. Melián-Cabrera and H. Heeres, *Appl. Catal., A*, 2012, **449**, 121.
75. A. Gutierrez, R. Kaila, M. Honkela, R. Slioor and A. Krause, *Catal. Today*, 2009, **147**, 239.
76. C. J. Freeman, S. B. Jones, A. B. Padmaperuma, D. M. Santosa, C. Valkenburg and J. Shinn, *Initial Assessment of US Refineries for Purposes of Potential Bio-based Oil Insertions*, Pacific Northwest National Laboratory (PNNL), Richland, WA, United States, 2013.
77. D. C. Elliott, T. R. Hart, G. G. Neuenschwander, L. J. Rotness, M. V. Olarte, A. H. Zacher and Y. Solantausta, *Energy Fuels*, 2012, **26**, 3891.
78. X. Xu, C. Zhang, Y. Zhai, Y. Liu, R. Zhang and X. Tang, *Energy Fuels*, 2014, **28**, 4611.
79. A. Ardiyanti, M. Bykova, S. Khromova, W. Yin, R. H. Venderbosch, V. A. Yakovlev and H. J. Heeres, *Energy Fuels*, 2016, **30**, 1544.

CHAPTER 2

Catalytic Processes and Catalyst Development in Biorefining

OXANA P. TARAN*, NIKOLAY V. GROMOV AND VALENTIN N. PARMON

Boreskov Institute of Catalysis, Department of Catalytic Methods for Sun Energy Conversion, Prospekt Akademika Lavrentieva, 5, Novosibirsk, 630090, Russia
*E-mail: oxanap@catalysis.ru

2.1 Introduction

We are entering a new period of technical progress when renewable plant materials and new technologies for their processing are becoming economical; they are used along with or even in competition with fossil organic materials (oil, natural gas, coal). Biomass first attracted attention as a feedstock for industry as long ago as the XIX century. At that time, industrial processes were being developed for the production of cellulose esters (nitrate and acetate), alcohols, organic acids, sugars, *etc.*,[1,2] and research works in those fields were constantly increasing in number. Since the beginning of the XXI century, interest in the catalytic processing of plant biomass has increased explosively (Figure 2.1).

Green Chemistry Series No. 56
Sustainable Catalysis for Biorefineries
Edited by Francesco Frusteri, Donato Aranda and Giuseppe Bonura

Published by the Royal Society of Chemistry, www.rsc.org

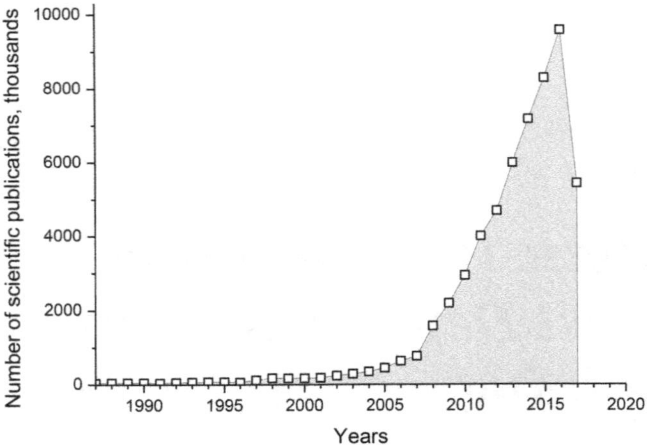

Figure 2.1 The number of scientific publications (papers, review papers, monographs, patents) devoted to catalytic processing of biomass in 1987–2017. The curve is plotted using Scopus, Science Direct, Springer, Royal Society of Chemistry databases. The keywords for inquiries are *biomass* + *catalysis.*

The great interest in biomass is mainly caused by the exhaustibility of fossil fuels (oil, coal, gas).[3] Moreover, the utilization of fossil fuels in petrochemical and chemical processes has become an environmental problem due to the emission of a huge amount of CO_2[4–8] that has provoked serious climate changes accompanied by economic troubles.[3,9] To overcome these problems, alternative solutions are being sought, and biomass utilization is one of them. Plant materials are renewable or even inexhaustible and, therefore, seem very promising. Biomass is formed from CO_2 and H_2O under Sun radiation in amounts of *ca.* 200 billion tons annually.[10] Lignocellulose, the main type of plant material, is up to 99% of biomass.[1] Numerous biomass sources are available, among which are wood and paper industrial wastes as well as agriculture wastes. The following resources of agricultural biomass can be considered for valorization: corn-cobs, straw, rice and sunflower husk, and energy crops (miscanthus, cane, millet). All these types of biomass comprise the same main components: cellulose, hemicelluloses and lignin, and they can all be referred to as lignocellulose. When the said biomass (agricultural wastes and energy crops) is used for the production of fuels and chemicals, there is no competition between valorization and human food supply, but economy and environmental safety are characteristic of biomass valorization methods.

One of the principal goals of modern research in the field of the biorefinery is the design of a series of novel chemical techniques based on the mentioned renewable feedstocks in order to produce valuable products from all the main components, not cellulose alone, of lignocellulose.[11] This is a serious problem due to the robust and very stable structure of two of the main components of biomass – cellulose and lignin. Catalytic methods of biomass transformation are most effective and promising. Of particular interest are catalytic one-pot technologies for direct synthesis of valuable chemicals or biofuels from biomass or its components to eliminate technological stages of semi-product isolation and purification. These technologies allow biofuels or platform molecules to be produced, for example levulinic acid, 5-hydroxymetilfurfural, furfural and furfuric alcohol, ethylene glycol, γ-valerolactone, and polyols (sorbitol, mannitol, *etc.*) from polysaccharides such as cellulose and hemicelluloses,[7,12–20] and aromatic compounds from lignin.[21,22]

Catalysts and catalytic processes involve several steps of a biorefinery; in general, these are (a) processes of biomass deconstruction to produce upgradable gaseous or liquid platforms; and (b) processes of upgrading of deconstructed biomass to useful fuels or chemicals. This Chapter gives a brief overview of the catalytic methods for biomass transformation and the progress in the field. We shall focus on the catalytic lignocellulose biorefinery for the production of fuels and base chemicals.

2.2 Lignocellulose Composition

The lignocellulose biomass consists of 95–99% lignocellulose and 1–5% extractive and ash compounds.[1] There are three main components of the lignocellulose: cellulose (33–60%), hemicellulose (6–30%), and lignin (9–30%).[23–25] Cellulose and lignin are inedible. Moreover, hemicelluloses and lignin do not have wide application in industry.

Cellulose is the key structural component of lignocellulose biomass. It is a polysaccharide built up by residues of glucose cross-linked by 1,4-glycoside bonds. The general chemical formula of cellulose can be written as $(C_6H_{10}O_5)_n$ or $(C_6H_7O_2(OH)_3)_n$. Wood and annual plants contain the highest and the lowest proportions of cellulose, respectively. The large concentration of hydroxyl groups in the cellulose chains allows the formation of numerous intra- and intermolecular hydrogen bonds and, as a result, the formation of high-ordered crystal domains in the cellulose structure. The degree of polymerization of natural cellulose depends on the plant species.[26]

Hemicelluloses are polysaccharides of different composition and structure. Hemicellulose components are hexoses (glucose, mannose, galactose, *etc.*) and pentoses (xylose, arabinose, *etc.*).[27] The exact composition of hemicelluloses depends strongly on the plant species. Xylan (polymer of xylose), arabinogalactan (polymer of galactose and arabinose) and mannans (galactoglucomannan or galactomannan, which are polymers of galactose, glucose and mannose) are most widespread in nature.[27,28] In contrast to cellulose, hemicelluloses are soluble in water and, therefore, capable of easier transforming. Processing of hemicelluloses does not need severe conditions: concentrated bases or acids, high temperature, pressure.

Lignin – the third component of lignocellulose – is an amorphous oxygen-containing organic polymer with a cross-linked structure and aromatic nature. The lignin structure is not uniform and identifiable; it can vary strongly even inside one plant. The main monomer units of lignin are phenylpropane species: guaiacols, syringols, and hydroxyphenols linked to one another through ether or carbon–carbon bonds. There is no regularity in the distribution of the lignin structural fragments.[26]

2.3 Catalytic Processes of Biomass Deconstruction to Produce Upgradable Gaseous and Liquid Platforms

At present, a number of catalytic methods are used for biomass processing. These methods allow different products to be manufactured from biomass (from fine chemicals to fuels and energy). The catalytic methods for lignocellulose transformation can be grouped as follows: (1) processes in the presence of chemical catalysts, and (2) biotechnologies based on the application of microorganisms and/or enzymes (Figure 2.2). The first group of methods is subdivided depending on the temperature of the reaction medium. They are divided into thermochemical and liquid-phase transformation techniques (Figure 2.2).

Thermochemical methods (catalytic combustion, gasification, liquefaction, and pyrolysis) need severe reaction conditions: very high temperature

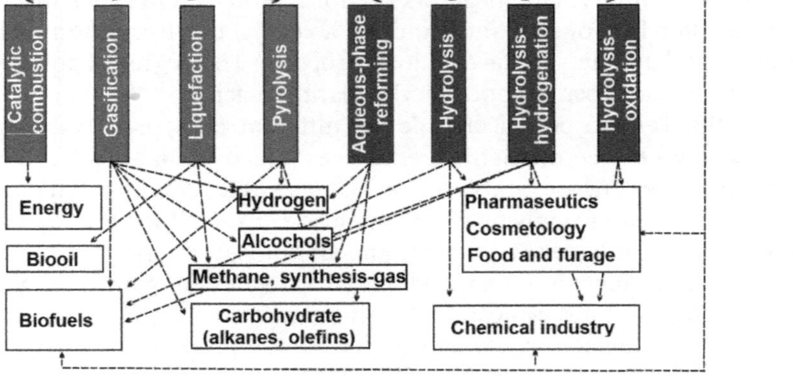

Figure 2.2 Catalytic ways of lignocellulose biomass transformation into chemicals and biofuels.

and pressure (Figure 2.2). Thermal methods for biomass transformations are used for manufacturing of gaseous (synthesis gas), liquid (bio-oil), and solid (biochar) products and energy generation (see Section 2.3.1).

The second subgroup of chemical catalytic methods for lignocellulose biomass conversion is the liquid-phase treatment, which is achieved in aqueous or water-organic media in the presence of catalysts at moderate temperature and pressure (not higher than 250 °C and 70 bar). These are hydrolysis, hydrolysis–hydrogenation, hydrolysis–oxidation, aqueous-phase reforming and biomass delignification (see Section 2.3.2). Since lignocellulose is, generally, insoluble in water, preliminary activation is necessary for the aqueous thermal treatment.[29–32] Chemical, physical and biological methods for the pretreatment of plant biomass materials have been developed.[30] Liquid-phase methods allow chemicals of different classes (monosaccharides, polyols, organic acids, furan and phenol derivatives, hydrocarbons) to be produced from lignocellulose. These methods are very attractive due to the possibility of one-pot implementation of plant biomass conversion.

2.3.1 Thermochemical Conversion of Biomass

Catalytic combustion of solid fuels including biomass in fluidized beds of solid catalysts for direct energy production was developed in the last century;[33] now, this is the technology for incineration and utilization of waste biomass containing high amounts of water.[34,35] The biomass conversion to fuel (bio-oil and biogas) seems more promising in spite of the very high process temperature and pressure. The advantage of the thermochemical approach compared to the liquid-phase treatment is the simultaneous processing of all components of biomass including lignin. Several reviewers[36–41] have discussed different aspects of thermochemical biomass valorization.

Thermochemical methods for the modification of plant biomass are low temperature pyrolysis (torrefaction, carbonization), high-temperature pyrolysis, liquefaction and gasification. They give three types of main products: biochar, bio-oil and gases. The product ratio depends on the feedstock type and, more strongly, on the temperature and time of the thermal treatment (see Table 2.1). The low temperature methods lead to the formation of solid products, mainly biochar. As the temperature rises, the proportion of the solid phase decreases but liquid and gaseous products are predominantly formed. Fast pyrolysis produces bio-oil in sufficiently large proportions; and gasification of plant biomass is the main method for the generation of synthesis gas (Table 2.1).

2.3.1.1 Pyrolysis and Liquefaction

Pyrolysis is the process of the oxygen-free thermal treatment of plant biomass to produce fuel and chemicals. Pyrolysis is among the most practiced methods for high-temperature conversion of biomass. It can be achieved

Table 2.1 Yields of thermal treatment products in different treatment methods according to ref. 42–46.

	Parameters		Approximate yields, %		
Method	Temperature, °C	Time	Biochar	Bio-oil	Gas
Torrefaction	<300	~1 h	80	0	20
Hydrothermal carbonization	~180–400	~1–16 h	50–80	5–20	5
Liquefaction[a]	~250–450	5–20 min	30	45	25
Slow pyrolysis	~400–700	>24 h	35	30	35
Fast pyrolysis	~500–1000	<1 min	10	75	15
Gasification	~700–1000	<1 min	10	5	85

[a]High pressure process, 50–200 bar.

at considerably variable parameters (temperature, treatment time, presence/absence of catalysts). Low temperature pyrolysis is conducted at 180–400 °C, and high-temperature pyrolysis at above 400 °C. Low temperature pyrolysis, when achieved at the bottom boundary of the temperature range, is called *torrefaction*,[47] and the transformations at the upper boundary are *carbonization*.[43] High-temperature pyrolysis can be *fast* and *slow* depending on the rate of biomass heating and treatment time. The estimated maximal temperature of the pyrolytic biomass processing is 800–1000 °C.[48] There are three successive stages of pyrolysis: (1) moist removal at below 200 °C; (2) heating up to the required process temperature; (3) treatment at constant temperature.[48] Fast pyrolysis favors the formation of liquid products; the effective processing of plant materials by fast pyrolysis needs biomass moisture to be of no more than 10% and a particle size less of than 2 mm.[43] The pyrolytic transformations can proceed in the presence of a catalyst.

Pyrolysis produces three types of products: char, gas and liquid. The pyrolysis conditions are the key factor affecting the product composition. Biochar is the main solid product and bio-oil is the main liquid product. The yields of gaseous products are usually no more than 20–25%. Generally, yields of the solid products decrease and yields of the liquid and gaseous products increase with temperature elevation. Biochar is the main product of low temperature pyrolysis (torrefaction and carbonization), and the yield can reach 80%, while the yields of the liquid products range from 0 to 10%. Fast pyrolysis is most promising to produce up to 75% of oil.[49] The optimal temperature range of pyrolytic bio-oil production is 480–520 °C with most feedstock types.[43]

Numerous catalysts were applied to facilitate the pyrolysis. Zeolites are most often investigated.[38–40] The particular interest is caused by their thermal stability, high selectivity to hydrocarbons and limited deactivation.[50] Note that zeolites are well-known catalysts for oil cracking. MCM-41 and metals supported on alumina have also been used as pyrolysis catalysts.[51,52] At the same time, pyrolysis in the presence of zeolite catalysts yields an organic liquid mixture (crude oil) that can contain up to 30% of aromatic products[39] and be used as a fuel.

Liquefaction is the second of the main methods for synthesis of bio-oil. It is essentially different from high-temperature pyrolysis. The plant biomass is treated at high pressure (50–200 bar) and not high temperature (250–450 °C).[46] Bio-oil synthesized by liquefaction contains less oxygen and, therefore, is more energetically capacious. A variety of liquefaction processes include hydrothermal processing (water or aqueous solvent), hydropyrolysis (no carrier liquid solvent), and solvolysis (reactive liquid solvent). The high-pressure characteristic of the liquefaction brings about technical problems and an increased capital cost.[46] A number of catalysts including alkali materials (alkaline oxides, carbonates, bicarbonate, hydroxides), salts of Zn, Cu, Co and Ni (formates, iodines, sulfides, zinc chlorides), and solid catalysts of Ni and Ru nanoparticles are used for liquefaction.[46] Catalysts containing nanostructures of metal are applied for liquefaction followed by hydrogenation.[46]

We should note that, generally, bio-oil is accepted as a promising feedstock for fuel synthesis. However, the bio-oil constituents such as water, oxygen-containing and mineral compounds and solid admixtures worsen its energetic value, making it more corrosive and hindering the economically effective application of bio-oil as a fuel, and this gives an impetus for development of stages of bio-oil upgrading.

2.3.1.2 *Gasification*

Gasification of biomass is a technology for the production of synthesis gas (syngas, or producer gas). Syngas is a mixture of gases (carbon monoxide, carbon dioxide and hydrogen) widely used in chemical industry. Nowadays, synthesis gas is mainly produced from fossil sources (natural gas or coal). However, it can also be produced from plant biomass *via* direct gasification. The following chemical equation describes the overall gasification process:

$$\underset{(\text{Biomass})}{CH_xO_y} + \underset{(\text{Pure or Air})}{O_2} + \underset{(\text{Steam})}{H_2O} \rightarrow CH_4 + CO + CO_2 + H_2 + \underset{(\text{Steam})}{H_2O} \qquad (2.1)$$

Synthesis gas can also be produced in two steps: (1) conversion of biomass to bio-oil; (2) gasification of the bio-oil.[53] Biomass conversion is achieved in gasifiers, which are bed or flow reactors.[45,46,53–59]

Direct biomass gasification is usually conducted over supported metal catalysts in the presence of water steam and small amounts of oxygen.[60] Steam is necessary for the control and uniform distribution of the supplied energy, as well as for increasing the H_2 proportion in the synthesis gas product. Pure oxygen is rather expensive but air gasification produces the gas product with a high concentration of nitrogen. The biomass gasification is a highly endothermic process conducted at temperatures higher than 700 °C.[53] The temperature of biomass gasification is a significant factor and plays an important role in controlling gas composition, tar concentration, reaction rate, and char and ash formation.[57] A high process temperature is reported to provide higher carbon conversion rates, which, ultimately, reduce the tar content and produce more combustible gases.[56,57] At above 750–800 °C, the endothermic nature of the reactions of H_2 production (steam reforming and water–gas reactions) results in an increase in the H_2 quantity and a decrease in the CH_4 content.[61–63] High temperature also favors destruction and reforming of tar to decrease its content and increase the gas yield.[61–63]

The catalysts for gasification must effectively remove tars and be capable of reforming methane, resistant to deactivation, and easy to regenerate.[64] Supported metals (mainly Fe, Ni, as well as Pd, Pt, Ru, Rh), alkali metal salts, alumina and some natural minerals (dolomites, olivines) can behave as catalysts for biomass gasification.[36,64] Among them, Ni seems most promising for biomass gasification, but the metal content, particle size, as well as the catalyst support still need optimization. González *et al.* showed that the use of $ZnCl_2$ and dolomite increases the yield of the gases upon gasification of wastes from the production of olive oil at 800–900 °C.[63] $Rh/CeO_2/SiO_2$ was a very effective noble-containing catalyst for reducing tar in synthesis gas.[65] Alkali salts decrease the tar formation but enhance the char yields.[46] Alkali metal salt catalysts seem unattractive for commercial use because of the poor carbon conversion, increased ash content, and the difficulty of recovering alkali metals.[46]

Processes of direct gasification are used in industry. Large plants for generation of synthesis gas are in operation in Germany, Italy, Sweden, Brasilia, USA, and the UK.[53,66] However, substantial drawbacks of direct gasification are logistic difficulties in product storage and transport. For this reason, development of another process is in progress: the two-stage production of synthesis gas from biomass. The first stage is the production of bio-oil as a storable, transportable and stable (under ambient conditions) feedstock. Then, bio-oil is used for the production of synthesis gas.[53] At present, gasification processes are commercially attractive due to the high price for fossil hydrocarbons.[66]

Another trend in gasification of plant materials is the catalytic processing of biomass in supercritical water.[53,55] This method may be economically effective when the feedstock humidity is high (the humidity may reach 95% in some types of wastes). It is also advantageous over traditional gasification owing to there being no need for the energy consumption of drying.[53,55] Modell *et al.* were first to suggest this technology in their patent in 1978.[67] They discovered that the processing of biomass in supercritical water led to a decrease in the formation of biochar during generation of synthesis gas. Over the last thirty years, there has been considerable progress in biomass gasification in supercritical water. Various types of reactors were suggested: tubular reactor, batch reactor, continuous stirred tank reactor, fluidized bed micro-reactor, and so on.[55] Temperature strongly influences the gasification process. The optimal temperature range is 650–800 °C in the absence of catalysts.[54] As the temperature rises, an increase in the hydrogen proportion in the product mixture, process rate and total of the reaction product yield is observed.[54,68] It is an interesting fact that pressure does not noticeably affect the gasification in supercritical water. Lu *et al.* studied the gasification of wood sawdust at 650 °C and at pressures ranging from 17 to 30 MPa. The total yield of gases was *ca.* 40 mol kg^{-1}.[69] The time of the substrate treatment is, typically, several seconds.[55,69,70] Lengthening of the treatment time results in an increase in the gas yield and in the proportion of H_2 and CH_4.[69,70]

Application of catalysts allows the process temperature to be reduced, as well as the hydrogen yield to be increased and the formation of tar to be suppressed during gasificaiton.[46] Three main types of catalysts were tried: activated carbon prepared from coal or biomass, metal and alkali based catalysts.[55,69] The Ru/TiO_2 catalyst favors decreasing the reaction temperature from 500–800 °C to 350–600 °C.[71] The nickel metal catalysts are usually used, although several noble metals also are applied. An additional advantage of the alkali catalysts, such as calcium hydroxide, is capturing carbon dioxide. Guo *et al.* found that the most effective catalyst depends on the type of biomass to be gasified.[55]

Along with traditional biomass gasification to synthesis gas, plant feedstock can also be used for the production of biogas, which is a gaseous product containing up to 50% methane. Biogas is produced by biotechnological enzymatic treatment of predominantly solid municipal and agricultural wastes. It is a kind of low-energy fuel that contributes only slightly to the power industry. At the same time, biogas production from wastes plays a part in environmental protection.

In conclusion, gasification of plant biomass to produce synthesis gas or biogas seems a promising trend in developing technologies for the processing of alternative raw materials. Of particular interest is the possibility of combining processes of gasification and traditional industrial processes utilizing synthesis gas, for example, the Fischer–Tropsch synthesis. The possibility of the use of plant-based gas as fuel also seems attractive. However, industrial implementation of plant biomass gasification meets a number of technological, economic and ecological challenges. Sansaniwal and co-authors[57] identified the following problems: (1) acceptable biomass composition for processing; (2) biomass pretreatment is an energy and labor intensive process and therefore requires attention to maintain overall economic feasibility; (3) dependence of the gasification process on numerous parameters including temperature and feedstock humidity; (4) synthesis gas conditioning and cleaning are the main challenges. Synthesis gas contains various undesirable substances and chemicals including tar, particulates, alkali metals, ammonia, sulfur and hydrochloric acid. The development of methods for impurity removal is an important challenge.[45,56]

2.3.2 Liquid-phase Methods

Considering the structure of all the lignocellulose components leads to the conclusion that hydrolysis is the main tool for chemical catalytic depolymerization of biomass. Indeed, hydrolytic depolymerization of lignocellulose biomass produces monomers (monosaccharides and aromatic compounds) from the lignocellulose components (Figure 2.3). The monosaccharides may undergo further transformations in the presence of the same solid acid catalysts as the ones used for the hydrolysis. Consecutive dehydration–hydration reactions transform sugars to some furan compounds (5-hydroxymethylfurfural (5-HMF), furfural) and carbonic acids, for example levulinic acid (LA).[14–16] These compounds are known as *platform molecules* and fall into the set of the most promising compounds for industrial development.[2,7,17–20] While these compounds can be produced in the presence of the same catalysts that favor hydrolytic depolymerization of lignocellulose and its main components, it seems possible to develop one-pot processes for the direct synthesis of platform molecules from biomass.[12,13] Advantages of the one-pot processes are well known. These are improvement of the economic and ecological efficiency, as well as an increase in the selectivity and product yields due to avoiding the stages of isolation, purification and pretreatment of intermediate compounds.

All potential products of the hydrolytic transformation of lignocellulose are either in current use in industry or may be used when the economically effective technologies for their production are developed. These products are used in the food and pharmaceutical industries (manufacturing of food supplements sorbitol, mannitol, ethyleneglycol), for manufacturing of octane number improving additives (based on 5-HMF, furfural, alkyl levulinates) to traditional gasolines, and as antifreeze (ethyleneglycol and propylene), biofuels (including bioethanol and biobutanol), some solvents (hexane, toluene, tetrahydrofuran, *etc.*), resins and polymer materials, *etc.*[7,72–74] In the European Union, the commercial implementation of

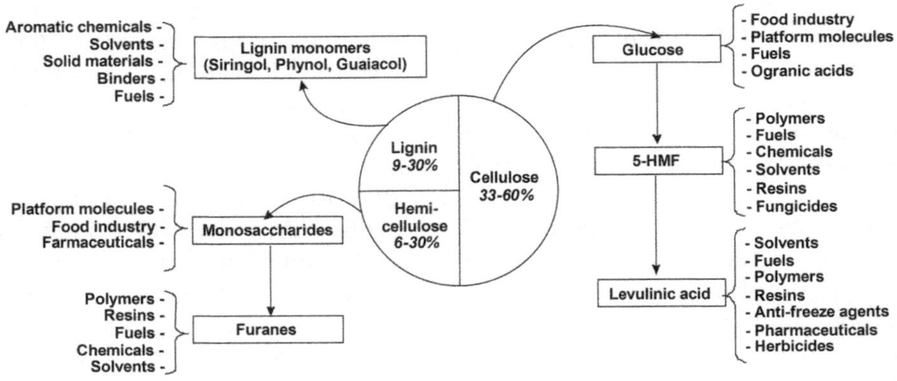

Figure 2.3 Practical applications of lignocellulose depolymerization derivatives.

technologies for manufacturing 5-HMF and levulinic acid from raw lignocellulose materials is in progress. For example, AVA Biochem (Germany) put into operation a commercial plant for the production of 5-hydroxymethylfurfural in 2014,[75] GFBiochemicals (Italy) started the commercial production of levulinic acid in 2015, *etc.*[76]

The present part of the Chapter is an overview of the liquid-phase catalytic methods for processing of lignocellulose and its components in the presence of catalysts, predominantly solid catalysts. More details on catalysts and catalytic processes for depolymerization of biomass are given in Chapter 3.

2.3.2.1 Hydrolytic Depolymerization of Cellulose and Hemicelluloses

Conversion of the polysaccharide components of biomass such as cellulose and hemicelluloses has been studied very intensively. Numerous research papers and reviews have been published since the pioneering research done in the XIX century.[24] Nowadays, hydrolysis is the main reaction applied for cleavage of glycoside bonds in polysaccharides.[24,77] The reaction is usually conducted in the presence of acid or base catalysts in the temperature range of 97–297 °C, acid hydrolysis being most common. Acids also catalyze the dehydration of sugars obtained by hydrolysis to produce furfural, 5-hydroxymethylfurfural, levulinic acid and some other chemicals. For example, hydrolysis of cellulose makes it possible to obtain glucose, 5-hydroxymethylfurfural, and levulinic acid. The conversion of xylan yields xylose and furfural. All the products are well-known platform molecules and can be used in industry.[19,78]

Biocatalysis or chemocatalysis are appropriate for the hydrolytic transformation of biomass. Biocatalytic processing of cellulose in the presence of an enzyme produces glucose at yields close to 80%.[79–81] However, biocatalytic processes need the activation of plant biomass. The disadvantages of biocatalysis, among which are low reaction rates, high costs of enzymes and consumption of sugars for feeding enzyme-producing microorganisms, have brought about a great interest in chemocatalytic hydrolysis.[1,3,9,82,83]

Polysaccharide hydrolysis in acid solutions has a long history. The process was developed and implemented in industry in the early XX century.[84–87] Sulfuric acid,[88] its derivatives[85] and sodium hydroxide[89] are commonly used soluble catalysts. Hydrochloric acid,[90,91] nitric acid[92] and phosphoric acid[88,92] are also catalytically active. Among the examined mineral acids, H_2SO_4 is the most active catalyst for cellulose hydrolysis.[93] Industrial methods allow glucose to be produced at very high yields (as high as 80%).[85] However, the strong corrosive impact of concentrated acids prevents general industrial applications of the technology.[32] A more general method is cellulose depolymerization by diluted acids on heating to 200 °C[85] to obtain 80–90% glucose.[88,94] Mineral acids catalyze not only cellulose hydrolysis but also synthesis of 5-HMF or levulinic acid from glucose.[72,95] However, this is undesirable when glucose is the target product.

Organic acids (acetic, formic, malic, oxalic acids) also catalyze biomass and cellulose hydrolysis but the yields of sugars are lower than in the presence of mineral acids.[85,87,96,97] Promising soluble catalysts for hydrolysis are heteropolyacids (HPAs),[98,99] which can be more effective than mineral acids[91] due to their stronger acidity.[98,99] Glucose yields of up to 60–75% may be achieved upon cellulose hydrolysis in the presence of HPAs.[91,100]

On the other hand, a new approach based on solid acid catalysts for polysaccharide hydrolysis is in rapid progress. This approach is attractive owing to the simple separation of the solid catalyst from the soluble products, the catalyst reusability, and the reduction in the risk of corrosion impact on the equipment and wastewater generation. A wide range of catalysts such as insoluble heteropolyacids,[101] modified carbons,[73,102–112] supported noble metals,[113] polymer resins,[104] oxides of aluminium, silicon, niobium,[104,114,115] and zeolites[115,116] were tried for biomass and cellulose hydrolysis and hydrolysis–dehydration.

The most common method of hemicelluloses hydrolysis is catalysis by diluted mineral acids (sulfuric acid,[117,118] HCl,[117] H_3PO_4,[118] trifluoroacetic,[118] maleic,[119] oxalic,[120] acitic,[121] acids and their mixtures[117]). Hemicelluloses hydrolysis needs milder conditions than cellulose hydrolysis or biomass delignification (100–220 °C).[117] Hemicelluloses can be removed directly from biomass without any activation.[118] Xylan can be hydrolyzed under milder conditions (90–160 °C) compared to hydrolysis of arabinogalactan (100 °C) or mannan (180–230 °C).[117,122]

Only a few studies have dealt with the hydrolysis of hemicelluloses and starch over solid acids. The zeolite catalyst LZ-M-8 applied for inulin hydrolysis to fructose at 100 °C afforded the fructose yield of 96% at almost 100% substrate conversion.[123] Polystyrene resins allowed starch to be hydrolyzed to give a yield of 100% glucose at 120 °C.[124] Zeolites H-Beta, H-Y, and H-ZSM-5 produce 5-HMF from starch with the yields of 9%, 9% and 12%, respectively.[125] Ti(IV) oxide appeared active to xylan hydrolysis–dehydration with the yield of furfural equal to 25%.[126]

Biphase reaction systems comprising water and organic solvents are effective to stabilize the reaction products, especially 5-HMF, and to prevent their decomposition at high temperature *via* extraction into the organic phase. Numerous papers are devoted to the synthesis of 5-HMF from monosaccharides in biphase water–organic systems.[7] The first works dealing with the conversion of polysaccharides to 5-HMF in biphase media were published in the 1980's. Now, this field has become more and more attractive. 5-HMF can be synthesized from inulin at a high yield (87%) in a water–butanol-2 mixture over a Ta_2O_5 catalyst.[127] The highest yield of 5-HMF from cellulose reported in the literature is 35% over TiO_2 in the water–MIBK system in a flow reactor.[128]

A method for the synthesis of levulinic acid from sugars was developed as long ago as the latter half of the XIX century[129] and remains the most practiced till now.[130] The process is catalyzed by soluble acid catalysts, preferably sulfuric acid.[46,72,130–132] BioMetics Inc. developed the biorefining process to produce levulinic acid at 50–70% yields from cellulosic feedstocks, including paper mill waste, wood waste, and agricultural residues, using dilute acid hydrolysis.[46]

2.3.2.2 Cellulose and Hemicelluloses Hydrolysis–Hydrogenation and Hydrolysis–Oxidation

Another way to convert polysaccharides into valuable chemicals is one-pot hydrolysis–hydrogenation in the presence of acid–base and reducing catalysts. This is a two-stage process. The first stage is hydrolysis of polysaccharides to give monosaccharides, and the second is hydrogenation of monosaccharides into their respective polyols. Two main approaches are suggested for the one-pot hydrolysis–hydrogenation of biomass produced polysaccharides. The first is the simultaneous use of two catalysts in one reactor. One catalyst is for hydrolysis and another for hydrogenation. In this case, mineral acids or heteropolyacids in combination with noble metal nanoparticles (Ru, Pt, *etc.*) are usually applied for hydrolysis and hydrogenation, respectively. Development of a bifunctional catalyst that bears both hydrolysis and hydrogenation sites is the second approach. These catalysts are prepared *via* immobilization of noble metal nanostructures on a support with acid and/or base surface groups. Hydrogen gas is used as a reductant for hydrolysis–hydrogenation. Propanol-2 can behave as a reducing agent instead of hydrogen gas; a high pressure is not required in this case.[133]

The pathway for the one-pot transformation of cellulose is as follows: Cellulose is hydrolyzed to glucose, then an aldehyde group of glucose is reduced under hydrogen pressure to alcohols (sorbitol and mannitol) as the main products.[17] Ruthenium and platinum are most promising among the noble metals (Ru, Pt, Au, Rh, Ir, Pd) for cellulose hydrolysis–hydrogenation.[134] The 2.5% Pt/γ-Al_2O_3 catalyst provides the highest yields of sorbitol and mannitol equal to 25% and 6%, respectively. Modifying Ru/C catalysts by phosphoric acid redoubles the catalyst activity.[135]

Ni-containing catalysts can also be used for hydrolysis–hydrogenation.[136] Sorbitol and mannitol yields equal to 50 and 6% can be observed over 3% Ni/C (190 °C and 60 bar H_2). Ni supported on carbon nanotubes produces 64% sorbitol and 7% mannitol.[137] It should be noticed that Ni catalysts are quite stable under the reaction conditions. However, in contrast to Ru catalysts, Ni/C promoted with phosphorous appeared unstable, and the sorbitol yield of 48–60% was only observed in the first reaction cycle.[138,139] Iridium was shown to improve the catalyst stability; bimetal Ir–Ni catalysts supported on mesoporous carbon allowed the sorbitol yield of *ca.* 58%.[137]

Ru/C in combination with mineral acids hydrolyzed cellulose at 70 bar H_2 to give 39% sorbitol.[140,141] The 73% yield of sorbitol was attained over Ru supported on carbon nanotubes in mineral acid solutions.[138] Mixing Ru/C with the soluble heteropolyacid $H_4SiW_{12}O_{40}$ increased the rate of cellulose hydrolysis–hydrogenation, and the sorbitol yield reached 85%.[142]

Another one-pot method for cellulose hydrolysis–hydrogenation is a combination of solid cesium salts of heteropolyacids (CsHPA) with Ru/C. The highest yields of sorbitol and mannitol equal to 59% and 12%, respectively, were obtained in the presence of the mechanical mixture $Cs_{2.5}H_{0.5}PW_{12}O_{40}$–Ru/C at 190 °C at 50 bar H_2. It should be noted that the combination of two solid catalysts appeared as effective as 3 times that of Ru/C with HPA solution.[143] Bifunctional catalysts were also developed for the process under discussion. 1% $Ru/Cs_2HPW_{12}O_{40}$ and 1% $Ru/Cs_3PW_{12}O_{40}$ produced sorbitol at the yield of 40% from cellulose at 160 °C.[144] The selectivity to sorbitol increased in the presence of the bifunctional catalyst compared to that over the mixed catalysts.

One-pot hydrolysis–hydrogenation of hemicelluloses into their corresponding polyols was also studied. Arabinogalactan was converted to arabitol and galacitol over Ru/MCM-48 at 185 °C and 20 bar H_2 in 24 h with a total yield of polyols equal to 25%.[28]

At higher temperatures and pressures of H_2, the hydrogenation process is accompanied by hydrogenolysis to form propylene glycol and ethylene glycol. These are industrially significant products synthesized usually from natural gas. Recent studies have led to the development of a method for direct production of propylene glycol and ethylene glycol from plant biomass.[145–147] An aqueous medium, 60–100 bar H_2, 230–250 °C and 0.5–4 h reaction time are typical reaction conditions for lignocellulose hydrolysis–hydrogenation.[148] The most effective catalysts for cellulose transformation to glycols were W-containing compounds and composites.[145,148] Tungsten carbide supported on carbon produced 73% ethylene glycol from cellulose.[149] Nanoparticles of platinum group metals and Ni supported on tungsten oxides (WO_3, H_2WO_4) and W-containing heteropolyacids were also studied as catalysts for glycol production,[145,149,150] Ni being the most active among them. Ru is very promising, while Pt, Pd, Rh, and Ir were less active.[145,149,150] A 50% yield of ethylene glycol from cellulose was observed with the 3% $Ru/NbOPO_4$ catalyst.[147]

Hydrolysis–oxidation of biomass and cellulose seems a particularly promising route for plant biomass valorization.[8,151] This method allows such valuable chemicals as gluconic acid,[3,152] furandicarbonic acid,[153] and formic acid[154] to be produced. Chemicals produced *via* biomass oxidation are valuable substances for chemical synthesis, and the polymer,[155] textile and chemical industries.[156] Formic acid can behave as the reducing agent or hydrogen source.[157,158]

Highly dispersed supported metal particles (Au, Pt, Pd) are the known catalysts for oxidation of saccharides to their corresponding acids, including oxidation of glucose to gluconic acid.[159–161] The one-pot synthesis of gluconic acid from cellulose over bifunctional acid-oxidative catalysts is very attractive.[151] Cellobiose hydrolysis–oxidation to gluconic acid was studied in the presence of gold and platinum nanoparticles supported on sulfonated activated carbon[160] and carbon nanotubes pretreated with nitric acid.[159] The yields of gluconic acid reached 80%. Onda and co-authors observed a synergetic effect with the bifunctional catalyst Au/AC-SO_3H; the yield of gluconic acid from starch was 40% over the bifunctional catalyst and only 10% in the presence of a mechanical mixture of Au/C and AC-SO_3H.[160] Another type of polysaccharide hydrolysis–oxidation catalyst is noble metal nanoparticles supported on solid HPAs. Such systems allow the production of gluconic acid from cellobiose at yields close to 95%.[162]

While both acid and oxidative properties are characteristic of Mo–V–P heteropolyacids, they are very good catalysts for the one-pot production of formic acid from biomass polysaccharides under mild conditions.[154,157] Mo–V–P HPA catalysts were first used for cellulose hydrolysis–oxidation.[163,164] Formic acid can be obtained from cellulose at yields as high as 66% at 150–160 °C and 20 bar air.[154] HPAs with different compositions (different acid and vanadium contents) were tested: $H_5PMo_{10}V_2O_{40}$ (HPA-2), $Co_{0.6}H_{3.8}PMo_{10}V_2O_{40}$ (Co-HPA-2), $H_{11}P_3Mo_{16}V_6O_{76}$ (HPA-6′) and $H_{17}P_3Mo_{16}V_{10}O_{89}$ (HPA-10′). The highest reaction rate was reached in the most acidic solutions, and the linear dependence of the initial reaction rate on H^+ concentration was observed. Both the yield of formic acid and the initial reaction rate decreased in the series: HPA-2 > Co-HPA-2 > HPA-6′ > HPA-10′.[154]

2.3.2.3 Aqueous-phase Reforming

The most traditional process for converting biomass to H_2 is high-temperature pyrolysis (see Part 2.3.1.1). Aqueous-phase reforming (APR) is another way to produce H_2 containing gas mixtures (H_2, CO and CO_2) to be transformed into pure gas and used for energy applications or the Fischer–Tropsch process.[165] Numerous studies dealt with APR of biomass and its components.[166] Catalysts were shown to play a crucial role in APR of polysaccharides. The yield of hydrogen and the gaseous product ratio depend strongly on the catalytic system. Pt, Ni or Sn catalysts are usual for the process conducted below 230 °C.[36,166,167] Pt-containing catalysts are highly selective to the formation of H_2.[166] On the other hand, high yields of methane over Ni provoke coke formation and cause very rapid catalyst deactivation. A combination of Ni and Sn improves the characteristics of Ni-containing systems and makes the activity of nickel catalysts close to that of Pt.[167] It should be emphasized that Pt/C is the catalyst most often used for APR.[168] The activity of metal particles supported on silica was shown to decrease in the series: Pt ~ Ni > Ru > Rh ~ Pd > Ir in the temperature range of 180–225 °C. The activity of the Pd catalyst can be improved by promoting the system with Fe (Pd_1Fe_9/Al_2O_3 as an example). Pt and Pd are more selective to hydrogen, while Ni, Ru and Rh are more selective to C_1–C_{15} products.[169]

2.3.2.4 Biomass Delignification

Delignification processes, which are traditional for the pulp and paper industry, are achieved in the presence of soluble acid–base catalysts. The currently used processes are, first, the most technologically effective and economic alkali methods including the Kraft process, and second, versions of the Sulfite process. They are used for the preparation of wood pulp (semicellulose, cellulose) from biomass; technologies based on these methods take 60% and 5%, respectively, of the global industrial production of cellulose materials.[170] The Kraft and Alkaline processes are distinguished among the alkali methods. In the Alkaline process, hydrazine sulfate or a mixture of sodium hydroxide and sulfide is used. In the Kraft process, the reagent is a mixture of sodium hydroxide and sulfide. The main reagents of the Sulfite process are sulfur dioxide, sulfurous acid, as well as bisulfates and sulfates. The ecologically dangerous traditional technologies for chemical wood cooking make it necessary to develop alternative environmentally friendly technologies.

The methods for delignification suggested to date can be classified into five groups:[171]

(1) Chemical treatment by acids, alkalis, or oxidants. Traditional technologies are also the chemical methods.
(2) Solvent fractionation with organic solvents such as methanol, ethanol, ethylene glycol, or acetone or a mixture of organic solvents and water.
(3) Biological treatment with microorganisms.
(4) Physical treatment (steam explosion, supercritical CO_2 explosion).
(5) Catalytic fractionation.

Catalytic fractionation, known also as the lignin-first strategy, is a new field.[172,173] In the narrowest sense of the term, lignin-first is often interpreted literally as *disassemble lignin prior to carbohydrate valorization*.[173] This interpretation indicates the strategy that combines biomass fractionation with tandem catalytic lignin depolymerization and stabilization of low molecular weight products. The catalytic hydrogenolysis of native lignin in wood and straw is a well-studied illustration of this strategy.[172–186] Although various names were given to this particular process (*e.g.*, Reductive Catalytic Fractionation (RCF),[168,174,175,179] Early-stage Catalytic Conversion of Lignin (ECCL),[172,179] Catalytic Upstream Biorefining (CUB),[172,179] hydrogenolysis of protolignin *etc.*), and small variations in the process exist, they follow the same basic mechanism: the solvolytic extraction of lignin is accompanied by instant lignin depolymerization and reductive stabilization of reactive intermediates. The depolymerization proceeds better in the presence of acid catalysts.[182] The reductive step requires the presence of a reductive catalyst (*e.g.*, Pd/C, Reney Ni, Ni/C, Ni/Al_2O_3)[175,178,179,184,185] in combination with a reducing agent, which can be pressurised H_2,[175,181] or a hydrogen donor (*e.g.*, iPrOH, MeOH, formic acid).[174,176,178] The product of this lignin-first approach is a highly depolymerized lignin oil comprising a set of monomers in close to the theoretical yield, and dimers and short oligomers.

Recently, Luterbacher *et al.* developed a new fractionation method that fits perfectly with the lignin-first strategy.[187] Formaldehyde was used as a coreagent during the acid-catalyzed delignification procedure, which forms stable acetals with lignin α- and γ-OH side-chain groups. The authors demonstrated that the reductive catalytic depolymerization of the isolated lignin (in a second step) results in the same close-to-theoretical monomer yield as obtained through depolymerization of native lignin (47%).[187]

Another approach that also fits the lignin-first strategy is oxidative catalytic fractionation. Kuznetsov *et al.* developed single-step processes of wood fractionation to cellulose and soluble lignin through oxidative depolymerization of lignin with hydrogen peroxide and oxygen in an acetic acid–water medium in the presence of different catalysts.[188–196] The peroxide catalytic oxidation of hardwood and softwood in the acetic acid–water medium under optimized mild conditions (≤100 °C, 1 MPa) enables the effective separation of wood biomass components. Solid microcrystalline cellulose and soluble products derived from lignin and hemicelluloses are produced by the peroxide catalytic oxidation.[191–195] The possibility of the successful replacement of the dangerous H_2SO_4 catalyst by a non-toxic and non-corrosive solid TiO_2 catalyst for peroxide fractionation of hardwood and softwood to microcrystalline cellulose and organosolv lignin under mild conditions was shown.[189,195] The catalytic oxidation with O_2 makes it possible to decompose the lignocellulosic biomass into aromatic aldehydes (mainly vanillin and syringaldehyde) and cellulose.[190,196]

However, implementation of the mentioned catalytic approaches of the lignin-first strategy is more complex compared to the conventional fractionation strategies and, therefore, more expensive. This extra cost can be critical for the economic viability of the future lignin-first biorefineries. Further research activities are still needed in order to comprehend the following aspects: (1) catalyst stability and reusability; (2) solvent stability and reusability; (3) energy requirements; (4) product purification; (5) reactor set-up and operational cost; (6) consumption of hydrogen or other reagents; (7) feedstock requirements and cost. Although a number of other challenges may also appear.[172,173]

2.4 Catalytic Processes for Upgrading Deconstructed Biomass to Useful Fuels and Chemicals

2.4.1 Synthesis Gas

Synthesis gas is the basis of a number of industrial processes,[56] for example the production of a variety of chemicals such as ammonia or methanol. Synthesis gas can be converted into synthetic petroleum as a fuel or lubricant *via* the Fischer–Tropsch synthesis using technologies that were originally developed in Germany during the Second World War and refined in South Africa when it was embargoed. Synthesis gas is an intermediate when the desired product is hydrogen, with the carbon monoxide component transformed into additional hydrogen through the water-gas shift reaction. Synthesis gas utilized in industry is traditionally produced from coal; the gas obtained from plant biomass can also find commercial application.[56,57]

Studies on synthesis gas processing are in progress. A number of new catalysts have been suggested to be used for conversion of synthesis gas to valuable products.[46,56,66]

Production of hydrogen, which is an important feedstock for energy generation and biomass processing, is one of the key methods for synthesis gas processing.[46] Two types of processes for the production of hydrogen from synthesis gas *via* the water-gas shift reaction over Fe or Cu-based catalysts are currently practiced in industry.[46] Zhang and co-authors designed a pilot plant for processing of switchgrass to hydrogen *via* the water-gas shift reaction over Fe_2O_3 and CuO catalysts.[197] Supported nanosized gold catalysts were shown to be very active in water-gas shift reactions.[46] Biological methods are available to achieve the water-gas shift reaction with photoheterotrophic bacteria at ambient temperature and pressure. However, this is a low efficiency process because of the low rate of H_2 production.[198]

Synthesis gas is also used to produce methanol through methanol synthesis over Cu/ZnO catalysts in the temperature range of 220–300 °C and pressures of 50–100 bar.[46] Gerber reported that Rh/Mn/Fe/SiO_2 is best among 10 catalysts for C2–C5 alcohol production.[56]

The Fischer–Tropsch (FT) process produces alkanes from synthesis gas in the presence of Fe, Co and Ru-based catalysts. Co and Fe are more practical to use because of the high cost of Ru. Co is also more expensive than Fe but more useful due to its higher activity and longer life. The process occurs at 20–40 bar and 180–250 °C. Many authors assumed that the selectivity to the product over catalysts on zeolite based supports can be narrowed.[199] The bifunctional catalysts such as Cr_2O_3–ZnO/ZSM-5 allow two simultaneous reactions (synthesis gas to methanol with the metal function and methanol to hydrocarbons with the acidic function) to take place.[200]

Another possibility to utilize synthesis gas is its enzymatic conversion to bioethanol, which is successfully used as a transportation fuel in many countries, notably in the USA and Brazil. Henstra *et al.* summarized the microorganisms used for the fermentation of synthesis gas.[201]

2.4.2 Bio-oil

Bio-oil is the main product of the fast pyrolysis of biomass. It can be formed at yields of 70–75% from lignocellulose and is known as a *crude pyrolysis liquid*.[43] Bio-oil is a viscous dark brown or black colored liquid consisting of a very complex mixture of organic and mineral substances. There are very high quantities of oxygen and sulfur, as well as moisture and macro- and microelements in the organic fraction. The oxygen content may reach 10–40%, and water content can reach 25%.[202] These are distinguishing features of bio-oil from fossil oil. The heating value of bio-oil is approximately 50% of the heating value of fuels, which makes it necessary to upgrade bio-oil.[53] The concept of bio-oil upgrading can be briefly described as *from bio-oil to biofuels*. Even though the bio-oil composition differs from the composition of fossil oil, the bio-oil upgrading may be based, generally, on the same processing stages and procedures to give motor fuels (diesel, gasoline, kerosene *etc.*). One of the principal purposes of bio-oil processing is deoxygenation, *i.e.* oxygen removal.

The process of enhancement of bio-oil properties starts with physical pretreatment stages, first of all, filtration of the bio-oil prepared through biomass pyrolysis to remove solid reaction products, as well as dilution of bio-oil using organic solvents in order to homogenize the mixture and to make it less viscous. At present, various methods for bio-oil upgrading are under development. These are cracking, decarbonylation, decarboxylation, hydrocracking, hydrodeoxygenation, hydrogenation, and polymerization. Most attention is paid to cracking and hydrodeoxygenation. Cracking implies removal of oxygen in the form of carbon dioxide (eqn (2.2)). The chemical equation of catalytic cracking has been suggested in a number of review papers:[43,202]

$$CH_{1.33}O_{0.43} + 0.26O_2 \rightarrow 0.65CH_{1.20} + 0.34CO_2 + 0.27H_2O. \quad (2.2)$$

Among the advantages of this method is the feasibility of the ambient pressure process without additional hydrogen sources. Catalysts based on zeolites and mesoporous materials, which are traditional cracking catalysts in petrochemistry, can be used for bio-oil deoxygenation through cracking, too. Catalysts based on ZSM-5, H-Y, MCM-41, SBA-15, as well as FCC catalysts, *etc.* operate at low temperatures (*ca.* 200 °C).[43,202–205] The low temperature processes were considered preferable because of the anticipated increase in the yield of gaseous products at high temperature. However, some researchers report bio-oil conversion under more severe conditions at 300–500 °C[206] or even 300–600 °C.[204] The temperature elevation is thought necessary for more effective bio-oil deoxygenation.

Another method for bio-oil deoxygenation is hydrodeoxygenation, which is more practical than cracking.[204] Hydrodeoxygenation is conducted in barosphere hydrogen that reduces the oxygen-containing organic constituents to form water. In terms of the process arrangement, hydrodeoxygenation of bio-oil is similar to the industrial hydrodesulfurization technology.[204] The chemical equation of bio-oil hydrodeoxygenation has been suggested in a number of review papers (eqn (2.3)):[43,202,204,206]

$$CH_{1.33}O_{0.43} + 0.77H_2 \rightarrow CH_2 + 0.43H_2O. \quad (2.3)$$

Similar to the industrial hydrodesulfurization, hydrodeoxygenation of bio-oil involves processes of hydrocracking, reduction, decarboxylation and decarbonylation, the pivotal reaction being hydrodeoxygenation. Bio-oil is hydrodeoxygenated at a hydrogen pressure varied over a wide range from 10 to 300 bar.[43,203,205,207] High hydrogen pressures enable hydrogen dissolution in the bio-oil thus increasing its concentration and, as a consequence, the rate and depth of hydrodeoxygenation. The deep and relatively fast hydrodeoxygenation favors a decrease in the coke formation.[202] Temperature of bio-oil upgrading usually ranges from 250 to 450 °C.[43,204] The conversion can be deepened by lengthening of the reaction time.[204] Eventually, the yield of naphta may be *ca.* 25% of the initial mass of the plant feedstock.[43] Historically, sulfides Co–MoS_2 and Ni–MoS_2 supported on aluminosilicates were the first catalysts for the processes under study.[43,202,204] They were chosen as the traditional catalysts for industrial hydrotreatment processes. However, some drawbacks of these catalysts were revealed regarding the bio-oil hydrodeoxygenation. For example, aluminosilicate supports were unstable in the presence of large quantities of water. The next generation of catalysts for hydrodeoxygenation were noble and some transition metals supported on stable oxides. Systems of Ni and Ni–Cu supported on ZrO_2, Al_2O_3, SiO_2 or CeO_2, and mixed oxides, which were suggested by Yakovlev *et al.* for processing of model compounds (anisole, guaiacol),[208–210] seem promising as catalysts for the hydrodeoxygenation. Ni–Cu had the potentiality to completely eliminate the oxygen from anisole. Unfortunately, this comparison gives only a vague idea about how one can compare the nickel based catalysts to other catalysts. Quantification of the activity and affinity of these catalysts to carbon formation against those of noble metal catalysts such as Ru/C and Pd/C or Co–MoS_2 would be interesting. Catalysts containing noble Ru, Rh, and Pd metals supported on oxides or carbon are another series of promising catalysts[211] under especially intensive studies. Zhao *et al.* demonstrated the potential of Ni, Co, Fe, W and Mo phosphides.[212] These catalysts were less active than noble metals but much more stable than the traditional sulfide catalysts. The suggested catalytic systems underwent comprehensive testing at 250–400 °C and up to 20 bar hydrogen during hydrodeoxygenation of organic substrates modeling bio-oil. Of particular interest is the search for and characterization of a promising support for the active component. The support must be highly stable. The insufficient stability of the support and the tendency toward coke formation were demonstrated earlier with the Ni–MoS_2/Al_2O_3 catalyst as an example.[213] The promising supports are carbon[214] and the oxides SiO_2, ZrO_2, and CeO_2.[208–210,213] Coke is formed in a smaller quantity on these supports because of their lower surface acidity compared to that of alumina. Generally speaking, new catalysts for hydrodeoxygenation of bio-oil are being intensively developed and bio-oil hydrodeoxygenation in the presence of these catalysts is accepted as the most prospective method for bio-oil upgrading. One of the challenges in the field is the development of a stable low-coking catalyst and increasing the reaction efficiency to 100%.

Different methods for bio-oil upgrading are under study along with cracking and hydrodeoxygenation. Among them is mild cracking that is conducted, unlike the classical method, in the presence of base catalysts to suppress the coke and gas formation. ZnO and MgO are suggested catalysts for this process.[43] A particular field of research is the catalytic reductive processing of bio-oil to synthesis gas. Catalysts containing noble metals, mainly Pt, and

Ni supported on oxides, have been tested with model substrates. The most important published papers in the field have been reviewed by Bridgwater and co-authors.[43]

2.4.3 Sugars

Monosaccharides formed from plant polysaccharides of cellulose and hemicelluloses may undergo further chemical catalytic transformations to obtain valuable chemicals and value-added products. One of the promising research trends is the conversion of di- and monosaccharides to furan compounds (5-HMF, furfural) and levulinic acid.

For example, 5-HMF can quite easily be prepared from hexoses such as glucose and fructose.[215,216] A high selectivity (up to 89%) to 5-HMF can be achieved by treating fructose in biphasic systems comprising diluted aqueous solutions of mineral acids, inorganic salts and organic solvents (DMSO, MIBK, THF, alcohols, ketones *etc.*).[217] Substitution of the solid acid catalyst (silicoaluminophosphate SAPO-44) for mineral acids provides high yields of 5-HMF from fructose (78%) and glucose (67%).[218] Different catalytic systems with ionic liquids as solvents were used to produce 5-HMF at high yields from fructose (75–92%)[219] and glucose (up to 88%).[220] The oxides Nb_2O_5, TiO_2, ZrO_2, WO_3, and Al_2O_3 with acid or acid–base properties are stable in hot water and seem to be the most promising solid catalysts. These oxides were successfully applied for the dehydration of monosaccharides (fructose, glucose) to 5-HMF.[221–223] Zirconia was active to isomerization of glucose to fructose[224,225] and to glucose dehydration into 5-HMF.[226,227] Niobium oxide is another potential catalyst for dehydration of monosaccharides into 5-HMF.[221–223,228] Nb_2O_5 bears a great array of acidic centers on the surface and seems to be a very promising additive to ZrO_2. Niobium oxide supported on zirconia and zirconium–niobium mixed oxides successfully catalyzed the dehydration of 2-prapanol,[229] glycerol[230] and 1,2-propanediol[231] due to the Brønsted acidity of polymeric niobium oxide species.

Since 5-HMF is an intermediate in the formation of levulinic acid (LA) from biomass, it is unsurprising that liquid acid catalysts were also used for decades to produce LA from sugars.[72,132] Generally, higher acid strengths and longer residence times are needed to produce LA as the final product rather than 5-HMF. The yields of LA are higher from fructose than from glucose. Heterogeneous catalysts are practical alternatives to homogeneous catalysts; they are more environmentally attractive due to their selective and easy to handle nature, reduced corrosive impact on equipment and relatively low cost because the catalyst can be easily separated and recycled. The other solid acid catalysts are zirconia and titania, stannous and stannic chloride, $SO_4^{2-}/ZrO_2–Al_2O_3$, $AlCl_3·6H_2O$, Sn-Mont, zirconium phosphate, ion-exchange resins and sulfonated polymers, *etc.*[6,72,132]

Saccharic acids are a valuable feedstock for pharmaceutics. Supported dispersed noble metals (Au, Pt, Pd) have been under long and intensive studies as catalysts for the oxidation of sugars to acids, for example the oxidation of glucose to glyconic acid.[161]

Monosaccharides prepared from plant polysaccharides can be treated to produce ethanol. There is the ever expanding application of ethanol as a transportation fuel. Although pure ethanol can be used in modified internal combustion engines, it is mostly attractive in 10% or 15% blends with gasoline to fuel regular internal combustion engines.[46,53] The progress in application of bioethanol as a fuel is most apparent in the USA, Brazil and EU. In these countries and worldwide, there has been a constant growth of bioethanol production, which was only interrupted by the global financial crisis (Figure 2.4A). In 2016, the USA, Brazil and EU produced 56, 27 and 5 billion liters of bioethanol, respectively, *i.e.* almost 90% of global production (Figure 2.4B). The stable growth of the bioethanol market in the USA has reached 7.7–9% per year. In fact, the production is expected to exceed 160 billion liters by 2020.[232] In the USA, State programs have been set up

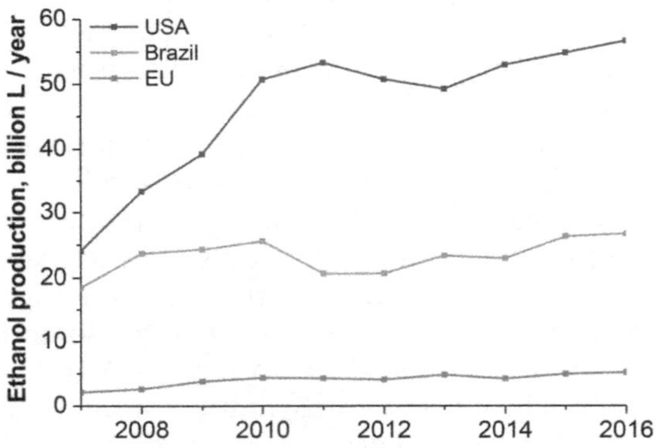

A

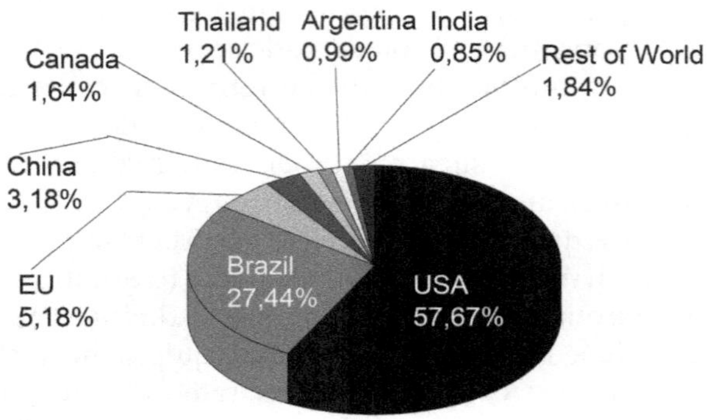

B

Figure 2.4 Growth of the bioethanol production in the USA, Brazil and EU in 2007–2016 (A) and the proportion of countries in the bioethanol production in 2016 (B). From the data of the Renewable Fuels Association (RFA) of USA.[233]

that imply expansion of bioethanol production up to 20% of the total fuel production by 2020.[232] Similar measures have been taken in the European Union.

The main method for bioethanol production is enzymatic treatment.[46] The target compounds are synthesized depending on the process type under the following conditions: fermentation of monosaccharides prepared by hydrolysis; fermentation with simultaneous saccharification of the substrate; the one-pot process of the same microorganism is responsible for the generation of hydrolytic enzymes and for the conversion of the formed sugars into the product.[234] A particular case is the process of the concurrent processing of C5 and C6 sugars (saccharification and cofermentation). There are a number of examples in the literature that describe the production of bioethanol by growing on media containing products of cellulose processing.[235] A yield as high as 100% of the theoretically possible bioethanol yield was observed with the enzymatic processing of saccharified biomass that underwent hydrothermal treatment at 200 °C in the presence of 0.5% sulfuric acid.[235] In addition to ethanol, 1-butanol and isobutanol are accepted promising biofuels due to their high octane numbers and hydrophobic behavior. They are mainly obtained by using selected genetically engineered strain-producers.[236–239]

Sugars extracted from plant biomass can be used for the production of hydrocarbons or hydrogen in the presence of solid catalysts. A mixture of hydrocarbons containing alkanes (methane, ethane, propane), alkenes (butene, pentene) and aromatic hydrocarbons (benzene, toluene) can be obtained from sugars over ZSM-5 zeolite catalyst at 510 °C and 1 bar, with gaseous products (CO, CO_2, H_2O) and coke also being formed.[46]

Hydrogen and light alkanes can be produced by aqueous-phase reforming of sugar or sugar-derived feeds with liquid water using a solid catalyst at temperatures of 200–260 °C and at 10–50 bar. The activity of metal catalysts for the process increases in the series: Ir $<$ Pd $\sim$ Rh $<$ Ru $<$ Ni $\sim$ Pt. The promising platinum catalyst can be made more active by adding Ni, Co or Fe additives to the metal.[240] The platinum and nickel catalysts can be used for the aqueous-phase hydrogenation of sugars and sugar model compounds (sorbitol, ethylene glycol) by H_2.[240] Hydrogen to be used in the process can be generated both by the individual aqueous-phase reforming and by the one-pot process combining aqueous-phase reforming and aqueous-phase reduction.[46]

2.4.4 Furfurals and Levulinic Acid

5-Hydroxymethylfurfural (5-HMF), furfural and levulinic acid (LA) are three of the most promising platform molecules owing to their convertibility into a variety of value-added chemicals and fuels by different catalytic reactions. These compounds are listed in the 10 top value-added bio-based chemicals.[1,7,17,72]

It can be seen in Figure 2.5 that a number of important C6 compounds can be formed through one common intermediate molecule, namely 5-HMF.

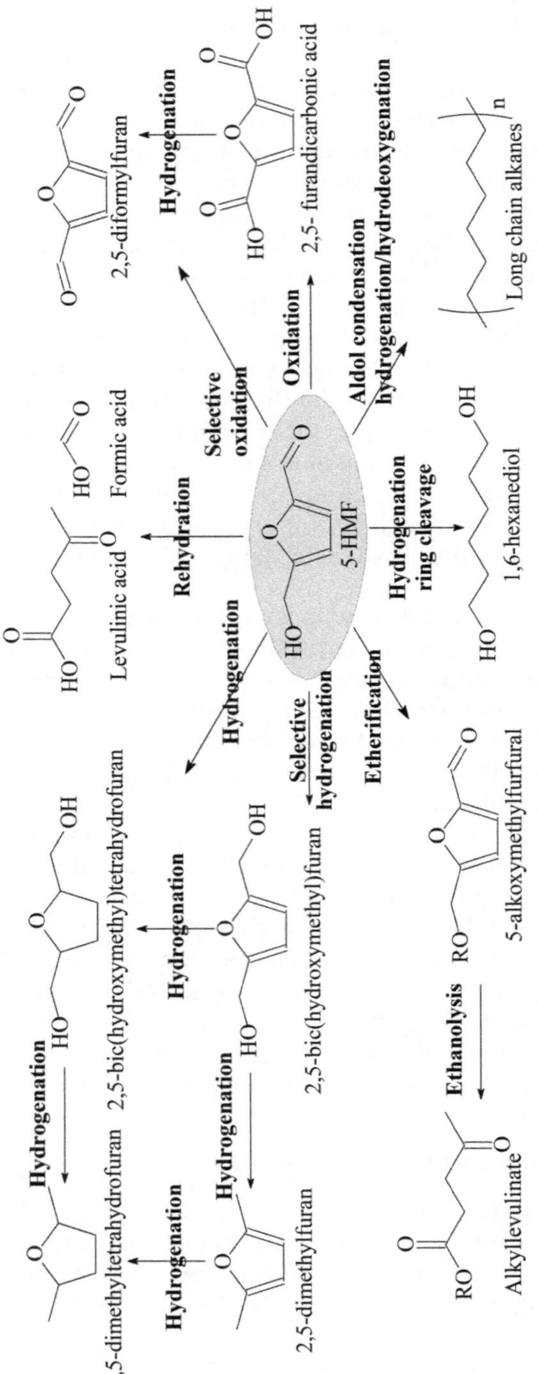

Figure 2.5 5-HMF as a platform chemical for the synthesis of other chemicals.

It can undergo oxidation over supported metal (Ru, Pt, Pd) catalysts promoted by bases to yield 2,5-furan dicarboxylic acid (used as an alternative to terephthalic acid in the production of polyamide, polyurethane and polyester),[241–243] or selective oxidation to 2,5-diformylfuran (used for synthesis of diamine and Schiff bases) in the presence of different catalysts (supported metals,[244] layered manganese-copper double hydroxides,[245] metal free acid-base carbon,[246] *etc.*). On the other hand, hydrogenation of 5-HMF produces various significant chemicals such as 2,5-dimethylfuran (used as a transportation fuel), 2,5-bis(hydroxymetyl)furan (used for manufacturing of polyurethane foam), 2,5-bis(hydroxymethyl)tetrahydrofuran, and 2,5-dimethyltetrahydrofuran (used for manufacturing of polyester) over supported metal catalysts (RuCu/C,[247] Ni/C,[248] Fe/C,[249] *etc.*). Furthermore, ring hydrogenation of 5-HMF in the presence of a mixture of copper chromite and Pd/C catalysts or Pd/zirconium phosphate produces 1,6-hexanediol (used for production of plastics and chemicals), which yields caprolactam (monomer of nylon-6) through further amination.[250,251] Rehydration of 5-HMF in the presence of acid catalysts yields equimolar amounts of formic acid and levulinic acid (used for synthesis of various chemicals).[72,252] Moreover, esterification of 5-HMF yields 5-alkoxymethylfurfural (a potential diesel additive), which produces further alkyl levulinates (fine chemicals, solvents, and additives) *via* ethanolysis.[253] It is also possible to produce long chain alkanes (used as transportation fuel) *via* aldol condensation of 5-HMF with suitable carbonyl reactants in the presence of acid or basic catalysts followed by hydrogenation/hydrodeoxygenation of the intermediates over supported noble metal catalysts.[254,255]

Another furan derivative, furfural, also has a wide range of industrial applications (Figure 2.6) described in several excellent reviews.[4,255–258] Its hydrogenation over cupper chromite, Ni or noble metal catalysts produces furfuryl alcohol (used for production of resin, adhesives, wetting agents; used as a solvent).[256,258] Tetrahydrofuran alcohol (used as a green solvent in agricultural applications, printing inks, and industrial and electronics cleaners) can be produced directly from furfural or furfural alcohol over Cu, Ni, and Cu–Cr, supported noble metal catalysts at higher H_2 pressures. On the other hand, furfural alcohol can be further converted to alkyl levulinates (used as plasticizing materials, solvents or odor substances and, potentially, as fuel additives) in alcohol solutions in the presence of soluble (HCl, H_2SO_4) or solid (mesoporous aluminosilicate Al-TUD-1, Amberlyst 46, Purolite MN500, ZSM-5 (SiO_2/Al_2O_3 = 30), $[MIMBS]_3PW_{12}O_{40}$) acidic catalysts.[257,259] Catalytic hydrogenation at higher temperature gives 2-methylfuran (used as an alternative fuel, flavoring agent) and 2-methyltetrahydrofuran (used as a biofuel and substitute for tetrahydrofuran to allow reactions at higher temperature). The catalytic decarbonylation of furfural on noble metal and/or acid–base catalysts at high temperature (300–500 °C) is expected to lead to the production of furan (used as a platform chemical) through the release of a molecule of carbon monoxide. Besides, production of tetrahydrofuran (used as a solvent, precursor for anionic polymerization) from furan

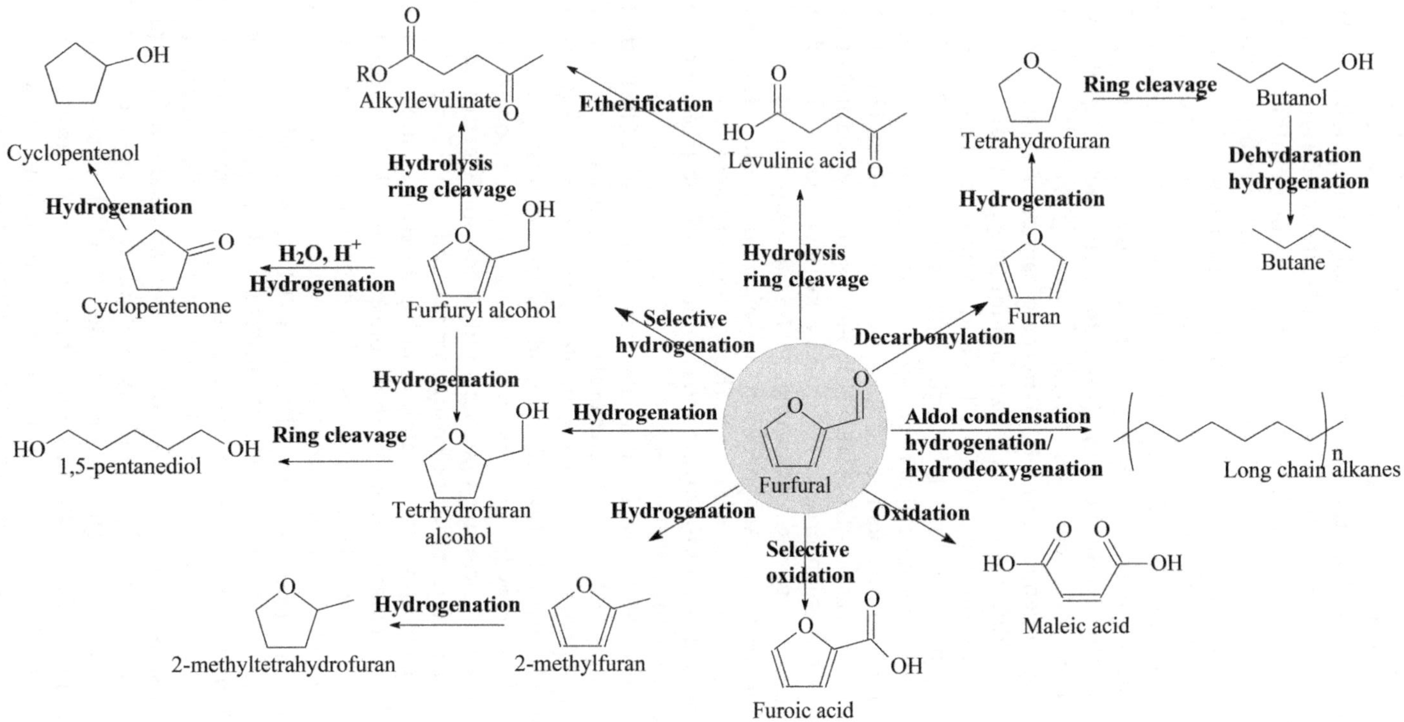

Figure 2.6 Furfural as a platform chemical for the synthesis of other chemicals.

(ultimately from furfural) is a highly promising alternative.[256] Furfural is also used in the synthesis of alkanes (used as transportation fuels) by aldol condensation with aldehydes and ketones in the presence of excess base (*e.g.*, $NaHCO_3$) followed by hydrogenation and/or hydrodeoxygenation.[257] These saturated compounds show better fuel properties (cetane number, pour point) than the condensation products obtained from furans and methyl furan. It can also be oxidized to yield furoic acid (used for nylon preparation, optic technologies, food products as flavoring agent, and preservative) and maleic acid (used for synthesis of several fine chemicals). Many other products (*e.g.*, 1-butanol, 1,5-pentanediol, cyclopentanol, cyclopentanone, dihydropyran and butane) can be produced by transformation of furfural over various catalysts.[255–258,260] In addition, hydrogenation of furfural to furfuryl alcohol can produce another important platform chemical - levulinic acid - *via* acid-catalyzed ring-opening in H_2O.[258]

Levulinic acid can be synthesized both from sugars including polysaccharides (cellulose, hemicelluloses and biomass) and from furan derivatives in the presence of soluble acid catalysts.[72,252,257–259,261] LA is a bifunctional gamma-keto-carboxylic acid; it has a rich chemistry and can be converted into a wide range of fuel additives and commodity chemicals (Figure 2.7).

γ-Valerolactone (GVL) is considered to be the most promising LA derived platform chemical that can be employed as a food additive, solvent, bio-oxygenate and platform for high value chemicals and fuels. GVL can also be used as a fuel additive that exhibits excellent fuel properties.[252,255,258,262] GVL is

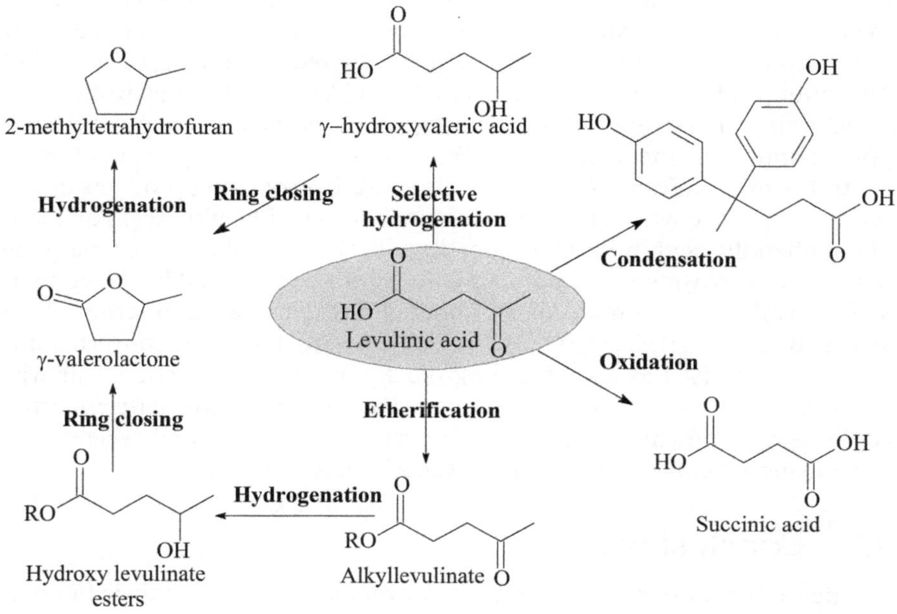

Figure 2.7 Levulinic acid as a platform chemical for the synthesis of other chemicals.

usually produced *via* selective hydrogenation of levulinic acid over different soluble metal (the Ru-derived homogeneous systems) and solid (supported Ru, Pd, Pt and inexpensive Raney Ni and CuO/Cr_2O_3) catalysts at moderate H_2 pressures (1–5.5 MPa) and temperatures (130–200 °C).[253,255] The γ-hydroxyvaleric acid intermediate forms first and is then subjected to ring closing. In the case of alkyl levulinates as reactants, this last step releases an equivalent of alcohol. A different approach was proposed in the literature by substituting H_2 for other hydrogen donors (catalytic transfer hydrogenation). An example is the Meerwein–Ponndorf–Verley type reduction using alcohols as the hydrogen source, which is very selective to carbonyl groups. The advantage here is that reactions can be performed in the absence of precious metals and catalyzed by simple metal oxides (ZnO_2, $Zr(OH)_4$, $CeZnO_2$, γ-Al_2O_3 *etc.*).[253,255] Non-alcohol hydrogen sources (formic acid or esters) have also been explored; they are always coproducts during the synthesis of LA or esters from carbohydrate reactants.[253] LA can be converted to fuels *via* dehydration/hydrogenation or esterification. Final products of these transformations are methyltetrahydrofuran and levulinic acid esters (alkyl levulinates, ALs). High octane numbers are characteristic of levulinic acid esters.[263] Mixtures containing 20% levulinate, 79% diesel fuel and 1% other additives can be used as motor fuels to minimize sulfur in exhaust gases of diesel engines.[46,252] Levulinic acid is converted to alkyl levulinates in the presence of alcohols and acid catalysts.[46,264,265] On the other hand, ALs can be produced from sugars, cellulose and biomass without intermediate isolation of LA. Soluble (H_2SO_4, 2-naphtanelesulfonic acid) or solid catalysts (zeolites, ion-exchanged resin, sulfonated TiO_2 or C, heteropolyacid salts) were used successively in these processes.[253,266] Methyltetrahydrofuran can be prepared by reduction of levulinic acid to γ-valerolactone followed by reduction to 1,4-pentendiol. The latter product is dehydrated at a high yield to methyltetrahydrofuran.[46]

Different homogenous Ru-containing catalysts, heterogeneous Cu composites and noble metal catalysts were reported for the production of methyltetrahydrofuran from LA.[252] Many valuable and promising products can be produced *via* catalytic transformation of LA or alkyl levulinates. For example, diphenolic acid (promising substitute for bisphenol A, the primary raw material of epoxy resin and polycarbonates) can be prepared by the condensation of LA with two molecules of phenol in the presence of strong Brønsted acid catalysts (HCl, H_2SO_4, $H_3PW_{12}O_{40}$/SBA-15). LA can be converted into succinic acid (used as a C4 building block for various fine chemicals with potential applications as fuel additives, solvents, food/cosmetics/pharmaceuticals, and polymers) by oxidation over V_2O_5. Recent good reviews[252,265] provide more details on the catalytic transformation of LA.

2.5 Conclusions

The development of processes for the conversion of lignocellulosic biomass to fuel and valuable chemicals has rapidly progressed in the past 10–15 years. The application of catalysts not only accelerates processes, but also allows researchers to increase the selectivity and yields of useful bio-based chemicals, to reduce the reaction temperature and pressure, and, consequently, to improve the process economy, and to change reaction routes and to obtain new valuable products.

There are two main approaches to the catalytic processing of lignocellulosic biomass. (1) Thermochemical catalytic treatment is conducted at severe reaction conditions (very high temperature and pressure). This approach includes: catalytic combustion, gasification, liquefaction, and pyrolysis, which are used for obtaining energy, gaseous (synthesis gas), liquid (bio-oil), and solid (biochar) products as well as the catalytic treatment of gas and liquid products to refine synthesis gas and bio-oil, reduce the content of oxygen and finally produce fuels and chemicals. (2) Liquid-phase catalytic processes are conducted at lower temperatures and pressures (up to 250 °C and 70 bar) in the presence of liquid or solid acid or bifunctional catalysts. This subgroup includes hydrolysis, hydrolysis–hydrogenation, hydrolysis–oxidation, biomass delignification and aqueous-phase reforming, which produce monosaccharides, furan and levulinic acid derivatives, polyols, organic acids, phenol derivatives, and hydrocarbons, respectively. The produced chemicals can be used by themselves or utilized for the production of other chemicals or fuel components *via* catalytic conversions in the liquid or vapor phase.

Heterogeneous catalysts for the processes under discussion are particularly interesting due to their easier separation from the reaction mixtures, reusability, and low corrosion activity. In recent years, many heterogeneous catalysts have been suggested. Meanwhile, in addition to the usual requirements for high activity and selectivity, special requirements can be formulated for heterogeneous catalysts owing to the specific characteristics of plant biomass as the substrate. (1) Biomass comprising polymeric molecules requires catalysts with large pores or even nonporous catalysts. (2) High contents of oxygen in lignocellulose, and therefore high water and organic acid contents in the reaction medium (especially for liquid-phase processing), require the catalysts to be highly stable against leaching in the aggressive reaction medium. Hence, the rational arrangement and optimization of the processes require the further improvement of catalyst morphology and stability.

As for industrial implementation of the methods under discussion, the thermochemical catalytic processes can be used for biomass treatment at specialized enterprises or at additional facilities of existing refineries, and the liquid-phase processes for the production of additional valuable chemicals can be applied at existing paper or bioethanol plants. However, implementation of the chemical biorefinery concept that integrates processes of biomass conversion to produce multiple fuels and chemicals from plant biomass needs further activities to improve and optimize the suggested catalytic methods and, in particular, methods for liquid-phase fractionation of lignocellulose and/or catalytic disintegration of the most robust components of biomass – cellulose and lignin.

Acknowledgements

The support from the Russian Academy of Science and Federal Agency of Scientific Organizations is acknowledged.

References

1. P. Gallezot, *Chem. Soc. Rev.*, 2012, **41**, 1538.
2. P. Gallezot, *ChemSusChem*, 2008, **1**, 734–737.
3. J. Geboers, S. Van de Vyver, R. Ooms, B. Op de Beeck, P. A. Jacobs and B. F. Sels, *Catal. Sci. Technol.*, 2011, **1**, 714–726.
4. P. Bhaumik and P. L. Dhepe, *Catal. Rev.*, 2016, **58**, 36.
5. D. Y. Murzin and T. Salmi, *Catal. Lett.*, 2012, **142**, 676.
6. S. Van de Vyver, J. Geboers, P. A. Jacobs and B. F. Sels, *ChemCatChem*, 2011, **3**, 82.
7. R.-J. van Putten, J. C. van der Waal, E. de Jong, C. B. Rasrendra, H. J. Heeres and J. G. de Vries, *Chem. Rev.*, 2013, **113**, 1499.
8. M. Besson, P. Gallezot and C. Pinel, *Chem. Rev.*, 2014, **114**, 1827.
9. S. Dutta and S. Pal, *Biomass Bioenergy*, 2014, **62**, 182.
10. M. L. Imhoff, L. Bounoua, T. Ricketts, C. Loucks, R. Harriss and W. T. Lawrence, *Nature*, 2004, **429**, 870.
11. C.-H. Zhou, X. Xia, C.-X. Lin, D.-S. Tong and J. Beltramini, *Chem. Soc. Rev.*, 2011, **40**, 5588.
12. P. Gallezot and A. Kiennemann, in *Handbook of Heterogeneous Catalysis*, ed. G. Ertl, Wiley-VCH Verlag GmbH & Co. KGaA, 2008, pp. 2447–2476.
13. A. Bruggink, R. Schoevaart and T. Kieboom, *Org. Process Res. Dev.*, 2003, **7**, 622.
14. T. Thananatthanachon and T. B. Rauchfuss, *Angew. Chem., Int. Ed.*, 2010, **49**, 6616.
15. V. E. Tarabanko, M. A. Smirnova, M. Yu. Chernyak, Russian Pat., 2363698 C1, 2009.
16. G. Fleche, A. Gaset, J.-P. Gorrichon, E. Truchot, P. Sicard, US Pat., 4339387, 1984.
17. T. Werpy and G. Petersen, in *Top Value Added Chemicals from Biomass*, ed. T. Werpy and G. Petersen, Energy Efficiency and Renewable Energy, U.S. Department of Energy, 2004, vol. 1, p. 63.
18. E. M. Rubin, *Nature*, 2008, **454**, 841.
19. P. Gallezot, *Chem. Soc. Rev.*, 2012, **41**, 1538.
20. J.A. Dumesic, J. M. Ribeiro Gallo, D. Alonso, US Pat., 8772515 B2, 2014.
21. J. van Haveren, E. L. Scott and J. Sanders, *Biofuels, Bioprod. Biorefin.*, 2008, **2**, 41.
22. J. Zakzeski, P. C. A. Bruijnincx, A. L. Jongerius and B. M. Weckhuysen, *Chem. Rev.*, 2010, **110**, 3552.
23. R. Deutschmann and R. F. H. Dekker, *Biotechnol. Adv.*, 2012, **30**, 1627.
24. T. Heinze, in *Polysaccharides: Structural Diversity and Functional Versatility*, ed. S. Dumitriu, Marcel Dekker, New York, 2nd edn, 2005, p. 551.

25. A. M. Ruppert, K. Weinberg and R. Palkovits, *Angew. Chem., Int. Ed.*, 2012, **51**, 2564.
26. N. V. Gromov, O. P. Taran, V. N. Parmon and S. Uthandi, in *Biocatalysis in Biomass to Bioproducts*, ed. K. Ramasamy, S. Karthikeyan and U. Sivakumar, Daya Publ. House, New Delhi, 2017, pp. 237–264.
27. I. Spiridon, in *Polysaccharides: Structural Diversity and Functional Versatility*, ed. S. Dimitriu, Marcel Dekker, New York, 2nd edn, 2005, pp. 475–487.
28. B. T. Kusema, L. Faba, N. Kumar, E. Díaz, S. Ordóñez, T. Salmi and D. Y. Murzin, *Catal. Today*, 2012, **196**, 26.
29. A. V. Pestunov, A. O. Kuzmin, D. A. Yatsenko, M. K. Pravdina and O. P. Taran, *J. Sib. Fed. Univ., Chem.*, 2015, **8**, 386.
30. N. V. Gromov, O. P. Taran, K. N. Sorokina, T. I. Mischenko, U. Sivakumar and V. N. Parmon, *Catal. Ind.*, 2016, **8**, 176–786.
31. R. Singh, A. Shukla, S. Tiwari and M. Srivastava, *Renewable Sustainable Energy Rev.*, 2014, **32**, 713.
32. Y. Zheng, J. Zhao, F. Xu and Y. Li, *Prog. Energy Combust. Sci.*, 2014, **42**, 35.
33. Z. R. Ismagilov and M. A. Kerzhentsev, *Catal. Today*, 1999, **47**, 339.
34. G. Pifer, *Pulp Pap.*, 2005, **79**, 54.
35. S. Ravelli, A. Perdichizzi and G. Barigozzi, *Prog. Energy Combust. Sci.*, 2008, **34**, 224.
36. D. A. Bulushev and J. R. H. Ross, *Catal. Today*, 2011, **171**, 1.
37. D. M. Alonso, J. Q. Bond and J. A. Dumesic, *Green Chem.*, 2010, **12**, 1493.
38. A. V. Bridgwater, *Appl. Catal., A*, 1994, **116**, 5.
39. T. R. Carlson, G. A. Tompsett, W. C. Conner and G. W. Huber, *Top. Catal.*, 2009, **52**, 241.
40. G. W. Huber and A. Corma, *Angew. Chem., Int. Ed.*, 2007, **46**, 7184.
41. M. Balat, *Energy Sources, Part A*, 2011, **33**, 674.
42. N. A. Qambrani, M. M. Rahman, S. Won, S. Shim and C. Ra, *Renewable Sustainable Energy Rev.*, 2017, **79**, 255.
43. A. V. Bridgwater, *Biomass Bioenergy*, 2012, **38**, 68.
44. M. J. Antal, in *Advances in Solar Energy: An Annual Review of Research and Development*, ed. K. W. Böer and J. A. Duffie, Springer US, Boston, MA, 1985, vol. 2, pp. 175–255, DOI: 10.1007/978-1-4613-9951-3_4.
45. T. B. Reed, in *Advances in Solar Energy: An Annual Review of Research and Development*, ed. K. W. Böer and J. A. Duffie, Springer US, Boston, MA, 1985, vol. 2, pp. 125–169.
46. G. W. Huber, S. Iborra and A. Corma, *Chem. Rev.*, 2006, **106**, 4044.
47. A. T. Tag, G. Duman, S. Ucar and J. Yanik, *J. Anal. Appl. Pyrolysis*, 2016, **120**, 200.
48. P. Fu, S. Hu and J. Xiang, *J. Anal. Appl. Pyrolysis*, 2010, **88**, 117.
49. T. Bridgwater, *J. Sci. Food Agric.*, 2006, **86**, 1755.
50. A. G. Gayubo, A. T. Aguayo, A. Atutxa, R. Prieto and J. Bilbao, *Energy Fuels*, 2004, **18**, 1640.
51. E. F. Iliopoulou, E. V. Antonakou, S. A. Karakoulia, I. A. Vasalos, A. A. Lappas and K. S. Triantafyllidis, *Chem. Eng. J.*, 2007, **134**, 51.

52. C. Wang, Q. Hao, D. Lu, Q. Jia, G. Li and B. Xu, *Chin. J. Catal.*, 2008, **29**, 907.
53. C. Briens, J. Piskorz and F. Berruti, *Int. J. Chem. React. Eng.*, 2008, **6**, R2 1.
54. M. J. Antal, S. G. Allen, D. Schulman, X. Xu and R. J. Divilio, *Ind. Eng. Chem. Res.*, 2000, **39**, 4040.
55. L. J. Guo, Y. J. Lu, X. M. Zhang, C. M. Ji, Y. Guan and A. X. Pei, *Catal. Today*, 2007, **129**, 275.
56. A. Kumar, D. Jones and M. Hanna, *Energies*, 2009, **2**, 556.
57. S. K. Sansaniwal, M. A. Rosen and S. K. Tyagi, *Renewable Sustainable Energy Rev.*, 2017, **80**, 23.
58. R. Luque and J. G. Speight, *Gasification for Synthetic Fuel Production*, Woodhead Publishing - Elsevier, 2015.
59. M. Miccio and F. Miccio, in *Fluidized Bed Technologies for Near-zero Emission Combustion and Gasification*, Woodhead Publishing, 2013, pp. 388–434.
60. S. Czernik, R. French, C. Feik and E. Chornet, *Ind. Eng. Chem. Res.*, 2002, **41**, 4209.
61. A. Kumar, K. Eskridge, D. D. Jones and M. A. Hanna, *Bioresour. Technol.*, 2009, **100**, 2062.
62. A. K. Gupta and W. Cichonski, *Environ. Eng. Sci.*, 2007, **24**, 1179.
63. J. F. González, S. Román, D. Bragado and M. Calderón, *Fuel Process. Technol.*, 2008, **89**, 764.
64. D. Sutton, B. Kelleher and J. R. H. Ross, *Fuel Process. Technol.*, 2001, **73**, 155.
65. K. Tomishige, M. Asadullah and K. Kunimori, *Catal. Today*, 2004, **89**, 389.
66. A. P. C. Faaij, *Energy Policy*, 2006, **34**, 322.
67. M. Modell, R. C. Reid, S. I. Amin, USA Pat., 4,113,446, 1978.
68. X. H. Hao, L. J. Guo, X. Mao, X. M. Zhang and X. J. Chen, *Int. J. Hydrogen Energy*, 2003, **28**, 55.
69. Y. J. Lu, L. J. Guo, C. M. Ji, X. M. Zhang, X. H. Hao and Q. H. Yan, *Int. J. Hydrogen Energy*, 2006, **31**, 822.
70. I.-G. Lee, M.-S. Kim and S.-K. Ihm, *Ind. Eng. Chem. Res.*, 2002, **41**, 1182.
71. D. C. Elliott, L. J. Sealock and E. G. Baker, *Ind. Eng. Chem. Res.*, 1993, **32**, 1542.
72. A. Mukherjee, M.-J. Dumont and V. Raghavan, *Biomass Bioenergy*, 2015, **72**, 143.
73. F. Guo, Z. Fang, C. C. Xu and R. L. Smith Jr, *Prog. Energy Combust. Sci.*, 2012, **38**, 672.
74. L. Hu, L. Lin, Z. Wu, S. Zhou and S. Liu, *Appl. Catal., B*, 2015, **174–175**, 225.
75. M. J. Antal and G. Varhegyi, *Ind. Eng. Chem. Res.*, 1995, **34**, 703.
76. K. Raveendran, A. Ganesh and K. C. Khilar, *Fuel*, 1995, **74**, 631.
77. J. N. Chheda, G. W. Huber and J. A. Dumesic, *Angew. Chem., Int. Ed.*, 2007, **46**, 7164.

78. J. C. Serrano-Ruiz, R. Luque and A. Sepulveda-Escribano, *Chem. Soc. Rev.*, 2011, **40**, 5266.
79. B. C. Saha, L. B. Iten, M. A. Cotta and Y. V. Wu, *Process Biochem.*, 2005, **40**, 3693.
80. A. K. Kivaisi and S. Eliapenda, *Renewable Energy*, 1994, **5**, 791.
81. F. Monlau, A. Barakat, J. P. Steyer and H. Carrere, *Bioresour. Technol.*, 2012, **120**, 241.
82. I. Delidovich, K. Leonhard and R. Palkovits, *Energy Environ. Sci.*, 2014, **7**, 2803.
83. P. Dhepe and A. Fukuoka, *Catal. Surv. Asia*, 2007, **11**, 186.
84. J. F. Saeman, *Ind. Eng. Chem.*, 1945, **37**, 43.
85. A. S. Amarasekara, in *Handbook of Cellulosic Ethanol*, ed. A. S. Amarasekara, John Wiley & Sons, Beverly, USA, 2013, ch. 7, pp. 247–281.
86. E. A. Immergut and B. G. Ranby, *Ind. Eng. Chem.*, 1956, **48**, 1183.
87. L. Kupiainen, J. Ahola and J. Tanskanen, *Ind. Eng. Chem. Res.*, 2012, **51**, 3295.
88. P. Lenihan, A. Orozco, E. O'Neill, M. N. M. Ahmad, D. W. Rooney and G. M. Walker, *Chem. Eng. J.*, 2010, **156**, 395.
89. V. Soldi, in *Polysaccharides: Structural Diversity and Functional Versatility*, ed. S. Dimitriu, Marcel Dekker, New York, 2nd edn, 2005, pp. 395–406.
90. G. Bustos, J. Ramírez, G. Garrote and M. Vázquez, *Appl. Biochem. Biotechnol.*, 2003, **104**, 51.
91. J. Tian, J. Wang, S. Zhao, C. Jiang, X. Zhang and X. Wang, *Cellulose*, 2010, **17**, 587.
92. M. A. Abeer, H. A. E. W. Zeinab, A. I. Atef and T. A. S. Mona, *Bioresour. Technol.*, 2010, **101**, 4446.
93. P. Laopaiboon, A. Thani, V. Leelavatcharamas and L. Laopaiboon, *Bioresour. Technol.*, 2010, **101**, 1036.
94. W. S. Mok, M. J. Antal and G. Varhegyi, *Ind. Eng. Chem. Res.*, 1992, **31**, 94.
95. H. E. van Dam, A. P. G. Kieboom and H. van Bekkum, *Starch/Stärke*, 1986, **38**, 95.
96. N. S. Mosier, A. Sarikaya, C. M. Ladisch and M. R. Ladisch, *Biotechnol. Prog.*, 2001, **17**, 474.
97. T. vom Stein, P. Grande, F. Sibilla, U. Commandeur, R. Fischer, W. Leitner and P. Dominguez de Maria, *Green Chem.*, 2010, **12**, 1844.
98. Y. Izumi, M. Ono, M. Kitagawa, M. Yoshida and K. Urabe, *Microporous Mater.*, 1995, **5**, 255.
99. C. L. Hill and C. M. Prosser-McCartha, *Coord. Chem. Rev.*, 1995, **143**, 407.
100. Y. Ogasawara, S. Itagaki, K. Yamaguchi and N. Mizuno, *ChemSusChem*, 2011, **4**, 519.
101. K.-I. Shimizu, H. Furukawa, N. Kobayashi, Y. Itaya and A. Satsuma, *Green Chem.*, 2009, **11**, 1627.
102. S. Suganuma, K. Nakajima, M. Kitano, D. Yamaguchi, H. Kato, S. Hayashi and M. Hara, *J. Am. Chem. Soc.*, 2008, **130**, 12787.

103. J. Pang, A. Wang, M. Zheng and T. Zhang, *Chem. Commun.*, 2010, **46**, 6935.
104. A. Onda, T. Ochi and K. Yanagisawa, *Green Chem.*, 2008, **10**, 1033.
105. A. Onda, T. Ochi and K. Yanagisawa, *Top. Catal.*, 2009, **52**, 801.
106. S. Dora, T. Bhaskar, R. Singh, D. Naik Viswanatha and D. K. Adhikari, *Bioresour. Technol.*, 2012, **120**, 318.
107. M. Kitano, D. Yamaguchi, S. Suganuma, K. Nakajima, H. Kato, S. Hayashi and M. Hara, *Langmuir*, 2009, **25**, 5068.
108. S. Li, Z. Gu, B. E. Bjornson and A. Muthukumarappan, *J. Environ. Chem. Eng.*, 2013, **1**, 1174.
109. K. Nakajima and M. Hara, *ACS Catal.*, 2012, **2**, 1296.
110. G. S. Foo and C. Sievers, *ChemSusChem*, 2015, **8**, 534.
111. A. H. Van Pelt, O. A. Simakova, S. M. Schimming, J. L. Ewbank, G. S. Foo, E. A. Pidko, E. J. M. Hensen and C. Sievers, *Carbon*, 2014, **77**, 143.
112. O. P. Taran, E. M. Polyanskaya, O. L. Ogorodnikova, C. Descorme, M. Besson and V. N. Parmon, *Catal. Ind.*, 2011, **2**, 381.
113. H. Kobayashi, T. Komanoya, K. Hara and A. Fukuoka, *ChemSusChem*, 2010, **3**, 440.
114. H. Wang, C. Zhang, H. He and L. Wang, *J. Environ. Sci.*, 2012, **24**, 473.
115. P. L. Dhepe and A. Fukuoka, *ChemSusChem*, 2008, **1**, 969.
116. K. Y. Nandiwale, N. D. Galande, P. Thakur, S. D. Sawant, V. P. Zambre and V. V. Bokade, *ACS Sustainable Chem. Eng.*, 2014, **2**, 1928.
117. P. Mäki-Arvela, T. Salmi, B. Holmbom, S. Willför and D. Y. Murzin, *Chem. Rev.*, 2011, **111**, 5638.
118. T. Marzialetti, M. B. Valenzuela Olarte, C. Sievers and T. J. C. Hoskins, *Ind. Eng. Chem. Res.*, 2008, **47**, 7131.
119. N. S. Mosier, C. M. Ladisch and M. R. Ladisch, *Biotechnol. Bioeng.*, 2002, **79**, 610.
120. A. Demirbas, *Energy Convers. Manage.*, 2009, **50**, 2782.
121. C. Blecker, C. Fougnies, J.-C. Van Herck, J.-P. Chevalier and M. Paquot, *J. Agric. Food Chem.*, 2002, **50**, 1602.
122. H.-J. Sun, S. Yoshida, N.-H. Park and I. Kusakabe, *Carbohydr. Res.*, 2002, **337**, 657.
123. A. E. Abasaeed and Y. Y. Lee, *Chem. Eng. Technol.*, 1995, **18**, 440.
124. L. Shuai and X. Pan, *Energy Environ. Sci.*, 2012, **5**, 6889.
125. Y. Yang, X. Xiang, D. Tong and C. Hu, *Bioresour. Technol.*, 2012, **116**, 302.
126. A. Chareonlimkun, V. Champreda, A. Shotipruk and N. Laosiripojana, *Fuel*, 2010, **89**, 2873.
127. F. Yang, Q. Liu, M. Yue, X. Bai and Y. Du, *Chem. Commun.*, 2011, **47**, 4469.
128. C. V. McNeff, D. T. Nowlan, L. C. McNeff, B. Yan and R. L. Fedie, *Appl. Catal., A*, 2010, **384**, 65.
129. J. Horvat, B. Klaid, B. Metelko and V. Sunjid', *Tetrahedron Lett.*, 1985, **26**, 2111.
130. B. V. Timokhin, V. A. Baransky and G. D. Eliseeva, *Russ. Chem. Rev.*, 1999, **68**, 73.

131. B. Girisuta, PhD Thesis Levulinic Acid from Lignocellulosic Biomass, University of Groningen, 2007.
132. D. W. Rackemann and W. O. S. Doherty, *Biofuels, Bioprod. Biorefin.*, 2011, **5**, 198.
133. T. Komanoya, H. Kobayashi, K. Hara, W.-J. Chun and A. Fukuoka, *Appl. Catal., A*, 2011, **407**, 188.
134. H. Kobayashi, H. Matsuhashi, T. Komanoya, K. Hara and A. Fukuoka, *Chem. Commun.*, 2011, **47**, 2366.
135. W. Deng, X. Tan, W. Fang, Q. Zhang and Y. Wang, *Catal. Lett.*, 2009, **133**, 167.
136. S. Van de Vyver, J. Geboers, M. Dusselier, H. Schepers, T. Vosch, L. Zhang, G. Van Tendeloo, P. A. Jacobs and B. F. Sels, *ChemSusChem*, 2010, **3**, 698.
137. J. Pang, A. Wang, M. Zheng, Y. Zhang, Y. Huang, X. Chen and T. Zhang, *Green Chem.*, 2012, **14**, 614.
138. L.-N. Ding, A.-Q. Wang, M.-Y. Zheng and T. Zhang, *ChemSusChem*, 2010, **3**, 818.
139. P. Yang, N. Kobatashi and A. Fukuoka, *Chin. J. Catal.*, 2011, **32**, 716.
140. C. Luo, S. Wang and H. Liu, *Angew. Chem., Int. Ed.*, 2007, **46**, 7636.
141. R. Rinaldi and F. Schuth, *Energy Environ. Sci.*, 2009, **2**, 610.
142. J. Geboers, S. Van de Vyver, K. Carpentier, K. de Blochouse, P. Jacobs and B. Sels, *Chem. Commun.*, 2010, **46**, 3577.
143. J. Geboers, S. Van de Vyver, K. Carpentier, P. Jacobs and B. Sels, *Green Chem.*, 2011, **13**, 2167.
144. M. Liu, W. Deng, Q. Zhang, Y. Wang and Y. Wang, *Chem. Commun.*, 2011, **47**, 9717.
145. N. Ji, T. Zhang, M. Zheng, A. Wang, H. Wang, X. Wang, Y. Shu, A. L. Stottlemyer and J. G. Chen, *Catal. Today*, 2009, **147**, 77.
146. N. Ji, T. Zhang, M. Zheng, A. Wang, H. Wang, X. Wang and J. G. Chen, *Angew. Chem., Int. Ed.*, 2008, **47**, 8510.
147. J. Xi, D. Ding, Y. Shao, X. Liu, G. Lu and Y. Wang, *ACS Sustainable Chem. Eng.*, 2014, **2**, 2355.
148. H. Li, Z. Fang, R. L. Smith Jr and S. Yang, *Prog. Energy Combust. Sci.*, 2016, **55**, 98.
149. Y. Zhang, A. Wang and T. Zhang, *Chem. Commun.*, 2010, **46**, 862.
150. X. Zhao, J. Xu, A. Wang and T. Zhang, *Chin. J. Catal.*, 2015, **36**, 1419.
151. W. Deng, Q. Zhang and Y. Wang, *Catal. Today*, 2014, **234**, 31.
152. W. Deng, Q. Zhang and Y. Wang, *Dalton Trans.*, 2012, **41**, 9817.
153. S. E. Davis, M. S. Ide and R. J. Davis, *Green Chem.*, 2013, **15**, 17.
154. N. V. Gromov, O. P. Taran, I. V. Delidovich, A. V. Pestunov, Y. A. Rodikova, E. G. Zhizhina and V. N. Parmon, *Catal. Today*, 2016, **278**, 74.
155. C. Moreau, M. N. Belgacem and A. Gandini, *Top. Catal.*, 2004, **27**, 11.
156. H.-J. Hustede and E. S. Haberstroh, in *Ulmann's Encyclopedia Industrial Chem*, ed. H. J. Arpe, Wiley-VCH Verlag GmbH & Co., Гамбург, 5th edn, 1989, vol. A12, pp. 449–456.
157. J. Li, D.-J. Ding, L. Deng, Q.-X. Guo and Y. Fu, *ChemSusChem*, 2012, **5**, 1313.

158. J. C. Serrano-Ruiz, D. J. Braden, R. M. West and J. A. Dumesic, *Appl. Catal., B*, 2010, **100**, 184.
159. X. Tan, W. Deng, M. Liu, Q. Zhang and Y. Wang, *Chem. Commun.*, 2009, **14**, 7179.
160. A. Onda, T. Ochi and K. Yanagisawa, *Catal. Commun.*, 2011, **12**, 421.
161. I. V. Delidovich, B. L. Moroz, O. P. Taran, N. V. Gromov, P. A. Pyrjaev, I. P. Prosvirin, V. I. Bukhtiyarov and V. N. Parmon, *Chem. Eng. J.*, 2013, **223**, 921.
162. D. An, A. Ye, W. Deng, Q. Zhang and Y. Wang, *Chem.–Eur. J.*, 2012, **18**, 2938.
163. D. V. Evtuguin, C. Pascoal Neto and J. D. Pedrosa De Jesus, *J. Pulp Pap. Sci.*, 1998, **24**, 133.
164. A. A. Shatalov, D. V. Evtuguin and C. Pascoal Neto, *Carbohydr. Polym.*, 2000, **43**, 23.
165. Y. Yao, B. Gao, H. Chen, L. Jiang, M. Inyang, A. R. Zimmerman, X. Cao, L. Yang, Y. Xue and H. Li, *J. Hazard. Mater.*, 2012, **209**, 408.
166. R. D. Cortright, R. R. Davda and J. A. Dumesic, *Nature*, 2002, **418**, 964.
167. G. W. Huber, J. W. Shabaker and J. A. Dumesic, *Science*, 2003, **300**, 2075.
168. G. Wen, Y. Xu, Z. Xu and Z. Tian, *Catal. Commun.*, 2010, **11**, 522.
169. G. Wen, Y. Xu, H. Ma, Z. Xu and Z. Tian, *Int. J. Hydrogen Energy*, 2008, **33**, 6657.
170. I. Groon and B. Swan, *Sven. Papperstidn.*, 1963, **66**, 812.
171. J. Zakzeski, P. C. A. Bruijnincx, A. L. Jongerius and B. M. Weckhuysen, *Chem. Rev.*, 2010, **110**, 3552.
172. M. V. Galkin and J. S. M. Samec, *ChemSusChem*, 2016, **9**, 1544.
173. T. Renders, S. Van den Bosch, S. F. Koelewijn, W. Schutyser and B. F. Sels, *Energy Environ. Sci.*, 2017, **10**, 1551.
174. Q. Song, F. Wang, J. Cai, Y. Wang, J. Zhang, W. Yu and J. Xu, *Energy Environ. Sci.*, 2013, **6**, 994.
175. T. Parsell, S. Yohe, J. Degenstein, T. Jarrell, I. Klein, E. Gencer, B. Hewetson, M. Hurt, J. I. Kim, H. Choudhari, B. Saha, R. Meilan, N. Mosier, F. Ribeiro, W. N. Delgass, C. Chapple, H. I. Kenttamaa, R. Agrawal and M. M. Abu-Omar, *Green Chem.*, 2015, **17**, 1492.
176. M. V. Galkin and J. S. M. Samec, *ChemSusChem*, 2014, **7**, 2154.
177. M. V. Galkin, A. T. Smit, E. Subbotina, K. A. Artemenko, J. Bergquist, W. J. J. Huijgen and J. S. M. Samec, *ChemSusChem*, 2016, **9**, 3280.
178. P. Ferrini and R. Rinaldi, *Angew. Chem., Int. Ed.*, 2014, **53**, 8526.
179. P. Ferrini, C. A. Rezende and R. Rinaldi, *ChemSusChem*, 2016, **9**, 3171.
180. W. Schutyser, S. Van den Bosch, T. Renders, T. De Boe, S. F. Koelewijn, A. Dewaele, T. Ennaert, O. Verkinderen, B. Goderis, C. M. Courtin and B. F. Sels, *Green Chem.*, 2015, **17**, 5035.
181. S. Van den Bosch, W. Schutyser, R. Vanholme, T. Driessen, S. F. Koelewijn, T. Renders, B. De Meester, W. J. J. Huijgen, W. Dehaen, C. M. Courtin, B. Lagrain, W. Boerjan and B. F. Sels, *Energy Environ. Sci.*, 2015, **8**, 1748.

182. T. Renders, W. Schutyser, S. Van den Bosch, S.-F. Koelewijn, T. Vangeel, C. M. Courtin and B. F. Sels, *ACS Catal.*, 2016, **6**, 2055.
183. T. Renders, S. Van den Bosch, T. Vangeel, T. Ennaert, S.-F. Koelewijn, G. Van den Bossche, C. M. Courtin, W. Schutyser and B. F. Sels, *ACS Sustainable Chem. Eng.*, 2016, **4**, 6894.
184. S. Van den Bosch, T. Renders, S. Kennis, S. F. Koelewijn, G. Van den Bossche, T. Vangeel, A. Deneyer, D. Depuydt, C. M. Courtin, J. M. Thevelein, W. Schutyser and B. F. Sels, *Green Chem.*, 2017, **19**, 3313.
185. E. M. Anderson, R. Katahira, M. Reed, M. G. Resch, E. M. Karp, G. T. Beckham and Y. Román-Leshkov, *ACS Sustainable Chem. Eng.*, 2016, **4**, 6940.
186. X. Huang, O. M. Morales Gonzalez, J. Zhu, T. I. Koranyi, M. D. Boot and E. J. M. Hensen, *Green Chem.*, 2017, **19**, 175.
187. L. Shuai, M. T. Amiri, Y. M. Questell-Santiago, F. Héroguel, Y. Li, H. Kim, R. Meilan, C. Chapple, J. Ralph and J. S. Luterbacher, *Science*, 2016, **354**, 329.
188. B. N. Kuznetsov, S. A. Kuznetsova, V. G. Danilov, I. A. Kozlov, V. E. Taraban'ko, N. M. Ivanchenko and N. B. Alexandrova, *Catal. Today*, 2002, **75**, 211.
189. B. N. Kuznetsov, S. A. Kuznetsova, V. G. Danilov and O. V. Yatsenkova, *React. Kinet. Catal. Lett.*, 2008, **94**, 311.
190. B. N. Kuznetsov, V. E. Taraban'ko and S. A. Kuznetsova, *Kinet. Catal.*, 2008, **49**, 517.
191. B. N. Kuznetsov, N. V. Chesnokov, O. V. Yatsenkova and V. I. Sharypov, *Russ. Chem. Bull.*, 2013, **62**, 1493.
192. B. N. Kuznetsov, V. I. Sharypov, L. I. Grishechko and A. Celzard, *Kinet. Catal.*, 2013, **54**, 344.
193. B. N. Kuznetsov, I. G. Sudakova, N. V. Garyntseva, L. Djakovitch and C. Pinel, *React. Kinet., Mech. Catal.*, 2013, **110**, 271.
194. B. N. Kuznetsov, N. V. Chesnokov, O. V. Yatsenkova, V. I. Sharypov, N. V. Garyntseva, N. M. Ivanchenko and V. A. Yakovlev, *Wood Sci. Technol.*, 2017, **51**, 1189.
195. B. N. Kuznetsov, I. G. Sudakova, N. V. Garyntseva, L. Djakovitch and C. Pinel, *React. Kinet., Mech. Catal.*, 2017, **120**, 81–94.
196. V. E. Tarabanko, K. L. Kaygorodov, E. A. Skiba, N. Tarabanko, Y. V. Chelbina, O. V. Baybakova, B. N. Kuznetsov and L. Djakovitch, *J. Wood Chem. Technol.*, 2017, **37**, 43.
197. R. Zhang, R. C. Brown and A. Suby, *Energy Fuels*, 2004, **18**, 251.
198. D. B. Levin, L. Pitt and M. Love, *Int. J. Hydrogen Energy*, 2004, **29**, 173.
199. M. E. Dry, *J. Chem. Technol. Biotechnol.*, 2002, **77**, 43.
200. J. Ereña, J. M. Arandes, J. Bilbao, M. Olazar and H. I. de Lasa, *J. Chem. Technol. Biotechnol.*, 1998, **72**, 190.
201. A. M. Henstra, J. Sipma, A. Rinzema and A. J. M. Stams, *Curr. Opin. Biotechnol.*, 2007, **18**, 200.
202. S. Morales-Delarosa and J. M. Camos-Martin, in *Advances in Biorefineries*, ed. K. Waldron, Elsevier, UK, 2014, pp. 152–198.

203. S. Czernik and A. V. Bridgwater, *Energy Fuels*, 2004, **18**, 590.
204. P. M. Mortensen, J. D. Grunwaldt, P. A. Jensen, K. G. Knudsen and A. D. Jensen, *Appl. Catal., A*, 2011, **407**, 1.
205. Q. Zhang, J. Chang, T. Wang and Y. Xu, *Energy Convers. Manage.*, 2007, **48**, 87.
206. M. Balat, M. Balat, E. Kırtay and H. Balat, *Energy Convers. Manage.*, 2009, **50**, 3147.
207. F. M. Mercader, M. J. Groeneveld, S. R. A. Kersten, N. W. J. Way, C. J. Schaverien and J. A. Hogendoorn, *Appl. Catal., B*, 2010, **96**, 57.
208. V. A. Yakovlev, S. A. Khromova, O. V. Sherstyuk, V. O. Dundich, D. Y. Ermakov, V. M. Novopashina, M. Y. Lebedev, O. Bulavchenko and V. N. Parmon, *Catal. Today*, 2009, **144**, 362.
209. M. V. Bykova, D. Y. Ermakov, S. A. Khromova, A. A. Smirnov, M. Y. Lebedev and V. A. Yakovlev, *Catal. Today*, 2014, **220**, 21.
210. M. V. Bykova, D. Y. Ermakov, V. V. Kaichev, O. A. Bulavchenko, A. A. Saraev, M. Y. Lebedev and V. A. Yakovlev, *Appl. Catal., B*, 2012, **113**, 296.
211. A. Gutierrez, R. K. Kaila, M. L. Honkela, R. Slioor and A. O. I. Krause, *Catal. Today*, 2009, **147**, 239.
212. H. Y. Zhao, D. Li, P. Bui and S. T. Oyama, *Appl. Catal., A*, 2011, **391**, 305.
213. A. Popov, E. Kondratieva, J. M. Goupil, L. Mariey, P. Bazin, J.-P. Gilson, A. Travert and F. Maugé, *J. Phys. Chem. C*, 2010, **114**, 15661.
214. S. Echeandia, P. L. Arias, V. L. Barrio, B. Pawelec and J. L. G. Fierro, *Appl. Catal., B*, 2010, **101**, 1.
215. N. V. Gromov, O. P. Taran, V. S. Semeykina, I. G. Danilova, A. V. Pestunov, E. V. Parkhomchuk and V. N. Parmon, *Catal. Lett.*, 2017, **147**, 1485.
216. R.-J. van Putten, J. C. van der Waal, E. de Jong, C. B. Rasrendra, H. J. Heeres and J. G. de Vries, *Chem. Rev.*, 2013, **113**, 1499.
217. Y. Román-Leshkov, J. N. Chheda and J. A. Dumesic, *Science*, 2006, **312**, 1933.
218. P. Bhaumik and P. L. Dhepe, *RSC Adv.*, 2013, **3**, 17156.
219. C. Moreau, A. Finiels and L. Vanoye, *J. Mol. Catal. A: Chem.*, 2006, **253**, 165.
220. Z. Fang, B. Liu, J. Luo, Y. Ren and Z. Zhang, *Biomass Bioenergy*, 2014, **60**, 171.
221. P. Carniti, A. Gervasini, S. Biella and A. Auroux, *Catal. Today*, 2006, **118**, 373.
222. X. Qi, M. Watanabe, T. M. Aida and R. L. Smith Jr, *Catal. Commun.*, 2008, **9**, 2244.
223. M. Kitano, K. Nakajima, J. N. Kondo, S. Hayashi and M. Hara, *J. Am. Chem. Soc.*, 2010, **132**, 6622.
224. W. Y. Qi, C. W. Hu, G. Y. Li, L. H. Guo, Y. Yang, J. Luo, X. Miao and Y. Du, *Green Chem.*, 2006, **8**, 183.
225. M. Watanabe, Y. Aizawa, T. Iida, C. Levy, T. M. Aida and H. Inomata, *Carbohydr. Res.*, 2005, **340**, 1931.

226. A. Chareonlimkun, V. Champreda and A. Shotipruk, *et al.*, *Bioresour. Technol.*, 2010, **101**, 4179.
227. M. Watanabe, Y. Aizawa, T. Iida, R. Nishimura and H. Inomata, *Appl. Catal., A*, 2005, **295**, 150.
228. F. Yang, Q. Liu, X. Bai and Y. Du, *Bioresour. Technol.*, 2011, **102**, 3424.
229. T. Onfroy, G. Clet and M. Houalla, *J. Phys. Chem. B*, 2005, **109**, 14588.
230. P. Lauriol-Garbey, J. M. M. Millet, S. Loridant, V. Bellière-Baca and P. Rey, *J. Catal.*, 2011, **281**, 362.
231. G. Peng, X. Wang, X. Chen, Y. Jiang and X. Mu, *J. Ind. Eng. Chem.*, 2014, **20**, 2641.
232. R. Bibi, Z. Ahmad, M. Imran, S. Hussain, A. Ditta, S. Mahmood and A. Khalid, *Renewable Sustainable Energy Rev.*, 2017, **71**, 976.
233. Renewable Fuels Association of the USA (RFA), *Ethanol industry statistics*, http://www.ethanolrfa.org/resources/industry/statistics/, accessed July 28, 2017.
234. K. N. Sorokina, O. P. Taran, T. B. Medvedeva, Y. V. Samoylova, A. V. Piligaev and V. N. Parmon, *ChemSusChem*, 2017, **10**, 562.
235. I. Cybulska, T. Chaturvedi, G. P. Brudecki, Z. Kádár, A. S. Meyer, R. M. Baldwin and M. H. Thomsen, *Bioresour. Technol.*, 2014, **153**, 165.
236. F. Collas, W. Kuit, B. Clement, R. Marchal, A. M. Lopez-Contreras and F. Monot, *AMB Express*, 2012, **2**, 45.
237. S. Atsumi, T.-Y. Wu, E.-M. Eckl, S. Hawkins, T. Buelter and J. Liao, *Appl. Microbiol. Biotechnol.*, 2010, **85**, 651.
238. B. Blombach, T. Riester, S. Wieschalka, C. Ziert, J. W. Youn, V. F. Wendisch and B. J. Eikmanns, *Appl. Environ. Microbiol.*, 2011, **77**, 3300.
239. S. S. Pang and R. G. Duggleby, *Biochem. J.*, 2001, **357**, 749–757.
240. G. W. Huber, J. W. Shabaker, S. T. Evans and J. A. Dumesic, *Appl. Catal., B*, 2006, **62**, 226.
241. R. Sahu and P. L. Dhepe, *React. Kinet., Mech. Catal.*, 2014, **112**, 173.
242. L. Zheng, J. Zhao, Z. Du, B. Zong and H. Liu, *Sci. China: Chem.*, 2017, **60**, 950.
243. N. K. Gupta, S. Nishimura, A. Takagaki and K. Ebitani, *Green Chem.*, 2011, **13**, 824–827.
244. Y. Nakagawa, M. Tamura and K. Tomishige, *J. Jpn. Pet. Inst.*, 2017, **60**, 1–9.
245. F. Neaţu, N. Petrea, R. Petre, V. Somoghi, M. Florea and V. I. Parvulescu, *Catal. Today*, 2016, **278**, 66.
246. P. V. Rathod, S. D. Nale and V. H. Jadhav, *ACS Sustainable Chem. Eng.*, 2017, **5**, 701.
247. Y. Roman-Leshkov, C. J. Barrett, Z. Y. Liu and J. A. Dumesic, *Nature*, 2007, **447**, 982–985.
248. M. S. Gyngazova, L. Negahdar, L. C. Blumenthal and R. Palkovits, *Chem. Eng. Sci.*, 2017, **173**, 455.
249. J. Li, J.-l. Liu, H.-y. Liu, G.-y. Xu, J.-j. Zhang, J.-x. Liu, G.-l. Zhou, Q. Li, Z.-h. Xu and Y. Fu, *ChemSusChem*, 2017, **10**, 1436.

250. T. Buntara, S. Noel, P. H. Phua, I. Melián-Cabrera, J. G. de Vries and H. J. Heeres, *Angew. Chem., Int. Ed.*, 2011, **50**, 7083.
251. J. Tuteja, H. Choudhary, S. Nishimura and K. Ebitani, *ChemSusChem*, 2014, 7, 96.
252. K. Yan, C. Jarvis, J. Gu and Y. Yan, *Renewable Sustainable Energy Rev.*, 2015, **51**, 986.
253. A. Démolis, N. Essayem and F. Rataboul, *ACS Sustainable Chem. Eng.*, 2014, **2**, 1338.
254. M. Balakrishnan, E. R. Sacia and A. T. Bell, *ChemSusChem*, 2014, **7**, 1078.
255. A. Bohre, S. Dutta, B. Saha and M. M. Abu-Omar, *ACS Sustainable Chem. Eng.*, 2015, **3**, 1263.
256. K. Yan, G. Wu, T. Lafleur and C. Jarvis, *Renewable Sustainable Energy Rev.*, 2014, **38**, 663.
257. J.-P. Lange, E. van der Heide, J. van Buijtenen and R. Price, *ChemSusChem*, 2012, **5**, 150.
258. R. Mariscal, P. Maireles-Torres, M. Ojeda, I. Sadaba and M. Lopez Granados, *Energy Environ. Sci.*, 2016, **9**, 1144.
259. I. Delidovich, K. Leonhard and R. Palkovits, *Energy Environ. Sci.*, 2014, 7, 2803.
260. Y. Nakagawa, M. Tamura and K. Tomishige, *Catal. Surv. Asia*, 2015, **19**, 249.
261. W. Deng, Q. Zhang and Y. Wang, *Sci. China: Chem.*, 2015, **58**, 29.
262. Z. Zhang, *ChemSusChem*, 2016, **9**, 156.
263. G. Shrivastav, T. S. Khan, M. Agarwal and M. A. Haider, *ACS Sustainable Chem. Eng.*, 2017, **5**, 7118.
264. Q. Fang and M. A. Hanna, *Bioresour. Technol.*, 2002, **81**, 187.
265. F. D. Pileidis and M.-M. Titirici, *ChemSusChem*, 2016, **9**, 562.
266. F. Rataboul and N. Essayem, *Ind. Eng. Chem. Res.*, 2011, **50**, 799.

CHAPTER 3

Catalysts for Depolymerization of Biomass

NIKOLAY V. GROMOV, OXANA P. TARAN AND VALENTIN N. PARMON*

Boreskov Institute of Catalysis SB RAS, Department of Nontraditional Catalytic Processes, Akademika Lavrentieva, 5, Novosibirsk, 630090, Russia
*E-mail: parmon@catalysis.ru

3.1 Introduction

There is an ever increasing worldwide interest in the processes of depolymerization of lignocellulose biomass, meeting the arising challenges. The queries *biomass depolymerization* and *lignocellulose depolymerization* in databases such as Science Direct, Wiley, Royal Society of Chemistry, American Chemical Society and Springer demonstrate the multiplication in the number of related scientific papers during the last three decades, especially during the last decade. The number of patents in the field has increased rapidly, too (Figure 3.1). The great interest is underlain by the progressive depletion of fossil natural gas and petroleum resources and the necessity to solve some ecological problems like, primarily, fighting global warming by bonding technogenically evolved carbon dioxide.[1–3]

Green Chemistry Series No. 56
Sustainable Catalysis for Biorefineries
Edited by Francesco Frusteri, Donato Aranda and Giuseppe Bonura

Published by the Royal Society of Chemistry, www.rsc.org

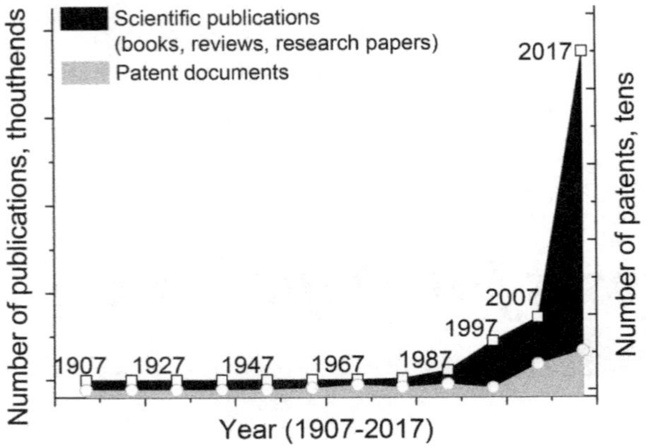

Figure 3.1 Evolution of the scientific interest in lignocellulose depolymerization.

An incredibly large variety of catalytic methods have been suggested for processing lignocellulose. Its depolymerization may be achieved under comparatively mild conditions at temperatures below 250 °C,[4–6] as well as with the use of stronger methods of pyrolysis, gasification, and fluidization[7–11] in either water or ionic liquids, organic solvents or mixtures, *etc.*[12–16] One may classify the methods for biomass transformation according to the origin of the catalyst used. For example, catalysts of biological or chemical nature (enzymes,[17,18] acids,[12,19–25] bases[19–21,26–28]) as well as catalysts operating in different physical states (soluble,[12,19–28] solid[4,5,29–38]) may be distinguished among these catalytic systems.

Depolymerization of lignocellulose under mild conditions is used mainly for the production of the so-called "platform molecules". Solid acid and acid–base catalysts, which demonstrate high selectivity[39] and activity in the depolymerization,[40] are of particular interest for the process implementation. The advantages of such processes are as follows:

- much easier separation of the catalyst from the reaction medium and reaction products;
- no corrosion to the facilities;
- high stability even at high temperatures;
- high rates of the initiated catalytic transformations;
- possibility of the catalyst recovery and recycling;
- generation of lower quantities of contaminants, waste products and wastewater.

At present, a large number of reviews devoted to the depolymerization of lignocellulose and its components are available.[4–6,12–14,29,30,32,33,35,37,41–49] In spite of considerable progress in processing lignocellulose, the authors of the research report some problems in the use of solid catalysts for the biomass depolymerization. Mass transfer between the solid substrate and solid catalyst, accessibility of the active acid and/or base catalytic center to large cellulose and lignin species,[4] elucidation of the detailed mechanism of depolymerization as well as of elementary acts of the catalytic transformation, and identification of the location of the catalytic transformation (surface, subsurface dielectric layer, solution) are among the serious challenges. As a result, the solid catalysts for the depolymerization may be less active and effective than the soluble catalysts.

This chapter is an attempt to provide an overview of the known methods for biomass depolymerization over solid acid catalysts and the progress in the application of these techniques. The main reaction in the depolymerization of lignocellulose and its components in the presence of solid acid catalysts is known as hydrolysis, which is usually carried out in an inert or gasless medium. The processes of lignin transformations in the presence of solid catalysts under an active redox atmosphere of O_2 or H_2 or in the presence of their chemical precursors, *e.g.* formic acid, are also referred to as *depolymerization processes* in the literature.[49–51] However, we shall consider these lignin transformations as one-pot processes of hydrolysis–oxidation and hydrolysis–reduction and leave them out of consideration in the Chapter. Since the process of lignin hydrolysis is less studied and less well understood now than the hydrolysis of polysaccharides, we shall focus our discussion on saccharides.

The main types of solid catalysts such as resins, carbons, zeolites, oxides, and heteropoly acids are considered in the Chapter. Under discussion are the known data on the catalyst activity towards lignocellulose depolymerization, as well as factors affecting the efficiency, and auxiliary methods favoring the lignocellulose depolymerization. Modern concepts on the solid substrate depolymerization in the presence of the solid acid catalyst will be discussed, too.

3.2 Solid Catalysts for the Depolymerization of Lignocellulose Biomass

It is known that the soluble acid catalysts are of limited industrial application due to the need for expensive non-corrodible equipment and difficulties in the separation of the catalyst from the dissolved substrate. These disadvantages could be overcome by creating high performance solid acid catalysts that can assist the hydrolysis of native biomass to valuable substances. Recently, a lot of articles aimed at solving this problem have been published. There are several types of promising catalysts: (1) ion-exchange resins and polymers (Amberlyst, Nafion) bearing Brønsted acid sites (-SO_3H); (2) carbon materials bearing acid sites; (3) nanodispersed metals of the platinum group (Pt, Pd, Ru, *etc.*) and Au on carbon or oxide supports. These catalysts are typically used for the combined hydrolysis–oxidation or hydrolysis–reduction processes.[41] They are also catalytically active to the hydrolysis of polysaccharides; (4) zeolites (Hβ, H-ZSM-5, *etc.*), which are aluminosilicate framework structures with Brønsted or Lewis type acid centers; (5) insoluble oxides (ZrO_2, TiO_2, SiO_2, *etc.*) containing Lewis and Brønsted acid sites; (6) insoluble heteropoly acids (HPA), for example, the salts CsHPA or AgHPA, which are acid catalysts with a strong Brønsted acidity.

3.2.1 Resins

Resins are rather effective catalysts for hydrolytic transformations of lignocellulose. These are organic polymers with a large surface area and capable of ion exchanging. The acidic properties of resins are accounted for by the presence of acid groups, *e.g.* $-SO_3H$, on the surface. The resins were among the first solid catalysts tested to hydrolyze plant biomass. Hartler and Hyllengren were first to apply polystyrene resins Amberlite-120 and Dowex-50 for cellulose hydrolysis in 1962.[52] We should emphasize that the styrene-based resins are till now most often used to study lignocellulose transformations. Application of polystyrene resins allows starch and cellulose to be hydrolyzed in hot water under mild conditions (120 °C) at 100% yield of the target product.[53]

The resin-based solid acid catalysts appear highly active and stable. For example, the polystyrene resin bearing surface –Cl and $-SO_3H$ groups allows glucose to be obtained from cellulose at yields as high as 95% in 10 hours of reaction under quite mild conditions (at 120 °C).[53] The results of the catalyst testing were successfully reproduced in several cycles.

Ion-exchange resin Amberlist®-15 is a typically used catalyst for hydrolysis of natural polymers; this resin is a phenylsulfoacid based on polyvinylbenzene (polystyrene). It catalyzes a large number of processes of organic synthesis including hydrolysis.[54] However, the published data on its catalytic activity in the hydrolysis of cellulose are contradictory. Some authors[55] observed rather high activity of this catalyst, while others[56] reported its only minor activity. A lower glucose yield in the latter case may be caused by a lower temperature of hydrolysis (130 *vs.* 150 °C, respectively) and by half as long contact times. The Amberlist®-15 resin becomes rather effective in the hydrolysis of cellulose pre-solved in an ionic liquid. Rinaldi *et al.* succeeded in reaching a 90% yield of glucose from cellulose at 100 °C after dissolving the polysaccharide in the 1-butyl-3-methylimidazolium chloride [BMIM] Cl ionic liquid.[57] A substantial disadvantage of the Amberlist resins is their deactivation at high temperatures as a result of leaching of the Brønsted acid sites.[58] Therefore, the maximal temperature of the catalyst operation must be not higher than 130–150 °C.[5]

Nafion resins (sulfated tetrafluoroethylene polymers) are more stable at high temperatures, and thus seem to be an alternative to the Amberlist catalysts. However, Nafion is far less active to cellulose hydrolysis than the polystyrene resin Amberlist. The yield of glucose upon cellulose hydrolysis was not higher than 9–16%, even though cellulose was preactivated by solving in ionic liquids.[59,60]

The search for the possibility of improving the activity of polymer resins is in progress. Hahn-Hägerdal *et al.* hydrolyzed cellobiose disaccharide in the presence of the Nafion resin and sodium chloride at 95 °C.[61] The glucose yields were 15, 50 and 80% at the salt loading equal to 0, 1 and 10%, respectively. The increase in the catalyst activity was accounted for by a stronger ion exchange between the resin and the salt that results in releasing protons to the reaction solution and in an increase in the acidity of the medium.

3.2.2 Carbon Based Catalysts

3.2.2.1 Modified Carbons

Comparative studies of different solid acid catalysts show that carbon materials with acid sites are among the most promising catalysts for the cellulose depolymerization.[55,62–64] Various carbons (*e.g.* activated carbons, pyrolytic carbons derived from monosaccharides such as glucose and sucrose, lignin, cellulose or phenol-containing organic waste) can be used for the preparation of acid catalysts.[4,62,64–68]

Carbon catalysts were shown to be highly active to cellulose hydrolysis. For example, Onda *et al.* compared a number of solid acid catalysts (oxides, zeolites, sulfonated carbon, Amberlyst resin). The catalytic activity of the sulfonated carbon was several times higher than the activity of the other solid acid catalysts. The yield of glucose over the carbon reached 40%.[55,64] Other researchers[63] observed a glucose yield equal to 62% from the mechanically activated microcrystalline cellulose under mild conditions at 150 °C in 24 hours. The authors of this chapter obtained a glucose yield as high as 74% from cellulose in a flow reactor in the presence of a sulfonated carbon catalyst based on the carbon material Sibunit at the reaction temperature of 200 °C.[69]

It should be noted that a non-modified carbon support does not exhibit any significant catalytic activity to polysaccharide depolymerization.[63,67,70] The most practiced method to increase the acidity of the carbon materials is their treatment with sulfuric acid to anchor sulfo groups onto the carbon surface.[55,62–64,71–75] Treatments with solutions of HNO_3,[63,68] H_2O_2,[68,70] oxidation by air,[68] wet air,[76] or NaOCl[76] also are appropriate, as well as a combination of different activation methods. For example, nitric acid was used[63,68] for surface oxidation and the formation of acid groups. The carbon catalysts treated successively with nitric and sulfuric acids were also tested. Activated carbon was modified using sulfuric acid and H_2O_2, as well as both simultaneously.[70] The oxidation by H_2O_2 led to a six-fold increase in the number of acid sites (phenol, lacto and carboxyl sites), while the catalyst activity increased only negligibly to afford a glucose yield as low as 2% at the selectivity of 8% (150 °C, 24 hours). The further sulfonation of the carbon by H_2SO_4 results in a five-fold increase in the yield of glucose at a selectivity as high as 26%. Similar results were obtained with carbon catalysts sulfonated at 100 and 200 °C.

The conditions (temperature, contact time and agent concentration) of the carbon pretreatment with sulfuric acid and/or with oxidants significantly influence the catalyst properties. Elevation of the sulfonation temperature provides a higher content of the surface sulfo groups and, as a result, a higher catalyst activity. However, activated carbons are destroyed at high temperatures due to deep oxidation processes.[63] The insufficient mechanical strength and stability of most activated carbons prevent them from practical use for processes of hydrothermal hydrolysis of cellulose.

Carbon materials treated with sulfuric acid or any oxidants are stable enough during the hydrothermal treatment.[68] Sulfonated carbon materials can be sufficiently stable to leaching of sulfo groups from the surface. Van Pelt *et al.* compared the stabilities of the activated carbons treated with various oxidants against leaching of acid groups from the surface after hydrothermal processing.[68] Carbon materials treated with sulfuric acid appeared to be more stable than the samples oxidized by other oxidants such as nitric acid, hydrochloric acid and hydrogen peroxide. Under the reaction conditions, sulfur is usually leached in negligible quantities from the surface: less than 0.03 mmol L^{-1} that corresponded to minor leaching of $-SO_3H$ to the reaction mixture (less than 0.7% of the initial sulfur content in the catalyst) but it may increase over a longer treatment time.[55] Thus, under hydrothermal conditions (200 °C) over long (24 hours) treatment times, the proportion of leached sulfo groups can reach 38% and the total acid groups decrease by 47%.[68] Testing the carbon catalysts during several reaction cycles under comparatively mild conditions demonstrated a high stability of the catalytic systems.[65,77–79] Thus, Yang *et al.* showed that the catalytic activity of the catalyst Fe_3O_4@C–SO_3H decreased after only 5 reaction cycles.[77]

The influence of salts on the activity of carbon catalysts in cellulose depolymerization was also a subject for studies.[66,80,81] In 1983, Hahn-Hägerdal *et al.* demonstrated an impact of sodium chloride upon the hydrolysis of cellobiose over the Nafion®-50 catalyst.[81] The yield of glucose increased by a factor of 2.5 from 15% to 75% in the presence of 100 g L^{-1} NaCl.[81] The influence of NaCl described by Hahn-Hägerdal *et al.* was explained by the exchange of H^+ ions in the solid catalyst for Na^+ of the salt that increased the reaction medium acidity (confirmed by pH) and consequently the product yields.[81] Shen *et al.* demonstrated a positive influence of NaCl, KCl, and KNO_3 salts on the cellulose depolymerization over pyrolytic carbon at 160 °C and 6 h reaction time. The yield of sugars was 39% in the absence of the additives. The sugar yields increased up to 55% upon addition of NaCl or KCl but slightly with potassium nitrate. The authors assigned the high activity of NaCl and KCl, firstly, to ion exchanges of H^+ in the solid catalyst for Na^+ and, secondly, to the destruction of hydrogen bonds in cellulose by chloride anions.[66,80] It is our opinion that the replacement of a noticeably high quantity of hydrogen on the solid catalyst surface by well soluble sodium and potassium ions is less probable, while the dispersion of cellulose under the action of concentrated salt solutions is described in commonly used manuals. Hence, the increased glucose yields in the presence of the mentioned salts can be accounted for by the destruction of hydrogen bonds in cellulose by chloride anions and cations.

The similar influence of Cl^- is known for the transformation of cellulose in some ionic liquids. The origin of the good solubility of cellulose in ionic liquids is the disappearance of hydrogen bonds –O···H–O– in the cellulose structure through the interaction of –O–H groups with halide anions (Cl^- or Br^-), which are strong electron acceptors. The influence of Cl^- ions is more pronounced.[82]

Thus, the carbon catalyst bearing both strong acid sites and Cl^- groups is promising for the processing of cellulose. Shen *et al.* obtained two types of carbon materials by pyrolysis of pure starch or by combined pyrolysis of starch with polyvinyl chloride. The amount of acid groups in the catalysts was the same (0.8 mmol g^{-1}). However, the sample containing C–Cl^- structures appeared two times more active than the sample obtained from starch only. The yields of reducing sugars were 95 and 40%, respectively.[80] The influence of chloride ions on the destruction of hydrogen bonds –O···H–O– was demonstrated by experimental studies of hydrolysis of cellobiose, which has no hydrogen bonds, when both catalysts showed close activities. The process seems to occur on or near the carbon surface and it involves chloride anions anchored to the surface (from the FT-IR spectrometric data).

To facilitate removal of the carbon catalyst from the reaction mixture containing saccharides and their dehydration products as well as non-reacted solid particles of cellulose, original catalysts containing a magnetic Fe_3O_4 core covered by a sulfonated carbonaceous material were developed. Yang *et al.* described the hydrolysis of cellulose in the presence of such a catalyst to give *ca.* 50% yields of glucose.[77] Pure Fe_3O_4 was catalytically inactive. Guo *et al.*[83] suggested the composite material carbon/Fe_3O_4 instead of the core–shell system but they did not test it in aqueous solutions. Studies with the use of an ionic liquid (1-butyl-3-methylimidazolium chloride, [BMIM]Cl) showed the activity of such a system to be much higher than the activities of the reference solid catalysts like sulfonated carbon and polymers Amberlist®-15 and Nafion® NR-50. However, the tested composition had low stability: the catalyst lost *ca.* 30% of its activity after three reaction cycles, presumably, because of the destruction of the poorly bound composite material.

3.2.2.2 Metals Supported on Carbon

Highly dispersed metal particles supported on carbons can also catalyze the hydrolysis of polysaccharides. The catalytic activity of nanodispersed metals was discovered by the team headed by Professor Atsushi Fukuoka from the Institute of Catalysis in Hokkaido (Hokkaido University, Japan).[84] Comparative studies were focused on the activities of supported highly dispersed Ru, Pt, and Pd metal particles in hydrolysis in an inert atmosphere at 230 °C.[85] The Ru-containing catalysts were most active. The yield of glucose was higher over the metal-containing catalyst than in the presence of pure carbon (31 and 16%, respectively). The yields of the hydrolysis products (glucose and soluble oligosaccharides) increased with the Ru content. Thus, the glucose yields reached 16, 20, 25 and 31% in the presence of catalysts containing 0, 2, 5 and 10% ruthenium, respectively. The Ru/C system was highly stable during 5 reaction cycles.[86,87] Importantly, the glucose yield was no more than 4% in the absence of any catalyst, which indicated the catalytic action of Ru/C and not hot water.[88] Fukuoka and co-workers explained the catalytic activity of nanodispersed metals as follows: the metal nanoparticles favor the *in situ* formation of H^+ and/or OH^- on their surface *via* dissociation of water.[84,89] Protons and hydroxide ions formed on the metal particles are stabilized on the carbon support surface. Then, water-soluble oligomers are adsorbed by the carbon support and/or metal particles of the catalyst, which promotes hydrolysis of the 1,4-β-glycosidic bond by protons and hydroxide ions to form glucose through the mechanisms characteristic of acidic or basic hydrolysis of saccharides. Eventually, glucose is desorbed back into the aqueous solvent from the catalyst surface.[84]

3.2.3 Zeolites and Silicates

Zeolites fall into a unique family of materials widely used as catalysts for oil cracking and, potentially, for a number of processes of organic synthesis.[4,5] These are structures built up by SiO_4 and AlO_4 tetrahedrons linked to one another *via* oxygen bonds. Exceptional stability under severe conditions, large surface area, large pore volume, extensive channel structure, and the presence of acid and base (Brønsted and Lewis) sites are characteristic of zeolites.[90] Among the advantageous features of zeolites are their highly ordered structures, and the possibility to control physicochemical properties during the catalyst preparation. There is also a possibility of the creation of hierarchical and 3D-structured modifications.[91] The number of Brønsted acid sites in zeolites is controlled by the Si/Al ratio; the smaller the ratio, the higher the acidity. Catalysts with the required acidity can be synthesized by varying that ratio.[4]

However, till now, zeolites have been considered to be only moderately active to the hydrolysis of cellulose in water. Thus, the yield of glucose from cellulose was 8–12% in the presence of H-beta and H-ZSM-5 zeolites.[64] Even though these results differ from the results of the blank run (*ca.* 4%), zeolites were far less active than carbon and resins. Onda *et al.*, based on the results of comparative studies of activities of acid catalysts, inferred that the reason for the low activity is diffusion limitations due to the small pore size that makes it impossible for large molecules of water-soluble saccharide oligomers to penetrate inside the catalyst granule to reach the active sites.[64] However, there is also another explanation of the low zeolite activity in aqueous media.[4] In terms of the suggested mechanism, the heterogeneous pathway characteristic of polysaccharide hydrolysis over zeolites implies that the dissolved oligomers enter the catalyst pores where they are transformed in the subsurface layer. Hence, the zeolite catalyst does not favor the direct conversion of cellulose to oligosaccharides. Cellulose does not dissolve in water and, therefore, oligosaccharides capable of transforming on the zeolite surface are not formed in sufficient quantities. As a result, the transformation rates and product yields decrease. The use of ionic liquids, which are capable of dissolving cellulose, instead of water makes zeolite-containing catalysts much more effective.[4] Cai and co-authors demonstrated a noticeable activity of HY zeolite to cellulose hydrolysis in the ionic liquid [BMIM]Cl. The yield of glucose was 50% at 130 °C and after a two hour reaction time.[4,92] The authors refer the high activity of the zeolite to the chemical ion-exchange interactions between the catalyst and the ionic liquid. During these, the [BMIM] ion of the liquid starts a chemical reaction with the acid site of the zeolite to release a proton that catalyzes the transformation of oligomers dissolved in the ionic liquid.

It should be emphasized that the researchers who used zeolites as catalysts for cellulose depolymerization in aqueous media were mainly aiming at studying the formation of glucose. However, the possibility of the direct one-pot conversion of cellulose to a valuable chemical, 5-HMF, is also under study.[93] The target product was synthesized from microcrystalline cellulose at a yield of 49% at 190 °C using the bimodal zeolite H-Z-5 with the ratio Si/Al = 30.15 prepared by treating H-ZSM-5 zeolite (Si/Al = 37.00) with sodium hydroxide. In addition, the formation of levulinic acid (a product of the 5-HMF hydration) in a rather high proportion was also observed. The said bimodal zeolite was stable during four reaction cycles. The H-Z-5 zeolite is most promising among the zeolite-containing catalysts for synthesis of 5-HMF from cellulose.

Zeolite catalysts may be effectively applied for depolymerization of plant hemicellulose materials to monosaccharides. Hydrolysis of polysaccharide inulin at 100 °C in the presence of the acid zeolite catalyst LZ-M-8 allows fructose to be obtained at yields as high as 96% at almost 100% conversion of the substrate.[94] Zeolites HUSY, H-Beta and HMOR catalyze the transformation of mixtures of xylan and arabinan hemicelluloses to their respective monosaccharides xylose and arabinose at yields of 35–40% at 170 °C.[95] Furane compounds (5-HMF, furfural) can be synthesized at moderate yields (up to 15%) in the presence of a number of zeolites (H-Beta, H-Y, H-ZSM-5, HUSY, HMOR) under conditions close to those of hemicellulose conversion to sugars.[95,96] High stabilities are characteristic of zeolites during hemicellulose transformations, their test results being reproducible during no less than five reaction cycles.[95]

Acid protonated zeolites, *e.g.* H-Y, can be used for lignin depolymerization. The maximal organosolv lignin conversion of 92.58% at a total of phenolic compound yield of 6.02% can be reached at 300 °C in an hour in the presence of H-Y zeolite with Si/Al = 10.[97]

Some silicate structures can also be used for the hydrolysis. Silicates are silica-based inorganic materials, either amorphous (SiO_2) or crystalline (for example, SBA-15 and MCM-41). They feature high stability, porosity, and a large surface area.[5] Their principal disadvantage is a very low acidity that makes additional pretreatment including supporting sulfo groups on the surface or the creation of composite materials necessary. Van de Vyver *et al.* suggested a sulfonated composite based on carbon and silicate.[45] The catalyst was used to synthesize glucose at 50% yield under comparatively mild conditions: 150 °C, 24 h reaction time. In 2014, Jiang and co-authors modified the MCM-41 silicate by introducing titanium and sulfonating with sulfuric acid.[98] The authors synthesized 5-HMF at yields of 2.0–7.5% from cellulose in a batch reactor at 190–230 °C, while the yields of reducing sugars were *ca.* 50%. Sulfonated silicates SO_4^{2-}/SBA-15[99] and SO_4^{2-}/MCM-41 allowed glucose to be synthesized at yields as high as 50%.[5]

A number of research teams have focused their activities on the development of an efficient catalyst based on zeolites and silicates. However, examples of the successful application of zeolite and silicate catalytic systems for low-temperature hydrolysis of lignin are not numerous. Zeolite or silicate based systems (H-ZSM-5, H-mordenite, H-Y, MCM) are more often used for the moderate temperature cracking of lignin in bio-oil at *ca.* 400 °C.[51]

3.2.4 Oxides

3.2.4.1 Non-promoted Oxides

Oxides of some metals, owing to their high mechanical and thermal stability, seem promising catalysts for depolymerization.[12] The most attractive objects are catalysts based on Al, Si, Zr, Ti, Nb, Co, V, and Ta oxides. The oxide catalysts are classified into mono-oxide systems and complex oxide systems containing several metals.

Niobium oxide is a potential catalyst for some processes of hydrolysis[100] and dehydration[4] due to the high content of surface acid sites. Catalysts based on oxides of niobium (Nb_2O_5), tantalum (Ta_2O_5) and zirconium (ZrO_2) are effective in the hydrolysis of soluble hemicellulosic polysaccharides and in the dehydration of fructose and glucose to 5-HMF.[101–104] The catalytic systems of niobium molybdate and tantalum molybdate ($HNbMoO_6$ and $HTaMoO_6$) were more active to sucrose and cellobiose hydrolysis than H_2SO_4.[105]

However, far fewer studies have dealt with the catalytic activity of oxide systems to cellulose depolymerization.[5] The papers are mainly devoted to hydrolysis of cellulose to glucose. The other valuable compounds, such as 5-HMF and levulinic acid, are either neglected or listed among the side products with minor yields.

Mixed Nb–W oxide demonstrated a low activity to the cellulose depolymerization process and the total yield of glucose and cellobiose was 8.5%.[106] Niobium molybdate $HNbMoO_6$ also showed only minor activity to cellulose depolymerization.[56] Low catalytic activities to hydrolysis of cellulose were demonstrated with ZrO_2, TiO_2, and Al_2O_3.[64,79,107,108] The activities of ZrO_2 and Al_2O_3 were close to the blank runs in pure water.[64] The low activity was accounted for either by an insufficient size of the catalyst pores, or by the low strength or low concentration of acid groups. Similar to zeolites, the oxide catalyst activity and yields of target reaction products increased upon hydrolysis in ionic liquids.[4]

We also studied the catalytic depolymerization of cellulose in the presence of zirconium oxide catalyst.[69] The reaction was tested in pure water at 180 °C under hydrothermal conditions. The yields of both glucose and 5-HMF reached 12–13%. ZrO_2 was promoted by niobium oxide to increase the surface acidity. We prepared samples of NbO_x/ZrO_2 catalyst, which contained 0.5, 1.1 and 2.8% Nb. The catalyst 2.8% Nb/ZrO_2 showed the highest activity and the yields of glucose and 5-HMF reached 22 and 16%, respectively.[69]

There are contradictory data on the activity of SiO_2 to cellulose hydrolysis. Comparative studies of the activities of oxides SiO_2, ZrO_2, TiO_2, and Al_2O_3 and H-ZSM-5 zeolites (Si/Al = 28 and 35) to the depolymerization of cellulose in aqueous media revealed that silica is several times more active than the other catalysts.[107] In the presence of SiO_2, a 50% yield of glucose at 73% conversion of the substrate and a total of 14.6% yield of all the side products including furan derivatives were observed; all the other catalysts afforded no more than a total of 9% yield of glucose and side products at a conversion of no more than 17%. The authors refer the high activity of this catalyst over the activities of the others to the much higher content of surface acid groups, higher strength of these groups and to a large enough pore size. On the other hand, pure silica is inactive to the hydrolysis of cellulose, which may be accounted for by the extremely low acidity of the catalyst.[79]

The activities of mixed TiO_2–ZrO_2 oxide and individual ZrO_2 and TiO_2 were studied regarding the depolymerization of cellulose and plant biomass to 5-HMF.[103] The highest yields of 5-HMF were observed with TiO_2–ZrO_2 and TiO_2 (14 and 13%, respectively). A higher efficiency was characteristic of the mono-oxide of Ti(IV) because a yield of 5-HMF equal to 13% was achieved at one fifth content of the surface acid groups compared to that of TiO_2–ZrO_2. Similar ratios of TiO_2–ZrO_2 and TiO_2 activities were observed during the hydrolysis of maize cobs. The highest yields of 5-HMF (the product of cellulose transformation) and furfural (hydrolysis–dehydration of xylan) were 8.6 and 10.3%. Besides, mixed oxide ZrO_2–Al_2O_3 seems promising for the depolymerization of lignin. The yield of liquid products from the transformation of Kraft lignin reached 85–88% at 200–400 °C and 0.5–8 h reaction time.[109]

Separation of an oxide catalyst from the reaction mixture containing solid particles of unreacted cellulose can be made easier if the catalyst includes magnetic Fe_3O_4 particles. Lai and co-authors[79] were first to use the catalytic system comprising a magnetic core covered by a sulfonated oxide shell, $Fe_3O_4@SiO_2–SO_3H$, for hydrolysis of cellulose and to show a high catalytic activity to hydrolysis of cellobiose and cellulose at 150 °C. The yields of glucose were 96 and 50% from di- and polysaccharide, respectively. Efficiency of this catalyst was halved when microcrystalline cellulose was used (the glucose yield decreased down to 28%). The main product of the long treatment of cellulose for 12 hours was levulinic acid (42% yield).[79] Takagaki and co-authors[110] developed an easy-to-separate oxide catalyst with the core–shell structure $CoFe_2O_4@SiO_2–SO_3H$. This catalytic system was active to the hydrolysis of disaccharides such as cellobiose and maltose to allow yields of glucose as high as 80–90%. However, the catalyst activity to the hydrolysis of polysaccharide (cellulose) was extremely low and comparable to the activity of the Amberlist-15 catalyst.[56,110]

An innovative approach to the hydrolytic processing of biomass is the use of two-phase systems. The systems have been actively utilized for the transformation of water-soluble saccharides.[14] This approach allows the direct hydrolytic synthesis of a valuable product – 5-HMF – from plant polysaccharides. Essentially, 5-HMF formed during the reaction is extracted from the aqueous phase to the organic phase in order to prevent it from further transformations and to improve the yield and the reaction selectivity. Among the most prospective organic solvents are methylisobutylketone (MIBK), dimethylsulfoxide (DMSO), tetrahydrofurane (THF), toluene, as well as propyl and butyl alcohols.[29,41,111] The positive effect of the addition of organic extractants was demonstrated; for example, alcohols form esters with furfurals and improve the extractability of the target product. At present, the highest yields of 5-HMF from fructose in water–organic systems reach 90% when butanol-2 or methylixobutylketone is used as the extractant and DMSO is used as the organic solvent of 5-THF esters.[14] The earliest works on the transformations of poly-, di- and oligosaccharides to 5-HMF were started in the 1980s by studying the synthesis of 5-HMF from inulin, starch, raffinose, cellobiose, and saccharose.[14,112] Yang and co-authors synthesized 5-HMF from inulin in the water–butanol-2 system in the presence of the Ta_2O_5 catalyst; the yield of 5-HMF was 87% at a substrate conversion of 95%.[101] Xylan, a hemicellulosic polysaccharide, can be converted to 5-HMF at a yield of 85% at 170 °C using a water–toluene mixture in the presence of the WO_3/SiO_2 catalyst.[95] Zeolites have also been examined for transformations of hemicelluloses in two-phase systems.[95] The solid catalyst TiO_2 can catalyze the hydrolysis of cellulose in a two-phase water–organic medium (water–MIBK) at 270 °C.[113]

3.2.4.2 *Sulfated Oxides*

While oxide catalysts bear fewer surface acid groups than, for example, sulfonated carbons (*cf.* 0.1–1 mmol g^{-1} in oxides and 1.5–3.5 mmol g^{-1} in sulfonated carbons[56,63,107,114]), it seems reasonable to add promoters to the acid catalysts. It is expected by analogy with sulfonated carbons that the treatment of transition metal oxides with sulfuric acid would improve their catalytic activity. Sulfated oxides such as SO_4^{2-}/Al_2O_3, SO_4^{2-}/TiO_2, SO_4^{2-}/ZrO_2, SO_4^{2-}/SnO_2 and SO_4^{2-}/V_2O_5 are most often used for studies in the field.[12]

Onda *et al.* synthesized glucose at the yield of 14.2% in the presence of sulfated zirconium(IV) oxide, *i.e.* a factor of 3.5 lower than in the presence of sulfonated carbon.[55,64] The fact that the total yield of reducing sugars over ZrO_2–SO_3H was only marginally lower than that over carbon indicates a low selectivity of the developed sulfated catalyst. A 24% yield of glucose can be reached when the reaction is catalyzed by sulfated silica supported on ferromagnetic particles of Fe_3O_4.[115]

Transformations of cellulose to 5-HMF in the presence of TiO_2, ZrO_2 and sulfated ZrO_2 prepared by treating zirconia with sulfuric acid at 70 °C were studied to show that the preparation conditions (chemical nature of the precursor, treatment temperature) considerably influence the properties of the resulting catalysts.[108] For example, the use of zirconyl and titanyl nitrates instead of the chlorides as precursors, as well as calcinations of the precursor at a lower temperature, makes it possible to prepare oxides with a larger surface area and, as a result, with a higher content of surface acid groups that increase their activity in the depolymerization of polysaccharide substrates. The promotion of ZrO_2 with sulfuric acid allows the catalyst activity to be improved, the maximal activity being observed with ZrO_2 containing 1.8 wt% sulfur. However, this catalyst was somewhat less active than the untreated titania; the latter allowed the authors to observe a 12% yield of 5-HMF from cellulose at 250 °C. The promotion with sulfuric acid of mixed oxide catalysts with different ratios of the oxide species prepared by co-hydrolysis of the precursors resulted in a 4–5 fold increase in their activity in the hydrolysis of cellobiose to provide a glucose yield up to 50%.[116] Sulfation of SiO_2 led to a considerable increase in the catalyst activity to produce glucose at 50% yield from cellulose.[79]

Not only mineral sulfuric acid but also organic sulfo acids can be used as the promoting agents.[114] 7.5 and 15% arenesulfonic and 15% propylsulfonic acids supported on SiO_2 catalyze hydrolysis of cellobiose to glucose and provide a high conversion (more than 80%) during as short a time as half an hour of the reaction.[114] Carboxylic acids supported on the oxide support do not demonstrate activity comparable to the activity of the sulfo acids.[114]

The leachability of the active component is a major drawback of the sulfated catalysts. Only a minor loss in the activity of the catalyst ZrO_2 is observed during several reaction cycles,[107,108] while the activity of sulfated ZrO_2 decreases progressively from cycle to cycle.[108] Sulfo groups are leached from the catalysts both during the reaction and during hydrothermal treatments. For example, hydrothermal treatment of sulfated ZrO_2–SiO_2 catalysts at 160 °C results in leaching of more than 90% of the active acid groups in 30 min and in their complete removal in an hour.[116]

3.2.4.3 Metals Supported on Oxides

Catalysts containing Ru and Pt nanoparticles supported not only on carbon supports (see Section 3.2.2) but also on oxides and zeolites of various natures (H-ZSM-5, HUSY, HY, Hβ, SiO_2, ZrO_2, Al_2O_3) were studied in the hydrolysis of polysaccharides.[117] The catalysts were rather active to hydrolysis–reduction of cellulose to produce polyols at yields of up to 30%. However, the observed extremely low activity of the supports to hydrolysis led the authors to conclude that not only the hydrolytic activity of the support but also the supported metals are responsible for the high yields of polyols. The mechanism of polysaccharide hydrolysis over metal catalysts suggested by Fukuoka and co-authors was discussed above in Section 3.2.2.2.

3.2.5 Heteropoly Acids

Heteropoly acids are a wide range of polyoxo compounds formed by fifth and/or sixth group metals of the Periodic system. The basis of HPA is a heteropoly anion of the general formula $[A_xB_yO_z]^{n-}$ where O is oxygen, A is a heteroatom (phosphorus, silicon, boron *etc.*), B is a metal (more often molybdenum, vanadium, tungsten, more rarely niobium, tantalum). HPA with similar structures are grouped as, for example, Keggin structures $AB_{12}O_{40}$, Dawson structures $A_2B_1O_{48}$, and Silverton structures $AB_{12}O_{42}$. Keggin structured HPA are most often used in catalysis.[29,118] A distinctive feature of HPA is their very strong Brønsted acidity, which is much higher than that of mineral acids (sulfuric, nitric, perchloric acids, hydrogen chloride *etc.*),[119,120] and this attracts much scientific interest in these catalysts. HPA were studied for a while as promising catalysts for hydrolytic processes.[121,122] HPA in solution are more effective catalysts for lignocellulose hydrolysis than mineral acids.[24] HPA with essentially different structures and compositions can demonstrate different catalytic activities. The systems based on solutions of P-W HPA are most active to hydrolysis of cellulose, being more active than Si-W HPA, and mineral sulfuric and phosphoric acids.[123]

Along with numerous advantages, all the drawbacks of soluble catalysts are characteristic of HPA. A promising approach may be to immobilize HPA, for example, as low soluble cesium salts of HPA (CsHPA). The concentration of acid groups is much higher in CsHPA than in all the solid catalysts under consideration in the Chapter. Hence, it is reasonable to conclude that CsHPA can be effectively used for hydrolysis of polysaccharides. Tian *et al.* were first to describe cellulose hydrolysis over CsHPA in 2010[24] to discover that $CsH_2PW_{12}O_{40}$ is promising among the $Cs_xH_{3-x}PW_{12}O_{40}$ salts (where x ranges between 1 and 3); the yield of glucose was 27% at 160 °C in 6 hours. The rates of depolymerization and cellulose conversion were inversely related to the content of Cs in CsHPA, probably, due to a decrease in the catalyst acidity with an increase in Cs content. However, a sharp increase in the selectivity to glucose was observed at x = 2.0–2.5 in CsHPA with the maximum of $S = 85\%$ at $x = 2.2$; at lower or higher cesium contents, the selectivity was *ca.* 50%.[24] The yield of glucose also increased sharply at $x = 2.0–2.5$ that, however, did not influence the total dependence of the yield on the catalyst acidity.

Later, a team of researchers[124] studied a close to optimal catalyst, $Cs_{2.5}H_{0.5}PW_{12}O_{40}$. The catalyst provided a yield of glucose equal to 21% at 89.9% selectivity to the monosaccharide (170 °C, 6 h). The developed catalyst appeared almost two times as active as the catalyst suggested by Tian *et al.*[24] Cheng *et al.*[124] observed the substrate conversion of 20% against 10% reported by Tian *et al.*[24] at close selectivities to glucose (89.9 and 80%, respectively) under identical conditions (160 °C, 6 h). This effect can be accounted for by different acidities of the catalysts with identical compositions. Comparative studies of cellulose hydrolysis in the presence of a number of solid acid catalysts (CsHPA, zeolites, sulfated carbons)[125] showed 30% conversion of the substrate over $Cs_2HPW_{12}O_{40}$ at 190 °C in an inert helium atmosphere, this being comparable with the results obtained with the other acid catalysts.

3.2.6 Micellar and Nanosized Catalysts

In considering the hydrolysis of lignocellulose in aqueous media in the presence of solid catalysts, it is necessary to notice that serious diffusion limitations are characteristic of the process due to the fact that both the catalyst and the substrate are solid.[84] However, the exposure of active catalytic sites can be enhanced using a highly dispersed catalyst based on micellar structures stabilized by surfactants.[124] Cheng and co-authors developed a micellar HPA catalyst, $H_3PW_{12}O_{40}$, stabilized with the surfactant hexadecyltrimethylammonium bromide $[C_{16}H_{33}N(CH_3)_3]Br$.[124] The resulting micellar catalyst $[C_{16}H_{33}N(CH_3)_3]H_2PW_{12}O_{40}$ was tested in the reaction of cellulose hydrolysis at 170 °C. The maximal yield of glucose was 39.3% (selectivity of 89.1%) after 8 h reaction that was one and a half times as high as that over the solid CsHPA catalyst $Cs_{2.5}H_{0.5}PW_{12}O_{40}$ but 39% lower than in the $H_3PW_{12}O_{40}$ solution. The TOF of the catalysts changed in the series: $H_3PW_{12}O_{40} > [C_{16}H_{33}N(CH_3)_3]H_2PW_{12}O_{40} > Cs_{2.5}H_{0.5}PW_{12}O_{40}$ at the ratio 3.6 > 1.3 > 1, respectively. The catalyst $[C_{16}H_{33}N(CH_3)_3]H_2PW_{12}O_{40}$ demonstrated its high stability during five cycles of the reaction.[124] Even though micellar catalysts are not solids, they behave as heterogeneous catalysts similar to solid acid catalysts.

It is known that hydrolysis reactions can be catalyzed not only by Brønsted acid sites but also by Lewis acid sites. The latter allows the selectivity and yields of sugars and 5-HMF from cellulose to be considerably improved due to suppression of the product degradation.[14,29] However, the activity of Lewis acid sites (electron acceptors) is only observed in a dry (anhydrous) medium, for example in ionic liquids, since these sites are hydrated in water to give the Brønsted type acidity.[126] Zeolites containing acid sites of both types are an example of such catalysts. At the same time, the use of ionic liquids, because of their high cost, makes the process rather expensive.[29,126] Zhao *et al.* studied a heterogeneous (but not solid) micellar system, $Cr[(DS)H_2PW_{12}O_{40}]_3$, where DS is dodecylsulfate surfactant that contains Lewis acid sites Cr^{3+} stabilized inside the organic surfactant linked, from its external side, to P-W HPA.[127] With this catalyst, one-pot hydrolysis–dehydration of cellulose in an aqueous medium at 150 °C gave 5-HMF at 53% yield in 2 hours that was as high as 1.6 times that in the presence of surfactant-free $Cr(H_2PW_{12}O_{40})_3$ and 5–10 times that in the presence of $CrCl_3$ and $Cr([DS])_3$. The $H_3PW_{12}O_{40}$ solution was inactive to the formation of 5-HMF.[126]

Along with micellar catalytic systems, colloidal nanosize catalysts can also be used for hydrolysis of polysaccharides. Unlike bulky solid catalysts, where oligomer sugars are adsorbed on the catalyst surface, nanosize catalysts favor adsorption of their catalytic sites on the solid cellulose surface to catalyze cleavage of glycoside bonds. Nanosize particles can be modified in the same way as solid catalysts by anchoring active catalytic centers (metal complexes, enzymes or organic acid species) on their surface.[4] Nanosize hydrotalcite activated with $Ca(OH)_2$ is an example of such a catalyst[128] providing a yield of glucose equal to 46.6% at the selectivity of 85.3%.

Micellar and nanosize catalysts are new objects of research in the area of processing of polysaccharides into valuable chemicals. Some researchers report the stability of these catalytic systems but one can expect that agglomeration of the nanostructures and adsorption on the reactor walls or on unreacted cellulose particles are serious impediments to the progress and application of the method.

3.2.7 Other Catalysts

Among the rarely used catalysts for lignocellulose hydrolysis are polymer catalysts based on metal–organic frameworks (MOF). Akiyama *et al.* described a sulfo group containing catalyst based on MOF MIL-101.[129] The catalyst was moderately active to cellulose depolymerization. At 120 °C, the yields of glucose and cellobiose were 1.4 and 1.2%, respectively.

Qian *et al.* attempted to develop a bifunctional catalyst for lignocellulose depolymerization.[130] The ceramics-based catalyst contained the ionic liquids poly(vinyl imidazolium chloride) and poly(styrene sulfonic acid) chemically immobilized on the surface. The role of the ionic liquid was to dissolve polysaccharide of cellulose and sulfo acid hydrolyzed oligosaccharides. The catalyst allowed the cellulose hydrolyzation to produce reducing sugars at the yields of 97 and 33% in the ionic liquid [EMIM]Cl and in water, respectively, at 130–140 °C.

The insoluble salts CaP_2O_6 and $Sr(PO_3)_2$ also catalyze the transformation of cellulose to 5-HMF under hydrothermal conditions.[131] The total yield of glucose and 5-HMF was 18% in the presence of CaP_2O_6, which indicates a low activity of the salt, and 35% in the presence of $Sr(PO_3)_2$ against 12% in the non-catalytic depolymerization of cellulose. Silicoaluminophosphate (SAPO-44) provides a high yield of 5-HMF equal to 68% from starch.[132]

3.2.8 Influence of Reaction Conditions and Target Products on the Choice of a Promising Catalyst

The table illustrates the comparative advantages and drawbacks of the main catalytic systems under consideration (Table 3.1). The tabulated data make it possible to find out about the application of one or another solid acid catalyst with regard to the reaction conditions and target products.

At present, catalysts based on resins and carbons are accepted as the most active among the known catalysts for lignocellulose depolymerization. The carbon catalysts and Nafion are noticeably stable at high reaction temperatures (<200 °C). Solid HPA systems are active to cellulose depolymerization. The observed activity of highly dispersed metal catalysts in lignocellulose hydrolysis makes them promising for direct application in one-pot processes, for example for hydrolysis–reduction or hydrolysis–oxidation of lignocellulose where metal nanoparticles are well known to be active. The moderate activities of catalysts such as zeolites, silicates, and oxides were mainly observed in aqueous media. Attempts to improve the activity by supporting sulfo groups on the surface were unsuccessful due to the low stability of the systems. However, there is no alternative to these catalysts, when one

Table 3.1 Comparison of some advantages and drawbacks of solid acid catalysts for hydrolytic depolymerization of lignocellulose.

Type of catalyst	Advantages	Disadvantages
Resins	+ High specific surface area + Ion-exchange properties	- Deactivation of Amberlyst at high temperatures (>130–150 °C)
Carbons	+ High activity + High mechanical stability of carbon material + Low leaching of sulfo and acid groups under soft conditions (<200 °C)	- Low acidity of pure carbons - Leaching of sulfo and acid groups under harsh conditions (>200 °C)
Metals supported on carbon or oxide	+ High stability + Can be directly applied in one-pot hydrolysis–oxidation or hydrolysis–hydrogenation of lignocellulose	- High cost of Au and metals of the platinum group (Pt, Ru, *etc.*)
Zeolites	+ High stability + Well-structured + High acidity + Effective synthesis techniques + High activity in ionic liquids	- Low activity in aqua medium
Silicates	+ High stability + Porosity and significant specific surface area	- Very low acidity - Additional modification of surface required
Oxides	+ High mechanical and thermal stability	
HPA	+ Strong acidity	- Low surface area

carries out biomass depolymerization under severe conditions (>200 °C) due to the exceptional stability of oxide structures. Again, zeolites and oxides are highly active when dissolved in ionic liquids that makes them promising for application in such a medium.

3.3 Reaction Mechanisms

Lignocellulose transformation during depolymerization is a complex process comprising numerous stages and the formation of a number of intermediates.[133] The product composition depends strongly on the reaction conditions. The principal diagram of transformations of cellulose and possible intermediates (Figure 3.2) encompasses all the data available in the literature.

An important problem when considering transformations of lignocellulose and cellulose, which are insoluble in water and in many other solvents, in the presence of solid acid catalysts is the mechanism of the catalytic action. Two types of hydrolysis mechanism can be assumed when acid groups ($-SO_3H$, –COOH, –OH, *etc.*) reside on the surface. The first is catalysis mediated by H^+ ions generated in solution *via* dissociation of acid groups. According to this mechanism, the catalyst surface does not affect the occurring processes but only bears the acid sites to facilitate separation of the catalytic sites from the reaction medium after the process is over. In the second mechanism, the catalyst surface is assumed to participate directly in the process. Hence, the catalyst structure is not of primary importance in the former mechanism but becomes critical in the latter.

Recent data on the influence of the catalyst surface on the catalytic transformation of cellulose indicate the heterogeneous mechanism of at least the stage of hydrolysis of water-soluble oligosaccharides.[62,66,70,78,80,84] During this, the size of the substrate molecules is an important parameter. A glucose molecule is *ca.* 0.5–0.6 nm in size.[144] Correspondingly, an oligosaccharide containing 10 glycoside units will be *ca.* 6 nm in linear size, and cellulose (100 units) will be *ca.* 60 nm in size. The size of cellulose particles has a crucial impact upon formation of the target products. Rinaldi *et al.* studied transformations of cellulose dissolved in ionic liquid [BMIM]Cl in the presence of catalyst Amberlist-15. Transformations of large or small particles led to the formation of oligomers or reducing sugars, respectively, as the main reaction products.[57,145] Based on the size of the oligosaccharide structures, Onda *et al.*[55,64] formulated requirements for the effective solid catalyst for lignocellulose hydrolysis.[80,89] They supposed that the efficient cellulose transformation needs catalysts with large surface areas and sufficiently large pores (meso- and macropores).[75,84] Evidence of the non-homogeneous reaction pathway in the presence of solid catalysts is the low activation energy of the process: 83 kJ mol^{-1} for sulfonated polystyrene resin, 110 kJ mol^{-1} for sulfated carbons[62] (*cf.* 170–180 kJ mol^{-1} for homogenous hydrolysis of

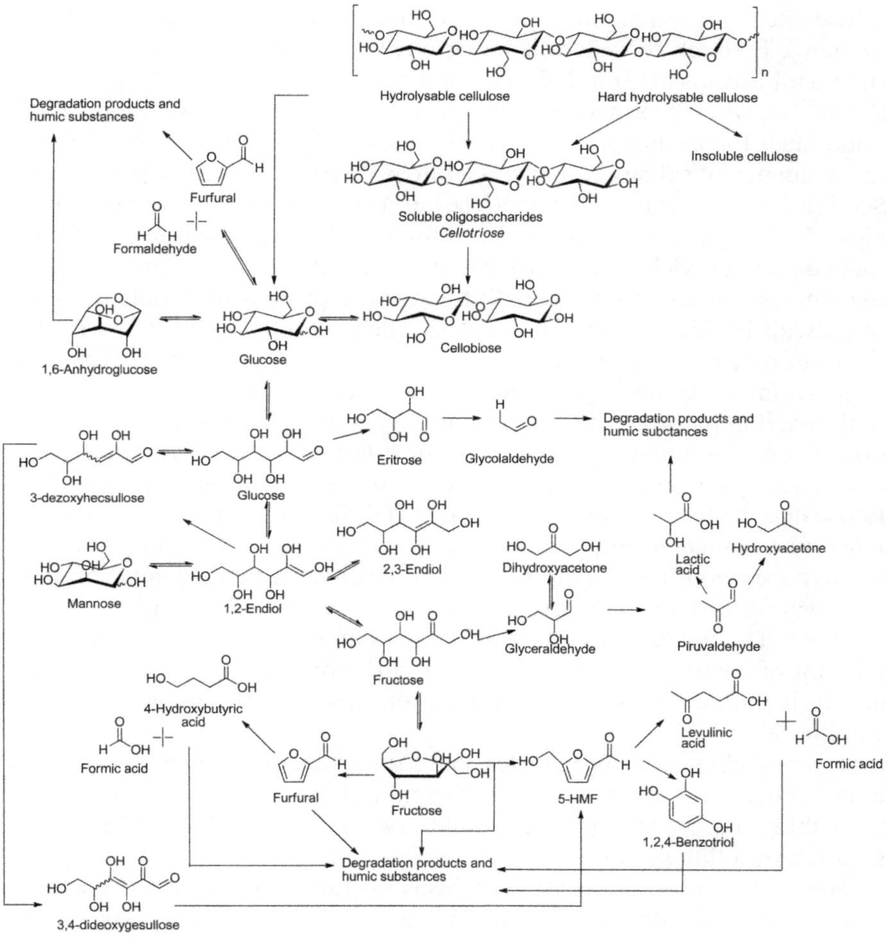

Figure 3.2 Putative paths of cellulose transformations.[134–143]

cellulose with sulfuric acid[146]). Suganuma *et al.* assigned the lower activation energy to higher acid concentrations in the catalyst surface layer caused by a smaller amount of water near the hydrophobic carbon surface. The uniform distribution of acidity through the carbon surface and in the solution indicates indirectly that the 1,4-β-glycosidic linkage is only cleaved on the surface of the solid carbonaceous catalyst.[62] However, some studies argue for the possibility of a mediated mechanism of hydrolysis of biopolymers and polysaccharide transformations in the solution bulk and not on the catalyst surface. It was demonstrated with ion-exchange resin Nafion as an example that the catalyst activity increases upon addition of salts to the reaction system.[61]

The effect of the salt addition was accounted for by strengthening the ion exchange between the salt and acid groups on the resin surface. As a result, the metal cations are bonded to the surface species to form structures such as $-SO_3Na$, while protons active to hydrolysis are transferred to the solution bulk. Such ion exchanges were confirmed by pH mesurements.[81] However, the influence of salts may lead to cellulose dispersing in their presence (see Section 3.2.2.1 above). The salt effect was also observed with the carbon catalysts.[66,80] It was shown that the addition of salts may result in a significant increase in the yields of hydrolysis products. Again, some anions providing cleavage of hydrogen bonds influence the process of depolymerization of polysaccharide (see Section 3.2.2.1). The proton transfer to the reaction solution may be responsible for an increase in the activity of zeolites upon substitution of an ionic liquid for water.[4,111] Our studies demonstrated that both reaction pathways may occur but the heterogeneous mechanism is predominant. The following facts argue for this assumption: (1) the activity of a solid catalyst decreases with a decrease in the specific surface area; (2) cellulose depolymerization in the presence of different catalysts yields different reaction products. Comparative studies were conducted in the presence of a Sibunit carbon based solid catalyst and of a sulfuric acid solution under identical conditions including the number of active acid sites in the reaction medium. The formation of 5-HMF as the main reaction product and a trace amount of levulinic acid was observed in the presence of the solid catalyst and of levulinic acid as the main reaction prooduct in the presence of the soluble catalyst.[69]

Water-soluble oligosaccharides are formed through mediated catalysis and through co-catalysis by hydrothermal water. Hyrolytic cleavage of water insoluble cellulose to glucose can follow two pathways. The first implies dissolution of cellulose and formation of soluble oligomers to be hydrolyzed to glucose in the solution. Another pathway to glucose is the direct abstraction of a monosaccharide from the polysaccharide upon dissolution (Figure 3.3). It was shown earlier that the former pathway is predominant, although it is undeniable that, statistically, the latter may occur.[140,147–149]

Solid acid catalysts can bear chemically different functional groups on the surface. When considering the heterogeneous mechanism, one should identify the influence of these groups on the process of lignocellulose depolymerization. Thus, there are differently strong functional groups on the surface of carbon materials treated with strong acids. The role of the different nature of acid groups ($-SO_3H$, $-COOH$, $-OH$) in the hydrolysis of cellulose was the subject of some studies.[62,74] The $-OH$ group was found to contribute significantly to the adsorption of soluble oligosaccharides *via* hydrogen bonding such as $-OH\cdots O-$ between a hydroxyl group on the carbon surface and oxygen of the β-1,4-glycosidic bond of the saccharide. Apparently, hydrolysis occurs on the catalyst surface at the Brønsted acid sites $-SO_3H$ and $-COOH$. The authors suggested the mechanism of cellulose hydrolysis in the presence of carbon catalysts involving depolymerization of cellulose to soluble oligosaccharides *via* a mediated

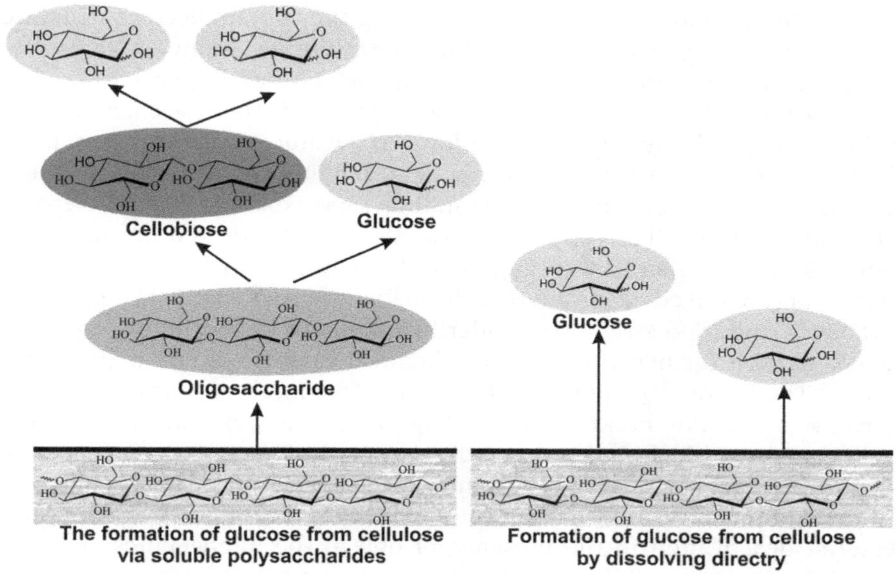

Figure 3.3 Two pathways of the formation of glucose from cellulose during polysaccharide solubilization.[69]

mechanism, adsorption of these sugars on the carbon surface and hydrolysis to glucose oligomers on the surface layer of the catalyst. There is another opinion that adsorption of oligo- and monosaccharides on the carbon surface is provided by the Van-der-Waals interactions at the hydrophobic surface regions, and hydrophilic groups such as hydroxyls are not involved in the adsorption process.[70,150] Different activation energies of the processes catalyzed by H_2SO_4 solution and solid sulfonated carbon argue for the heterogeneous mechanism of cellulose hydrolysis. It should be emphasized that the identical natures of the active site (proton) formed by dissociation of the –S–O—H bond are characteristic of both soluble sulfuric acid and sulfonated carbon. Similar reasoning can be related to the roles of different natured groups on the surface of polymer resin-based catalysts.[4] For example, functional groups –Cl and $-SO_3H$ on the surface of polystyrene resins behave as follows: the chlorides catch oligosaccharide molecules on the catalyst surface, while sulfo groups catalyze hydrolysis of the glycoside bonds.[53]

3.4 Auxiliary Methods for Lignocellulose Depolymerization

Very strong structures are characteristic of cellulose and lignin, two of the three main components of lignocellulose. The physical and chemical properties of cellulose are determined by a large number of hydrogen bonds that make cellulose molecules capable of forming the supramolecular structure from bundles of molecules with various crystallinities. From XRD data, the cellulose structure in lignocellulose comprises alternating crystalline and non-crystalline (amorphous) domains where the order of molecules is determined and kept by hydrogen interactions. In the amorphous domains, there is no long-range ordering but only general longitudinal orientation of the chains. High-ordered domains (crystallites) are considerably different in size depending on the type of cellulose; they can be 10–200 nm in length and 2–5 nm in crosswise size. Lignin is a branched polymer where aromatic base units are linked to one another through numerous chemical bonds. The branched and rigid structures hinder hydrolysis of lignocellulose constituents and their further processing by chemical catalytic or enzymatic methods.[151] The factor impeding the processing of lignocellulose is the need for effective interaction between the solid lignocellulose substrate and the solid acid catalyst for hydrolysis. Again, development of an effective process for depolymerization of lignocellulose may face the problem of pre-separation of the components and isolation of one of them. This approach can cause catalyst deactivation in the presence of one of the lignocellulose components, limited accessibility of the target component to be processed in the plant biomass structure, or the necessity of decreasing the number of side processes and products.

Three hindering factors – the rigid structure, solid–solid interaction, and the necessity of component separation – make it necessary to use auxiliary methods for the processing. The auxiliary methods are used for the following purposes. First, to reduce the crystallinity of cellulose.[17,18,152–154] The polysaccharide and lignin undergo partial depolymerization.[17,18,155,156] Second, a co-catalyst can be used for depolymerization in order to form soluble oligomers to be transformed over solid catalysts.[157] And third, one or two components of lignocellulose can be isolated, for example, through delignification.[17,18,152,155,156] Changes in the lignocellulose structure are illustrated schematically in Figure 3.4.[44]

All the auxiliary methods for cellulose processing can be grouped according to the time of their application. The first group methods are used before the target depolymerization process. They can be referred to as methods of activation. The second group methods are used concurrently with hydrolysis of lignocellulose; these are co-catalysts of the main reaction. It is interesting to note that the use of auxiliary components can be avoided when the biomass is hydrolyzed under hydrothermal conditions. Water can behave as the co-catalyst at high temperature owing to an increase in the ionic product of water.[158] The catalytic effect of water can be detected during blank runs.[64] Heated water starts influencing the hydrolysis at above 150 °C (amorphous cellulose) or 180 °C (crystalline cellulose).[39,88] The short-time treatment of cellulose with superheated water (sub- and supercritical conditions, 320–380 °C) is a simple procedure for cellulose dissolution.

All the auxiliary pretreatment (activation) methods can be classified into three groups based on the nature of the impact on the biomass substrate.

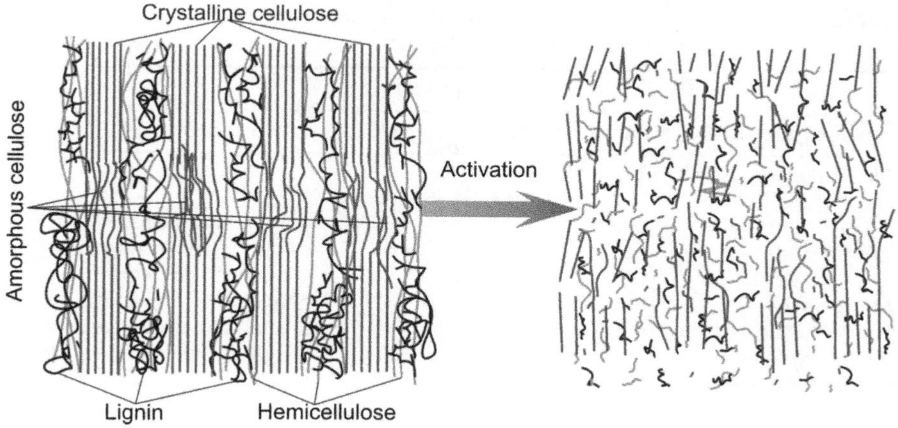

Figure 3.4 Changes in the lignocellulose structure during auxiliary impacts (Reproduced with permission from ref. 44).

These are physical,[17,18] chemical[17,18,27,159] and biological[17,18] methods. Physical methods of the activation are used for the pretreatment of lignocellulose biomass. They include: mechanical grinding,[160,161] microwave exposure,[162] thermal treatment[163] including steam explosion,[159,164–166] γ-irradiation,[167,168] microwave radiation,[169] and ultrasonic treatment.[170–175] The chemical methods are treatment of lignocellulose with concentrated acids, diluted acids, alkalis, and hydrogen peroxide. High- and low-temperature ammonolysis, ammonia autohydrolysis, ozonolysis, and treatment with ionic liquids and organic solvents can be classified as chemical (physicochemical) methods. Chemical pretreatment techniques have been reviewed in depth.[4,17,18,27,84,153,154,159,176–182] Biological methods imply delignification by laccase enzymes.[183] These auxiliary methods can be used both before and during catalytic depolymerization. All the activation methods, and their advantages and drawbacks are briefly generalized in Table 3.2.

We put special focus on some of the most effective and widely used methods for auxiliary treatment of cellulose. Mechanical milling and ultrasonic treatment are of special interest among the physical methods. They are especially effective for destruction of the lignocellulose structure, reduction of cellulose crystallinity, and improvement of the solubility of plant biomass components.[155,156,159,188,189] Pretreatment of pure cellulose and lignocellulose materials in a roll mill followed by fermentation allows the glucose yield to be 12 and 17 times increased, respectively.[190] The ultrasonic treatment of biomass enlarges the surface area of the sample, loosens cellulose fibers, diminishes the crystallinity and degree of polymerization, and increases hydrophilicity.[171,184] Steam explosion and hydrothermal treatment, which is also known as hydrothermolysis or liquid fractionation, seems most effective for both separation of biomass into three main components and improvement of the cellulose reactivity.[159] A specific feature of all these methods is

Table 3.2 Auxiliary methods of lignocellulose depolymerization (+++ significant effect, ++ partial effect, + negligible).

Auxiliary method	Component degradation			Advantages and disadvantages
	Lignin	Hemi	Cell	
Physical methods[155,156,159,163,184,185]				
Mechanical	+++	+++	+++	
Ultrasonic	+++	+++	+++	√ No or small waste water
Thermal	++	+++	++	√ High reaction rate and conversion degree
Microwave	+++	+++	++	√ No chemical pollutions
Hydrothermolysis	+++	+++	+	√ Small times of treatment
Steam explosion	+++	+++	+	- High energy consumption
Gamma radiation	+	+	++	
Chemical methods[17,18,27,153,154,159,176–181]				
Acid	++	+++	++	√ Rather simple, relatively low energy consumption
Alkali	+++	++	+	√ Cheap activation agent
Ammonolysis	+++	+++	+	√ Availability of activation agent
Ammonia explosion	+++	+++	+	- Negative impact on the environment - Corrosion
Organic solvent	+++	+	+	- Formation of large amounts of waste water
Ionic liquid	+	++	+++	√ Environmentally friendly
Ozonolysis	++	+++	+	- High cost of ionic liquids
Biological methods[183,186,187]				
Enzymes	+++	+	+	√ Low energy consumption - High cost - Partial consumption of sugars for growth of bacteria - Low reaction rates

the high energy consumption in comparison to that of chemical and biological methods. That is why the physical methods are only of limited use for the biomass activation.

Intensive research has addressed chemical acidic activation with H_2SO_4, HNO_3, HCl, H_3PO_4 and CH_3COOH.[44] In this case, the H^+ ion attacks intermolecular and intramolecular bonds of cellulose, hemicelluloses and lignin to hydrolyze them. On activation of lignocellulose materials with acids, hemicellulose is removed and cellulose crystals are destroyed,[17] and the substrate porosity increases.[4] Concentrated and diluted acids can be used for the acidic activation. However, concentrated solutions of inorganic acids cause the corrosion effect and environmental damage.[21] Therefore, the activation by diluted acid solutions at ambient temperature is preferable.

These conditions allow the loss of saccharides *via* the dehydration and retro-aldol cleavage reactions to be prevented. Application of the carbon catalyst in combination with hydrochloric or sulfuric acid produces glucose at yields of up to 35%.[157] The yields of target products increase significantly if the carbon catalyst is added at the step of mechanical activation of cellulose. Thus, the glucose yield was 69 and 88% at the substrate conversion of 95 and 98% in the presence of carbon–sulfuric acid and carbon–hydrochloric acid, respectively.

The most commonly used reagents for alkali processing are aqueous solutions of NaOH, KOH, and NH_3, as well as lime $Ca(OH)_2$. Alkali processing of biomass for isolation of holocellulose (cellulose and hemicelluloses) seems less expensive and more effective than the acid treatment.[159,177] Diluted 1–2% NaOH and KOH solutions, as well as lime $Ca(OH)_2$, can be used at moderate temperatures (*ca.* 100 °C) for removal of 70% of lignins and reduction of cellulose crystallinity in substrates such as cotton footstalk, wheat straw, rice husk, of rice, beat bagasse, and maize cob.[18,177]

Application of ionic liquids is of special interest among the chemical methods.[154,181,191] Ionic liquids can dissolve biomass completely and reduce the content of lignin in biomass in the temperature range of 80 to 130 °C.[181] Ionic liquids 1-butyl chloride and 1-allyl-3methylimidazole chloride ([AMIM]Cl and [BMIM]Cl), and 1-ethyl-3-methylimidazole acetate ([EMIM]OAc) are the most effective solvents of cellulose; [AMIM]Cl and [EMIM]OAc are promising ionic liquids for transformation of lignocellulose biomass.[154] Unfortunately, ionic liquids are rather expensive to be used for industrial chemical processes even at the moderate scale.

3.5 Conclusions

In this Chapter, we discussed and summarized the available data on solid acid catalysts for hydrolytic depolymerization of lignocellulose biomass and on some similar heterogeneous systems (micellar and nanosized catalysts). Auxiliary methods for the transformation of renewable substrates were considered, and the reaction pathways and mechanisms were discussed. A special emphasis was put on the transformation of cellulose as the main component of plant biomass. The progress in the fields of depolymerization of lignocellulose and its components in the presence of acid catalysts such as polymer resins, carbons, zeolites, silicates, oxides, supported dispersed metals *etc.* was described. The main purpose of most of the research works is the synthesis of primary monosaccharides from polysaccharides and aromatic compounds from lignin. Possibilities to synthesize 5-HMF and levulinic acid have also been studied.[14,29] There are numerous remarkable advantages of heterogeneous catalysts (including solid catalysts) over soluble homogeneous catalysts; these are the possibilities of easy separation from the reaction mixtures, recycling, and environmental compatibility. In order to develop an effective catalyst for the hydrolysis of plant biomass, the catalytic system has to meet the following requirements: (1) stability in the solvent including water as the most used solvent for hydrolysis. (2) Stability under severe reaction conditions (high temperature and pressure). (3) High concentration of strong Brønsted acid groups on the surface. (4) Large surface area and large size pores. Enhancement of the hydrolytic process needs some auxiliary methods for the substrate preactivation. By now, a number of impressive results have been reported but the development and scaling-up of the effective methods for depolymerization of lignocellulose are still a relevant goal. The future studies should be aimed at improvement of the catalyst stability and activity, selectivity to hydrolytic depolymerization as well as the development of ecologically and economically effective auxiliary methods for lignocellulose processing.

Acknowledgements

The support from projects of the Russian Academy of Science and Federal Agency of Scientific organizations is acknowledged (project V.47.1.12).

References

1. A. J. Ragauskas, C. K. Williams, B. H. Davison, G. Britovsek, J. Cairney, C. A. Eckert, W. J. Frederick, J. P. Hallett, D. J. Leak, C. L. Liotta, J. R. Mielenz, R. Murphy, R. Templer and T. Tschaplinski, *Science*, 2006, **311**, 484.
2. J. N. Chheda, G. W. Huber and J. A. Dumesic, *Angew. Chem., Int. Ed.*, 2007, **46**, 7164.
3. D. M. Alonso, J. Q. Bond and J. A. Dumesic, *Green Chem.*, 2010, **12**, 1493.
4. F. Guo, Z. Fang, C. C. Xu and R. L. Smith Jr, *Prog. Energy Combust. Sci.*, 2012, **38**, 672.
5. L. Hu, L. Lin, Z. Wu, S. Zhou and S. Liu, *Appl. Catal., B*, 2015, **174–175**, 225.
6. N. V. Gromov, O. P. Taran, V. N. Parmon and S. Uthandi, in *Biocatalysis in Biomass to Bioproducts*, ed. K. Ramasamy, S. Karthikeyan and U. Sivakumar, Daya Publ. House, New Delhi, 2017, pp. 237–264.
7. D. M. Alonso, J. Q. Bond and J. A. Dumesic, *Green Chem.*, 2010, **12**, 1493.
8. A. V. Bridgwater, *Appl. Catal., A*, 1994, **116**, 5.
9. T. R. Carlson, G. A. Tompsett, W. C. Conner and G. W. Huber, *Top. Catal.*, 2009, **52**, 241.
10. G. W. Huber and A. Corma, *Angew. Chem., Int. Ed.*, 2007, **46**, 7184.
11. M. Balat, *Energy Sources, Part A*, 2011, **33**, 674.
12. A. S. Amarasekara, *Handbook of Cellulosic Ethanol*, John Wiley & Sons Inc., 2013, pp. 247–281.
13. S. Dutta and S. Pal, *Biomass Bioenergy*, 2014, **62**, 182.
14. R.-J. van Putten, J. C. van der Waal, E. de Jong, C. B. Rasrendra, H. J. Heeres and J. G. de Vries, *Chem. Rev.*, 2013, **113**, 1499.

15. R. P. Swatloski, S. K. Spear, J. D. Holbrey and R. D. Rogers, *J. Am. Chem. Soc.*, 2002, **124**, 4974.
16. H. Zhao, J. E. Holladay, H. Heather and Z. C. Zhang, *Science*, 2007, **316**, 1597.
17. R. Singh, A. Shukla, S. Tiwari and M. Srivastava, *Renewable Sustainable Energy Rev.*, 2014, **32**, 713.
18. Y. Zheng, J. Zhao, F. Xu and Y. Li, *Prog. Energy Combust. Sci.*, 2014, **42**, 35.
19. J. F. Saeman, *Ind. Eng. Chem.*, 1945, **37**, 43.
20. E. A. Immergut and B. G. Ranby, *Ind. Eng. Chem.*, 1956, **48**, 1183.
21. V. Soldi, in *Polysaccharides. Structural Diversity and Functional Versatility*, ed. S. Dimitriu, Marcel Dekker, New York, 2nd edn, 2005, pp. 395–406.
22. P. Lenihan, A. Orozco, E. O'Neill, M. N. M. Ahmad, D. W. Rooney and G. M. Walker, *Chem. Eng. J.*, 2010, **156**, 395.
23. G. Bustos, J. Ramírez, G. Garrote and M. Vázquez, *Appl. Biochem. Biotechnol.*, 2003, **104**, 51.
24. J. Tian, J. Wang, S. Zhao, C. Jiang, X. Zhang and X. Wang, *Cellulose*, 2010, **17**, 587.
25. M. A. Abeer, H. A. E. W. Zeinab, A. I. Atef and T. A. S. Mona, *Bioresour. Technol.*, 2010, **101**, 4446.
26. G. De Wit, A. P. G. Kieboom and H. van Bekkum, *Carbohydr. Res.*, 1979, **74**, 157.
27. J. S. Kim, Y. Y. Lee and T. H. Kim, *Bioresour. Technol.*, 2016, **199**, 42.
28. P. Mäki-Arvela, I. L. Simakova, T. Salmi and D. Y. Murzin, *Chem. Rev.*, 2014, **114**, 1909.
29. P. Bhaumik and P. L. Dhepe, *Catal. Rev.*, 2016, **58**, 36.
30. J. H. Clark, *Acc. Chem. Res.*, 2002, **35**, 791.
31. W. Deng, Y. Wang, Q. Zhang and Y. Wang, *Catal. Surv. Asia*, 2012, **16**, 91.
32. Y.-B. Huang and Y. Fu, *Green Chem.*, 2013, **15**, 1095.
33. T. Okuhara, *Chem. Rev.*, 2002, **102**, 3641.
34. R. Rinaldi and F. Schüth, *Energy Environ. Sci.*, 2009, **2**, 610.
35. R. Rinaldi and F. Schüth, *ChemSusChem*, 2009, **2**, 1096.
36. K.-I. Shimizu and A. Satsuma, *Energy Environ. Sci.*, 2011, **4**, 3140.
37. L. Vilcocq, P. C. Castilho, F. Carvalheiro and L. C. Duarte, *ChemSusChem*, 2014, **7**, 1010.
38. C.-H. Zhou, X. Xia, C.-X. Lin, D.-S. Tong and J. Beltramini, *Chem. Soc. Rev.*, 2011, **40**, 5588.
39. O. Bobleter, in *Polysaccharides. Structural Diversity and Functional Versatility*, ed. S. Dimitriu, Marcel Dekker, New York, 2nd edn, 2005, pp. 893–913.
40. P. Laopaiboon, A. Thani, V. Leelavatcharamas and L. Laopaiboon, *Bioresour. Technol.*, 2010, **101**, 1036.
41. M. Besson, P. Gallezot and C. Pinel, *Chem. Rev.*, 2014, **114**, 1827.
42. P. Gallezot, *Chem. Soc. Rev.*, 2012, **41**, 1538.
43. J. Geboers, S. Van de Vyver, R. Ooms, B. Op de Beeck, P. A. Jacobs and B. F. Sels, *Catal. Sci. Technol.*, 2011, **1**, 714.

44. N. V. Gromov, O. P. Taran, K. N. Sorokina, T. I. Mischenko, U. Sivakumar and V. N. Parmon, *Catal. Ind.*, 2016, **8**, 176.
45. S. Van de Vyver, J. Geboers, P. A. Jacobs and B. F. Sels, *ChemCatChem*, 2011, **3**, 82.
46. K. D. O. Vigier and F. Jérôme, in *Carbohydrates in Sustainable Development II*, ed. A. P. Rauter, P. Vogel and Y. Queneau, Springer Berlin Heidelberg, Berlin, Heidelberg, 2010, pp. 63–92.
47. P. Yang, N. Kobatashi and A. Fukuoka, *Chin. J. Catal.*, 2011, **32**, 716.
48. A. Mukherjee, M.-J. Dumont and V. Raghavan, *Biomass Bioenergy*, 2015, **72**, 143.
49. N. Mahmood, Z. Yuan, J. Schmidt and C. Xu, *Renewable Sustainable Energy Rev.*, 2016, **60**, 317.
50. M. P. Pandey and C. S. Kim, *Chem. Eng. Technol.*, 2011, **34**, 29.
51. J. Zakzeski, P. C. A. Bruijnincx, A. L. Jongerius and B. M. Weckhuysen, *Chem. Rev.*, 2010, **110**, 3552.
52. N. Hartler and K. Hyllengren, *J. Polym. Sci.*, 1962, **56**, 425.
53. L. Shuai and X. Pan, *Energy Environ. Sci.*, 2012, **5**, 6889.
54. R. Pal, T. Sarkar and S. Khasnobis, *ARKIVOC*, 2012, **2012**, 570.
55. A. Onda, T. Ochi and K. Yanagisawa, *Top. Catal.*, 2009, **52**, 801.
56. A. Takagaki, C. Tagusagawa and K. Domen, *Chem. Commun.*, 2008, 5363.
57. R. Rinaldi, R. Palkovits and F. Schüth, *Angew. Chem., Int. Ed.*, 2008, **47**, 8047.
58. R. Karinen, K. Vilonen and M. Niemelä, *ChemSusChem*, 2011, **4**, 1002.
59. S.-J. Kim, A. A. Dwiatmoko, J. W. Choi, Y.-W. Suh, D. J. Suh and M. Oh, *Bioresour. Technol.*, 2010, **101**, 8273.
60. J. Hegner, K. C. Pereira, B. DeBoef and B. L. Lucht, *Tetrahedron Lett.*, 2010, **51**, 2356.
61. B. Hahn-Hägerdal, K. Skoog and B. Mattiasson, *Eur. J. Appl. Microbiol. Biotechnol.*, 1983, **17**, 344.
62. S. Suganuma, K. Nakajima, M. Kitano, D. Yamaguchi, H. Kato, S. Hayashi and M. Hara, *J. Am. Chem. Soc.*, 2008, **130**, 12787.
63. J. Pang, A. Wang, M. Zheng and T. Zhang, *Chem. Commun.*, 2010, **46**, 6935.
64. A. Onda, T. Ochi and K. Yanagisawa, *Green Chem.*, 2008, **10**, 1033.
65. M. Toda, A. Takagaki, M. Okamura, J. N. Kondo, S. Hayashi, K. Domen and M. Hara, *Nature*, 2005, **438**, 178.
66. S. Shen, C. Wang, Y. Han, B. Cai and H. Li, *Fuel*, 2014, **134**, 573.
67. K. Fukuhara, K. Nakajima, M. Kitano, H. Kato, S. Hayashi and M. Hara, *ChemSusChem*, 2011, **4**, 778.
68. A. H. Van Pelt, O. A. Simakova, S. M. Schimming, J. L. Ewbank, G. S. Foo, E. A. Pidko, E. J. M. Hensen and C. Sievers, *Carbon*, 2014, **77**, 143.
69. N. V. Gromov, Catalytic methods of cellulose transformation in pure water into valuable chemical substances, PhD Thesis, University of Bordeaux and Boreskov Institute of Catalysis, 2016.
70. G. S. Foo and C. Sievers, *ChemSusChem*, 2015, **8**, 534.
71. H. Guo, X. Qi, L. Li and R. L. Smith Jr, *Bioresour. Technol.*, 2012, **116**, 355.

72. S. Dora, T. Bhaskar, R. Singh, D. Naik Viswanatha and D. K. Adhikari, *Bioresour. Technol.*, 2012, **120**, 318.
73. S. Li, Z. Gu, B. E. Bjornson and A. Muthukumarappan, *J. Environ. Chem. Eng.*, 2013, **1**, 1174.
74. M. Kitano, D. Yamaguchi, S. Suganuma, K. Nakajima, H. Kato, S. Hayashi and M. Hara, *Langmuir*, 2009, **25**, 5068.
75. K. Nakajima and M. Hara, *ACS Catal.*, 2012, **2**, 1296.
76. O. P. Taran, E. M. Polyanskaya, O. L. Ogorodnikova, C. Descorme, M. Besson and V. N. Parmon, *Catal. Ind.*, 2011, **2**, 381.
77. Z. Yang, R. Huang, W. Qi, L. Tong, R. Su and Z. He, *Chem. Eng. J.*, 2015, **280**, 90.
78. S. Suganuma, K. Nakajima, M. Kitano, D. Yamaguchi, H. Kato, S. Hayashi and M. Hara, *Solid State Sci.*, 2010, **12**, 1029.
79. D.-M. Lai, L. Deng, J. Li, B. Liao, Q.-X. Guo and Y. Fu, *ChemSusChem*, 2011, **4**, 55.
80. S. Shen, B. Cai, C. Wang, H. Li, G. Dai and H. Qin, *Appl. Catal., A*, 2014, **473**, 70.
81. B. Hahn-Hägerdal, K. Skoog and B. Mattiasson, *Eur. J. Appl. Microbiol. Biotechnol.*, 1983, **17**, 344.
82. R. P. Swatloski, S. K. Spear, J. D. Holbrey and R. D. Rogers, *J. Am. Chem. Soc.*, 2002, **124**, 4974.
83. H. Guo, Y. Lian, L. Yan, X. Qi and R. L. Smith, *Green Chem.*, 2013, **15**, 2167.
84. P. L. Dhepe and A. Fukuoka, *ChemSusChem*, 2008, **1**, 969.
85. H. Kobayashi, T. Komanoya, K. Hara and A. Fukuoka, *ChemSusChem*, 2010, **3**, 440.
86. T. Komanoya, H. Kobayashi, K. Hara, W.-J. Chun and A. Fukuoka, *Appl. Catal., A*, 2011, **407**, 188.
87. H. Kobayashi, T. Komanoya, S. K. Guha, K. Hara and A. Fukuoka, *Appl. Catal., A*, 2011, **409–410**, 13.
88. Y. Yu, X. Lou and H. Wu, *Energy Fuels*, 2008, **22**, 46.
89. T. M. Yurieva, L. M. Plyasova, O. V. Makarova and T. A. Krieger, *J. Mol. Catal. A: Chem.*, 1996, **113**, 455.
90. J. Perez-Ramirez, C. H. Christensen, K. Egeblad, C. H. Christensen and J. C. Groen, *Chem. Soc. Rev.*, 2008, **37**, 2530.
91. K. A. Sashkina, V. S. Labko, N. A. Rudina, V. N. Parmon and E. V. Parkhomchuk, *J. Catal.*, 2013, **299**, 44.
92. H. Cai, C. Li, A. Wang, G. Xu and T. Zhang, *Appl. Catal., B*, 2012, **123–124**, 333.
93. K. Y. Nandiwale, N. D. Galande, P. Thakur, S. D. Sawant, V. P. Zambre and V. V. Bokade, *ACS Sustainable Chem. Eng.*, 2014, **2**, 1928.
94. A. E. Abasaeed and Y. Y. Lee, *Chem. Eng. Technol.*, 1995, **18**, 440.
95. P. L. Dhepe and R. Sahu, *Green Chem.*, 2010, **12**, 2153.
96. Y. Yang, X. Xiang, D. Tong and C. Hu, *Bioresour. Technol.*, 2012, **116**, 302.
97. W. Wanmolee, P. Daorattanachai and N. Laosiripojana, *Energy Procedia*, 2016, **100**, 173.
98. C.-W. Jiang, X. Zhong and Z.-H. Luo, *RSC Adv.*, 2014, **4**, 15216.

99. D.-M. Lai, L. Deng, Q.-x. Guo and Y. Fu, *Energy Environ. Sci.*, 2011, **4**, 3552.
100. K. Tanabe, *Catal. Today*, 2003, **78**, 65.
101. F. Yang, Q. Liu, M. Yue, X. Bai and Y. Du, *Chem. Commun.*, 2011, **47**, 4469.
102. F. Yang, Q. Liu, X. Bai and Y. Du, *Bioresour. Technol.*, 2011, **102**, 3424.
103. A. Chareonlimkun, V. Champreda and A. Shotipruk, *et al.*, *Bioresour. Technol.*, 2010, **101**, 4179.
104. M. Watanabe, Y. Aizawa, T. Iida, R. Nishimura and H. Inomata, *Appl. Catal., A*, 2005, **295**, 150.
105. C. Tagusagawa, A. Takagaki, K. Takanabe, K. Ebitani, S. Hayashi and K. Domen, *J. Catal.*, 2010, **270**, 206.
106. C. Tagusagawa, A. Takagaki, A. Iguchi, K. Takanabe, J. N. Kondo and K. Ebitani, *Angew. Chem., Int. Ed.*, 2010, **49**, 1128.
107. H. Wang, C. Zhang, H. He and L. Wang, *J. Environ. Sci.*, 2012, **24**, 473.
108. A. Chareonlimkun, V. Champreda, A. Shotipruk and N. Laosiripojana, *Fuel*, 2010, **89**, 2873.
109. T. Yoshikawa, T. Yagi, S. Shinohara, T. Fukunaga, Y. Nakasaka, T. Tago and T. Masuda, *Fuel Process. Technol.*, 2013, **108**, 69.
110. A. Takagaki, M. Nishimura, S. Nishimura and K. Ebitan, *Chem. Lett.*, 2011, **40**, 1195.
111. D. Murzin and T. Salmi, *Catal. Lett.*, 2012, **142**, 676.
112. F. Yang, Q. Liu, M. Yue, X. Bai and Y. Du, *Chem. Commun.*, 2011, **47**, 4469.
113. C. V. McNeff, D. T. Nowlan, L. C. McNeff, B. Yan and R. L. Fedie, *Appl. Catal., A*, 2010, **384**, 65.
114. J. A. Bootsma and B. H. Shanks, *Appl. Catal., A*, 2007, **327**, 44.
115. D.-M. Lai, L. Deng, Q.-X. Guo and Y. Fu, *Energy Environ. Sci.*, 2011, **4**, 3552.
116. V. Degirmenci, D. Uner, B. Cinlar, B. H. Shanks, A. Yilmaz, R. A. van Santen and E. J. M. Hensen, *Catal. Lett.*, 2011, **141**, 33.
117. A. Fukuoka and P. L. Dhepe, *Angew. Chem., Int. Ed.*, 2006, **45**, 5161.
118. M. T. Pope, *Heteropoly and Isopoly Oxometalates*, Springer-Verlag, Berlin, 1983, p. 180.
119. I. V. Kozhevnikov, *J. Mol. Catal. A: Chem.*, 2007, **262**, 86.
120. M. N. Timofeeva, *Appl. Catal., A*, 2003, **256**, 19.
121. Y. Izumi, M. Ono, M. Kitagawa, M. Yoshida and K. Urabe, *Microporous Mater.*, 1995, **5**, 255.
122. C. L. Hill and C. M. Prosser-McCartha, *Coord. Chem. Rev.*, 1995, **143**, 407.
123. K.-I. Shimizu, H. Furukawa, N. Kobayashi, Y. Itaya and A. Satsuma, *Green Chem.*, 2009, **11**, 1627.
124. M. Cheng, T. Shi, H. Guan, S. Wang, X. Wang and Z. Jiang, *Appl. Catal., B*, 2011, **107**, 104.
125. F. Chambon, F. Rataboul, C. Pinel, A. Cabiac, E. Guillon and N. Essayem, *Appl. Catal., B*, 2011, **105**, 171.

126. S. Zhao, M. Cheng, J. Li, J. Tian and X. Wang, *Chem. Commun.*, 2011, **47**, 2176.
127. S. Zhao, M. Cheng and J. Li, *Chem. Commun.*, 2011, **47**, 2176.
128. Z. Fang, F. Zhang, H.-Y. Zeng and F. Guo, *Bioresour. Technol.*, 2011, **102**, 8017.
129. G. Akiyama, R. Matsuda, H. Sato, M. Takata and S. Kitagawa, *Adv. Mater.*, 2011, **23**, 3294.
130. X. Qian, J. Lei and S. R. Wickramasinghe, *RSC Adv.*, 2013, **3**, 24280.
131. P. Daorattanachai, P. Khemthong, N. Viriya-empikul, N. Laosiripojana and K. Faungnawakij, *Carbohydr. Res.*, 2012, **363**, 58.
132. P. Bhaumik and P. L. Dhepe, *RSC Adv.*, 2013, **3**, 17156.
133. M. S. Feather and J. F. Harris, in *Advances in Carbohydrate Chemistry and Biochemistry*, ed. R. S. Tipson and H. Derek, Academic Press, 1973, vol. 28, pp. 161–224.
134. M. Sasaki, B. Kabyemela, R. Malaluan, S. Hirose, N. Takeda, T. Adschiri and K. Arai, *J. Supercrit. Fluids*, 1998, **13**, 261.
135. D. Klinger and H. Vogel, *J. Supercrit. Fluids*, 2010, **55**, 259.
136. D. A. Cantero, M. D. Bermejo and M. J. Cocero, *ChemSusChem*, 2015, **8**, 1026.
137. B. M. Kabyemela, T. Adschiri, R. M. Malaluan and K. Arai, *Ind. Eng. Chem. Res.*, 1999, **38**, 2888.
138. M. J. Antal Jr, W. S. L. Mok and G. N. Richards, *Carbohydr. Res.*, 1990, **199**, 91.
139. T. M. Aida, K. Tajima, M. Watanabe, Y. Saito, K. Kuroda, T. Nonaka, H. Hattori, R. L. Smith Jr. and K. Arai, *J. Supercrit. Fluids*, 2007, **42**, 110.
140. W. S. Mok, M. J. Antal and G. Varhegyi, *Ind. Eng. Chem. Res.*, 1992, **31**, 94.
141. A. A. Peterson, F. Vogel, R. P. Lachance, M. Froling, J. M. J. Antal and J. W. Tester, *Energy Environ. Sci.*, 2008, **1**, 32.
142. E. Anet, *Aust. J. Chem.*, 1965, **18**, 240.
143. I. V. Delidovich, A. N. Simonov, O. P. Taran and V. N. Parmon, *ChemSusChem*, 2014, **7**, 1833.
144. L. P. Ramos, *Quim. Nova*, 2003, **26**, 863.
145. R. Rinaldi, N. Meine, J. vom Stein, R. Palkovits and F. Schüth, *ChemSusChem*, 2010, **3**, 266.
146. B. Girisuta, L. P. B. M. Janssen and H. J. Heeres, *Ind. Eng. Chem. Res.*, 2007, **46**, 1696.
147. S. E. Jacobsen and C. E. Wyman, *Appl. Biochem. Biotechnol.*, 2000, **84**, 81.
148. G. SriBala and R. Vinu, *Ind. Eng. Chem. Res.*, 2014, **53**, 8714.
149. Q. Xiang, J. S. Kim and Y. Y. Lee, *Appl. Biochem. Biotechnol.*, 2003, **106**, 337.
150. M. Yabushita, H. Kobayashi, J. Y. Hasegawa, K. Hara and A. Fukuoka, *ChemSusChem*, 2014, **7**, 1443.
151. S. Perez and K. Mazeau, in *Polysaccharides. Structural Diversity and Functional Versatility*, ed. S. Dimitriu, Marcel Dekker, New York, 2nd edn, 2005, pp. 41–64.
152. Y. Y. Lee, P. Iyer and R. W. Torget, *Adv. Biochem. Eng./Biotechnol.*, 1999, **65**, 93.

153. W. Xiao and W. Clarkson, *Biodegradation*, 1997, **8**, 61.
154. P. Maki-Arvela, I. Anugwom, P. Virtanen, R. Sjöholm and J. P. Mikkola, *Ind. Crops Prod.*, 2010, **32**, 175.
155. M. Ishiguro and T. Endo, *Bioresour. Technol.*, 2014, **153**, 322.
156. A. I. Yeh, Y. C. Huang and S. H. Chen, *Carbohydr. Polym.*, 2010, **79**, 192.
157. H. Kobayashi, M. Yabushita, T. Komanoya, K. Hara, I. Fujita and A. Fukuoka, *ACS Catal.*, 2014, **3**, 581.
158. W. L. Marshall and E. U. Franck, *J. Phys. Chem. Ref. Data*, 1981, **10**, 295.
159. N. Mosier, C. Wyman, B. Dale, R. Elander, Y. Y. Lee, M. Holtzapple and M. Ladisch, *Bioresour. Technol.*, 2005, **96**, 673.
160. P. Gosh and A. Singh, *Adv. Appl. Microbiol.*, 1993, **39**, 295.
161. S. B. Baryshnikov, V. I. Sharypov, N. G. Beregovtseva, O. P. Taran, V. E. Agabekov and B. N. Kuznetsov, *J. Sib. Fed. Univ., Chem.*, 2014, **7**, 455.
162. E. Nakayama and K. Okamura, *Mokuzai Gakkaishi*, 1989, **35**, 251.
163. L. Dai, L.-Y. Wang, T.-Q. Yuan and J. He, *Polym. Degrad. Stab.*, 2014, **99**, 233.
164. L. P. Ramos, C. Breuil and J. N. Saddler, *Appl. Biochem. Biotechnol.*, 1992, **34–35**, 37.
165. W. Schwald, C. Breuil, H. H. Brownell, M. Chan and J. N. Saddler, *Appl. Biochem. Biotechnol.*, 1989, **20–21**, 29.
166. H. H. Brownell and J. N. Saddler, *Biotechnol. Bioeng.*, 1987, **29**, 228.
167. W. G. Glasser, *For. Prod. J.*, 1981, **31**, 24.
168. A. P. Sinitsyn, A. V. Gusakov and E. Y. Vlasenko, *Appl. Biochem. Biotechnol.*, 1991, **30**, 43.
169. A. Whittaker and D. Mingos, *J. Microwave Power*, 1994, **29**, 195.
170. N. Kardos and J.-L. Luche, *Carbohydr. Res.*, 2001, **332**, 115.
171. Y. Wu, Z.-H. Wu, X.-J. Zhang, J.-L. Zhang and X.-X. Yan, *Key Eng. Mater.*, 2014, **609–610**, 526.
172. J. Zawadzki, A. Radomski, T. Zielenkiewicz, A. Wysocka-Robak, P. Przybysz and P. Maksimowski, *Wood Res.*, 2012, **57**, 279.
173. X. Wang, G. Fang, C. Hu and T. Du, *J. Appl. Polym. Sci.*, 2008, **109**, 2762.
174. V. G. Yachmenev, N. R. Bertoniere and E. J. Blanchard, *J. Chem. Technol. Biotechnol.*, 2002, **77**, 559.
175. V. G. Yachmenev, E. J. Blanchard and A. H. Lambert, *Ultrasonics*, 2004, **42**, 87.
176. B. C. Saha, L. B. Iten, M. A. Cotta and Y. V. Wu, *Process Biochem.*, 2005, **40**, 3693.
177. M. Balat, *Energy Convers. Manage.*, 2011, **52**, 858.
178. T. H. Kim, J. S. Kim, C. Sunwoo and Y. Y. Lee, *Bioresour. Technol.*, 2003, **90**, 39.
179. V. Balan, B. Bals, S. S. Chundawat, D. Marshall and B. Dale, in *Biofuels: Methods and Protocols*, ed. J. R. Mielenz, Humana Press, Totowa, New Jersey, USA, 2009, ch. 5, vol. 581, pp. 61–77.
180. T. H. Kim, R. Gupta and Y. Y. Lee, in *Methods in Biofuels: Methods and Protocols*, ed. J. R. Mielenz, Humana Press, Totowa, New Jersey, USA, 2009, ch. 5, vol. 581, pp. 79–91.

181. I. Kilpeläinen, H. Xie, A. King, M. Granstrom, S. Heikkinen and D. S. Argyropoulos, *J. Agric. Food Chem.*, 2007, **55**, 9142.
182. D. Ishii, D. Tatsumi and T. Matsumoto, *Carbohydr. Res.*, 2008, **343**, 919.
183. W. Yao and S. E. Nokes, *Biomass Bioenergy*, 2014, **62**, 100.
184. D. V. Pinjari and A. B. Pandit, *Ultrason. Sonochem.*, 2010, **17**, 845.
185. S. Zhou, L. Liu, B. Wang, F. Xu and R. Sun, *Process Biochem.*, 2012, **47**, 1799.
186. U. Moilanen, M. Kellock, S. Galkin and L. Viikari, *Enzyme Microb. Technol.*, 2011, **49**, 492.
187. G. M. Scott, M. Akhtar, R. E. Swaney and C. J. Houtman, in *Progress in Biotechnology*, ed. L. Viikari and R. Lantto, Elsevier, 2002, vol. 21, pp. 61–71.
188. G. G. Silva, M. Couturier, J. G. Berrin, A. Buleon and X. Rouau, *Bioresour. Technol.*, 2012, **103**, 192.
189. M. Hall, P. Bansal, J. H. Lee, M. J. Realff and A. S. Bommarius, *FEBS J.*, 2010, **277**, 1571.
190. T. Tassinary and C. Macy, *Biotechnol. Bioeng.*, 1977, **19**, 1321.
191. M. E. Zakrzewska, E. Bogel-Łukasik and R. Bogel-Łukasik, *Chem. Rev.*, 2011, **111**, 397.

CHAPTER 4

Advances in Catalytic Processes of Microalgae Conversion into Biofuels and Chemicals

RONGHONG LIN[a], CONGJUN WANG[a,b], PING WANG[a], YUHUA DUAN[a], DIRK LINK[a], BRYAN MORREALE[a] AND FAN SHI*[a,b]

[a]National Energy Technology Laboratory, Pittsburgh, Pennsylvania 15236, USA; [b]AECOM, Pittsburgh, Pennsylvania 15236, USA
*E-mail: Fan.shi@netl.doe.gov

4.1 Introduction

In the United States, approximately 90% of the energy consumed within the transportation sector is petroleum, and this usage reflects nearly 70% of the oil processed nationally.[1] In addition, the large consumption of petroleum-based transportation fuels accounts for roughly one-third of CO_2 emissions — thought to be a major contributor to global climate change.[1] With increasing demands but limited fossil fuel resources, the production of alternative transportation fuels *via* renewable routes is becoming more important for addressing future energy needs to enhance the economic, environmental, and social benefits. In addition, diversifying the energy portfolio to include more renewable energy resources can assist in overcoming impacts associated with increased greenhouse gas concentrations in the atmosphere.[1-3]

Green Chemistry Series No. 56
Sustainable Catalysis for Biorefineries
Edited by Francesco Frusteri, Donato Aranda and Giuseppe Bonura

Published by the Royal Society of Chemistry, www.rsc.org

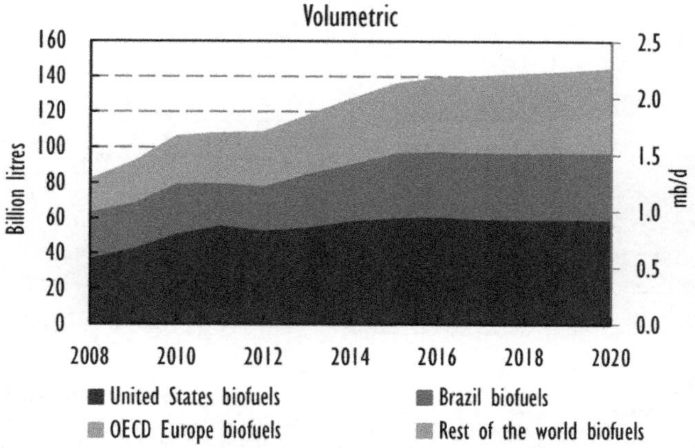

Figure 4.1 Projected biofuels production worldwide.[4] (© OECD/IEA 2015, *Medium-term Renewable Energy Market Report 2015: Market Analysis and Forecasts to 2020*, IEA Publishing. Licence: http://www.iea.org/t&c).

Current trends show that the United States biofuel production for transportation will continually increase over the next few decades, as shown in Figure 4.1.[4] In the United States, the Energy Independence and Security Act (EISA) of 2007 requires production of 36 billion gallon per year (bgy) of renewable transportation fuels by 2022.[1]

Algae based biomass has been considered as a potential carbon-neutral alternative feedstock for the production of transportation fuels.[5–8] The ability of algae to accumulate significant amounts of lipids makes it particularly well suited for conversion to hydrocarbon-based fuels, such as renewable diesel and jet fuel. Microalgae's potentials of fast growth rates, high oil yield (1000–6500 gal/acre/year *vs.* 48 gal/acre/year for soybean), the use of non-arable land for algae cultivation, growth in a variety of water sources, and the benefits associated with large-scale CO_2 mitigation[8–16] have attracted significant attention. Furthermore, microalgae based biofuels do not compete with food crop production, unlike other conventional biofuels, which typically use fertile land and edible oils in their production cycle. Research supported by the U.S. Department of Energy is studying the potential use of microalgae to produce biodiesel with CO_2 captured from point sources, including coal fired power plants.[17–24] Microalgae are able to produce more than 50% dry weight of bio-crude oil with the potential to yield 100 times higher oil production than conventional crops.[16,25]

The three major components of microalgae are carbohydrates, lipids and proteins.[11] The lipid content of dry biomass of various microalgae species varies from 4% (such as in *spirulina*) to as high as 80% (such as in *botyococcus braunii*).[14] Depending on the species of microalgae, three major pathways can be used to convert it to transportation fuel, including whole algae conversion *via* thermochemical reactions, lipid (algal oil)

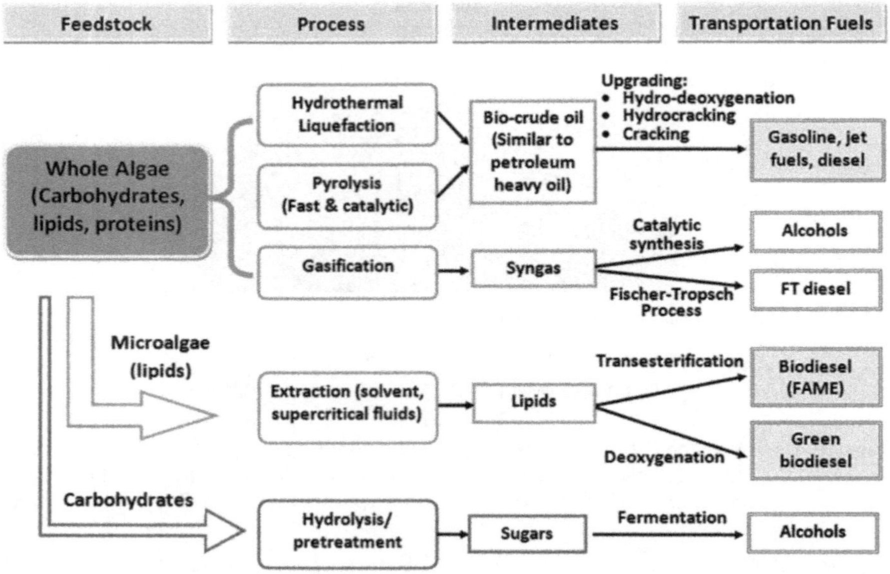

Figure 4.2 Pathways for converting microalgae to biofuels.[1,24]

extraction followed by catalytic reactions, and hydrolysis of carbohydrates *via* biochemical processes, as shown in Figure 4.2.[1,26] Because the aim of this chapter focuses on the microalgae based catalytic processes, the hydrolysis of carbohydrates and fermentation of sugar to ethanol will not be discussed further.

- Whole microalgae can be processed into fuels directly *via* gasification, pyrolysis, and hydrothermal liquefaction (HTL).[26–29] The gasification or partial oxidation of microalgae in the presence of steam produces an intermediate synthesis gas (H_2 and CO) that can be further converted *via* thermocatalytic pathways to liquid fuels and chemicals.[1,28,30,31] Microalgae can also be converted thermochemically in the absence of oxygen, where the pyrolysis (fast pyrolysis, up to 450–550 °C at a heating rate of 10^3–10^4 °C s^{-1}) generates bio-oil with a lower oxygen content and higher heating values than lignocellulosic based bio-oil,[32,33] which is upgraded to transportation fuels by thermocatalytic processes.[28,31,34–36] High water content (as high as 92%)[37] in harvested microalgae makes HTL processes more economically attractive than pyrolysis, gasification or extraction methods: all of which require an energy-intensive pre-drying procedure.[28,38,39] The resulting bio-crude oil is subsequently fed into the refinery stream and upgraded *via* conventional petroleum catalytic processes to transportation fuel through cracking, hydrocracking, and hydrodeoxygenation using commercial catalysts.

- Traditional biodiesel fuels, *i.e.*, fatty acid methyl esters (FAMEs), are generally produced *via* transesterification reactions in solvents using homogeneous catalysts.[30,31,34–36,40] However, due to undesirable oxygen groups, FAMEs suffer from the propensity of having lower energy density, poor low-temperature properties, absorption of water, and they are also susceptible to microbial fouling.[41–43] The selective conversion of triacylglycerides and free fatty acids (FFAs) to transportation fuels from microalgae has received extensive attention due to environmental concerns and the high demand for transportation fuels. A new generation of "green diesels" produced from microalgae[44–46] is alkane hydrocarbons similar to those found in conventional fossil fuel-derived fuels, thus making them fully compatible with existing engines and infrastructure. The composition of typical microalgal oils includes higher molecular weight species ranging from C_{14} to C_{26} and often contains carboxylic acid. Microalgae-based oil also typically contains 20–50% free fatty acids.[41,47] While many examples of the growth and extraction of the resultant microalgae are available in the literature,[36,48–54] instances discussing the upgrading or catalytic conversion of microalgal lipids to alkane-based transportation fuels has not been discussed extensively.

In the following discussion, we will investigate the catalytic conversion of microalgae to transportation fuels and valuable chemicals, and we will provide a fundamental understanding of the processes. Finally, some perspectives and concluding remarks will be given.

4.2 Hydrothermal Liquefaction (HTL) of Microalgae to Bio-crude Oil

HTL has been considered as a promising technology to generate liquid fuel from wet biomass due to its low energy consumption.[55,56] This technology is a representation of the natural geological processes in the formation of petroleum fossil fuels.[16] In the 1970s, pioneering work on the HTL of cellulosic biomass using a sodium carbonate catalyst to produce a heavy oil was conducted by researchers at the Pittsburgh Energy Research Center (now, the National Energy Technology Laboratory).[57] A pilot plant capable of treating 3 tons per day of biomass was demonstrated in Albany, Oregon.[58] In 1994, Dote *et al.* conducted the pioneering work on HTL for microalgae (*Botryococcus braunii*) with high lipid content.[59] In the last five years, research on HTL of whole algae has been tremendously expanded. HTL has been applied to microalgae with a wide range of lipid contents (<1–50% lipid contents of total microalgae),[60–66] microalgae residues (*e.g.* from post-lipid extraction),[56,67] and macroalgae.[68–70] Studies on HTL of microalgae have been moved from laboratory batch reactors to pilot scale continuous flow reactor systems.[71,72] Multi-step HTL processes have recently been developed to recover high-value products[73] and to produce bio-crude with low nitrogen content.[74] HTL of whole algae was reviewed by Chen *et al.*[75] In addition, Guo *et al.*[76] covered fundamental chemical conversion and Elliott[77] focused on the continuous flow system. This section mainly discusses catalytic hydrothermal liquefaction of microalgae to liquid fuels.

4.2.1 Development of HTL of Microalgae to Liquid Fuel

HTL generally converts biomass to bio-crude oil in subcritical water (below the water critical point of 374 °C and 22 MPa).[38,78] It is typically operated in a temperature range between 200 and 370 °C, at pressures from 12 to 20 MPa, for 5 to 60 minutes in subcritical water with or without catalysts (alkaline or metal catalyst) under CO and H_2 or inert gases, such as N_2 and He. A slurry of typically 10–20 wt% dry solids of algae is directly fed to a HTL reactor.[77] A mechanical method is usually used to dewater the algae grown in 0.1 wt% dry solids of algae. During the HTL process, oxygen in raw microalgae (oxygen content of 31.7 wt% on a dry organic basis)[63] is removed in the form of water by dehydration or carbon dioxide by decarboxylation.[38] The bio-crude oil as the main product is produced along with byproducts of biochar, aqueous phase and gases. The bio-crude is similar to a petroleum-derived product (diesel, jet, and gasoline) but with a higher viscosity and oxygen content, typically 10–20 wt% compared to <1% in conventional petroleum.[38] It can be upgraded to the entire distillate range of petroleum-derived fossils fuels by catalytic hydrotreating to remove oxygenates and reduce molecular weight.[16,79]

HTL of microalgae has mainly been investigated in laboratory batch reactors (10–1000 ml). Recently, pilot scale continuous flow reactor systems have been developed. In one of the proposed continuous HTL processes (Figure 4.3),[80] a high slurry concentration of up to 35 wt% dry solids is fed to the HTL reactor. The HTL conditions are 348 °C and 20 MPa for 15 min. After phase separation removing the solid phase, the bio-crude and aqueous phase can be separated. The bio-crude is upgraded to transportation fuels by catalytic hydrotreatment to produce hydrocarbons followed by catalytic cracking in a process similar to fossil crude oil upgrading.[77] Aqueous byproduct (water stream) of the HTL process contains dissolved organic materials in the range of 25–40% carbon in the algae feedstock. The dissolved organic material is converted to gas products (CO_2, CH_4, *etc.*) for process energy using catalytic hydrothermal gasification (CHG). The water after the CHG is recycled to the algae growth ponds. The catalyst is ruthenium metal, 7.8% on a partially graphitized carbon extrudate. In addition, for aqueous byproduct with rich nutrient content (N, P), the nutrient may be recycled through the precipitation of solids, allowing phosphorous recovery and the transformation of nitrogen into ammonium that can be fed back into the algae pond.[77,81]

Multi-step HTL processes have recently been developed to recover high-value products and to produce bio-crude with low nitrogen content. For HTL

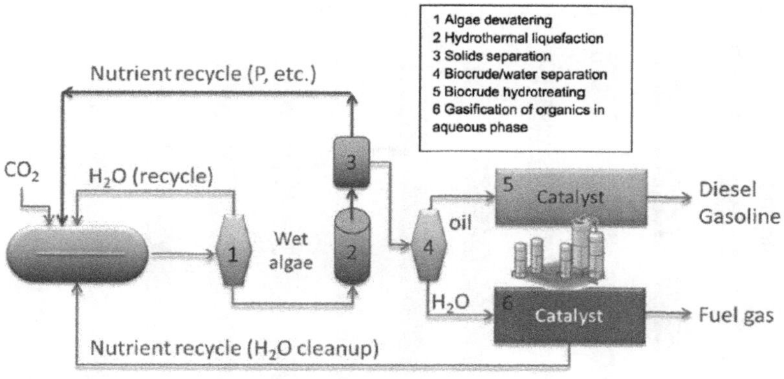

Figure 4.3 Process flow diagram for hydrothermal liquefaction of whole algae. Reprinted from *Algal Research*, **13**, D. C. Elliott, Review of recent reports on process technology for thermochemical conversion of whole algae to liquid fuels, 255–263, Copyright 2016, with permission from Elsevier.[77]

of a high protein microalgae, a two-stage HTL process is proposed to reduce the nitrogen concentration in the bio-crude.[72] In stage I, the algae is pre-treated to extract nitrogen from the algae under mild hydrothermal conditions of 100–200 °C for 5–15 min. In stage II, the residue produces bio-crude under high temperature HTL at 250–350 °C for 10 min. Compared to direct HTL, two-stage HTL achieves lower nitrogen content in the bio-crude (3.5% compared to 7.5% in direct HTL) but the total bio-crude yield decreases to 12.7% from 24.7% of the direct HTL process.

4.2.2 Microalgae Conversion Under HTL Conditions

Microalgae have three major components: carbohydrates, lipids and proteins.[11] In the HTL process, all the lipids, proteins and carbohydrates are converted to oil. The raw microalgae have an initial crude lipid content of 20.5 wt% on an organic basis, and they produce 37 wt% oil yield on an organic basis, which is higher than the microalgae crude lipid content.[38] In addition, the maximum oil yield is 64% from HTL of the microalgae *Botryococcus braunii* as the microalgae starting material has a crude lipid content of 50%.[59] These two studies also show that the higher the lipid content, the higher the bio-oil yield. Biller and Ross[61] further confirmed this observation by using model compounds of proteins (albumin, soya protein, and amino acids of asparagines and glutamine), carbohydrates (starch and glucose) and lipids (triglyceride from sunflower oil) and microalgae strains with high (*Chlorella vulgaris*, *Nannochloropsis occulata*) and low (*Porphyridium cruentum* and the *cyanobacteria Spirulina*) lipid contents. Figure 4.4 shows the yields of products of the model compounds and microalgae based on a dry, ash-free basis in sodium carbonate at 350 °C for 60 min. Bio-crude oil is generated from each biochemical component. This is a unique advantage of HTL

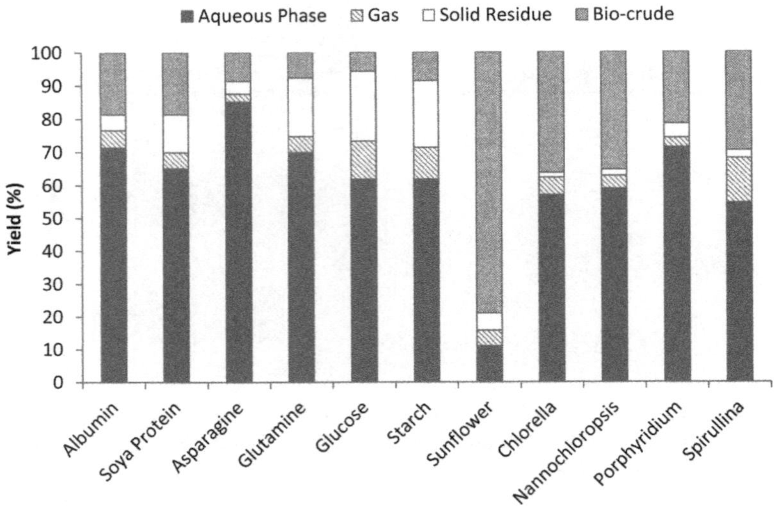

Figure 4.4 Yields of products from hydrothermal processing of microalgae and model compounds at 350 °C for 60 min. (Data from ref. 61).

compared with conventional lipid extraction and conversion methods, and it speaks to the conversion efficiencies and yields that are possible with the HTL process. Furthermore, the results of these studies using model compounds and microalgae oil also demonstrate that the relative oil contributions are lipids > proteins > carbohydrates.

Algal lipids mainly consist of simple fatty acid triglycerides (TAGs), glycolipids and a phospholipid.[11] The lipids can function as the structural membranes of the cells. High contents of unsaturated fatty acids, 50% of those with a carbon number fewer than C_{18}, are typical of algal lipids and may need to be hydrogenated to improve their potential fuel properties.[11] Under the HTL conditions, TAGs are hydrolyzed to fatty acids and glycerol, which are not converted to an oil phase but rather to the water-soluble compounds in HTL. The free fatty acids are relatively stable, but they partially degrade to produce long-chain hydrocarbons for transportation usage.[38,82] Adding NaOH or KOH can increase the decomposition of fatty acids.[38] In addition, hydrocarbons and lipids may be formed from small organic materials by Fischer–Tropsch reactions with water acting as the hydrogen source under the HTL conditions.[38]

Carbohydrates are monosaccharide polymers. Under HTL conditions (subcritical water system), starch is rapidly hydrolyzed to monosaccharides with glucose as one of the main products. The glucose is readily converted to fructose, an isomer of glucose.[82] This fructose subsequently undergoes degradation to generate fragmentation products, *e.g.* glycolaldehyde and glyceraldehyde. Further fragmentation and dehydration produce complex products, *e.g.* acetic acid, formic acid, and aromatic compounds.[38,82]

Most proteins consist of linear polymers of amino acids that have both structural and metabolic functions.[11] The peptide bond C–N links the amino acids together between the carboxyl and amine groups. The bond is hydrolyzed under hydrolysis conditions and results in the production of amino acids with optimal yields at ~250 °C.[38] The peptide bonds of proteins are more stable than the glycosidic bonds in starch, and peptide bond hydrolysis occurs slowly below 230 °C.[82] The amino acids rapidly undergo decarboxylation and deamination and produce mainly hydrocarbons, amines, aldehydes and acids. Some of these products are the same as those from the hydrolysis of carbohydrate. Thus, similar condensation may take place in protein degradation alone or together with carbohydrate degradation. In addition, a considerable fraction of nitrogen in protein is incorporated in the bio-oil during the HTL process. The nitrogen in oil causes NO_x emission when it is combusted.

Algal composition of lipid, protein, and carbohydrate influences HTL conversion efficiency. Leow *et al.*[83] showed that HTL bio-crude yield varies from 33–68 wt% for *Nannochloropsis* sp. cultivated with varying composition before processing. Li *et al.*[84] reported that HTL of microalgae with higher total lipid content and lower protein content yields higher energy recovery. Potential production strains of algae have high nitrogen contents (4–10%). High protein content results in higher nitrogen concentration in the bio-crude, requiring significant denitrification during upgrading to limit nitrogen-oxide emissions.[85]

Although HTL of microalgae with higher lipid content produces the higher bio-oil yield, utilization of low lipid content algae may have potential for large-scale production, as well as potential environmental advantages because high-lipid algae usually have lower growth productivity compared to low-lipid algae. Algae that are typically used for wastewater treatment, and those species that form algal blooms in nature, usually have low lipid contents.[56,60,86,87] In addition, HTL of low lipid algae decreases capital and operation costs compared to processes requiring high lipid yield algae and the extraction of the lipid from the algae.[16]

4.2.3 Effect of Catalysts and HTL Conditions on Bio-crude Oil Properties and Yields

Water plays an important role in the HTL processes. It serves as a solvent (vehicle) and catalyst for the reactions, supplies hydrogen for addition to the substrate, and hydrolyzes high molecular weight reactants such as carbohydrates and proteins.[44,64] As the water temperature increases, for example from 25 °C to 300 °C, the relative permittivity decreases from 78.85 to 19.66, indicating that ordinarily, very polar water molecules become relatively non-polar, which allows the water to show greater affinity for the non-polar organic hydrocarbons.[68] In addition, the dissociation of water dramatically increases with the increase of temperature. For example, the water dissociation constant at 300 °C is about 500 times higher than that at 25 °C, under atmospheric pressure.[68] Water is split into H^+ and OH^- ions *via* hydrolysis or dissociation, which is reversible and rapid. In that way, the rate of both acid- and base-catalyzed reactions is increased.

Table 4.1 lists the key parameters of temperature, time and catalysts used in the HTL process, the optimum conditions where maximum bio-oil yield is obtained, maximum bio-oil yields, and heating value. The table shows that feedstocks strongly affect bio-oil yield and properties, as optimum conditions are different for different feedstocks. Microalgae with high lipid content tend to generate a high oil yield. As the temperature increases, the viscosity of the oil decreases, its heating value increases slightly, the carbon and hydrogen contents increase, and the oxygen content decreases.[63] HTL of *Microcystis viridis* strain was investigated using a similar method.[65] The maximum oil yield was obtained at 340 °C for 30 min with 5 wt% Na_2CO_3 and the oil had the maximum energy yield (the ratio of the weight of C and H in oil products after liquefaction to the weight of C and H in the feedstock).[65] In addition, higher temperature increases both the oil yield and nitrogen content in the oil of *Desmodesmus* sp.[60] and *Chlorella pyrenoidosa*.[66]

Effects of temperature above the water critical temperature of 374 °C, *i.e.*, up to 550 °C, have been investigated on the maximum oil yields. It was found that the maximum oil yields are obtained under subcritical conditions (Table 4.1).[60,62,88] Brown *et al.* conducted tests using the microalgae *Nannochloropsis* in the processing temperature range of 200 to 500 °C and the reaction time of 60 minutes.[62] The highest bio-oil yield (dry basis) was 43 wt% at 350 °C in subcritical water liquefaction. The oil yields decrease modestly as temperature increases from 400 °C to 450 °C, and at 500 °C, it is nearly half of the yields achieved in supercritical water. In addition, the chemical composition of the oil and gas fractions changes with varying temperatures. The oils generated at lower temperatures consist of fatty acids, alkenes, sterol-related compounds, and heterocyclic *N*-containing compounds, with CO_2 being the main compound in the gas phase. As the temperature increases, the oils show more polycyclic aromatic hydrocarbons (PAHs), and the gas phase contains some lighter hydrocarbons (such as CH_4, C_2H_4 and C_2H_6) in addition to CO_2.

Ross *et al.*[64] investigated the effect of alkali catalysts and organic acids along with temperature on the yield and properties of produced oil. The alkalis include potassium hydroxide and sodium carbonate. The organic acids, as hydrogen donors, are acetic acid and formic acid. The microalgae *Chlorella vulgaris*, which was found to have a low lipid content and high protein content, was selected as the feedstock. The HTL was performed in a batch reactor at 300 and 350 °C for 1 hour. The oil yield based on organic content follows the trend $Na_2CO_3 > CH_3COOH > KOH > HCOOH$. The oil yield on a dry basis follows the trend $CH_3COOH > HCOOH > KOH > Na_2CO_3$. The bio-crude oils using the alkali catalysts have a higher heating value, but the oils generated from HTL using organic acids have a lower boiling point, as measured using thermal gravimetric analysis (TGA), and better flow properties.

Table 4.1 Summary of catalysts and process conditions in hydrothermal liquefaction of microalgae to bio-crude oil.

Feedstocks	Lipid (wt%)	Catalyst/Organic acid	Temperature (°C)	Time (min)	Optimum conditions	Max. oil yield (wt%)[b]	Oil HHV (MJ Kg^{-1})	References
Botryococcus braunii	50	0, 5 wt% Na_2CO_3[a]	200, 300, 340	60	300 °C 60 min 5% Na_2CO_3	64	49	37 and 59
Dunaliella tertiolecta	20.5	0, 5 wt% Na_2CO_3	250, 300, 340	5, 60	300 °C 5 min	43.8	34	63
Spirulina	12	0, $Fe(CO)_5$–S	350	60	350 °C 60 min	61	26	89
Microcystis viridis		0, 5 wt% Na_2CO_3	300, 340	30, 60	340 °C 30 min 5% Na_2CO_3	33	31	65
Spirulina, platensis	13.1	0, 5 wt% Na_2CO_3	350	30–60	350 °C 60 min 5% Na_2CO_3	45.06	29.32	86
Mixed algae[d]	6.7	NiO				17.46	35.18	
Dunaliella tertiolecta	22.17	NA	300–380	10–90	360 °C 30 min	36.9[c]	26.62	78
Nannochloropsis sp.	28	NA	200–550	60	350 °C 60 min	43[c]	39	62
Chlorella vulgaris	25	1M Na_2CO_3	300, 350	60	350 °C 60 min 1M Na_2CO_3	27.3	37.1	64
Spirulina	5	1M KOH, 1M CH_3COOH, 1M HCOOH				20	34.8	
Dunaliella tertiolecta cake[e]	2.87	5 wt% Na_2CO_3	280–380	10–90	360 °C 50 min 5% Na_2CO_3	25.8	30.74	67
Chlorella vulgaris	25	0, 1M Na_2CO_3	350	60	350 °C 60 min	35.8	35.1	61
Nannochloropsis oculata	32	1M HCOOH				34.3	34.5	
Porphyridium cruentum	8					20	35.7	
Spirulina	5					29	36.8	
Chlorella pyrenoidosa	<1	NA	100–300	10–120	280 °C 120 min	39.4[c]	35.42	66 and 87
Desmodesmus sp.	10–14	NA	175–450	5–60	375 °C 5 min	49[c]	35.4	60 and 88
Scenedesmus	13	NA	300	30		45	35.5	56 and 90
Spirulina	5					36	35.8	
Defatted scenedesmus	<1					31	35.3	

(*continued*)

Table 4.1 (*continued*)

Feedstocks	Lipid (wt%)	Catalyst/Organic acid	Temperature (°C)	Time (min)	Optimum conditions	Max. oil yield (wt%)[b]	Oil HHV (MJ Kg^{-1})	References
Nannochloropsis sp.	28	0, Pd/C, Pt/C, Ru/C	350	60	350 °C 60 min Pd/C	57	38	91
		Ni/SiO_2–Al_2O_3						
		CoMo/γ–Al_2O_3						
		Zeolite						
Cyanidioschyzon merolae	1.99	0	180–300	30	300	16.98	32.22	92
		KOH				22.67	33.68	
		NaOH				21.78	32.89	
		H_2SO_4				16.98	33.76	
		CH_3COOH				21.23	33.36	

[a]Weight percentage of catalyst to dry solid in algal cells.
[b]Maximum oil-yield on a dry organic in algal cells (dry ash free basic, daf) = (weight of oil/dry weight of organic in algal cells) × 100.
[c]Maximum oil-yield on a dry matter in algal cells (dry basis, db) = (weight of oil/dry weight in algal cells) × 100.
[d]Grown in open pond with waste water.
[e]Residues after recovering β-carotene as a bioactive compound.

The oils, analyzed using GC-MS, contain aromatic hydrocarbons, substituted phenols, nitrogen heterocycles, fatty acids and fatty acid amides. Elemental analysis of the algae oil shows that it has a significantly lower oxygen content compared with the biomass pyrolysis oil. The overall elemental balance of nitrogen indicates that a large portion of the algae nitrogen (up to 50%) is transferred to the aqueous phase in the form of ammonium (NH_4^+). The nitrogen content of the oils is typically 4–6%. Such a high nitrogen content in the oil is undesirable because it increases NO_x emission as the fuel is combusted, and because of the technical challenges for upgrading bio-oil to remove nitrogen.[56,60,68]

Alkalis such as sodium carbonate have been widely used in the HTL of algae. The weak organic acids as hydrogen donors positively affect bio-crude oil yield and boiling point distribution of the oils, but they may either decompose to gaseous products (such as CO and H_2) or react with the microalgae degradation compounds in the HTL process.[64] To address homogeneous catalyst recovery and reuse issues following the HTL process, metal catalysts have been applied to the HTL process (Table 4.1).[86,89,93] The catalysts affect the oil yield and oil properties. The HTL of *Nannochloropsis* sp. with Pd/C leads to a higher bio-crude oil yield and lower O/C ratio in the oil than the oil from the uncatalyzed process.[91] However, the metal catalyst function and catalytic mechanism are not clearly understood and more research is needed.

The effect of catalyst loading along with temperature and reaction time on the yield and properties of the produced oil including viscosity and ultimate analysis was studied by Minowa *et al.*[63] In their study, *Dunaliella tertiolecta* (microalgae) with 78.4% moisture content in a common algae loading of about 10–20%[60,86] is converted in a 100 ml autoclave. The process conditions are temperatures of 250, 300 and 340 °C, residence times of 5 and 60 min with high pressure (about 10 MPa), and a Na_2CO_3 catalyst loading of 0 and 5 wt% of the dry algae. The products of HTL include a bio-oil fraction (extracted by dichloromethane), a water soluble fraction, gases, and solid residues. The oil yields on an organic basis are 31–44 wt% (average 37 wt%), while the microalgae have 76.4 wt% dry basis non-ash organic content (daf). The oil obtained at 340 °C for 60 min has a heating value of 36 MJ kg^{-1} and a viscosity of 150–330 mPa s, which is comparable to fuel oil (about 40 MJ kg^{-1}, 50–1000 mPa s). The gas yield is about 10% on an organic basis and the gas contains mainly CO_2. The solid residue is negligible and the inorganic matter in the microalgae is mostly soluble in the aqueous phase. The catalyst, temperature, and reaction time affect the oil properties, but they do not significantly affect the oil yield. At the maximum oil yield obtained at 300 °C for 5 min, the viscosity of the oil with the catalyst is 500 mPa s, which is much lower than the viscosity of 14 000 mPa s for oil produced without the catalyst.

HTL research has been investigated to achieve a more detailed characterization of products, and to determine the relationship between composition and HTL process conditions.[60,65,88,90,94] The understanding of bio-oil chemical composition is required for improving the upgrading pathways for bio-oil to achieve higher quality products. Yang *et al.*[65] studied HTL conditions for the *Microcystis viridis* strain using a method similar to that used by Monowa (Table 4.1).[63] The liquefied oil is separated into four fractions using thin layer chromatography and analyzed with a flame ionization detector (TLC-FID).[65] The fractions included saturated compounds, aromatic compounds, resin, and asphalt. Furthermore, gas chromatography with mass spectrometry (GC-MS) is used to analyze the low molecular weight saturated and aromatic hydrocarbon compounds. The saturated compounds in the oil are mainly *n*-alkanes of the C_{17}–C_{18} hydrocarbon range, and the aromatic compounds contain *n*-naphthalene and *n*-dibenzothiophene, which are typical components in petroleum-based heavy oil. Therefore, the liquefied oil is similar to heavy oil. The gas consists primarily of CO_2 and methane. The solid residues decrease with increasing temperature and the addition of sodium carbonate. The aqueous phase has a total nitrogen concentration ranging from 998 to 1157 mg l^{-1} and a total phosphate concentration from 2.47 to 5.38 mg l^{-1}. Half of the total nitrogen is detected as ammonia (NH_3–N). Therefore, it is suggested that the final process water will require further treatment to remove nitrogen and phosphate prior to disposal. On the other hand, microalgae growth requires nitrogen and phosphate. So, the nitrogen and phosphate in the water could potentially be recycled as nutrients for microalgae growth.[60,66,95] At the same time, this approach would have an environmental benefit. Microalgae grown in wastewater from secondary treated sewage or from the carpet industry have been converted to bio-crude oil using HTL.[37,86] Nitrogen distributions in the products of HTL of microalgae and macroalgae have been evaluated for future nitrogen control and recycling studies.[60,68]

HTL has been considered as a promising technology to generate liquid fuel from whole algae. The research on the HTL processes has advanced from small laboratory reactor to pilot scale continuous reactor systems.

4.3 Catalytic Conversion of Microalgae Extracts

Most common microalgae species can produce lipids in the range of 20–50% dry weight.[96] Under certain conditions, lipid contents in some species can reach up to 90% dry weight.[96] Microalgal lipids are mainly composed of C_{14}–C_{20} fatty acids and can be converted to renewable biodiesel *via* catalytic or non-catalytic transesterification reactions.

4.3.1 Catalytic Transesterification of Microalgal Lipids to Produce Biodiesel

Catalytic transesterification is the conventional and commercially available method for producing biodiesel from various triglyceride feedstocks such as vegetable oils, animal fats and waste cooking oils. In the general

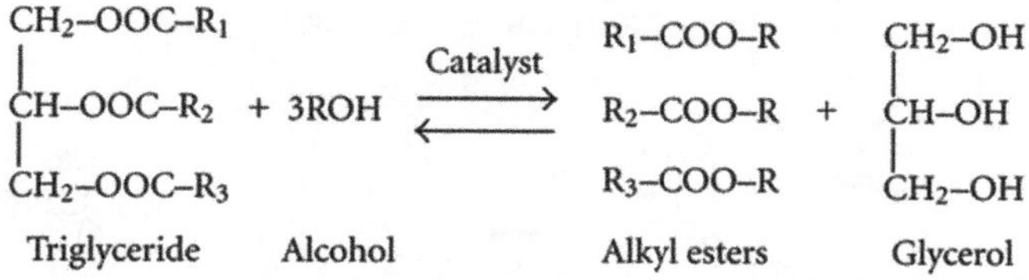

Figure 4.5 Overall transesterification reaction to produce biodiesel.

transesterification reaction, one mole of triglyceride reacts with three moles of alcohol, such as methanol or ethanol, to produce three moles of fatty acid esters and one mole of glycerol as a byproduct (Figure 4.5). Various catalysts have been reported for catalyzing the reaction including homogeneous and heterogeneous catalysts. Each category includes both acid and base catalysts. In addition, biocatalysts have also been reported for use in biodiesel product processes.[97]

4.3.1.1 Homogeneous Catalytic Transesterification

Traditional homogeneous catalysts, such as NaOH, H_2SO_4 and HCl, have been reported for the direct transesterification of algae oil to biodiesel.[98–100] A few non-traditional homogeneous catalysts such as Lewis acids have also been reported for this application.[101–103] Homogeneous base-catalyzed transesterification of triglycerides follows the reaction pathway:[104] (1) production of the active species, RO^-, (2) nucleophilic attack on the carbonyl group of the triglyceride by RO^- to form a tetrahedral intermediate, (3) intermediate breakdown, and (4) regeneration of the RO^- active species, as illustrated in Figure 4.6. Homogeneous acid-catalyzed transesterification of triglycerides follows a different reaction pathway:[104] (1) protonation of the carbonyl group of the triglyceride by the acid catalyst, (2) nucleophilic attack by an alcohol to form a tetrahedral intermediate, and (3) proton migration and breakdown of the intermediate, as illustrated in Figure 4.7. Both sequences are repeated to convert diglycerides and monoglycerides.

Both base and acid catalysts have advantages and disadvantages. The main advantages of homogeneous base catalysts are: (1) low cost, (2) mild reaction conditions (low temperature and atmosphere pressure), and (3) high biodiesel yield within a reasonably short period of time.[97] The major limitation of using base catalysts is that they cannot tolerate high free fatty acid (FFA) and water contents due to soap formation, which leads to catalyst deactivation. This issue can be addressed by the acid catalysts because they are insensitive to the presence of FFAs and can catalyze esterification and transesterification simultaneously.[97] However, acid-catalyzed transesterification requires much longer reaction times and more demanding operating conditions such as high temperature and high alcohol to oil ratios, among other issues. The selection of base or acid catalysts for use in a biodiesel

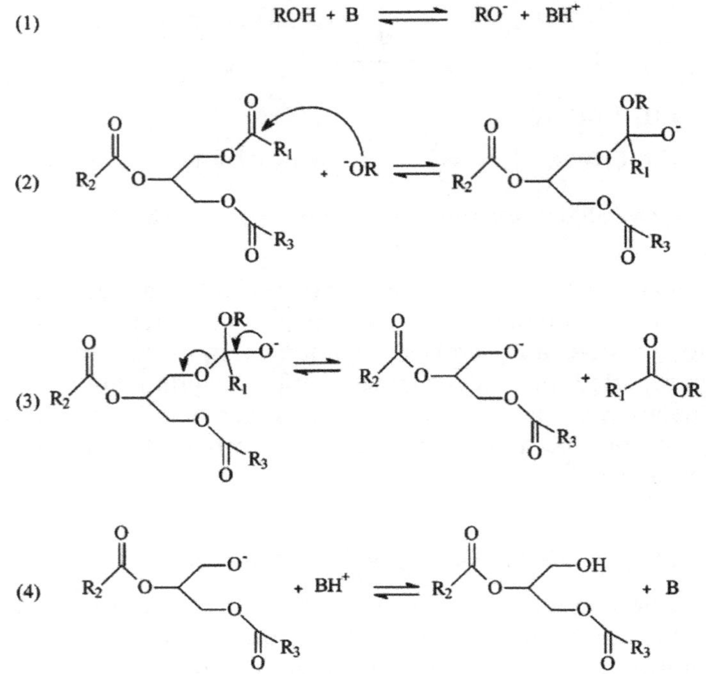

Figure 4.6 Proposed reaction mechanism for homogeneous base catalyzed transesterification of triglycerides. B: base catalyst. R1, R2, R3: carbon chain of the fatty acids. R: alkyl group of the alcohol. Reprinted with permission from E. Lotero, Y. Liu, D. E. Lopez, K. Suwannakarn, D. A. Bruce and J. G. Goodwin, *Ind. Eng. Chem. Res.*, 2005, **44**, 5353. Copyright 2005 American Chemical Society.[104]

production process will depend on the characteristics of the algae biomass and the final downstream processes.[98]

Since homogeneous acid and base catalysts exhibit different advantages and limitations, strategies have been proposed to overcome the limitations by combining the advantages of these two types of catalysts. For example, Suganya *et al.*[99] proposed a two-step transesterification process for converting *Enteromorpha compressa* algal oil with high FFA content. The FFA content was first reduced from 6.3% to 0.34% *via* acid esterification under optimized conditions: 1.5% H_2SO_4, 12:1 methanol/oil ratio, 60 °C and 90 min of reaction time, followed by base-catalyzed transesterification under optimized conditions: 1% NaOH, 9:1 methanol/oil ratio, 60 °C and 70 min, leading to the maximum biodiesel yield of 90.6%. Teo and Idris[100] examined the one-step and two-step transesterification of *Nannochloropsis* sp. wet biomass under microwave irradiation and found that the two-step process (NaOH and H_2SO_4) increased the biodiesel yield by 2.3 fold (water bath heating) and 2.87 fold (microwave heating) compared with the one-step NaOH catalyzed process.

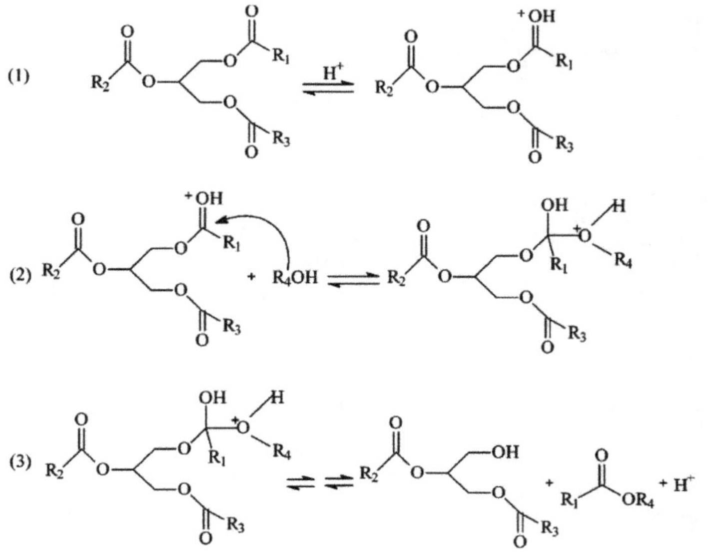

Figure 4.7 Proposed reaction mechanism for homogeneous acid catalyzed transesterification of triglycerides. R1, R2, R3: carbon chain of the fatty acids. R4: alkyl group of the alcohol. Reprinted with permission from E. Lotero, Y. Liu, D. E. Lopez, K. Suwannakarn, D. A. Bruce and J. G. Goodwin, *Ind. Eng. Chem. Res.*, 2005, **44**, 5353. Copyright 2005 American Chemical Society.[104]

4.3.1.2 Heterogeneous Catalytic Transesterification

To overcome the disadvantages of homogeneous catalysts, a wide variety of heterogeneous base and acid catalysts have been studied for biodiesel production from microalgal lipids. Among the heterogeneous base catalysts are metal oxides and mixtures such as CaO, MgO and CaO–MgO mixed oxides, supported or unsupported. The Al_2O_3 supported CaO, MgO and CaO–MgO are found to result in higher biodiesel yields than the unsupported ones under the same conditions.[105,106] The results indicate that both the basic site density and the basic strength of the catalysts are important to achieve a high biodiesel yield.[105] Also, the high contents of Ca^{2+} and Mg^{2+} precipitated on an Al_2O_3 supported CaO–MgO catalyst tend to increase the total basicity and provide more active sites for transesterification reaction.[106] Another benefit of using these supported catalysts is improved moisture resistance, which is favorable for transesterification of high water content microalgae oil.[106] Other heterogeneous base catalysts also show high conversion of microalgal lipids to biodiesel including SrO,[107] Mg–Zr,[108] KOH/La–Ba–Al_2O_3,[109] $Ca(OCH_3)_2$,[110] Li_4SiO_4,[111] and functionalized Fe_3O_4/silica.[112] The general reaction mechanism for CaO catalysis is:[113] (1) abstraction of a proton from methanol by the basic sites to form a methoxide anion, (2) attack on the carbonyl carbon by the methoxide anion to form

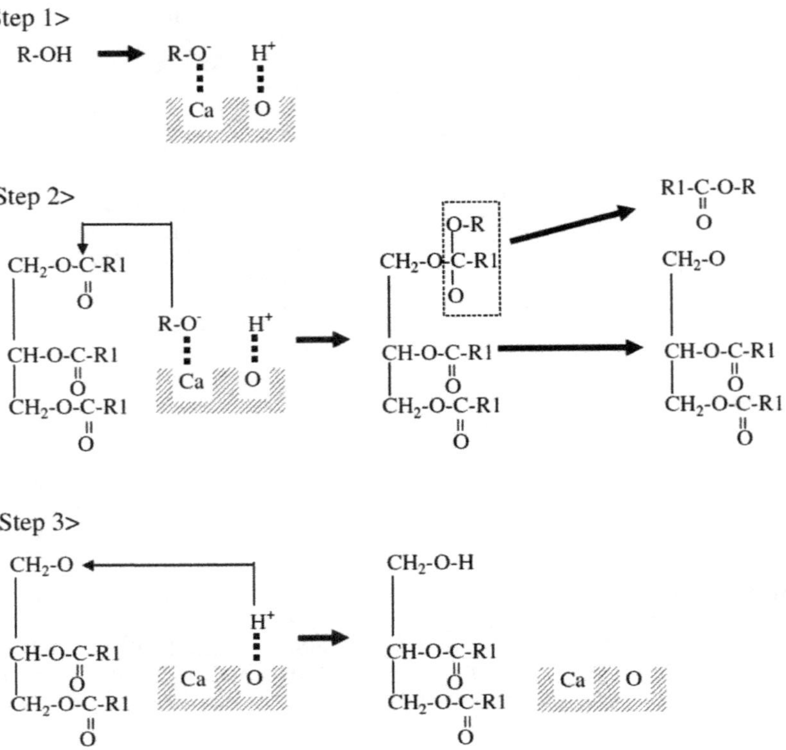

Figure 4.8 Proposed reaction mechanism for CaO catalyzed transesterification of triglycerides. Reprinted from *Fuel*, **87**, M. Kouzu, T. Kasuno, M. Tajika, Y. Sugimoto, S. Yamanaka and J. Hidaka, Calcium oxide as a solid base catalyst for transesterification of soybean oil and its application to biodiesel production, 2798–2806, Copyright 2008, with permission from Elsevier.[113]

an alkoxycarbonyl intermediate, (3) transformation of the alkoxycarbonyl intermediate into a fatty acid ester and an anion of diglyceride, and (4) attack on the anion of diglyceride by the methoxide cation to form diglyceride, as illustrated in Figure 4.8. One major issue with CaO catalyst is calcium soap formation by reacting with FFAs, leading to a reduced catalyst lifetime and deterioration in biodiesel quality.[113] Therefore, a pretreatment step is necessary to remove FFAs in microalgal oils prior to transesterification using CaO catalysts.

There has been increasing interest in developing solid acid catalysts to replace conventional liquid acid catalysts for transesterification reactions because solid acid catalysts offer many benefits such as easy separation from the reaction medium, easy regeneration and recycling and reduced corrosion problems.[97,114] Both inorganic acid catalysts and organically-functionalized acid catalysts have been studied.[114] A few solid acid catalysts have been reported for transesterification of microalgal lipids

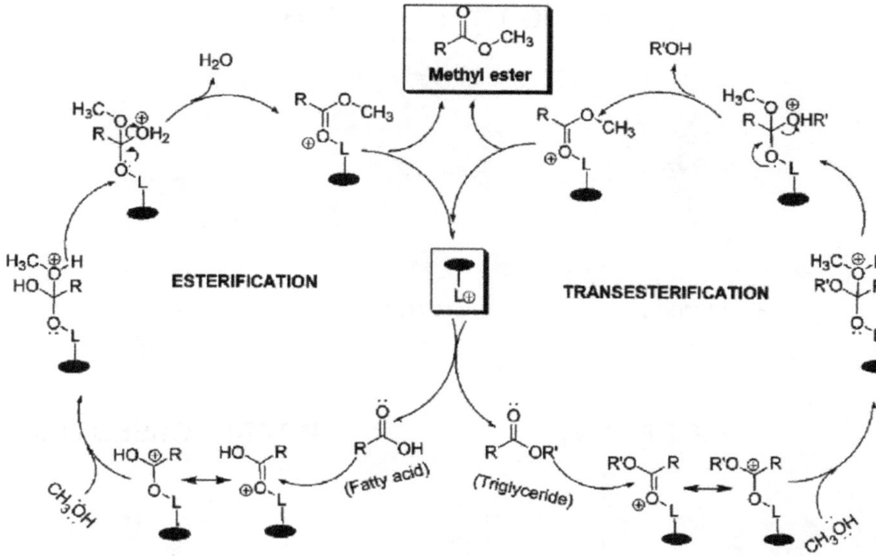

Figure 4.9 Proposed reaction mechanism of simultaneous esterification and trans-esterification using solid Lewis acid catalysts. L⊕: acid site on the catalyst surface; R: alkyl group of fatty acid; R′: alkyl esters of triglyceride.[122]

including zeolites,[115] TiO_2,[116] graphene oxide,[117] carbon-based catalysts,[118] WO_3/ZrO_2,[119] Amberlyst-15[120] and $Cr_2O_3/Na_2Cr_2O_3/Al_2O_3$.[121] Figure 4.9 illustrates the reaction mechanism of simultaneous esterification and trans-esterification using solid Lewis acid catalysts.[122] The reaction starts with the interaction of the carbonyl oxygen of a free fatty acid or monoglyceride with the acidic site of the catalyst to form a carbocation. The carbocation is then attacked by an alcohol to produce a tetrahedral intermediate, followed by the transformation of the tetrahedral intermediate to form biodiesel.

4.3.1.3 *Biocatalytic Transesterification*

Enzymes such as lipases obtained from various sources can be used as biocatalysts for transesterification reactions.[97,123–125] Both extracellular lipases and intracellular lipases have been reported for this application.[124] Immobilized lipases rather than free lipases are preferred mainly due to easy recovery and reuse.[125] Three basic types of immobilization techniques have been developed including binding to support materials, confinement and inter-crosslinking, as illustrated in Figure 4.10.[126] The mechanism for biocatalytic transesterification follows the ping–pong bi–bi mechanism and three reaction pathways have been proposed: (1) direct alcoholysis of glycerides, (2) hydrolysis of glycerides followed by esterification and (3) simultaneous alcoholysis and hydrolysis followed by esterification.[125] The reactions occur

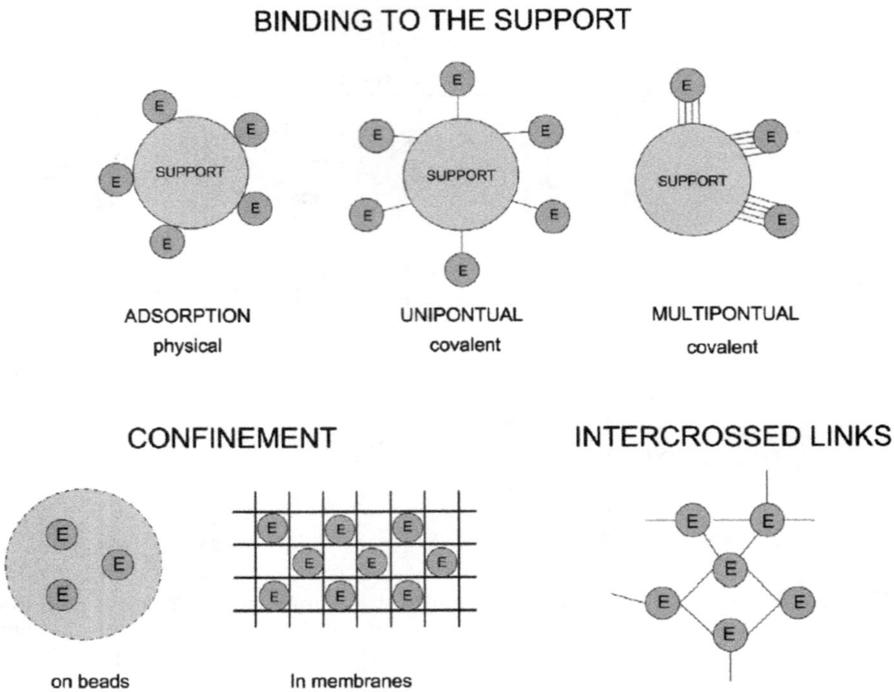

Figure 4.10 Three basic types of immobilization techniques. Reprinted from *Biotechnology Advances*, **33**, J. K. Poppe, R. Fernandez-Lafuente, R. C. Rodriguez and M. A. Z. Ayub, Enzymatic reactors for biodiesel synthesis: present status and future prospects, 511–525, Copyright 2015, with permission from Elsevier.[126]

at the interface of the lipid–water two-phase system and are mass-transfer limited. Adsorption of lipases onto the interface initiates the reactions,[123] followed by transport of substrates to the lipase surface, catalysis and release of products, as illustrated in Figure 4.11.[126] Organic solvents such as tert-butanol[127] and hexane,[128–130] ionic liquids[127] and supercritical CO_2[131] have been reported for use in biocatalytic transesterification. The advantages of using such solvents include the (1) easy control of water contents, (2) improved lipase recovery and reuse, (3) enhanced byproduct recovery, and (4) increased lipase stability.[123]

A number of lipases have been reported for use in the catalytic transesterification of microalgal lipids. They include *Candida antarctica* lipase B (Novozym 435),[127,131–135] *Penicillium expansum* lipase (PEL),[127] *Burkholderia* lipase,[128,129] *M. miehei* lipase,[133] *T. lanuginosus* lipase,[133] *C. rugose* lipase,[136] *Rhizomucor miehei* lipase,[130] *Fusarium heterosporum* lipase[137] and lipase NS81006.[135] Novozym 435 is the most commonly used and commercially available immobilized lipase for this application. The limited research on lipase-catalyzed conversion of microalgal lipids has been focused on

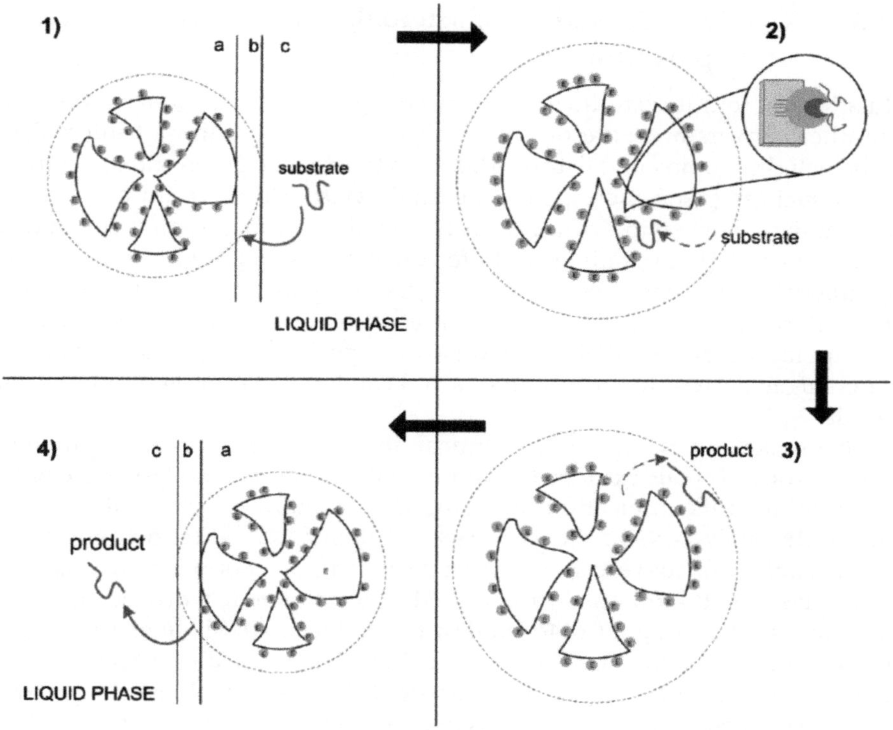

Figure 4.11 Mass transfer limited lipase-catalyzed transesterification process. Reprinted from *Biotechnology Advances*, **33**, J. K. Poppe, R. Fernandez-Lafuente, R. C. Rodriguez and M. A. Z. Ayub, Enzymatic reactors for biodiesel synthesis: present status and future prospects, 511–525, Copyright 2015, with permission from Elsevier.[126]

addressing the impacts of various operating parameters on biodiesel yield and quality mainly including reaction temperature,[128–134] water content,[128–130,135,137] lipase loading,[128–133] reaction duration,[128,129,132,134] and acyl acceptor/lipid ratio.[128–130,132–135] Issues of catalyst reuse[128–132,134,137] and process intensification and optimization[130,131,134,137] have also been discussed by several groups. Compared with conventional acid and base catalytic transesterification, the biocatalytic transesterification for biodiesel production offers many advantages such as mild reaction temperatures, low alcohol to oil ratios, high tolerance to water and free fatty acids, high specificity toward substrates, wide substrate variation, improved product quality, no saponification, and no generation of wastewater.[125] However, the disadvantages of this method such as high enzyme cost, slow reaction rates, enzyme inhibition and loss of activity limit its application in commercial biodiesel production. Future investigations should be directed to the research and development of novel enzymes with reduced production cost and improved activity and stability.

4.3.2 Catalytic Upgrading of Microalgal Oil to Produce Green Transportation Fuels

In addition to transesterification of microalgal lipids to produce biodiesel, another conversion route of interest is catalytic upgrading of microalgal oils including lipids and hydrocarbon oils to produce green transportation fuels such as green diesel, gasoline, and aviation fuel. In an early study on catalytic conversion of algal oil to gasoline, Milne *et al.*[138] concluded that the catalytic upgrading route for extracted lipids might be both cost competitive and have system advantages compared with transesterification. Depending on the type of microalgal oils to be treated, microalgal oil upgrading can be divided into two categories: catalytic upgrading of microalgae derived hydrocarbon oils and catalytic upgrading of microalgal lipids.

Hydrocracking is the main technical approach for catalytically upgrading hydrocarbon oils extracted from microalgae. Several catalysts have been reported for this application including CoMo catalysts,[139,140] zeolites,[141,142] and PtRe catalysts.[143] In the early 1980s, Hillen *et al.*[139] first demonstrated the feasibility of converting *Botryococcus braunii* oil (Bot-oil) to transportation fuels. The Bot-oil was hydrocracked at 400 °C and a hydrogen gas pressure of 20 MPa using a CoMo catalyst to produce a distillate comprising a 67% gasoline fraction, 15% aviation fuel fraction, 15% diesel fuel fraction and 3% residue oil. Tracy *et al.*[142] reported the hydrocracking of squalene ($C_{30}H_{50}$) extracted from *Botryococcus braunii* using H-BEA zeolite as a catalyst to produce gasoline-range hydrocarbons. The ASTM testing indicated that the resulting gasoline-range fraction had a high research octane number of 98 and many other desirable properties. Kitazato *et al.*[141] studied hydrocracking of botryococcenes ($C_{34}H_{58}$) also extracted from *Botryococcus braunii* using H-zeolite as a catalyst to produce an organic liquid containing 62% gasoline-range hydrocarbons. They found that the reaction mechanism involved pathways of consecutive cracking of botryococcenes and subsequent aromatization. Garciano *et al.*[140] studied the catalytic upgrading of squalene using $CoMo/Al_2O_3$–SiO_2 catalysts and found that these catalysts could induce cyclisation, hydrogenation, hydrocracking and esterification depending on the Al : Si ratio of the catalysts. Murata *et al.*[143] investigated hydrocracking of Bot-oil (mainly $C_{34}H_{58}$) and squalene using $PtRe/SiO_2$–Al_2O_3 catalysts. The resulting organic liquids contained 66.9% and 68.8% C_{10}–C_{20} hydrocarbons for Bot-oil and squalene, respectively. The results indicated that the synergism of Pt and Re could play a critical role in the hydrocracking process, although these two metals were present independently on the catalyst surface. The reaction pathway mainly involved the C–C bond scission reaction followed by hydrogenation of the C=C and C–C bonds and partial aromatization, as illustrated in Figure 4.12. These studies suggest that long-chain hydrocarbons derived from microalgae can be converted to green transportation fuels through catalytic upgrading. The reaction pathways and product

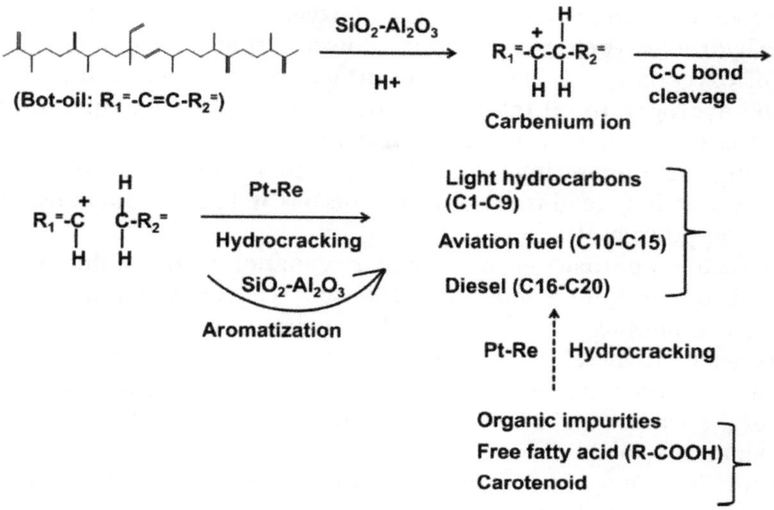

Figure 4.12 Proposed reaction pathway for hydrocracking of Bot-oil. Reprinted with permission from K. Murata, Y. Liu, M. M. Watanabe, M. Inaba and I. Takahara, *Energy Fuels*, 2014, **28**, 6999. Copyright 2014 American Chemical Society.[143]

distribution (diesel-range, gasoline-range or jet fuel-range), however, rely heavily on the type of catalysts employed. Future research should focus on catalyst design and development to address selectivity, regenerability, and long-term stability of the catalysts.

Different approaches have been reported for the catalytic upgrading of microalgal lipids.[144] The first approach is hydrotreating using the conventional CoMo and NiMo sulfide catalysts. Hydroprocessing generally offers high conversion of triglyceride oils and high yields of diesel-range hydrocarbons.[144] Zhou and Lawal[145] reported nearly complete conversion of *Nannochloropsis salina* oil at 360 °C and a H_2 pressure of 3.4 MPa using sulfided NiMo/γ-Al_2O_3 catalysts. The resulting organic fraction contained 56.2% diesel-range (C_{13}–C_{20}) hydrocarbons. Verma *et al.*[146] also studied algal lipid upgrading by a single-step route using sulfided NiMo catalysts supported on hierarchical mesoporous H-ZSM-5 zeolite. However, a high yield of 77% for the jet fuel-range hydrocarbons (C_9–C_{15}) with moderately high isomerization selectivity and low sulfur, olefin and aromatic contents was obtained. The resulting fuel product met all the basic requirements of freezing point, density, flash point, heat of combustion and viscosity for aviation fuel. Such a big difference in product distributions obtained from the studies by Zhou and Lawal[145] and Verma *et al.*[146] is likely caused by different catalyst supports because ZSM-5 is known for favoring the formation of gasoline-range hydrocarbons.[144] Deactivation is one major issue with any catalyst. Zhou and Lawal[145] showed that NiMo/γ-Al_2O_3 was deactivated due to the accumulation of oxygenated intermediates formed in the hydrodeoxygenation reaction. However, they found that the adsorption of oxygenates on the catalyst surface could be reduced by increasing the hydrogen to oil ratio, residence time, reaction temperature, and pressure. Consequently, the catalyst activity and life could be improved by tuning these operating parameters. Another issue with the sulfide catalysts is that they lead to contamination of the fuel products by organosulfur compounds.[147]

The second approach is catalytic deoxygenation using sulfide-free metal catalysts under mild conditions (260–300 °C) compared with the first approach (360–400 °C). Peng *et al.*[148] reported the conversion of crude microalgal oil to diesel-range hydrocarbons using 10 wt% Ni/H-β (Si/Al = 180) as a catalyst and obtained a 78% yield of liquid alkanes (containing 60 wt% yield of C_{18} octadecane) at 260 °C and 4 MPa H_2. They found that Ni catalyzes hydrogenation, hydrogenolysis, and decarbonylation reactions, while the acid function catalyzes dehydration, hydroisomerization, and hydrocracking reactions. Accordingly, an overall reaction pathway for catalytic upgrading of microalgae oil over the bifunctional Ni/H-β catalysts was proposed (Figure 4.13). The reaction pathway mainly involves initial hydrogenation

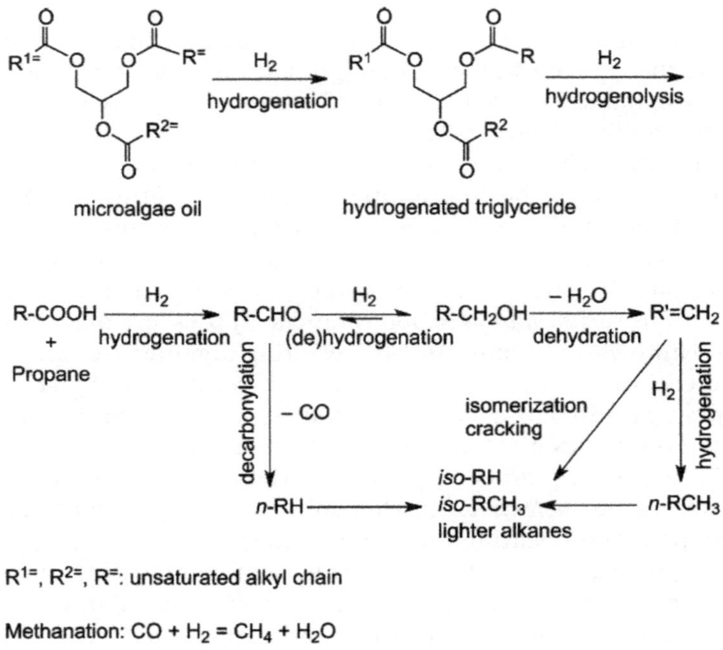

Figure 4.13 Proposed reaction pathway for catalytic conversion of microalgae oil to alkanes over bifunctional Ni/H-Beta catalysts. Reprinted with permission from ref. 148, Copyright © 2012 WILEYVCH Verlag GmbH & Co. KGaA, Weinheim.[148]

of unsaturated triglycerides followed by hydrogenolysis of saturated triglycerides to produce fatty acids and hydrogenation of the fatty acids to form aldehyde (R–CHO). The aldehyde further undergoes either hydrodeoxygenation to form *n*-RCH_3 (C_{18} octadecane) or decarbonylation to form *n*-RH (C_{17} heptadecane). Hydrodeoxygenation is the major pathway with the Ni/H-β catalysts. Peng *et al.*[147] further studied conversion of the same microalgal oil using Ni/ZrO_2 catalysts and found that the major alkane product was *n*-heptadecane (68%), suggesting that the hydrogenolysis/hydrogenation–decarbonylation route is the main reaction pathway with Ni/ZrO_2 catalysts. They further concluded that this route would be superior to the hydrodeoxygenation route considering the lower H_2 consumption. Santillan-Jimenez *et al.*[149] examined a series of oxide-supported Ni catalysts for deoxygenation of model and microalgal lipids. They found that using Ni–Al layered double hydroxides (LDHs) led to a higher yield of diesel-range hydrocarbons than using Ni/Al_2O_3, Ni/ZrO_2, and $Ni/La–CeO_2$ catalysts and the Ni–Al LDHs showed no significant loss of activity within 72 h. They also concluded that the deoxygenation reactions followed mainly the decarboxylation/decarbonylation pathway rather than the hydrodexygenation pathway.

Besides Ni catalysts, Fe, Pt, Rh, Ru and Pd catalysts have also been reported for the catalytic deoxygenation of microalgal lipids. Kandel *et al.*[150] found that iron nanoparticles supported on mesoporous silica nanoparticles (Fe-MSN) could catalyze the conversion of crude microalgal oil into diesel-range hydrocarbons. Their results indicated that the Fe-MSN catalyst was likely to be involved in a reverse Mars–Van Krevelen mechanism and offered a high selectivity for hydrodeoxygenation over decarbonylation and hydrocracking ($C_{18}/C_{17} = 6.4:1$). Yang *et al.*[151] demonstrated the capability of multiwall carbon nanotube supported Pt catalysts (Pt/MWCNT) for hydrothermal decarboxylation of fatty acids with 97% selectivity toward heptadecane. The Pt/MWCNT catalysts showed higher activity than the Pt/C and Ru/C catalysts. Fu *et al.*[152] demonstrated the direct conversion of microalgal lipids in water using Pt/C catalysts without H_2 and concluded that among the tested solvents, water was most favorable for the selective decarboxylation of methyl stearate. Robota *et al.*[153] also reported a multi-step process using Pd/C and Pt/US-Y catalysts to convert microalgal oil to diesel and jet fuels. Zhou and Lawal[154] evaluated Pt/Al_2O_3, Rh/Al_2O_3 and sulfide $NiMo/Al_2O_3$ catalysts for the catalytic deoxygenation of *Nannochloropsis salina* oil to green diesel. Their study suggested that the Pt and Rh catalysts favor the decarboxylation/decarbonylation route and the selectivity was independent of the experimental conditions. The selectivity of NiMo catalysts, however, was shifted from hydrodehydration to decarboxylation/decarbonylation as the H_2/oil ratio and residence time were reduced. They also found that coke accumulation decreased in the order NiMo > Pt > Rh. It can be concluded from these studies that catalytic deoxygenation is a promising, feasible approach for producing green fuels (mainly green diesel) from microalgal lipids. However, challenges remain in the development of cost-effective catalysts for commercial applications.

4.3.3 Catalytic Upgrading of Bio-crude Oil

Bio-crude oil produced from microalgae through thermochemical conversion methods cannot be directly used due to its poor quality. The physical and chemical properties of bio-crude oil are significantly different from conventional petroleum fuels and depend largely on the conversion routes.[155] Bio-crude oil generally has a higher water content, higher viscosity, higher oxygen content, higher nitrogen content, higher ash content, higher corrosiveness (acidity) and lower heating value compared with conventional petroleum fuels.[156,157] These undesired properties restrict its applications. Therefore, upgrading of bio-crude oil is necessary to improve its quality for various applications, especially as a transportation fuel for combustion engines. Depending on the target applications of bio-crude oil, various technologies have been developed for bio-crude oil upgrading including hydrotreating, hydrocracking, supercritical fluid processing, solvent addition/esterification, emulsification, steam reforming, and solvent extraction.[157] The discussion below focuses on catalytic upgrading of bio-crude oil *via* hydrotreating and hydrocracking to produce transportation fuels. Catalytic upgrading of bio-crude oil involves complex reactions including but not limited to cracking, decarbonylation, decarboxylation, hydrocracking, hydrodeoxygenation, hydrogenation, and polymerization.[158] Figure 4.14 shows examples of reactions associated with catalytic upgrading of bio-crude oil.

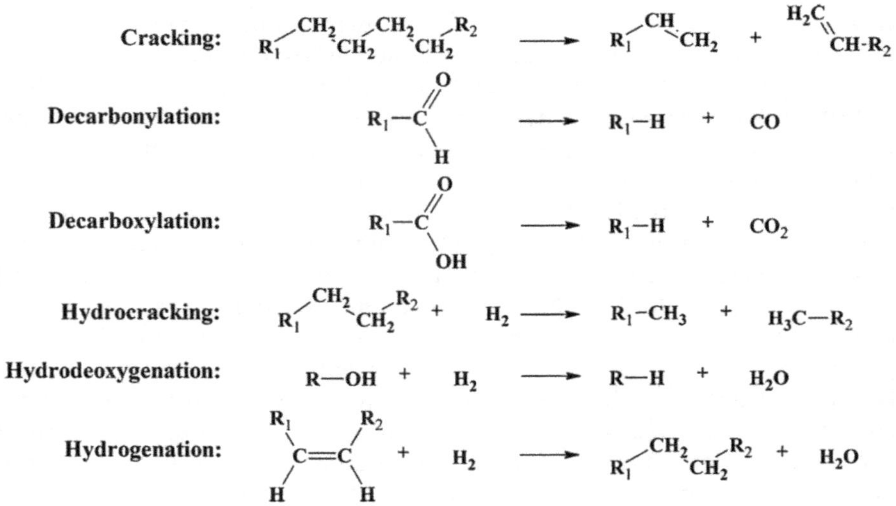

Figure 4.14 Examples of reactions associated with catalytic upgrading of bio-crude oil. Reprinted from *Applied Catalysis A: General*, **407**, P. M. Mortensen, J. D. Grunwaldt, P. A. Jensen, K. G. Knudsen and A. D. Jensen, A review of catalytic upgrading of biooil to engine fuels, 1–19, Copyright 2011, with permission from Elsevier.[158]

The chain length of hydrocarbons obtained from biological feedstocks is typically in the range of C_{14}–C_{22}, and those hydrocarbons also contain large amounts of heteroatoms. In order to produce light fractions from heavy hydrocarbons and to reduce heteroatom contents, hydrogenation, hydrocracking and hydroisomerization processes are necessary. Hydrocracking is commonly practiced in the petroleum industry to process oil residue. Typical catalysts used for bio-crude oil hydrocracking include mainly transition metal based catalysts and zeolites. Biller *et al.*[159] carried out hydroprocessing of bio-crude oil from HTL of Chlorella at 350 and 405 °C using NiMo and CoMo catalysts. They found that the upgraded bio-oils were completely free flowing at room temperature and light in colour compared with raw bio-crude oil. The yields of upgraded bio-oil were 93% and 89% at 350 °C and reduced to 41% and 69% at 405 °C for NiMo and CoMo catalysts, respectively. Despite a reduction in oil yield, a higher reaction temperature resulted in reductions in N, O and S contents and an increase in HHV from 42 to 45 MJ kg^{-1} for both catalysts. Elliott *et al.*[160] studied hydrotreating of algae bio-crude oil in a two-stage continuous flow fixed-bed reactor system using the CoMo catalyst for both stages. The bio-crude oil was first hydrotreated at 125–170 °C and the process then proceeded directly into the high-temperature stage at 405 °C. Both stages were operated at the same pressure of 13.6 MPa. The results showed that catalytic hydrotreating produced a nearly oxygen-free hydrocarbon blend with substantially lower levels of nitrogen and sulfur contents. Simulated distillation results further indicated that the upgraded oil product was primarily in the diesel-range and 80–85% of the hydrotreated bio-crude oil would be blendable into the conventional diesel fuel facility. Bai *et al.*[161] also reported a two-step hydrotreating process: non-catalytic pretreatment at 350 °C followed by catalytic hydrotreating at 400 °C. They examined the activities of 5% Pt/C, 5% Pd/C, 5% Ru/C, 5% Pt/C, Mo_2C, MoS_2, alumina, CoMo/c-Al_2O_3, Ni/SiO_2–Al_2O_3, H-ZSM-5, activated carbon, and Raney-Ni for hydrothermal hydrodeoxygenation and hydrodenitrogenation of algal bio-crude oil. They found that the Ru/C and Raney-Ni catalysts showed the best performances for deoxygenation and denitrogenation, respectively. The upgraded bio-oil had 97 wt% of material that boiled below 400 °C, contained only 2 wt% O and 2 wt% N, and had a heating value of 45 MJ kg^{-1}. They finally concluded that a two-step hydrotreating process with a two-component catalyst bed for the second step would be an attractive option for upgrading of algal bio-crude oil.

Zeolite catalysts are often used in hydrocracking processes. Li and Savage[162] studied hydrocracking of bio-crude oil produced from HTL of *Nannochloropsis* sp using a H-ZSM-5 catalyst at high H_2 pressure and determined the effects of reaction temperature (400–500 °C), reaction time (0.5–4 h), and catalyst loading (5–50 wt%) on bio-oil yield and composition. The produced oil had reduced heteroatom (N, O, and S) contents. Hydrocracking at 400 °C resulted in a paraffinic oil that contained 95 wt% C and H and retained 87% of the energy content of the crude bio-oil. Hydrocracking at 500 °C, however, produced a freely flowing liquid mixture composed of aromatic hydrocarbons in 44 wt% yield. The results indicated that H-ZSM-5 could be an effective catalyst for hydrotreating algal bio-crude oil and by tuning the operating conditions, a wide variety of fuels and chemicals could be produced. High temperatures will result in a better removal of oxygen but also enhance the degradation of the bio-crude oil to light gases and carbon.[158] Therefore, it is crucial to control the operating conditions for the desired products.

4.3.4 Catalytic Reforming of Glycerol

Glycerol is a key byproduct of biodiesel production processes. Economically converting glycerol to other valuable chemicals and fuels would indirectly reduce the biodiesel production costs and thus make the microalgae-to-biodiesel technical route economically more feasible. Extensive research work has been accomplished to address the technological issues of catalytic reforming of glycerol to produce a broad spectrum of valuable intermediates and products. A summary of the major reactions, key valuable intermediates and products and representative catalysts is given in Table 4.2. Major reaction pathways for catalytic reforming of glycerol include (1) selective oxidation of glycerol to produce dihydroxyacetone, glyceraldehyde, tartronic acid, hydroxyethanoic acid, glyceric acid, hydroxypyruvic acid, mesooxalic acid, mesoxalic acid, oxalic acid, formic acid, *etc.*, (2) selective hydrogenolysis of glycerol to produce propanediols and ethylene glycol, (3) dehydration of glycerol to produce acetol, acrolein, acrylic acid, acrylonitrile, *etc.*, (4) steam and aqueous-phase reforming of glycerol to produce mainly hydrogen, (5) transesterification or esterification of glycerol to produce monoglycides, α-monbenzoyl glycerol, glycerol dimethacrylate, *etc.*, (6) etherification of glycerol to produce monoethers, diethers and triethers, (7) oligomerization and polymerization of glycerol to produce, for example, glycerol 1-monoethers and polyglycerol methacrylates, respectively, (8) carboxylation of glycerol to produce glycerol carbonates, (9) acetalization of glycerol to form acetals and ketals, and (10) chlorination of glycerol to produce dichloropropanol.[163,164] Supported noble-metal and non-noble metal based catalysts play critical roles in oxidation, hydrogenolysis, steam forming and aqueous-phase reforming reactions. Metal oxide catalysts find important applications in transesterification, esterification, etherification, carboxylation and acetalization reactions. Zeolites have shown catalytic activity in dehydration, etherification, oligomerization and polymerization reactions. Other representative catalysts that have demonstrated potential useful applications in glycerol reforming include organometallic compounds, salts, MCM-41, and immobilized ionic liquids, to name a few.

Although various reaction pathways have been reported for catalytic reforming of glycerol to produce valuable chemicals and fuels, the vast majority of scientific research regarding glycerol conversion and utilization has been focused on renewable hydrogen production from glycerol. A recent study on techno-economic and environmental performances of glycerol reforming for hydrogen and power production suggested that glycerol

Table 4.2 Catalytic reforming of glycerol: major reactions, key valuable intermediates and products, and representative catalysts.[163,164]

Reactions	Valuable intermediates and products	Representative catalysts
Oxidation	Dihydroxyacetone, glyceraldehyde, tartronic acid, hydroxyethanoic acid, glyceric acid, hydroxypyruvic acid, mesooxalic acid, mesoxalic acid, oxalic acid, formic acid	Supported noble metal catalysts, organometallic compounds
Hydrogenolysis	1,2-Propanediol, 1,3-propanediol, ethylene glycol	Supported noble metal catalysts, Cu-based catalysts
Dehydration	Acetol, acrolein, acrylic acid, acrylonitrile	H_2SO_4, H_3PO_4, CsPW, zeolites, acidic binary metal oxides
Steam/aqueous phase reforming	Hydrogen	Supported Ni-, Pt-, Rh-, Ru-, Ir-, and Co-based catalysts
Transesterification, esterification	Monoglycides, α-monbenzoyl glycerol, glycerol dimethacrylate	MgO, CeO_2, La_2O_3, ZnO, Al–Mg hydrotalcites, Cs-exchanged sepiolite, and base-modified mesoporous MCM-41
Etherification	Monoethers, diethers, triethers	$CsHCO_3$, metal oxides, HY zeolites, SiO_2–SO_3H, Pd/C
Oligomerization, polymerization	Glycerol 1-monoethers, polyglycerol methacrylates	Zeolites, MCM-41
Carboxylation	Glycerol carbonate	Oxides (CaO, La_2O_3), mixed oxides (Mg/Al/Zr), KF modified hydroxyapatite, salts ($ZnSO_4$, $ZnCl_2$), immobilized ionic liquids
Acetalization	Acetals, ketals	MoO_3/SiO_2, SO_4^{2-}/ZrO_2
Chlorination	Dichloropropanol	$H_3PW_{12}O_{40}$

reforming is a promising concept for high energy efficiency processes with low CO_2 emissions.[165] Steam reforming is one of the major processes that convert glycerol to hydrogen and it has gained increasing interest in the scientific community. This process generally involves at least two reactions: (1) decomposition of glycerol to CO and H_2, and (2) the water–gas shift reaction to form additional H_2.[164] The overall endothermic reaction can be represented by: $C_3H_8O_3 + 3H_2O \rightarrow 7H_2 + 3CO_2$ ($\Delta H = 128$ kJ mol^{-1}).[166] The detailed reaction mechanisms and pathways, however, are very complicated and depend heavily on the catalysts used. Supported Ni-, Co-, Pt-, Rh-, Ru- and Ir-based catalysts are among the most active catalysts for steam reforming of glycerol. Although supported noble metal catalysts appear more stable and active than Ni-based catalysts, the latter become more attractive due to their lower cost.[164] The typical reaction conditions when supported Ni catalysts are used are: $T = 600–800$ °C, $P = 0.1$ MPa, H_2O : glycerol (molar ratio) = 3–16, and weight hourly space velocity = 1–10 h^{-1}.[164] Ni-based catalysts, however, suffer from deactivation caused by coke formation and Ni particle sintering. Recent studies have focused on developing novel catalysts for steam reforming of glycerol to address these issues.[167–170] The aqueous-phase reforming process is another relatively new process for glycerol reforming. It generally operates at lower temperatures (200–300 °C) but higher pressures (3–6 MPa) than the steam reforming process.[171] Other methods reported for producing H_2 from glycerol include supercritical water reforming, partial oxidation, and autothermal reforming.[172]

4.4 Computational Simulation of Model Feedstock

During upgrading of bio-crude-oil for biodiesel production, the catalysts play critical roles.[41,49–53] Computational modeling is an effective tool to help understand and optimize the conversion processes involved in algal biofuel production, such as growth and culture, transesterification, hydrolysis, decarboxylation, hydrodeoxygenation, hydrocracking, *etc.* Table 4.3 summarizes

Table 4.3 Computational modeling of various properties in each stage of biodiesel production from algae.

Major stages of production	Properties that can be investigated by computational methods
(a) Transesterification and hydrolysis of algae oil to biodiesel	(1) Reaction rate and kinetics (2) Catalytic sites and deactivation (3) Mechanism of catalytic reaction (4) Regulating saponification reactions (5) Process modeling to estimate the overall cost
(b) Decarboxylation and hydrodeoxygenation of biodiesel to hydrocarbons	(1) Reaction pathways (2) Catalysts selection and active sites (3) Kinetics and reaction rates (4) Mechanisms of catalytic reaction (5) Catalyst poisoning
(c) Hydrocracking and hydroisomerization of long-chain compounds to light hydrocarbons	(1) Catalysts screening to identify optional active sites (2) Kinetics modeling to get the reaction rate (3) Optimizing operating conditions to obtain specific products (4) Catalysis mechanism and deactivation (5) Acidity effects (6) Process modeling
(d) Byproduct glycerol reforming	(1) Oxidation mechanism (2) Optimal steam reforming conditions (3) Catalyst selection and catalysis mechanisms

the main properties that can be obtained using computational simulations. Since in this chapter, the focus is on the catalytic conversion of algae oil to liquid fuels, we only summarize the progress of simulations on processing and upgrading biodiesel and omit the algae growth and byproduct reforming.[23,24]

4.4.1 Transesterification and Hydrolysis of Algae Oil to Biodiesel

Many microalgae species can be induced to produce and accumulate substantial amounts of lipids; it has been reported that some species produce between 1 and 70% lipids and under certain conditions, some species can even reach 90% dry weight. These lipids are typically composed of glycerides and fatty acids.[36,49]

Usually, triglycerides (TGs) in algal lipids can be converted to biodiesel (mixture of fatty acid methyl esters (FAMEs)) through the transesterification or hydrolysis process. The transesterification reactions of triglycerides and free fatty acids with methanol or ethanol to biodiesel can be base catalyzed, acid catalyzed, and enzyme catalyzed.[36,173–177] Usually, base-catalyzed transesterification takes shorter times and occurs at lower temperatures while acid-catalyzed transesterification needs more time to complete and also requires heat. For example, Freedman *et al.*[178] reported that when using base catalyst (NaOH), the transesterification of vegetable oil was 99% complete at 32 °C in 4 hours, whereas at 60 °C, the reaction was completed in one hour. Comparing with alkali catalysis, they also observed that the transesterification by acid catalysis was much slower. However, solid acid catalysts can carry out transesterification and esterification simultaneously using low cost feedstocks without multiple reaction and post-treatment steps, greatly improving the economics of biodiesel production.[179] In the literature, computational modeling of transesterification focuses on the kinetics and the mechanism of the reactions and product design.[180]

The overall reaction scheme of transesterification of TG with methanol to give biodiesel (FAME) and byproduct glycerol (GL) is shown in Figure 4.15. Freedman *et al.*[174] proposed a three-step reaction scheme for the overall transesterification reaction that includes reaction intermediates, monoglyceride (MG) and diglyceride (DG), as shown in Figure 4.15(a). By examining the effects of the molar ratio of methanol to soybean oil, the type and amount of catalyst, and the reaction temperature on rate constants and kinetic order, they found that the forward reactions appear to be pseudo-first order or second order depending upon conditions used, while the reverse reactions appear to be second order.

Triglycerides and free fatty acids could react with a base catalyst to form soap as shown in Figure 4.15(b) and (c). The saponification leads to a lower production of biodiesel and higher cost of the purification process. Based on the work done by Freedman *et al.*,[174,178] Komers *et al.*[175] explored the kinetics and mechanism of the KOH-catalyzed methanolysis of rapeseed oil for biodiesel production by a model consisting of two sequences of consecutive

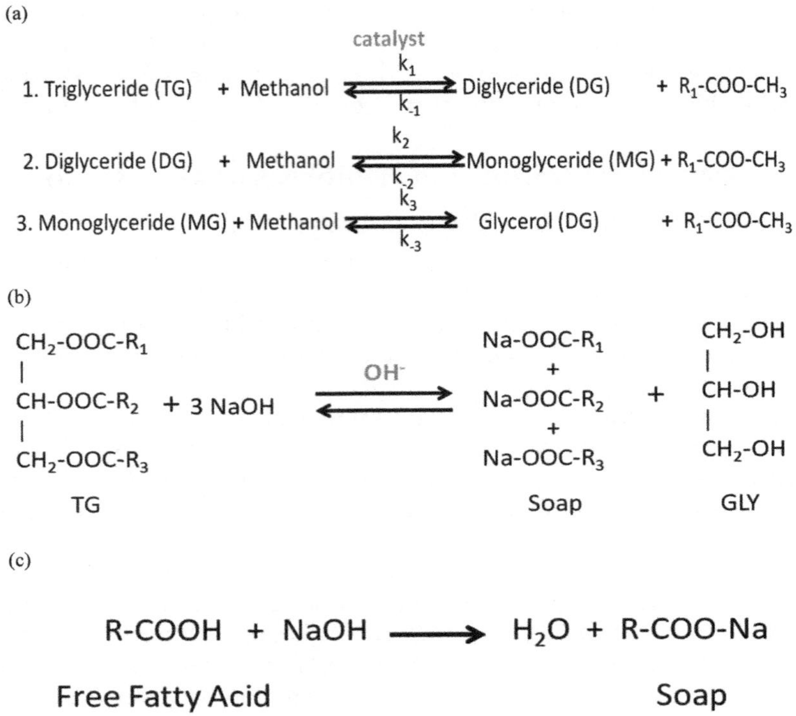

Figure 4.15 Transesterification of triglyceride: (a) stepwise reaction scheme; (b) saponification of triglyceride with NaOH; (c) saponification of free fatty acid with NaOH.[180]

competitive reactions (methanolysis and saponification). Their modeling results can describe the real cases with a probability of about 78% at given conditions. This model shows that all reaction rates are proportional to the amount of catalyst, which varies over time and is consumed in the competing saponification reactions.

When the starting lipids contain more than 0.5% (wt) free fatty acids (FFAs), a preliminary stage of acidity reduction is necessary by means of an esterification reaction of the FFAs to avoid their saponification. Tesser *et al.*[176] explored the kinetics of fatty acid esterification on acid exchange resins. As catalysts, the active site in acid ionic exchange resins is the sulphonic group, which exchanges its hydrogen ion with the components involved in the reaction after they are adsorbed on the resin surface. Their results showed that the rate determining step is the surface reaction of protonated fatty acid with methanol, which leads to the formation of protonated methyl ester. This model can provide a good description of the kinetic behavior, in particular for runs performed at different catalyst concentrations.

Hydrolysis and transesterification are two competing reactions that can occur during the synthesis of biodiesel. Early in 1997, Matsui *et al.*[89] investigated the liquefaction of microalgae with an iron catalyst ($Fe(CO)_5$–S) and obtained a 66.9(wt)% oil yield at 350 °C. They also found that liquefaction in toluene gave oil fractions of high carbon content and lower oxygen content while liquefaction in water caused a lower carbon content and higher oxygen content with an oil yield of 78.3(wt)% at 350 °C. Cheirsilp *et al.*[173] developed three kinetic models of the transesterification of palm oil fatty acids to ethanol using an immobilized lipase catalyst. Their results showed that the rate constants for alcoholysis of palm oil with ethanol are much higher than those for the hydrolysis reaction. The simulated results also showed that increasing initial ethanol concentration produces an increase in the initial production rate and yield of fatty acid ethyl ester and lowers the final concentration of free fatty acid. Using a tungstated zirconia catalyst, Ngaosuwan *et al.*[179] investigated the mechanistic pathways for hydrolysis (with water) and transesterification (with methanol) of tricaprylin and found that upon increasing the concentration of tricaprylin, the reaction rates for both hydrolysis and transesterification increased under all conditions. However, water inhibited the reaction rate of hydrolysis by poisoning the active sites. With a low water-to-tricaprylin ratio, the rate determining step for hydrolysis is the adsorption of tricaprylin while the rate determining step for transesterification is the surface reaction at low methanol-to-tricaprylin ratio. Based on quantum mechanical calculations, Zhang *et al.*[181] investigated the mechanisms of the conversion of methyl lactate (ML) over sodium tripolyphosphate. Their calculated results indicated that the conversion of ML over the catalyst is mainly through the direct decomposition of ML to methyl acrylate and methanol, and through the decarbonylation of ML to acetaldehyde, methanol and CO *via* stepwise mechanisms.

4.4.2 Conversion to Hydrocarbons by Decarboxylation and Hydrodeoxygenation

After transesterification or hydrolysis, the obtained biodiesel (FAMEs, FFAs) may still contain unsaturated C–C bonds and carboxyl groups, depending on the algal lipid composition, which reduce the combustion heat. In some cases, further treatments may be desirable in order to upgrade the biodiesel by decarboxylation and hydrodeoxygenation processes. In the literature, there are many experimental investigations for production of diesel fuel hydrocarbons from unsaturated renewable feedstocks. Snåre *et al.*[182,183] and Kubičková *et al.*[184] investigated the liquid-phase catalytic deoxygenation reaction of unsaturated renewables (oleic acid, linoleic acid, tristearine, methyl oleate) on Pd/C catalyst and the decarboxylation reaction of ethyl stearate to produce high carbon number hydrocarbons. Their schematic deoxygenation network of ethyl stearate to diesel products is depicted in Figure 4.16. With 5(wt)% Pd/C catalyst, Immer *et al.*[185] explored the catalytic reaction pathways in the liquid-phase deoxygenation of C_{18} FFAs, such as stearic, oleic, and linoleic acids, and found that the main liquid products are *n*-heptadecane and heptadecenes. FFA deoxygenation can occur *via* decarboxylation ($C_{17}H_{35}COOH \rightarrow CO_2 + n\text{-}C_{17}H_{36}$) and decarbonylation ($C_{17}H_{35}COOH \rightarrow CO + H_2O + C_{17}H_{34}$) pathways. Thermodynamics of both

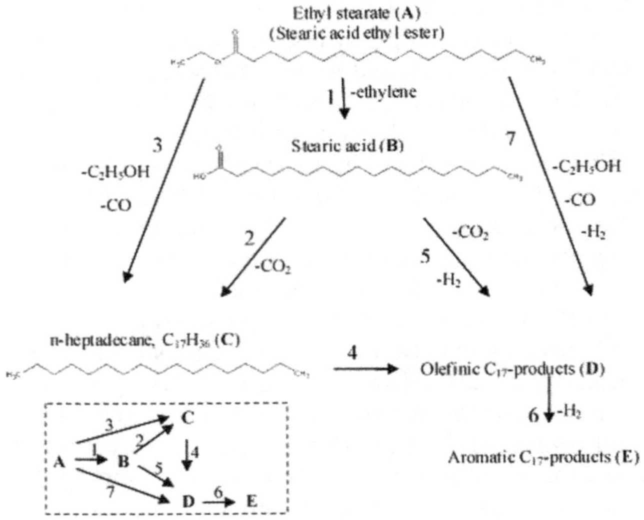

Figure 4.16 Schematic and simplified deoxygenation network of ethyl stearate to diesel products. The decarboxylation of fatty acid ester (A) proceeds *via* its corresponding fatty acid (B), which is subsequently decarboxylated to paraffin *n*-heptadecane (C). C is simultaneously dehydrogenated to unsaturated olefins (D) and aromatics (E). Furthermore, the direct formations of paraffin and olefin from fatty acid ester (A → C, A → D) have been detected from experiments. Adapted from *Chemical Engineering Journal*, **134**, M. Snåre, I. Kubičková, P. Mäki-Arvela, K. Eränen, J. Wärnå and D. Yu. Murzin, Production of diesel fuel from renewable feeds: kinetics of ethyl stearate decarboxylation, 29–34, Copyright 2007, with permission from Elsevier.

reactions are favorable at 300 °C: $\Delta G_{rxn} = -83.5$ kJ mol^{-1} for decarboxylation and -17 kJ mol^{-1} for decarbonylation. Their results showed that under 10% H_2, the hydrogenation of heptadecenes *via* hydrogen transfer from dodecane solvent leads to a 98% *n*-heptadecane yield.

Triglycerides can be directly converted to hydrocarbons without transesterification. Morgan *et al.*[186] conducted a study on the deoxygenation of triglycerides (tristearin, triolein, soybean oil) to hydrocarbons over supported metal catalysts (Ni/C, Pd/C and Pt/C). The formation of hydrocarbons spans a range of carbon chain lengths as shown in the following reaction:

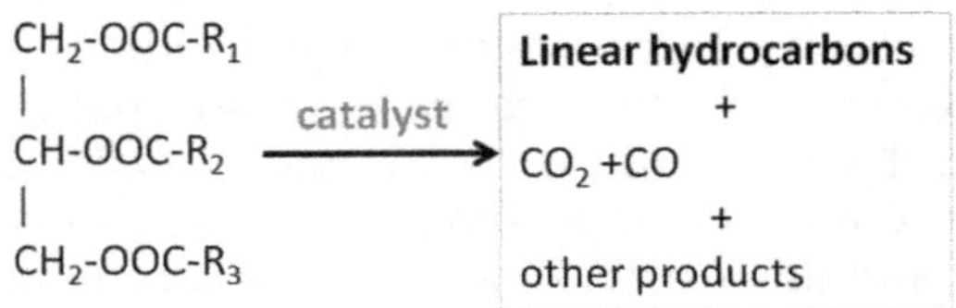

With Ni catalyst, the conversion of triglyceride produced high yields of linear C_5 to C_{17} alkanes and alkenes, while small amounts of light alkanes

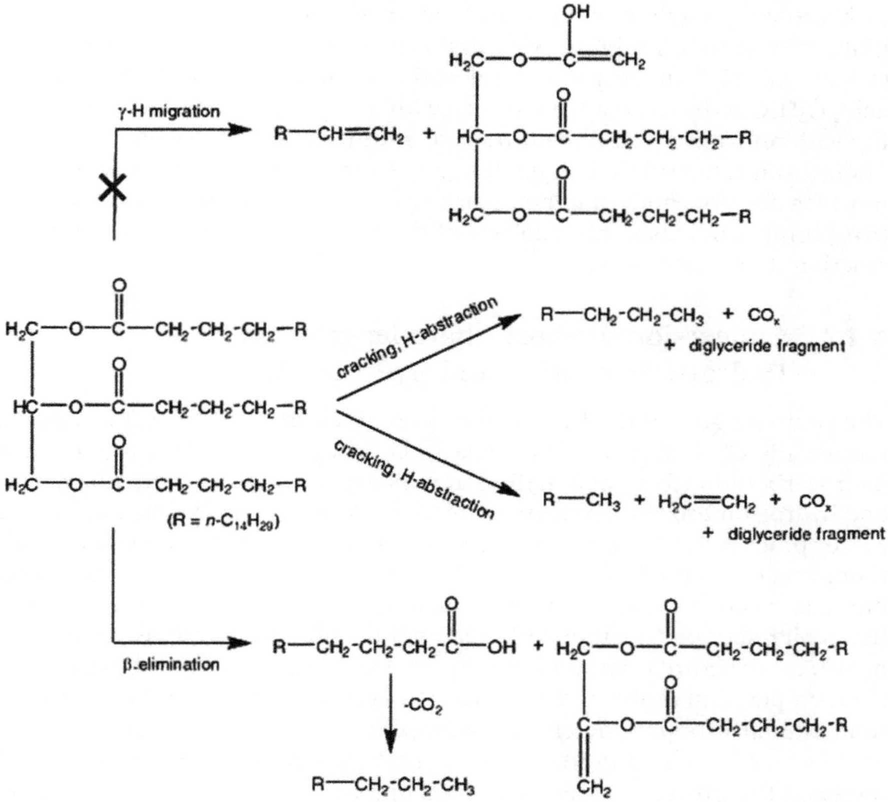

Figure 4.17 Simplified reaction scheme for tristearin deoxygenation. Reprinted from *Topics in Catalysis*, Conversion of triglycerides to hydrocarbons over supported metal catalysts, 53, 2010, 820–829, T. Morgan, D. Grubb, E. Santillan-Jimenez and M. Crocker, (Copyright © 2010, Springer Nature) with permission of Springer.[186]

(C_1–C_4) and H_2 were also formed. Figure 4.17 shows the simplified reaction scheme for tristearin deoxygenation, which involves liberation of fatty acids *via* C–O bond scission and concomitant H transfer, followed by elimination of CO_2 from the acids in a later step. Compared to Ni, their results showed that catalysts containing Pd or Pt supported on activated carbon showed lower activity for both triglyceride deoxygenation and for cracking of the fatty acid chains.

Snåre *et al.*[183] studied the kinetics of ethyl stearate decarboxylation over Pd/C catalyst. Their kinetic modeling and estimated parameters of temperature dependence (300–360 °C) of ethyl stearate decarboxylation fit the experimental results very well with a 99.28% degree of explanation. In the literature, we have not found studies involving theoretical modeling of the decarboxylation and deoxygenation of algal oil at an atomistic level. Due to limited computer resources, instead of long chain FFAs from algae transesterification or hydrolysis, only model fatty acids or small molecules are used for such kinds of research. Recently, with first-principles density functional theory, Geatches *et al.*[187] investigated the decarboxylation reaction of a model fatty acid, C_2H_5COOH, to an alkane (C_2H_6) with a clay mineral catalyst (pyrophillite with an isomorphic substitution of aluminum in the tetrahedral layer). Their results showed that an uncharged system with a sodium counterion is most feasible for catalyzing the decarboxylation reaction in an Al-substituted pyrophillite and, also, that analysis of the orbitals is a better indicator of a reaction than charge alone.

4.4.3 Conversion to Short Chain-length Fuel by Hydroisomerization and Hydrocracking

The chain-lengths of the hydrocarbons obtained from biological feedstocks are usually C_{11} and above (particularly, C_{14}–C_{22}). In order to produce light hydrocarbons (such as jet fuel JP-8) from heavy hydrocarbons, hydrocracking and hydroisomerization processes are used. Currently, hydrocracking is commonly practiced in the petroleum refinery industry to treat oil residue. In the literature, direct modeling work has not been published for hydrocracking of biofuels. Ancheyta *et al.*[188] reviewed the kinetic modeling of hydrocracking of heavy oil fractions. The models based on the lumping technique, continuous mixtures, structure oriented lumping, and single event were fully discussed and compared in terms of the capacity to predict detailed product composition, difficulty for parameter estimation, dependency of rate coefficient with feed properties, and required experimental data. Recently, Martinez *et al.*[189] reviewed the process aspects and modeling of an ebullated bed reactor for hydrocracking of heavy oils. Some key factors for applying these reactors to hydrocracking of heavy petroleum fractions, such as sediment formation, catalyst attrition and catalyst deactivation, were clearly discussed. Here, we summarize the progress in the theoretical modeling of the mechanisms of catalyzed hydrocracking of heavy compounds.

Based upon the elementary steps of carbenium-ion chemistry, Froment[190] developed a detailed kinetic model of the acid-catalyzed oil-refining process. In his approach, the computer calculation of the number of events in an elementary step requires the knowledge of the configuration of the transition state, which could be determined through quantum chemical calculations. Krishna and Balasubramanian[191] provided a general analytical solution for the full stoichiometry based discrete lumped kinetic equations in hydrocracking of heavier petroleum fractions. Their results revealed that the full stoichiometry is the best performing model for hydrocracking of heavier petroleum fractions and the product distribution in a hydrocracker is governed by the ordinary differential equations. Thybaut and Marin[192] investigated the kinetic modeling of the conversion of complex hydrocarbon feedstocks by acid catalysts and found that the number of elementary reaction families and hence, of the corresponding kinetic parameters, is rather limited and the variations in acid strength have a significant effect on the activation energy of the acid-catalyzed rearrangement steps. Using the principles of the single-event microkinetic concept, Mitsios *et al.*[193] investigated the hydrocracking and hydroisomerization of long-chain paraffin on an amorphous $Pt/SiO_2{\cdot}Al_2O_3$ catalyst. Their results showed that the invariant rate constants do not depend on the feedstock composition or the process configuration. From the experimental program involving liquid-phase pure *n*-hexadecane hydrocracking, a complete set of only nine rate constants for paraffin hydrocracking and hydroisomerization was obtained and gave a satisfactory fit of the experimental data.

Zeolite catalysts are often used in hydrocracking processes. According to the bifunctional reaction scheme, the kinetic modeling for hydrocracking of alkanes on Pt/US-Y zeolites has been carried out by Martens *et al.*[194] and Thybaut *et al.*[195] The differences in both the number of acid sites and the average acid length determine the activities of the zeolites. On catalyst, the reactivity of alkanes increased with carbon number, which was related to three phenomena: (1) the physisorption of heavier molecules was more favorable; (2) the reaction network and the number of parallel reactions became larger with larger molecules, and (3) the stabilization of alkylcarbenium ions and, hence, their concentration, increased with increasing size and electron donating property of alkyl-substituents. In order to explore the mechanisms of hydrocarbon conversion in zeolites, several atomistic level simulations have been done.[196–199] Based on *ab initio* quantum chemical calculations, Kazansky *et al.*[198] investigated the isobutene cracking on zeolites for three main elementary steps: (1) chain initiation resulting in formation of adsorbed carbenium ions by protonation of olefins, by protolytic cracking of paraffins, and by protolytic dehydrogenation of paraffins; (2) chain termination represents the decomposition of the *tert*-butoxide group into isobutene and bridging hydroxyl; and (3) chain propagation, which can involve hydride transfer. For each elementary step, at least one transition state was identified. Their calculated heats of reaction and activation energies of the main elementary steps are comparable with available experimental and other theoretical results. Using an *ab initio* method at Hartree-Fock level with the SCF-3-21G/6-31G* basis set and MP2 perturbation theory, Rigby *et al.*[199] further studied the mechanism of hydrocarbon conversion in zeolites and found that the intermediates are covalent alkoxide species. In their work, the influence of differences in the acid strength on the activation energies was explored. As shown in Figure 4.18, variations in acid strength were found to only have minor effects on the alkoxide stability and little effect on the activation energies for D/H exchange in methane and for olefin chemisorption, which have covalent transition states. However, acid strength was discovered to have a significant effect on the activation energy of the acid-catalyzed rearrangement steps, which have ionic transition states. It can be seen from Figure 4.18 that all of these activation energies are strongly reduced with increasing acidity.

Based on first-principles density functional theory, Frash *et al.*[196] explored the mechanism of the β-scission reaction in zeolites and found

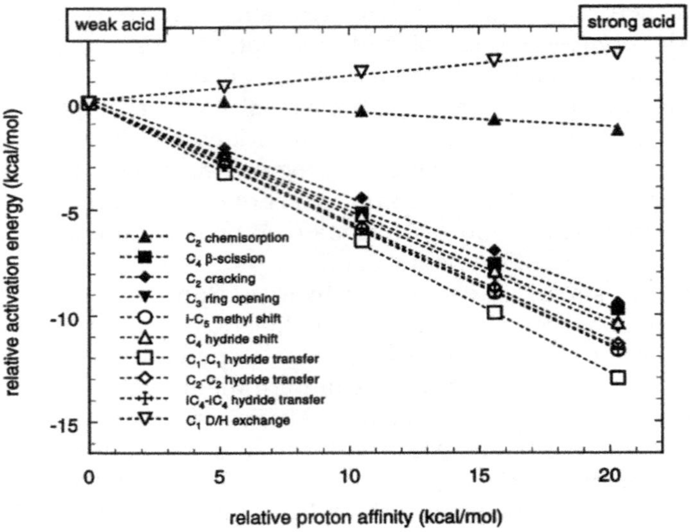

Figure 4.18 The calculated variation with absolute acid strength of the activation energies of the reaction steps of interest for hydrocarbon conversion. Reprinted from *Journal of Catalysis*, **170**, A. M. Rigby, G. J. Kramer, R. A. van Santen, Mechanisms of hydrocarbon conversion in zeolites: a quantum mechanical study, 1–10, Copyright 1997, with permission from Elsevier.

that the potential energy surface of this reaction is very complex. Three reaction paths were identified: (1) one-step *via* the "ringlike" transition state; (2) two-step β-scission reaction *via* a "hydrogen-bonded" transition state and substituted cyclopropane; and (3) one-step β-scission reaction *via* a "hydrogen-bonded" transition state. Among them, path (2) has the lowest activation barrier. Combining experimental measurements with molecular dynamical simulations, Isoda *et al.*[197] explored the mechanism of hydrocracking of pyrenes (Py) over nickel-supported Y-zeolite catalysts and the corresponding reaction pathway is depicted in Figure 4.19. Their results showed that pyrene was cracked into 1- and 2-ring aromatics by partial hydrogenation of aromatic rings. Molecular dynamical simulation results indicated that diffusion of the polycyclic aromatic products into the zeolite pore is not easy and that these molecules are cracked on the surface of the zeolite rather than in the pore.

4.5 Conclusions

Increasing global energy demands, limited oil reserves, combined with enhanced interest in developing more sustainable fuel and chemical sources have resulted in a continued focus on oil-rich algae as a potential

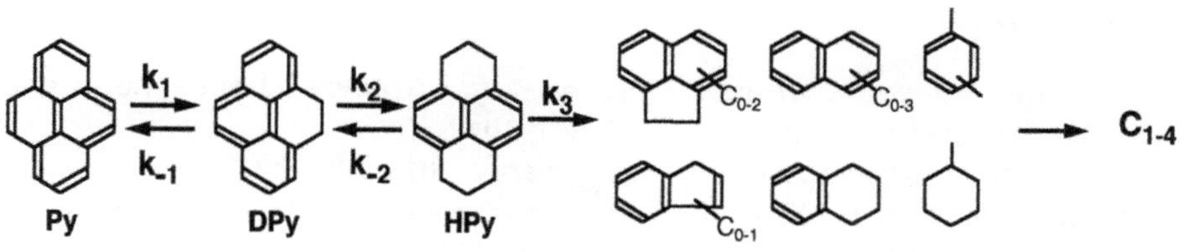

Figure 4.19 Hydrocracking reaction pathway of pyrene over metal supported Y-zeolite. Reprinted with permission from T. Isoda, S. Maemoto, K. Kusakabe and S. Morooka, *Energy & Fuels*, 1999, **13**, 617–623. Copyright 2014 American Chemical Society.[197]

solution. In this chapter, the progress and current status of the catalytic conversion of algae to biofuels and chemicals are discussed. The potential of converting algae to produce a variety of valuable fuels and chemicals is evident and is supported by the substantial technical advances as well as investment in research in this field. Future research should be focused on resolving key challenges to enable large-scale, practical deployment of this technology. Some of the challenges include substantial reduction of the process cost, by discovering catalysts based on inexpensive, abundant materials, by substantially improving the activity and stability of the catalysts, by developing processes that allow larger scale deployment, and by innovative computational approaches that can help understand the reaction mechanism, elucidate catalyst structure–property relationships, and ultimately predict the catalyst activity. Solving these challenges will have a revolutionary impact not only on algae conversion to biofuels and chemicals, but also on the extremely important field of catalysis in general.

Disclaimer

This report was prepared as an account of work sponsored by an agency of the United States Government. Neither the United States Government nor any agency thereof, nor any of their employees, makes any warranty, express or implied, or assumes any legal liability or responsibility for the accuracy, completeness, or usefulness of any information, apparatus, product, or process disclosed, or represents that its use would not infringe privately owned rights. Reference herein to any specific commercial product, process, or service by trade name, trademark, manufacturer, or otherwise does not necessarily constitute or imply its endorsement, recommendation, or favoring by the United States Government or any agency thereof. The views and opinions of authors expressed herein do not necessarily state or reflect those of the United States Government or any agency hereof.

References

1. EERE, *Biomass Multi-year Program Plan*, http://www1.eere.energy.gov/biomass/pdfs/mypp_april_2012.pdf, 2012.
2. R. Pate, G. Klise and B. Wu, *Appl. Energy*, 2011, **88**, 3377.
3. A. F. Clarens, H. Nassau, E. P. Resurreccion, M. A. White and L. M. Colosi, *Environ. Sci. Technol.*, 2011, **45**, 7554.
4. I. E. A. (IEA), *Medium-term Renewable Energy Market Report 2015: Market Analysis and Forecasts to 2020*, 2015.
5. EIA, *How Dependent Are We on Foreign Oil*, http://205.254.135.7/energy_in_brief/foreign_oil_dependence.cfm, 2011.
6. DOE-EERE, *Algae: 2011 Platform Review Report*, http://www1.eere.energy.gov/biomass/pdfs/2011_algae_review.pdf, 2011.
7. EPA, *Regulation of Fuels and Fuel Additives: Changes to Renewable Fuel Standard Program*, http://www.gpo.gov/fdsys/pkg/FR-2010-03-26/pdf/2010-3851.pdf, 2010.
8. H. Herzog, E. Drake and E. Adams, *CO_2 capture, Reuse, and Storage Technologies for Mitigating Global Climate Change. DOE Order No. DE-AF22–96PC01257*, 1997.
9. Y. Ghasemi, S. Rasoul-Amini, A. T. Naseri, N. Montazeri-Najafabady, M. A. Mobasher and F. Dabbagh, *Appl. Biochem. Microbiol.*, 2012, **48**, 126.
10. M. S. Wigmosta, A. M. Coleman, R. J. Skaggs, M. H. Huesemann and L. J. Lane, *Water Resour. Res.*, 2011, **47**, W00H04.
11. P. J. L. Williams and L. M. L. Laurens, *Energy Environ. Sci.*, 2010, **3**, 554.
12. D. H. Lee, *Bioresour. Technol.*, 2011, **102**, 43.
13. P. H. Pfromm, V. Amanor-Boadu and R. Nelson, *Bioresour. Technol.*, 2011, **102**, 1185.
14. A. Singh, P. S. Nigam and J. D. Murphy, *Bioresour. Technol.*, 2011, **102**, 10.
15. D. J. Stepan, R. E. Shockey, T. A. Moe and R. Dorn, *U.S. Department of Energy Report 2002-EERC-02-03*, 2002.
16. DOE, *National Algal Biofuels Technology Review. US. Department of Energy, Office of Energy Efficiency and Renewable Energy*, Bioenergy Technologies Office, 2016.
17. G. Andrich, U. Nesti, F. Venturi, A. Zinnai and R. Fiorentini, *Eur. J. Lipid Sci. Technol.*, 2005, **107**, 381.
18. G. Andrich, A. Zinnai, U. Nesti, F. Venturi and R. Fiorentini, *Acta Aliment.*, 2006, **35**, 195.
19. A. P. Carvalho, L. A. Meireles and F. X. Malcata, *Biotechnol. Prog.*, 2006, **22**, 1490.
20. Y. Chisti, *Biotechnol. Adv.*, 2007, **25**, 294.
21. N. Kosaric and J. Velikonja, *FEMS Microbiol. Rev.*, 1995, **16**, 111.
22. Q. Hu, M. Sommerfeld, E. Jarvis, M. Ghirardi, M. Posewitz, M. Seibert and A. Darzins, *Plant J.*, 2008, **54**, 621.
23. Y. Duan and F. Shi, in *Reactor and Process Design in Sustainable Energy Technology*, ed. F. Shi, Elsevier, Amsterdam, 2014, pp. 27–60.

24. F. Shi, P. Wang, Y. H. Duan, D. Link and B. Morreale, *RSC Adv.*, 2012, **2**, 9727.
25. M. P. Mansour, D. M. F. Frampton, P. D. Nichols, J. K. Volkman and S. I. Blackburn, *J. Appl. Phycol.*, 2005, **17**, 287.
26. L. Brennan and P. Owende, *Renewable Sustainable Energy Rev.*, 2010, **14**, 557.
27. H. H. Khoo, P. N. Sharratt, P. Das, R. K. Balasubramanian, P. K. Naraharisetti and S. Shaik, *Bioresour. Technol.*, 2011, **102**, 5800.
28. V. Patil, K. Q. Tran and H. R. Giselrod, *Int. J. Mol. Sci.*, 2008, **9**, 1188.
29. S. Amin, *Energy Convers. Manage.*, 2009, **50**, 1834.
30. P. Bondioli, *Top. Catal.*, 2004, **27**, 77.
31. A. Demirbas, *Energy Convers. Manage.*, 2009, **50**, 2782.
32. Z. Y. Du, Y. C. Li, X. Q. Wang, Y. Q. Wan, Q. Chen, C. G. Wang, X. Y. Lin, Y. H. Liu, P. Chen and R. Ruan, *Bioresour. Technol.*, 2011, **102**, 4890.
33. R. Li, Z. Zhong, B. Jin and A. Zheng, *Energy Fuels*, 2012, **26**, 2996.
34. H. Fukuda, A. Kondo and H. Noda, *J. Biosci. Bioeng.*, 2001, **92**, 405.
35. F. Ma and M. A. Hanna, *Bioresour. Technol.*, 1999, **70**, 1.
36. B. Smith, H. C. Greenwell and A. Whiting, *Energy Environ. Sci.*, 2009, **2**, 262.
37. S. Sawayama, S. Inoue, Y. Dote and S. Y. Yokoyama, *Energy Convers. Manage.*, 1995, **36**, 729.
38. A. A. Peterson, F. Vogel, R. P. Lachance, M. Froling, M. J. Antal and J. W. Tester, *Energy Environ. Sci.*, 2008, **1**, 32.
39. B. C. Wu, S. Ramon, C. A. DeLuca and E. K. Payne, US Patent No. 0050502, 2010.
40. M. Koberg and A. Gedanken, *Energy Environ. Sci.*, 2012, **5**, 7460.
41. N. H. Tran, J. R. Bartlett, G. S. K. Kannangara, A. S. Milev, H. Volk and M. A. Wilson, *Fuel*, 2010, **89**, 265.
42. R. O. Idem, S. P. R. Katikaneni and N. N. Bakhshi, *Fuel Process. Technol.*, 1997, **51**, 101.
43. A. Leung, D. G. B. Boocock and S. K. Konar, *Energy Fuels*, 1995, **9**, 913.
44. T. Kalnes, T. Marker and D. R. Shonnard, *Int. J. Chem. React. Eng.*, 2007, **5**, A48.
45. S. Lestari, I. Simakova, A. Tokarev, P. Maki-Arvela, K. Eranen and D. Y. Murzin, *Catal. Lett.*, 2008, **122**, 247.
46. J. Myllyoja, P. Aalto, P. Savolainen, V. Purola, V. Alopaeus and J. Gronqvist, US Patent US2007/010682 A1, 2007.
47. S. V. Ghadge and H. Raheman, *Biomass Bioenergy*, 2005, **28**, 601.
48. R. Luque, J. C. Lovett, B. Datta, J. Clancy, J. M. Campelo and A. A. Romero, *Energy Environ. Sci.*, 2010, **3**, 1706.
49. T. M. Mata, A. A. Martins and N. S. Caetano, *Renewable Sustainable Energy Rev.*, 2010, **14**, 217.
50. M. F. Demirbas, *Appl. Energy*, 2009, **86**, S151.
51. D. E. Resasco and S. Crossley, *AIChE J.*, 2009, **55**, 1082.
52. S. A. Scott, M. P. Davey, J. S. Dennis, I. Horst, C. J. Howe, D. J. Lea-Smith and A. G. Smith, *Curr. Opin. Biotechnol.*, 2010, **21**, 277.

53. V. A. Yakovlev, S. A. Khromova, O. V. Sherstyuk, V. O. Dundich, D. Y. Ermakov, V. M. Novopashina, M. Y. Lebedev, O. Bulavchenko and V. N. Parmon, *Catal. Today*, 2009, **144**, 362.
54. S. K. Hoekman, A. Broch, C. Robbins, E. Ceniceros and M. Natarajan, *Renewable Sustainable Energy Rev.*, 2012, **16**, 143.
55. S. Sawayama, T. Minowa and S. Y. Yokoyama, *Biomass Bioenergy*, 1999, **17**, 33.
56. D. R. Vardon, B. K. Sharma, G. V. Blazina, K. Rajagopalan and T. J. Strathmann, *Bioresour. Technol.*, 2012, **109**, 178.
57. H. R. Appell, Y. C. Fu, E. G. Illig, F. W. Steffgen and R. D. Miller, *Report of Investigations 8013*, Bureau of Mines, Washington, DC, 1975.
58. S. M. Kohan, *Biomass Conversion Processes for Energy and Fuels*, Plenum Press, New York, 1981.
59. Y. Dote, S. Sawayama, S. Inoue, T. Minowa and S. Yokoyama, *Fuel*, 1994, **73**, 1855.
60. L. G. Alba, C. Torri, C. Samori, J. van der Spek, D. Fabbri, S. R. A. Kersten and D. W. F. Brilman, *Energy Fuels*, 2012, **26**, 642.
61. P. Biller and A. B. Ross, *Bioresour. Technol.*, 2011, **102**, 215.
62. T. M. Brown, P. G. Duan and P. E. Savage, *Energy Fuels*, 2010, **24**, 3639.
63. T. Minowa, S. Yokoyama, M. Kishimoto and T. Okakura, *Fuel*, 1995, **74**, 1735.
64. A. B. Ross, P. Biller, M. L. Kubacki, H. Li, A. Lea-Langton and J. M. Jones, *Fuel*, 2010, **89**, 2234.
65. Y. F. Yang, C. P. Feng, Y. Inamori and T. Maekawa, *Resour., Conserv. Recycl.*, 2004, **43**, 21.
66. G. Yu, Y. H. Zhang, L. Schideman, T. Funk and Z. C. Wang, *Energy Environ. Sci.*, 2011, **4**, 4587.
67. S. P. Zou, Y. L. Wu, M. D. Yang, K. Imdad, C. Li and J. M. Tong, *Energy*, 2010, **35**, 5406.
68. K. Anastasakis and A. B. Ross, *Bioresour. Technol.*, 2011, **102**, 4876.
69. D. M. Li, L. M. Chen, D. Xu, X. W. Zhang, N. H. Ye, F. J. Chen and S. L. Chen, *Bioresour. Technol.*, 2012, **104**, 737.
70. D. Zhou, L. A. Zhang, S. C. Zhang, H. B. Fu and J. M. Chen, *Energy Fuels*, 2010, **24**, 4054.
71. D. C. Elliott, P. Biller, A. B. Ross, A. J. Schmidt and S. B. Jones, *Bioresour. Technol.*, 2015, **178**, 147.
72. C. Jazrawi, P. Biller, A. B. Ross, A. Montoya, T. Maschmeyer and B. S. Haynes, *Algal Res.*, 2013, **2**, 268.
73. M. Chakraborty, C. Miao, A. McDonald and S. L. Chen, *Fuel*, 2012, **95**, 63.
74. K. Prapaiwatcharapan, S. Sunphorka, P. Kuchonthara, K. Kangvansaichol and N. Hinchiranan, *Bioresour. Technol.*, 2015, **191**, 426.
75. Y. Chen, Y. L. Wu, D. R. Hua, C. Li, M. P. Harold, J. L. Wang and M. D. Yang, *RSC Adv.*, 2015, **5**, 18673.
76. Y. Guo, T. Yeh, W. H. Song, D. H. Xu and S. Z. Wang, *Renewable Sustainable Energy Rev.*, 2015, **48**, 776.

77. D. C. Elliott, *Algal Res.*, 2016, **13**, 255.
78. S. P. Zou, Y. L. Wu, M. D. Yang, C. Li and J. M. Tong, *Energy Environ. Sci.*, 2010, **3**, 1073.
79. P. G. Duan and P. E. Savage, *Bioresour. Technol.*, 2011, **102**, 1899.
80. D. C. Elliott, T. R. Hart, A. J. Schmidt, G. G. Neuenschwander, L. J. Rotness, M. V. Olarte, A. H. Zacher, K. O. Albrecht, R. T. Hallen and J. E. Holladay, *Algal Res.*, 2013, **2**, 445.
81. T. Selvaratnam, A. K. Pegallapati, H. Reddy, N. Kanapathipillai, N. Nirmalakhandan, S. Deng and P. J. Lammers, *Bioresour. Technol.*, 2015, **182**, 232.
82. S. S. Toor, L. Rosendahl and A. Rudolf, *Energy*, 2011, **36**, 2328.
83. S. Leow, J. R. Witter, D. R. Vardon, B. K. Sharma, J. S. Guest and T. J. Strathmann, *Green Chem.*, 2015, **17**, 3584.
84. H. Li, Z. D. Liu, Y. H. Zhang, B. M. Li, H. F. Lu, N. Duan, M. S. Liu, Z. B. Zhu and B. C. Si, *Bioresour. Technol.*, 2014, **154**, 322.
85. N. Sudasinghe, B. Dungan, P. Lammers, K. Albrecht, D. Elliott, R. Hallen and T. Schaub, *Fuel*, 2014, **119**, 47.
86. U. Jena and K. C. Das, *Production of Biocrude Oil from Microalgae via Thermochemical Liquefaction Process*, Amer. Soc. Agricultural & Biological Engineers, St Joseph, 2009.
87. G. Yu, Y. Zhang, L. Schideman, T. L. Funk and Z. Wang, *Trans. ASABE*, 2011, **54**, 239.
88. C. Torri, L. G. Alba, C. Samori, D. Fabbri and D. W. F. Brilman, *Energy Fuels*, 2012, **26**, 658.
89. T. O. Matsui, A. Nishihara, C. Ueda, M. Ohtsuki, N. O. Ikenaga and T. Suzuki, *Fuel*, 1997, **76**, 1043.
90. D. R. Vardon, B. K. Sharma, J. Scott, G. Yu, Z. C. Wang, L. Schideman, Y. H. Zhang and T. J. Strathmann, *Bioresour. Technol.*, 2011, **102**, 8295.
91. P. G. Duan and P. E. Savage, *Ind. Eng. Chem. Res.*, 2011, **50**, 52.
92. T. Muppaneni, H. K. Reddy, T. Selvaratnam, K. P. R. Dandamudi, B. Dungan, N. Nirmalakhandan, T. Schaub, F. Omar Holguin, W. Voorhies, P. Lammers and S. Deng, *Bioresour. Technol.*, 2017, **223**, 91.
93. P. Duan and P. E. Savage, *Ind. Eng. Chem. Res.*, 2011, **50**, 52.
94. P. J. Valdez, J. G. Dickinson and P. E. Savage, *Energy Fuels*, 2011, **25**, 3235.
95. T. Minowa and S. Sawayama, *Fuel*, 1999, **78**, 1213.
96. T. M. Mata, A. A. Martins and N. S. Caetano, *Renewable Sustainable Energy Rev.*, 2010, **14**, 217.
97. M. K. Lam, K. T. Lee and A. R. Mohamed, *Biotechnol. Adv.*, 2010, **28**, 500.
98. S. B. Velasquez-Orta, J. G. M. Lee and A. Harvey, *Fuel*, 2012, **94**, 544.
99. T. Suganya, N. Nagendra Gandhi and S. Renganathan, *Bioresour. Technol.*, 2013, **128**, 392.
100. C. L. Teo and A. Idris, *Bioresour. Technol.*, 2014, **174**, 281.
101. H. L. Tran, Y. J. Ryu, D. H. Seong, S. M. Lim and C. G. Lee, *Biotechnol. Bioprocess Eng.*, 2013, **18**, 242.
102. B. Jin, P. Duan, Y. Xu, B. Wang, F. Wang and L. Zhang, *Bioresour. Technol.*, 2014, **162**, 341.

103. Q. Guan, H. Shang, J. Liu, J. Gu, B. Li, R. Miao, Q. Chen and P. Ning, *Appl. Energy*, 2016, **164**, 380.
104. E. Lotero, Y. Liu, D. E. Lopez, K. Suwannakarn, D. A. Bruce and J. G. Goodwin, *Ind. Eng. Chem. Res.*, 2005, **44**, 5353.
105. E. S. Umdu, M. Tuncer and E. Seker, *Bioresour. Technol.*, 2009, **100**, 2828–2831.
106. S. H. Teo, Y. H. Taufiq-Yap and F. L. Ng, *Energy Convers. Manage.*, 2014, **88**, 1193.
107. E. O. Naor, M. Koberg and A. Gedanken, *Renewable Energy*, 2017, **101**, 493.
108. Y. Li, S. Lian, D. Tong, R. Song, W. Yang, Y. Fan, R. Qing and C. Hu, *Appl. Energy*, 2011, **88**, 3313.
109. X. Zhang, Q. Ma, B. Cheng, J. Wang, J. Li and F. Nie, *J. Nat. Gas Chem.*, 2012, **21**, 774.
110. S. H. Teo, A. Islam and Y. H. Taufiq-Yap, *Chem. Eng. Res. Des.*, 2016, **111**, 362.
111. Y. M. Dai, K. T. Chen and C. C. Chen, *Chem. Eng. J.*, 2014, **250**, 267.
112. Y. D. Chiang, S. Dutta, C. T. Chen, Y. T. Huang, K. S. Lin, J. C. S. Wu, N. Suzuki, Y. Yamauchi and K. C. W. Wu, *ChemSusChem*, 2015, **8**, 789.
113. M. Kouzu, T. Kasuno, M. Tajika, Y. Sugimoto, S. Yamanaka and J. Hidaka, *Fuel*, 2008, **87**, 2798.
114. J. A. Melero, J. Iglesias and G. Morales, *Green Chem.*, 2009, **11**, 1285.
115. A. Carrero, G. Vicente, R. Rodríguez, M. Linares and G. L. Del Peso, *Catal. Today*, 2011, **167**, 148.
116. B. J. Krohn, C. V. McNeff, B. Yan and D. Nowlan, *Bioresour. Technol.*, 2011, **102**, 94.
117. J. Cheng, Y. Qiu, R. Huang, W. Yang, J. Zhou and K. Cen, *Bioresour. Technol.*, 2016, **221**, 344.
118. X. Fu, D. Li, J. Chen, Y. Zhang, W. Huang, Y. Zhu, J. Yang and C. Zhang, *Bioresour. Technol.*, 2013, **146**, 767.
119. A. Guldhe, P. Singh, F. A. Ansari, B. Singh and F. Bux, *Fuel*, 2017, **187**, 180.
120. M. Veillette, A. Giroir-Fendler, N. Faucheux and M. Heitz, *Chem. Eng. J.*, 2017, **308**, 101.
121. A. Guldhe, C. V. R. Moura, P. Singh, I. Rawat, E. M. Moura, Y. Sharma and F. Bux, *Renewable Energy*, 2017, **105**, 175.
122. M. G. Kulkarni, R. Gopinath, L. C. Meher and A. K. Dalai, *Green Chem.*, 2006, **8**, 1056.
123. C. C. Akoh, S.-W. Chang, G.-C. Lee and J.-F. Shaw, *J. Agric. Food Chem.*, 2007, **55**, 8995.
124. A. Gog, M. Roman, M. Toşa, C. Paizs and F. D. Irimie, *Renewable Energy*, 2012, **39**, 10.
125. B. Norjannah, H. C. Ong, H. Masjuki, J. Juan and W. Chong, *RSC Adv.*, 2016, **6**, 60034.
126. J. K. Poppe, R. Fernandez-Lafuente, R. C. Rodrigues and M. A. Z. Ayub, *Biotechnol. Adv.*, 2015, **33**, 511.

127. J. Q. Lai, Z. L. Hu, P. W. Wang and Z. Yang, *Fuel*, 2012, **95**, 329.
128. D. T. Tran, C. L. Chen and J. S. Chang, *Bioresour. Technol.*, 2013, **135**, 213.
129. D. T. Tran, K. L. Yeh, C. L. Chen and J. S. Chang, *Bioresour. Technol.*, 2012, **108**, 119.
130. J. Huang, J. Xia, W. Jiang, Y. Li and J. Li, *Bioresour. Technol.*, 2015, **180**, 47.
131. H. Taher, S. Al-Zuhair, A. H. Al-Marzouqi, Y. Haik and M. Farid, *Biochem. Eng. J.*, 2014, **90**, 103.
132. R. S. Alavijeh, F. Tabandeh, O. Tavakoli, A. Karkhane and P. Shariati, *J. Oleo Sci.*, 2015, **64**, 69.
133. L. F. Bautista, G. Vicente, A. Mendoza, S. González and V. Morales, *Energy Fuels*, 2015, **29**, 4981.
134. B. Castillo López, L. Esteban Cerdán, A. Robles Medina, E. Navarro López, L. Martín Valverde, E. Hita Peña, P. A. González Moreno and E. Molina Grima, *J. Biosci. Bioeng.*, 2015, **119**, 706.
135. X. Tian, L. Dai, M. Liu, D. Liu, W. Du and H. Wu, *Catal. Commun.*, 2016, **84**, 44.
136. G. Bayramoglu, A. Akbulut, V. C. Ozalp and M. Y. Arica, *Chem. Eng. Res. Des.*, 2015, **95**, 12.
137. J. Amoah, S. H. Ho, S. Hama, A. Yoshida, A. Nakanishi, T. Hasunuma, C. Ogino and A. Kondo, *Biochem. Eng. J.*, 2016, **105**, 10.
138. T. A. Milne, R. J. Evans and N. Nagle, *Biomass*, 1990, **21**, 219.
139. L. Hillen, G. Pollard, L. Wake and N. White, *Biotechnol. Bioeng.*, 1982, **24**, 193.
140. L. O. Garciano, N. H. Tran, G. K. Kannangara, A. S. Milev, M. A. Wilson and H. Volk, *Chem. Eng. Sci.*, 2014, **107**, 302.
141. H. Kitazato, S. Asaoka and H. Iwamoto, *J. Jpn. Pet. Inst.*, 1989, **32**, 28.
142. N. I. Tracy, D. W. Crunkleton and G. L. Price, *Biomass Bioenergy*, 2011, **35**, 1060.
143. K. Murata, Y. Liu, M. M. Watanabe, M. Inaba and I. Takahara, *Energy Fuels*, 2014, **28**, 6999.
144. C. Zhao, T. Brück and J. A. Lercher, *Green Chem.*, 2013, **15**, 1720.
145. L. Zhou and A. Lawal, *Energy Fuels*, 2014, **29**, 262.
146. D. Verma, R. Kumar, B. S. Rana and A. K. Sinha, *Energy Environ. Sci.*, 2011, **4**, 1667.
147. B. Peng, X. Yuan, C. Zhao and J. A. Lercher, *J. Am. Chem. Soc.*, 2012, **134**, 9400.
148. B. Peng, Y. Yao, C. Zhao and J. A. Lercher, *Angew. Chem.*, 2012, **124**, 2114.
149. E. Santillan-Jimenez, T. Morgan, R. Loe and M. Crocker, *Catal. Today*, 2015, **258**, 284.
150. K. Kandel, J. W. Anderegg, N. C. Nelson, U. Chaudhary and I. I. Slowing, *J. Catal.*, 2014, **314**, 142.
151. C. Yang, R. Nie, J. Fu, Z. Hou and X. Lu, *Bioresour. Technol.*, 2013, **146**, 569.
152. J. Fu, C. Yang, J. Wu, J. Zhuang, Z. Hou and X. Lu, *Fuel*, 2015, **139**, 678.
153. H. J. Robota, J. C. Alger and L. Shafer, *Energy Fuels*, 2013, **27**, 985.
154. L. Zhou and A. Lawal, *Catal. Sci. Technol.*, 2016, **6**, 1442.

155. P. Duan, B. Wang and Y. Xu, *Bioresour. Technol.*, 2015, **186**, 58.
156. J. A. Ramirez, R. J. Brown and T. J. Rainey, *Energies*, 2015, **8**, 6765.
157. S. Xiu and A. Shahbazi, *Renewable Sustainable Energy Rev.*, 2012, **16**, 4406.
158. P. M. Mortensen, J. D. Grunwaldt, P. A. Jensen, K. G. Knudsen and A. D. Jensen, *Appl. Catal., A*, 2011, **407**, 1.
159. P. Biller, B. K. Sharma, B. Kunwar and A. B. Ross, *Fuel*, 2015, **159**, 197.
160. D. C. Elliott, T. R. Hart, A. J. Schmidt, G. G. Neuenschwander, L. J. Rotness, M. V. Olarte, A. H. Zacher, K. O. Albrecht, R. T. Hallen and J. E. Holladay, *Algal Res.*, 2013, **2**, 445.
161. X. Bai, P. Duan, Y. Xu, A. Zhang and P. E. Savage, *Fuel*, 2014, **120**, 141.
162. Z. Li and P. E. Savage, *Algal Res.*, 2013, **2**, 154.
163. C. H. Zhou, J. N. Beltramini, Y. X. Fan and G. Q. Lu, *Chem. Soc. Rev.*, 2008, **37**, 527.
164. C. H. Zhou, H. Zhao, D. S. Tong, L. M. Wu and W. H. Yu, *Catal. Rev.: Sci. Eng.*, 2013, **55**, 369.
165. A. M. Cormos and C. C. Cormos, *Int. J. Hydrogen Energy*, 2017, **42**, 7798.
166. P. D. Vaidya and A. E. Rodrigues, *Chem. Eng. Technol.*, 2009, **32**, 1463–1469.
167. K. D. Dewoolkar and P. D. Vaidya, *ChemCatChem*, 2016, **8**, 3499–3509.
168. C. Dang, H. Wang, H. Yu and F. Peng, *Appl. Catal., A*, 2017, **533**, 9.
169. M. Menor, S. Sayas and A. Chica, *Fuel*, 2017, **193**, 351–358.
170. L. Pastor-Pérez and A. Sepúlveda-Escribano, *Fuel*, 2017, **194**, 222–228.
171. R. D. Cortright, R. Davda and J. A. Dumesic, *Nature*, 2002, **418**, 964–967.
172. S. Adhikari, S. D. Fernando and A. Haryanto, *Energy Convers. Manage.*, 2009, **50**, 2600–2604.
173. B. Cheirsilp, A. H. Kittikun and S. Limkatanyu, *Biochem. Eng. J.*, 2008, **42**, 261–269.
174. B. Freedman, R. O. Butterfield and E. H. Pryde, *J. Am. Oil Chem. Soc.*, 1986, **63**, 1375–1380.
175. K. Komers, F. Skopal, R. Stloukal and J. Machek, *Eur. J. Lipid Sci. Technol.*, 2002, **104**, 728–737.
176. R. Tesser, L. Casale, D. Verde, M. Di Serio and E. Santacesaria, *Chem. Eng. J.*, 2010, **157**, 539–550.
177. Anon, *Prof. Eng.*, 2007, **20**, 45.
178. B. Freedman, E. H. Pryde and T. L. Mounts, *J. Am. Oil Chem. Soc.*, 1984, **61**, 1638–1643.
179. K. Ngaosuwan, X. H. Mo, J. G. Goodwin and P. Praserthdam, *Top. Catal.*, 2010, **53**, 783–794.
180. A. F. Chang and Y. A. Liu, *Ind. Eng. Chem. Res.*, 2010, **49**, 1197–1213.
181. Z. Q. Zhang, Y. X. Qu, S. I. Wang and J. D. Wang, *J. Mol. Catal. A: Chem.*, 2010, **323**, 91–100.
182. M. Snåre, I. Kubickova, P. Maki-Arvela, D. Chichova, K. Eranen and D. Y. Murzin, *Fuel*, 2008, **87**, 933–945.
183. M. Snåre, I. Kubickova, P. Maki-Arvela, K. Eranen, J. Warna and D. Y. Murzin, *Chem. Eng. J.*, 2007, **134**, 29–34.

184. I. Kubičková, M. Snare, K. Eranen, P. Maki-Arvela and D. Y. Murzin, *Catal. Today*, 2005, **106**, 197–200.
185. J. G. Immer, M. J. Kelly and H. H. Lamb, *Appl. Catal., A*, 2010, **375**, 134–139.
186. T. Morgan, D. Grubb, E. Santillan-Jimenez and M. Crocker, *Top. Catal.*, 2010, **53**, 820–829.
187. D. L. Geatches, S. J. Clark and H. C. Greenwell, *J. Phys. Chem. A*, 2010, **114**, 3569–3575.
188. J. Ancheyta, S. Sanchez and M. A. Rodriguez, *Catal. Today*, 2005, **109**, 76–92.
189. J. Martinez, J. L. Sanchez, J. Ancheyta and R. S. Ruiz, *Catal. Rev.: Sci. Eng.*, 2010, **52**, 60–105.
190. G. F. Froment, *Catal. Today*, 1999, **52**, 153–163.
191. P. C. Krishna and P. Balasubramanian, *Ind. Eng. Chem. Res.*, 2009, **48**, 6608–6617.
192. J. W. Thybaut and G. B. Marin, *Chem. Eng. Technol.*, 2003, **26**, 509–514.
193. M. Mitsios, D. Guillaume, P. Galtier and D. Schweich, *Ind. Eng. Chem. Res.*, 2009, **48**, 3284–3292.
194. G. G. Martens, G. B. Marin, J. A. Martens, P. A. Jacobs and G. V. Baroni, *J. Catal.*, 2000, **195**, 253–267.
195. J. W. Thybaut, G. B. Marin, G. V. Baron, P. A. Jacobs and J. A. Martens, *J. Catal.*, 2001, **202**, 324–339.
196. M. V. Frash, V. B. Kazansky, A. M. Rigby and R. A. van Santen, *J. Phys. Chem. B*, 1998, **102**, 2232–2238.
197. T. Isoda, S. Maemoto, K. Kusakabe and S. Morooka, *Energy Fuels*, 1999, **13**, 617–623.
198. V. B. Kazansky, M. V. Frash and R. A. vanSanten, *Appl. Catal., A*, 1996, **146**, 225–247.
199. A. M. Rigby, G. J. Kramer and R. A. vanSanten, *J. Catal.*, 1997, **170**, 1–10.

CHAPTER 5

Catalysts for Biofuels Production

LEONE FRUSTERI[a], GIUSEPPE BONURA[b] AND SIGLINDA PERATHONER*[a]

[a]University of Messina and INSTM/CASPE (Lab. of Catalysis for Sustainable Production and Energy), Department of Electronic Engineering, Industrial Chemistry and Engineering, Contrada di Dio, 98166 Messina, Italy; [b]CNR-ITAE, Istituto di Tecnologie Avanzate per l'Energia "Nicola Giordano", Via S. Lucia sopra Contesse 5, 98126 Messina, Italy
*E-mail: perathon@unime.it

5.1 Novel Catalytic Technologies for Biofuels Production

The development of biofuels has been promoted through incentive policies in many parts of the world. Europe is a good example: directive 2003/30/EC set targets for the renewable energy share in transportation fuels to be at 2% in 2005 and 5.75% in 2010. The climate-energy directive 2009/28/EC targets a share of 20% of renewable energy in the EU energy end-use and a share of 10% of this type of energy in each member state in the transport sector by the year 2020. Sustainability criteria have also been put in place in European legislation to encourage the use of resources that do not compete with the food chain and avoid undesirable effects such as deforestation or the use of environmentally sensitive land.

Green Chemistry Series No. 56
Sustainable Catalysis for Biorefineries
Edited by Francesco Frusteri, Donato Aranda and Giuseppe Bonura

Published by the Royal Society of Chemistry, www.rsc.org

Finally, objectives of GHG emissions reduction have been put in place for the biofuels: 35% GHG emissions saving by 2017 and 50% by 2018 for existing plants (commissioned before July 2014) and 60% GHG emissions saving for new plants (commissioned after July 2014). The shift towards biomass as a feedstock for fuels and chemicals requires the development of new catalytic chemistry.[1] Most refinery processes are run at high temperatures and with specific catalysts to activate the petroleum compounds. However, biomass comprises a complex mixture of multi-functionalized compounds, and thus the conversion into applicable fuels or chemicals usually requires removal or defunctionalization of most or all of these functionalities, depending on feedstock type.

The different possibilities for producing diesel oils from biomass resources have undergone substantial research over the years, and a number of catalytic methods have now been established. Fats and oils can either be transesterified with alcohol to form fatty acid alkyl esters,[2–7] deoxygenated with hydrogen to form hydrocarbons in the range of liquid fuels,[8–16] or cracked at elevated temperatures to form a hydrocarbon mixture.[14,17–19] Carbohydrates may be converted by aqueous-phase-reforming to form unfunctionalized or monofunctional hydrocarbons.[20–23] Biomass in general may be gasified to syngas and undergo Fischer–Tropsch-synthesis,[24–26] or bio-oils can be produced by flash pyrolysis and upgraded by cracking or hydrogenation.[27–32]

5.2 Transesterification of Vegetable Oils

Biodiesel is usually produced by transesterification of vegetable oils or animal fats with chemical catalysts, especially in the presence of strong acidic or basic solutions, such as hydrochloric acid, sulfuric acid, sodium hydroxide, sodium methoxide and potassium hydroxide. Homogeneous alkali catalysts can convert triglycerides to their corresponding fatty acid methyl esters (FAMEs) with high yield, less time and low cost. However, separating the catalyst from the product mixture for recycling is technically difficult. After reaction, the catalyst should be neutralized or removed with a large amount of hot water, which will produce a large amount of industrial wastewater. Typical plant oils, such as soybean oil, rapeseed oil and palm oil, are the main edible oils. They are not suitable as raw materials, particularly in developing countries, due to limited supply and high cost. Therefore, low-cost lipids, such as non-edible oils (*e.g.*, Jatropha oil), animal fats and waste oils, are used as ideal feedstocks. Such oils usually contain some water and free fatty acids (FFAs) that will form soap when homogeneous base catalysts are used. On the other hand, homogenous acid catalysts are corrosive to equipment. Solid heterogeneous catalysts are used to overcome these problems, because they are noncorrosive, non-toxic, and easily separated for recycling. As a rule, however, the presence of a multi-phase heterogeneous system (solid catalyst–alcohol–oil) will lead to diffusional phenomena that will lower the reaction rate.[33–35] Additional problems faced by solid catalysts are a low number

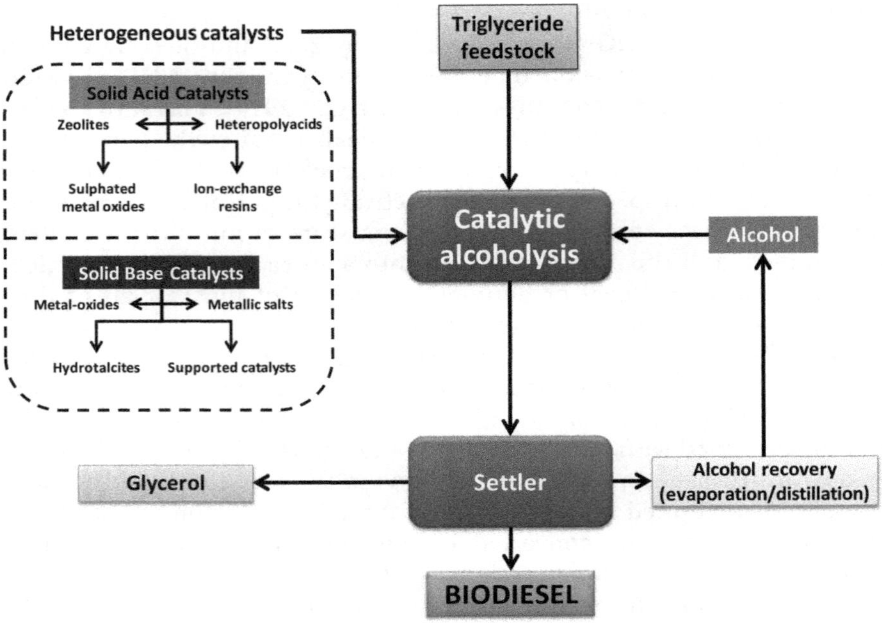

Figure 5.1 Heterogeneous biodiesel production by catalytic alcoholysis of triglyceride feedstock.

of active sites, micro porosity and leaching of the active phase.[36–38] Hence, in order to produce an excellent solid acid catalyst, the catalyst must comprise suitable textural/surface/morphological properties along with optimal mechanical strength.[35] Reusability of heterogeneous catalysts makes continuous fixed-bed operation possible. Such continuous process can minimize product separation and purification costs, making it economically viable to compete with commercial petroleum-based diesel fuel (see Figure 5.1). This chapter describes solid heterogeneous catalysts for biodiesel production and their typical catalytic mechanism.

In a laboratory-scale experiment, heterogeneous processes could be run in a continuous mode with a packed-bed continuous flow reactor. Heterogeneous catalysts were easily separated from the products, and water-washing processes and neutralization steps were avoided. Contaminated water from this process was greatly reduced, and the sewage treatment fees were also minimized. New types of heterogeneous catalysts have mushroomed and developed in recent years.

5.2.1 Solid Acid Catalysts

Acid catalysts can simultaneously catalyze both esterification and transesterification, showing a much higher tolerance to FFAs and water than basic homogeneous catalysts (*e.g.*, NaOH and KOH). Solid acids contain a variety of acid sites with different Brönsted or Lewis strength, compared to the homogenous acid catalysts. Despite the fact that a wide range of solid catalysts is available for the transesterification and esterification reactions, most of them suffer from one or more drawbacks. For example, zeolites are stable, robust and have good acidity but their small pore size causes diffusional limitations resulting in slow rates.[39] Amberlyst-15 or -70, Nafion and resin-based catalysts possess high acidity and pore size but have to operate under mild reaction conditions to avoid the catalyst degradation due to their polymeric matrix structure. In addition, when using an organic catalyst, the swelling capacity becomes a critical issue since it controls the substrate accessibility to the acid sites. Thus, inorganic solid acid catalysts are more desirable when high reaction temperatures are needed to enhance the catalyst performance. The inorganic acid catalysts operate at higher reaction temperatures than base catalysts because of their lower activity for transesterification. The main applications of the solid acid catalysts are described in the following sections: (1) ion-exchange resins; (2) zeolites; (3) heteropoly acid catalysts; (4) sulfated metal oxides.

5.2.1.1 Ion-exchange Resins

Ion-exchange resins are widely used in important industrial processes for both separation and reaction applications. Currently, sulfonated cation-exchange resins have been found to be very effective for esterification of FFA due to their high concentration of acid sites.[40–45] The activity of acid ion-exchange resins for the esterification reaction is influenced by the accessibility of reactants to the matrix. Shibasaki-Kitakawa *et al.*[46] found that the catalytic activity of ion-exchange resins correlated with cross-linking degree and particle size. Ion-exchange resins are classified into two types on the basis of the matrix structure: gelular or macroreticular. In fact, mass-transfer restriction represents a key factor affecting catalytic activity. In particular, internal diffusion was found to cause mass-transfer restriction being rate-limiting for regular resins. Most of the active sites are embedded in the gel matrix, so the resins with macropores have high catalytic activity. Furthermore, the catalytic activity usually decreases when the cross-linking degree of the polymeric matrix increases. Kouzu *et al.*[47] performed a study on the performance of gelular (Amberlyst-31) and macroreticular (Amberlyst-15) matrices in the esterification of soybean oil. It was found that the gelular resin has a higher catalytic activity compared to macroreticular resin. The uniqueness of these gelular resins lies in their textural properties (*i.e.*, the swelling behaviour). This swelling capacity controls the accessibility of acid sites in the catalyst, simultaneously affecting the overall reactivity. Once the small pores become macropores, they create more accessible sites for the acid functional group. Furthermore, the accessibility of long hydrocarbon chains of fatty acid molecules into the acid sites becomes easier and hence improves the catalytic activity. However, the internal mass transfer of gelular resins appears much slower compared to macroreticular resins. Similarly, Russbueldt and

Hoelderich[43] reported that the catalytic performance of a gelular sulfonic resin was superior to the macroporous catalysts investigated. Other studies dealing with the esterification of waste cooking oil in the presence of macroreticular Amberlyst-type resins and gelular matrices showed a higher FFA conversion for Amberlyst-15.[42]

Frusteri *et al.*[48] found a significantly higher turnover frequency of Amberlyst-15 in respect of other studied acid systems in the catalytic etherification of glycerol by *tert*-butyl alcohol to produce oxygenated additives for diesel fuel (see Figure 5.2), due to the wide pore diameter of such a resin that allows easier accessibility of the reagent molecules.

Reusability is an important evaluation index for industrial applications of resins. Mechanical strength and thermo-stability are decisive properties for large-scale applications of resins in biodiesel production. Bianchi *et al.*[49] studied the de-acidification of animal fats by using several types of macroreticular ion-exchange resins as catalysts. From this study, more than 90% FFA conversion was successfully achieved in the presence of Amberlyst-70 (wet), which also exhibited the possibility to be re-used at high conversion rate. Park *et al.*[50] found that the pore volume of macroreticular ion-exchange catalysts significantly affects the catalytic activity, owing to a different influence on the water formed during the esterification process. Caetano *et al.*[51] studied esterification of palmitic acid with methanol using poly(vinyl alcohol) cross-linked with sulfosuccinic acid (SSA) resin at 60 °C, reporting a conversion rate of 90% after 2 h. Only about 5% sulfosuccinic

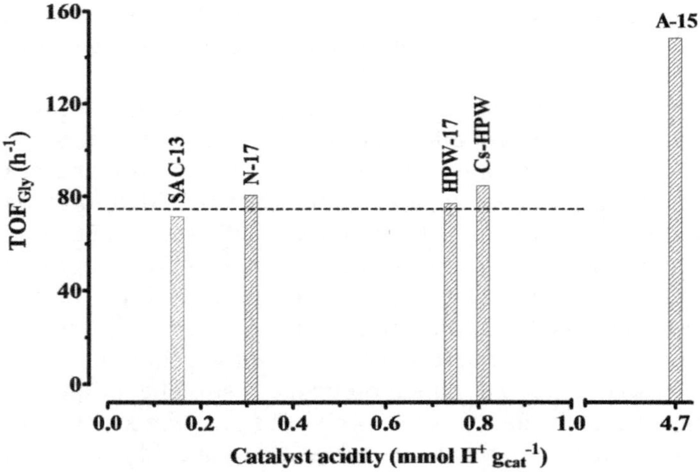

Figure 5.2 TOF of glycerol as a function of the catalyst acidity: T_R = 343 K; P_R = 0.1 MPa; $R_{Alc/Gly}$ = 4.0; reaction time: 2 h. Reprinted from *Applied Catalysis A: General,* **367**, F. Frusteri, F. Arena, G. Bonura, C. Cannilla, L. Spadaro and O. Di Blasi, Catalytic etherification of glycerol by *tert*-butyl alcohol to produce oxygenated additives for diesel fuel, 77–83, Copyright (2009), with permission from Elsevier.

acid was leached after 7 recycles. The activity of NKC-9 resin even slightly increased in the first 10 runs, due to the breakdown of resin particles under mechanical agitation.[52] In addition, the composition of fatty acids in vegetable oil has an impact on the catalytic activity of the ion-exchange resin, where a higher content of shorter-chain fatty acids in the oil composition is preferred when an ion-exchange resin is used.[53] However, thermal stability is the main concern when ion-exchange resins are used at higher temperatures. In esterification and transesterification reactions, a high temperature is usually required for higher reaction rates and catalyst regeneration. Fu *et al.*[54] examined the influence of impurities (*i.e.*, metal ions, organic acids, water, *etc.*) contained in used cooking oil (UCO) in the biodiesel production carried out in the presence of Amberlyst 15 and Amberlyst BD20 as solid acid catalysts. The results indicated that metal ions (Fe^{3+}) exert a significant deactivation effect on the resin catalyst by cationic exchange, also exhibiting catalytic activity as a Lewis acid when in free form. Moreover, the organic acids, highly active in the esterification reaction, showed a synergistic effect with the cation-exchange resin.

From a technological point of view, the combination of a fixed-bed reactor with supercritical CO_2 may enable a continuous process suitable for massive biodiesel production. Regarding catalyst deactivation, it is usually ascribed to water-swelling or salt contaminants. Considering that temperature has a negligible effect on water-swelling, the water absorbed on the resin surface can be readily extracted by light alcohols.[55] Catalytic active sites on acidic resins can exchange with salt ions contained in oil, leading to a continuous activity loss[43] that can be restored by acid washing.

5.2.1.2 Zeolites

Zeolites are crystalline alumino-silicates that can be synthesized with variable crystal structures, pore sizes, framework Si/Al ratios, surface hydrophobicity and proton-exchange levels. Zeolites have extremely high internal surface areas (600–800 $m^2\ g^{-1}$) and thermal stability (up to 1000 °C), while acidity is induced by aluminum atoms and ions in the skeletons and porosity. Based on these peculiar features, zeolites represent suitable solid catalysts for biodiesel production, although internal diffusion resistances are considered to significantly influence the reaction rate. Pérez-Pariente *et al.*[56] studied the selective synthesis of fatty monoglycerides with zeolites, reporting that the catalyst properties have more effects on monoglyceride yield than reaction parameters. Moreover, in such solids, the acidity must be thoroughly tuned, since a too high level will lead to catalyst deactivation along with the occurrence of side reactions and by-products, which result in additional costs for process separation. On the other hand, the reaction may not proceed at a reasonable rate in the presence of an acidity level that is too low. Zeolite β is a high silica zeolite with both Lewis-acid sites and Brønsted-acid sites, containing an intersecting three-dimensional structure of 12-membered ring channels.[57] Lewis-acid sites are mainly present in the microporous walls. In contrast, Brønsted-acid sites are present on the internal and external surface. Zeolite β does not exhibit high activity in transesterification, but it can be used for selective removal of FFAs in waste oil.[58] Jothiramalingam and Wang[59] claimed that the conversion of FFA increases with the increase of Si/Al ratio, indicating that the reactions are greatly influenced by surface hydrophobicity and acid site strength. In contrast, another study by Carmo *et al.*[60] found that the highest conversion of FFA is obtained with a decrease of Si/Al ratio, and methanol was found to be the best solvent (among other alcohols) in the esterification of palmitic acid with aluminum containing mesoporous MCM-41. Sasidharan and Kumar[61] found that large-pore zeolites such as Y, mordenite and β showed higher activity (biodiesel yield 92%) than the medium-pore ZSM-5 and Al-MCM-41 (biodiesel yield < 30%), owing to the active sites being more accessible to the bulky triglyceride molecules. However, H–β zeolite was tested in the transesterification of crude *Pongamia pinnata* oil with a yield as low as 59% at a long reaction time (24 h).[62] Bifunctional catalysts can usually be prepared by combining active catalytic sites on acid zeolites. Shu *et al.*[57] introduced La ions into zeolite β, so obtaining higher conversion and stability than the "bare" zeolite β as a result of there being more external Brønsted acid sites available for the reactants.

5.2.1.3 *Heteropoly Acid Catalysts*

Heteropoly acids (HPAs) have been extensively studied worldwide as solid acid catalysts for different reaction applications. A HPA consists of a particular combination of hydrogen and oxygen with certain metals (*i.e.*, tungsten, molybdenum and vanadium) and non-metals (*i.e.*, silicon, phosphorus).[63] There are many types of heteropoly acids, and the Keggin ($H_nXM_{12}O_{40}$) and Dawson ($H_nX_2M_{18}O_{62}$) structures are two of the better known groups.[64] HPAs are characterized by high stability and proton mobility, being typical strong Brønsted acids with an acidity level much higher compared to conventional acids, such as sulfuric acid, Al_2O_3–SiO_2, zeolites and acidic resins. HPAs can catalyze a wide range of reactions, including esterification and transesterification.[65] In general, HPAs are soluble in a polar solvent, but such a solubility can be changed *via* alkali-exchange, and modified HPAs exhibit significantly higher activity.[66]

Several studies have also reported on the catalytic activity of different solid heteropoly acids, particularly in the transesterification of virgin and waste oils. Alsalme *et al.*[67] studied the intrinsic catalytic activity of Keggin HPAs, indicating that the activity of HPAs is significantly higher than that of the conventional acid catalysts in (trans)esterification. Their acid strength in descending order is as follows:

$$H_3PW_{12}O_{40} > Cs_{2.5}H_{0.5}PW_{12}O_{40} > H_4SiW_{12}O_{40} > 15\%H_3PW_{12}O_{40}/Nb_2O_5, 15\%H_3PW_{12}O_{40}/ZrO_2, 15\%H_3PW_{12}O_{40}/TiO_2 > H_2SO_4 > \text{HY, H-Beta} > \text{Amberlyst-15}.$$

Kulkarni *et al.*[68] found that the catalytic activity of the HPA catalyst was highly dependent on the reactor temperature rather than the methanol to oil molar ratio or the FFA content. Cao *et al.*[69] investigated the production of biodiesel from waste frying oil and they found that the tungstophosphoric acid (HPW) catalyst shows higher activity under the optimized reaction conditions compared with the conventional homogeneous catalyst sulfuric acid. The final conversion was nearly 90%, however, the reaction required a high methanol to oil molar ratio and a long reaction time *i.e.* 70 : 1 and 14 h, respectively. HPA as a homogeneous catalyst was known to have several disadvantages, such as low surface area and low thermal stability. In order to overcome these limitations, HPA is now used together with a support material, such as hydrous zirconia,[68] activated carbon,[70] MCM-41,[71,72] silica,[73] niobia,[74] clays[75] and alumina powders.[68,76] These supports usually create a high surface area, providing a high thermally stable material even under reaction conditions of 200 °C, and simultaneously increase the catalytic performance. Most of the supported HPA catalysts are in the range of superacid catalysts and their properties vary with different porous structure and water solubility. Studies from several authors[77,78] showed that a salt of HPA with large monovalent ions (*e.g.*, Cs^+, NH_4^+ and Ag^+) has gained more interest nowadays in biodiesel research due to its excellent performance as a water tolerant catalyst with a large surface area. This combination of HPA with monovalent ions creates a microporous solid acid catalyst that is highly effective in most of the catalytic reactions including isomerization, hydration, alkylation, hydrolysis and esterification.[77,79–82] The proton replacement has similar effects on activity as cation concentration increase. $Cs_xH_{3-x}PW_{12}O_{40}$ (x = 0.9–3), one kind of insoluble Keggin HPA, offers excellent performance in (trans)esterification.[82] The catalytic activity of Cs-salts decreases as the content of Cs in HPW grows, due to the decrease of pH and the increase of conductivity of colloidal solutions in direct relation with the acidity of surface layers of primary particles. Chai *et al.*[83] and Hamad *et al.*[84] found that the $Cs_{2.5}H_{0.5}PW_{12}O_{40}$ and $Cs_2HPW_{12}O_{40}$ catalysts perform effectively in the transesterification of different vegetable oils. The catalysts were found to be less affected by the presence of high FFA and water contents. Such catalysts were also reported to have high activity, yielding 99% biodiesel at a low catalyst concentration (0.00185 : 1 weight of catalyst to oil), low methanol to oil ratio (5.3 : 1), low reaction temperature (338 K) and relatively short reaction time (<1 h).[83] The regeneration of the catalyst was also feasible and its capability for re-use was reported to be for a minimum of 6 times.[83] Zhang *et al.*[85] found that a zirconium dodecatungstophosphate ($Zr_{0.7}H_{0.2}PW_{12}O_{40}$) catalyst with a nanotube structure exhibited higher water tolerance and acid properties compared to $Cs_{2.5}H_{0.5}PW_{12}O_{40}$ and $H_3PW_{12}O_{40}$ in the transesterification of waste cooking oil. A comparison study between potassium and cesium salts of $H_3PW_{12}O_{40}$ in the transesterification of castor oil was also investigated. In spite of experiencing the formation of a colloidal dispersion during the reaction process, it was found that potassium salts resulted in higher activity compared to cesium salts.[86] Often, the reactions involving supported HPA catalysts were found to be affected by leaching of the active phase, as a result of the weak bonding between HPA and the support surface, especially for reactions involving a polar solvent.[87] Oliveira *et al.*[88] discovered a minor leaching of the zirconia-supported HPW catalyst (8 wt% relative to the initial mass of catalyst loading) during the esterification of oleic acid with methanol. Xu *et al.*[89] claimed that the combination of tantalum oxide and tungsten can create a strong affinity and this could potentially reduce the leaching of active species. However, the outcome showed only a moderate catalytic activity in the transesterification of tripalmitine and the catalyst was easily poisoned by other species over the active sites.

5.2.1.4 Sulfated Metal Oxides

Studies on the transesterification of vegetable oils using sulfated metal oxides have been carried out progressively. The structure of the sulfated metal oxide catalyst is a combination of a sulfate group with metal oxide centers. These catalysts were found to show excellence performance in the esterification and transesterification of triglyceride-containing feedstocks. Catalysts containing sulfate anions such as $SO_4^{=}/Al_2O_3$, $SO_4^{=}/TiO_2$, $SO_4^{=}/ZrO_2$, $SO_4^{=}/SnO_2$ and $SO_4^{=}/V_2O_5$[90] can act as acidic systems.[91,92] Such solid acids are usually prepared by impregnating the hydroxides from ammonia precipitation of the corresponding metal salt solutions with aqueous sulfuric acids followed by calcination. In addition to acid amount and acid strength adjustment, the catalysts are also active in a heterogeneous liquid–solid system, being recoverable and reusable. Jitputti *et al.*[93] reported the use of sulfated zirconia in the transesterification of crude palm oil and coconut oil with methanol. It was found that sulfated zirconia produced 90.3 and 86.3 wt% FAME from crude palm kernel oil and crude coconut oil, respectively. However, this catalyst deactivated after its first run due to catalytic site poisoning and leaching of sulfated ions from the porous support, resulting from the hydrolysis process. The sulfate group can be lost as H_2SO_4 or HSO_4^-,[94] which leads to a homogeneous reaction, interfering with the heterogeneous activity. The same findings, *i.e.* deactivation and leaching of sulfated zirconia catalyst, were reported by other authors.[90,95]

The preparation method for the sulfated zirconia catalyst was reported to have a significant effect on the catalytic performance.[90] Indeed, conventional sulfated zirconia displayed a very low catalytic activity for the transesterification of soybean oil, whereas sulfated zirconia that was synthesized using a free-solvent method achieved approximately a 99.8% conversion after 1 h reaction. However, the catalyst deactivated rapidly with only 59% conversion after being re-used once, which dropped to a 39% conversion after being re-used twice. Several studies also focused on the improvements of the catalyst preparation method to overcome the leaching problem. Yadav and Murkute[96] found a new route to prepare $SO_4^{=}/ZrO_2$ catalyst with a higher sulfate loading and resistance to leaching. It was done by dissolving a chlorosulfonic acid precursor in an organic solvent, instead of the conventional impregnation of sulfuric acid. This novel catalyst showed better catalytic activity than the conventional prepared $SO_4^{=}/ZrO_2$; no leaching was observed and it was able to retain its activity for further reutilization. Similar work done by Fu and co-workers[97] observed the stability of sulfated zirconia in the transesterification of waste cooking oil and yellow grease with methanol. They evidenced negligible loss of acid sites by analyzing the remaining sulfur content in the used catalyst. Thus, they concluded that no sulfate groups leached from the support throughout the process.

Another superacidic sulfated metal oxide that has been used in the transesterification reaction is sulfated titania. Although it was claimed to be less active than sulfated zirconia, the catalytic activity appears to be a linear function of the surface area. A novel silica supported sulfated titania was used in the simultaneous esterification and transesterification of oleic acid-refined cotton seed oil mixture with methanol.[98] The catalyst achieved a very high catalytic activity, yielding over 90% FAME at 473 K, 3 wt% catalyst loading and 9 : 1 methanol to oil molar ratio. Only small catalyst activity decay was found after four consecutive reutilization runs with the used catalysts. Melero *et al.*[99] found that the sulfated titania was proven to have higher stability compared to sulfated zirconia, due to several possible reasons: (1) a stronger interaction between titania and sulfate groups and (2) the presence of silica as the stabilizing agent. Still, sulfated tin oxide exhibited interesting performance in the production of biodiesel, owing to the higher acid surface compared to sulfated zirconia.[100] The mixed system of alumina/silica with sulfated tin oxide was seen to increase the catalytic activity in the transesterification process of waste cooking oil, by achieving a 92.3% FAME yield. Nevertheless, because of the high cost and difficulties posed during the catalyst separation process, the sulfated tin oxide catalysts become less favourable in industrial scale production. On the whole, specific surface area, pore size, pore volume and active site concentration on the surface of the catalyst are effective factors in catalytic activity. Stability of the active sites is important for their industrial applications.

5.2.2 Heterogeneous Base Catalysts

The transesterification of vegetable oils or animal fats to biodiesel by chemical catalysts, especially in the presence of a strong basic solution, such as sodium hydroxide and potassium hydroxide, has been widely used in industrial production of biodiesel. Such basic solutions can transform triglycerides to their corresponding FAMEs with higher yields at lower temperatures and shorter times than those by acid catalysts. However, separating the catalysts from the products is technically difficult. Moreover, natural vegetable oils and animal fats usually contain small amounts of FFAs and water, which can have significant negative effects on the transesterification of glycerides with alcohols, and also hinder the separation of FAMEs and glycerol due to saponification of FFAs. Compared with basic solutions, solid base catalysis is preferred due to easy separation. Detailed investigations of different types of base heterogeneous catalysts such as rare earth oxides, various base metal compounds supported on alumina or zeolite, hydroxides, alkoxides, hydrotalcites and ion-exchange resins have been carried out by several researchers.[46,101–105] Solid base catalysts are reported to have high catalytic activities but low stability.[39] They refer mainly to solids with Brønsted basic and Lewis basic activity centers that can supply electrons (or accept protons) for (or from) reactants. Heterogeneous base-catalyzed transesterification for biodiesel synthesis has been studied intensively over the last decade. Low-qualified oil or fat with FFAs and water can be used. However, the catalytic efficiency of conventional heterogeneous base catalysts is relatively low and needs to be improved. Heterogeneous base catalysts, such as hydrotalcites, metal oxides, metallic salts, and supported base catalysts, are discussed below.

5.2.2.1 Hydrotalcites

Hydrotalcites (HTs) are a class of anionic and basic clays known as layered double hydroxides. Hydrotalcites consist of positively charged brucite-like layers and interstitial layers formed by $CO_3^{=}$ anions, and water molecules compensate the positive charge resulting from the substitution. Hydrotalcites have gained more interest for the transesterification process due to their capability to be modified, producing catalysts with different basicity and surface area. The modification process can be done either by varying the chemical composition or by adopting different preparation methods. HTs have strong alkali sites and high stability, but they are characterized by low surface area, affecting the catalytic activity. The most well-known hydrotalcites are the Mg/Al-based ones as they appear to have consistent activity in the transesterification process.[106,107] Mg/Al mole ratio and calcination temperature are the determining factors for the base-catalyzed activities. The initial study by Cantrell *et al.*[108] on biodiesel synthesis with HTs indicated magnesium content has obvious effects on catalytic activity. A few studies have been conducted to investigate the best Mg/Al molar ratio for biodiesel application, and a ratio of 3 was claimed to be the best.[108–111] Decomposition of HTs after calcination yields a high surface area of the Mg/Al mixed oxide, which presumably exposes strong Lewis basic sites. During the calcination process, first, the interlayer water is lost, followed by dehydroxylation and decomposition of interlayer carbonate to CO_2, which generate a porous structure and specific surface area ranging from 150 to 300 $m^2\ g^{-1}$.[112] Modification of the calcination temperature also leads to better thermal stability as well as higher catalyst basicity.[113] Liu *et al.*[107] studied the transesterification of poultry fat with methanol using Mg/Al hydrotalcite derived catalysts and they achieved 67% conversion of FAME after 9 h of reaction, with a 15 : 1 methanol oil molar ratio, 333 K reaction temperature, 7.5 wt% catalyst and a molar ratio of 3 : 1 of Mg/Al. Hydrotalcites also showed good results in the transesterification reaction of cottonseed oil (9.5 wt% FFA) and animal fat oil (45 wt% water) with methanol.[114] After 3 h of reaction, both feedstocks resulted in 99% triglycerides conversion in the presence of 1 wt% catalyst, 6:1 methanol to oil molar ratio, 473 K and 1 wt% FFA content. Georgogianni *et al.*[115] have investigated the transesterification of rapeseed oil using homogeneous and heterogeneous catalysts and Mg/Al hydrotalcite shows the highest catalytic activity with 97% FAME conversion. The performance of the catalyst was found to be related to the basic strength of the hydrotalcite sample. Furthermore, Mg^{2+} can be replaced by other cations in the hydrotalcite structure. HTs substituted with copper have a relatively uniform porous structure with decreased specific surface area. For iron substituted HTs, microporosity developed, pore size decreased, while specific surface area increased.

Apart from the impregnation of additional metal in the hydrotalcite structure, substitution of Al^{3+} ions with other metal ions has also been investigated. Macala *et al.*[116] found that by doping Fe^{3+} ions into the Mg/Al structure, they could increase the catalytic activity of the hydrotalcites.

It is known that HTs present a lamellar structure and thus they do not pose accessibility restrictions of vegetable oil molecules to catalyst sites. However, the improvement of specific surface area becomes necessary to obtain high catalytic activity. The main advantage of hydrotalcites, instead, is their capability to operate in the presence of feedstocks with high FFA and water contents. Deng *et al.*[117] synthesized a series of nanosized HTs by a modified co-precipitation method. Variables of temperature, solution pH and ageing time have a strong influence on the final basicity of the mixed oxides. Mg/Al ratio in the precursor HT depends on the basic properties of these sites. In the transesterification experiment using Mg/Al-HT catalysts, 95% biodiesel yield was achieved from *Jatropha* oil in 1.5 h. Pre-mixing of HTs with methanol is essential to optimize catalyst activity to avoid lagging in reaction activity due to mass diffusion. Xi *et al.*[118] tested the influence of water on the activity and stability of activated Mg/Al-HTs. In the presence of a certain amount of interlayer water, Brønsted base sites were active. However, a high degree of hydration caused rapid deactivation of the catalyst. Mg/Al-HT shows relatively robust activity in the presence of water or FFAs, which is an attractive feature for biodiesel production.

5.2.2.2 *Metal Oxides*

Metal oxides are promising materials for base catalyzed transesterification, especially in biodiesel production. Metal oxides are composed of Lewis-acid cations and Brønsted base anions. They are classified as single metal oxides (*e.g.*, MgO, CaO and SrO) and mixed metal oxides [A–B–O type metal oxides, where A is an alkaline-earth metal (Ca, Ba, Mg), alkaline metal (Li), or rare earth metal (La) and B is a transition metal (Ti, Mn, Fe, Zr, Ce)].[119–122] Early studies on heterogeneously catalyzed transesterification were focused on the catalysis by single metal oxides. The basicity of oxides (especially, basic sites) directly depends on reaction rate. A comparison of several metal oxides (MgO, CeO_2, La_2O_3 and ZnO) indicated that the most basic one is La_2O_3, followed by MgO, CeO_2 and ZnO.[123] The order of activity among alkaline-earth oxide catalysts is BaO > SrO > CaO > MgO. CaO is the most frequently applied metal oxide catalyst for biodiesel preparation, due to its cheap price, relatively high basic strength and lower environmental impact. Among all the alkaline single metal oxide catalysts, calcium-derived oxides are claimed to be the most promising catalysts in biodiesel transesterification because they are cheap, easily available, have a long catalyst lifetime and are less toxic. Reddy *et al.*[103] used CaO as a solid base catalyst in the transesterification of soybean oil, reporting a biodiesel yield as low as 2%. However, the intrinsic basicity of nano-CaO is much higher, leading to a 99% biodiesel yield. The appropriate water level in the reaction process was found to be one of the factors that could increase the activity of CaO.[121] Indeed, O^{2-} ions on the catalyst surface can coordinate H^+ ions from water molecules, forming OH^- species that, if combining with other H^+ ions from the solvent (*e.g.*, methanol), produce alkoxide anions, which are among the most favourable species for the transesterification process. Nevertheless, too much water (more than 2.8 wt% of the oil) could result in saponification due to hydrolysis of fatty acids. Other studies showed that a 95% biodiesel yield was obtained using CaO as catalyst in the presence of about 2 wt% water.[118] Several studies have also been conducted to improve the catalytic properties of CaO. Studies on the improvement of CaO activity have been carried out by modifying the surface basicity.[124] Thus, by placing CaO in an ammonium carbonate solution, followed by calcination of the catalyst at 1173 K, a FAME yield of 94% was achieved, operating at a 9 : 1 molar ratio, 1.5 wt% catalyst loading and 343 K. The hydration–dehydration approach has also been used by Yoosuk *et al.*[125] to create a simple and flexible calcined catalyst. This new catalyst was found to have a higher surface area and a higher amount of basic sites. The FAME yield increased from 75.5 to 93.9% using the conventionally calcined and novel improvised CaO, respectively. Bai *et al.*[126] focused on the improvement of catalyst morphology by developing a highly porous microsphere of CaO to increase the catalytic performance. This study successfully reached nearly 99% FAME conversion. The main problem in the presence of CaO was the leaching of the active sites from the catalyst surface. This condition leads to the formation of homogeneous $Ca(OCH_3)_2$ catalysts, thus involving a pseudo-homogeneous reaction. SrO has basic sites stronger than H_0 = 26, making it suitable for applications in the transesterification reactions. Liu *et al.*[120] claimed a high catalytic activity of SrO to convert soybean oil into biodiesel with a yield over 95% at a temperature below 70 °C for 30 min. MgO has also shown a good catalytic activity for biodiesel synthesis. MgO has weak basic strength and low solubility in methanol. It is usually produced by direct heating of magnesium carbonate or magnesium hydroxide. Veljković *et al.*[127] obtained approximately a 98% biodiesel yield using MgO as catalyst. For this catalyst, the FAME yield is largely dependent on the methanol to oil molar ratio as well as the reaction temperature.[128]

Several studies have also been conducted using other types of metal oxides. An ultrasonic-assisted transesterification with different types of alkali oxide catalyst (CaO, SrO and BaO) has been carried out by Mootabadi *et al.*[129] They found that BaO gives the highest yield (95%) within an hour. However, the catalytic activity dropped during the subsequent reusability study, due to catalyst dissolution during the reaction process. The capability of transition-metal oxides as biodiesel catalysts, like zirconium oxide (ZrO_2) or zinc oxide (ZnO), also suggests huge promise, due to their acidic properties. Jitputti *et al.*[93] examined the performance of zinc oxide and zirconium oxide during the methanolysis of palm oil under supercritical conditions (50 bar under nitrogen atmosphere, 473 K). The FAME yield was found to be 86.1% for zinc oxide and 64.5% for zirconium oxide after 1 h of reaction time, using 3 wt% catalyst and a 6 : 1 methanol to oil molar ratio.

To increase the basic strength of a single metal oxide, mixed metal oxides have been synthesized too. The introduction of the third lattice within the original structure of the single metal oxide increase the catalyst's durability and stability. Kawashima *et al.*[119] studied the combination of A–B–O type metal oxides, where A is an alkaline-earth metal, alkaline metal or rare earth metal, such as calcium, barium, magnesium, or lanthanum, while B is a transition metal, such as titanium, manganese, iron, zirconium or cerium. Catalytic activity tests were performed for CaO–TiO_2, CaO–MnO_2, $(CaO)_2$–Fe_2O_3, CaO–ZrO_2 and CaO–CeO_2 samples, obtaining approximately 90% biodiesel yields.[119] Furthermore, active sites of metal oxides are easily blocked by adsorbing intermediates (diglyceride, monoglyceride) or products. Deactivated catalysts can be recovered to nearly the initial value through calcination.

Frusteri *et al.*[130] assessed the activity and selectivity of $MnCeO_x$ and $MnZrO_x$ systems, prepared *via* a redox-precipitation route, in biodiesel production by sunflower oil transesterification reaction with methanol. Weak Brønsted basic sites were indicated as not active in methyl ester formation, whereas medium and strong basic sites were reported as decisive for the activation of the methanol molecule. Anyhow, only a proper balance of Lewis acid and basic sites was claimed to define the catalytic performance of the investigated systems and, in synergy with active sites accessibility, it controls the conversion of the TG molecules and accordingly the methyl ester formation.

Supported materials have been widely used in transesterification and esterification catalysis due to their availability to enhance the performance of the catalyst, mainly in terms of stability and durability. Sodium, potassium, lithium, barium and magnesium were frequently used either in the ionic form of halide, hydroxide, carbonate, or nitrate or in the metallic form. These ionic and metallic metals were impregnated on various metal supports, such as Al_2O_3, ZnO, MgO and CaO, to develop various types of metal supported catalysts. In the attempt to enhance the catalytic activity of CaO, Yan and co-workers[131] prepared several systems by impregnation of CaO on various metal oxide supports, such as manganese (MgO), silica (SiO_2), alumina (Al_2O_3) and zeolite HY. The best condition was obtained with the CaO/MgO system and the optimum conversion was *ca.* 92% at 337.5 K. However, the catalyst activity dropped after the fourth run of reutilization due to the blockage of the active sites. Macedo *et al.*[132] examined the alcoholysis of soybean oil with Al_2O_3–SnO and Al_2O_3–ZnO as catalysts. Under the optimum conditions, an 80% yield of FAME was achieved after 4 h of reaction time, with 5 wt% catalyst at 333 K. The main difficulty faced with this type of catalyst is the catalyst dissolution in the reaction medium. In most cases, catalysts suffer from leaching of active phase, thus complicating the separation process due to the occurrence of pseudo-homogeneous reactions.

5.2.2.3 *Metallic Salts*

Inorganic solid bases, such as sodium silicate,[133] vanadyl phosphate,[134] calcium zincate[135] and calcium methoxide,[136] are low-cost and easy-to-use heterogeneous catalysts. Reports on metallic salt catalyzed conversion in biodiesel preparation are rare. Here, only sodium silicate, vanadyl phosphate (VOP) and calcium zincate are reviewed. Sodium silicate was used as the starting material to synthesize β-zeolite, NaY zeolite, and NaX zeolite. Guo *et al.*[133] used sodium silicate to catalyze the transesterification reaction for the first time. It catalyzed soybean oil to biodiesel with a yield of almost 100% under the conditions of 3.0 wt% sodium silicate, a molar ratio of methanol/oil of 7.5 : 1, a reaction time of 60 min, a reaction temperature of 60 °C, and a stirring rate of 250 rpm. In addition to high catalyst activity, sodium silicate also has other similar characteristics to supported solid base catalysts. Most of the basic sites were in the interior of the solid catalyst due to the low surface area and high density of the basic sites. The calcined sodium silicate could tolerate 4.0 wt% water or 2.5 wt% FFAs contained in soybean oil. The water tolerance is related to its special crystal and porous structure.

As a result, Si–O–Si bridges are hydrolyzed and H_4SiO_4 monomers are sequentially released. Such a series of reactions not only produces OH^-, but also avoids the formation of soap. Furthermore, sodium silicate could also be used to catalyze the dehydration of glycerol. Long *et al.*[137] used sodium silicate as the catalyst for transesterification of rapeseed oil for several recycles, and subsequently, the used sodium silicate without any modification catalyzed the hydrothermal production of lactic acid from glycerol at 300 °C. A yield of 80.5% lactic acid and only minor amounts of formic, acetic acid and acrylic acid were produced. Previous applications of VOP were mainly in hydrocarbon oxidation, dehydration and isomerization.[138] Di Serio *et al.*[134] confirmed that VOP-based catalysts were very active in the transesterification of vegetable oil with methanol despite their low specific surface area. VOP was deactivated due to a progressive reduction of vanadium (V) species from V^{5+} to V^{4+} and V^{3+} by methanol. Because the deactivation is reversible and catalyst activity can easily be restored by calcination. Rubio-Caballero *et al.*[135]

investigated the use of calcium zincate in the methanolysis of sunflower oil for biodiesel production. The activated calcium zincate at 400 °C is stable against lixiviation, attributed to its strong interaction with a much less soluble zinc oxide. But, calcium zincate is more sensitive to water (>0.2 wt%) than FFAs. Calcium methoxide has a moderate surface area, relative broad particle size distribution, narrow pore size distribution, strong basicity, long catalyst lifetime and better stability in organic solvent.[136] It has tremendous potential to replace some homogeneous catalysts.

5.2.2.4 *Supported Base Catalysts*

Alkali metals (Li, Na, K) and alkaline-earth metals (Mg, Ca, Ba) are the most common sources of super basicity, and they have been selected as the active species of supported catalysts for biodiesel synthesis. They are frequently used in the metallic form or as various ionic forms of hydroxide, halide, carbonate and nitrate, such as K^+, Li^+, La^{3+}, KOH, NaOH, KF, K_2CO_3, and KNO_3.[57,139,140] Alumina, silica, zinc oxide, zirconium oxide and zeolites were used as supports for these catalysts. Surface basicity is the primary determinant of catalyst activity, then the specific surface area and pore volume.[139] During the preparation of such catalysts, the mechanical intensity and surface area of carriers can be adjusted to obtain different basic intensities and activity sites. Almost all supported base catalysts were synthesized *via* loading of active species on carriers by covalent bonds, ionic bonds or physical adsorption. Despite the formation of the M–O–carrier (*e.g.*, Al–O–K, Si–O–Na and Ca–O–K), other possible interactions of the alkali species with supports include formation of solid solutions and acid–base reactions. Hydroxyl groups introduced to the surface of solids play an important role in the transesterification reaction.[141] The hydroxyls with alkali species enhance the catalytic activity. As the most popular carrier, Al_2O_3 has almost all noteworthy properties such as high temperature resistance, high surface area, high porosity, low density and a transition crystalline phase existing in a wide temperature range. Furthermore, it serves as a carrier with both a solid acid and base. Most super basicity sources can be well dispersed on the Al_2O_3 support in the form of a monolayer at a low loading. Furthermore, alumina is more resistant than other supports (*e.g.*, SiO_2, CaO and zeolite) to alkali species. Taking KNO_3/Al_2O_3 as an example, it is usually prepared by impregnation and subsequent calcination at 500 °C.[141] K^+ ions replaced protons of isolated hydroxyl groups to form Al–O–K groups. The Al–O–K groups and K_2O derived from KNO_3 are active basic species. The base strength could be tentatively denoted as $15 < H_0 < 18.4$ by using the Hammett indicator. Basic strength of KNO_3/Al_2O_3 was influenced by KNO_3 loading and temperature. The 35% KNO_3/Al_2O_3 sample calcined at 500 °C had the highest basicity. However, the sample prepared at 700 °C was most stable. Some of the potassium species are lost during a solid–solid reaction leading to formation of spinels or penetration into the subsurface. Aends and Sheldon[142] indicated that such a kind of catalyst is unstable during reaction, mainly due to M–O–Al decomposing in the presence of methanol. Arzamendi *et al.*[143] confirmed that NaOH reacted with the support to form aluminates during preparation of $NaOH/Al_2O_3$. Leaching of sodium species from Al_2O_3 was also found. Furthermore, problems of high cost, difficult preparation and easy poisoning by absorption of H_2O and CO_2 should be solved. The supported solid base catalysts are excellent for transesterification of triglyceride, but a higher temperature is needed.

5.3 Hydrotreating of Bio-oils

The hydrotreating of vegetable oils as an alternative to transesterification with methanol was developed to overcome the limits in terms of oxidative stability, solvency, energy density, distillation curve, density, H_2O solubility and in part cold properties shown by fatty acid methyl esters (FAMEs), *e.g.* biodiesel obtained by vegetable oil transesterification. More specifically, some issues dealing with FAMEs are the following:

- Corrosion issues, in particular when distributed in the existing pipeline network. Therefore, it should be transported *via* rail or truck, resulting in higher costs and higher energy consumption for transport.
- FAME density is higher than most diesel specifications allow, with implications in the type of components that may be added to the diesel pool.
- FAME shows low blend limits in diesel (about 5–7 vol%), due to stability and decomposition issues.

However, the main limit of FAMEs is the difficult integration within current refinery schemes (currently all production of FAMEs is done by small companies, and mostly with old 1st generation technologies), while the hydrotreating of vegetable oils may be suitably utilized, for example, in the conversion of old refineries to new "green" refineries, as applied by eni for their refinery in Porto Marghera, Venice (Italy) (about 500 kt per year production) and planned also for their refinery in Gela (Italy).[144]

For these motivations, Neste Oil and UOP/eni developed in parallel the hydrotreating process of vegetable oils, the NEXBTL and Ecofining™ technologies, respectively. Some other alternative processes exist, such as the Vegan™ technology developed by IFP Energies Nouvelles and commercialized by Axens, but the first two are the main technologies used. Often, hydrotreated vegetable oils are indicated with the acronym HVOs, or in general with the term Green Diesel.[145–151] Sometimes, the term Hydroprocessed Esters and Fatty Acids (HEFA) is also used.

Green Diesel (HVO) can be produced from different types of vegetable oils and fats. Therefore, a higher flexibility in feedstock with respect to transesterification processes is expected. It is possible, in fact, to feed triglycerides and fatty acids from vegetable oils, (*e.g.*, rapeseed, soybean and corn oil), tall-oil (a co-product from the pulp and paper industry) and animal fats. However, hydrogen is necessary to remove the oxygen from the triglyceride in the HVO process and produce a mix of linear alkanes, CO_2 and water. While methanol should be fed in the transesterification process with production of raw glycerol as the byproduct (with the need of further expensive purification for most of the uses), propane is the main byproduct of the HVO process. The alkane can find various uses inside the refinery, differently from glycerol. However, consumption of H_2 is a major issue in the HVO process.

After the deoxygenation step, the product is hydroisomerized, in order to branch the linear chains for improving the cold flow properties of the final products. The HVO process thus give rise to a mixture of paraffinic hydrocarbons, free of sulfur and aromatics, and with a very high cetane number. HVO offers a number of benefits over FAMEs, such as reduced NO_x emissions, better storage stability and better cold properties. In addition, it can be used in diesel engines without blend walls or the modifications required for biodiesel. Another key point is that jet biofuel, bionaphta and biopropane can be co-produced.

Furthermore, it should be noted that there are a number of companies co-processing vegetable oils with fossil fuels, like Cepsa (Spain), Preem (Sweden) and Repsol (Spain). Several Companies (Neste Oil, BP, PetroBras, Syntroleum/Tyson Foods, ConocoPhilipps, Haldor Topsoe, Nippon Oil, Axens and eni/UOP) have developed proprietary technologies for HVO to green diesel and green jet. Neste Oil is the leading renewable diesel producer in the world, with four facilities producing high-quality renewable diesel with a total capacity of 2 Mt per year. The feed is mainly vegetable oils, especially crude palm oil, but its amount is progressively decreasing with increasing use of waste and residue-based raw materials including animal and fish fats, used cooking oil and various residues generated during vegetable oil refining, such as palm fatty acid distillate (PFAD) and technical corn oil. These raw materials accounted for 62 percent of Neste Oil's renewable inputs in 2014 (52 percent in 2013, 35 percent in 2012).

Petroleum diesel (baseline) requires 1.27 MJ of input energy to yield 1 MJ of diesel fuel. FT syndiesel requires higher energy inputs throughout the life cycle, regardless of feedstock, even if natural gas FT syndiesel is better than coal syndiesel. Rapeseed oil biodiesel and green diesel both have a slightly lower CED than FT syndiesel. When palm oil is used (still the main vegetable oil used for green diesel), the CED is substantially higher than for petroleum diesel. The FED of petroleum diesel is very close to the CED value, while coal and natural gas syndiesel have significantly higher FED values compared with petroleum diesel. All of the biofuels have very similar FED values. Therefore, the use of renewable diesel leads to higher total energy consumption, but lower demand in fossil fuel consumption. As a result, greenhouse (GHG) emissions are lower than for petroleum diesel. Overall, green diesel emits less GHG than biodiesel for all feedstocks. Oil palms are responsible for fewer emissions in cultivation compared with rapeseed.[152,153]

5.3.1 Catalysts for Vegetable/Algal Oil Hydroconversion

The catalyst characteristics are one of the most critical factors determining the yield and composition of liquid products, such as naphtha (C_5–C_{10}), jet fuel (C_{11}–C_{13}), diesel (C_{14}–C_{20}) and liquid petroleum gas (LPG), in the hydroprocessing of triglycerides. A severe hydrocracking catalyst produces a higher amount of naphtha, whereas a hydroisomerization catalyst produces mainly diesel. The severity of operations (mainly the reaction temperature) also determines the yield and quality of hydroprocessed oils, but also the stability of the operations. In general, diesel selectivity decreases with increasing reaction temperature, becoming more relevant to the hydrocracking activity. High temperatures and strong acid catalysts maximize naphtha, while moderate temperatures and catalysts with mild acid sites produce middle distillates (diesel).[154] Intermediate reaction conditions and catalyst acidity would maximize the jet fraction, which is the most valuable, but to effectively improve this characteristic, catalysts with selective hydrocracking properties have to be developed.[155–157]

The yield of gasoline can also be increased by using a two-step process, *i.e.* hydrotreating followed by hydrocracking. In the first step, oxygen is removed from biomass as water. The deoxygenated product is then separated by distillation, and the heavier fractions are further brought to the second step to convert them into (lighter) molecules within the boiling range of naphtha. Gasoline is maximized using NiMo/zeolite catalysts at 300–320 °C. Egia *et al.*[158] reported, for example, the hydrocracking properties of Ni, Mo and Ni–Mo sulfides supported on an ultrastable HY zeolite. Their cracking conversions are significantly higher than that of the HY support. The strong acid sites in the zeolite enhance the cracking of the large molecules in the vegetable oil into naphtha; however, middle distillates (C_{13}–C_{19}) were also obtained. Hydrotreatment of jatropha oil over sulfided NiMo/SiO_2–Al_2O_3 and NiMo/ZSM-5–Al_2O_3 catalysts[159] evidenced that the rates of decarboxylation and/or decarbonylation increased with increasing Si/(Si + Al) ratio, while the ratio of hydrodeoxygenation decreased with increasing Si/(Si + Al) ratio. These results evidence that the total acidic sites of the catalyst have a positive effect on the decarboxylation and/or decarbonylation pathways. NiMo/SiO_2–Al_2O_3 catalysts showed a much higher isomerization/cracking ratio than NiMo/ZSM-5–Al_2O_3 catalysts, likely because the isomerization reaction is favourable for weak and middle acidic sites while cracking reactions are favourable for strong acidic sites.

In general, in these reactions, nanoscale intimacy in bifunctional catalysts plays a relevant role. Zecevic *et al.*,[160] however, recently remarked that the so-called intimacy criterion, *e.g.* the maximum distance between the two types of sites (metal and acid sites) is not true. The authors showed that for a bifunctional catalyst comprising an intimate mixture of zeolite Y and alumina binder, and with platinum metal controllably deposited on either the zeolite or the binder, the closest proximity between metal and zeolite acid sites can even be detrimental. They showed that the cracking selectivity of large hydrocarbon feedstock molecules for high-quality diesel production is optimized with the catalyst that contains platinum on the binder, *i.e.*, with a nanoscale rather than closest intimacy of the metal and acid sites. Thus, cracking of the large and complex hydrocarbon molecules, as for vegetable oil or algal oil, should benefit especially from bifunctional catalysts that avoid locating platinum on the zeolite (the traditionally assumed optimal location). Spatial organization of different active sites at the nanoscale will benefit the further development and optimization of the emerging generation of multifunctional catalysts. While the results are interesting, they are not fully convincing under more optimized catalysts (with the right balance between metal and acid sites) and practical reaction conditions.

Various non-sulfide zeolite catalysts, such as noble metals supported on ZSM-5, HZSM-5, SAPO-11, beta- zeolite, SBA-15 and mesoporous-Y zeolite, also showed considerable activity for biofuel conversion.[161] A catalyst with high surface area and pore volume tends to be more active. Zeolite X, HY and faujasite-type zeolites are large-pore zeolites that show good activity for the transesterification reaction. However, zeolites with a small pore volume are better for esterification, but diffusion is limited; therefore, this type of system is not good for the transesterification reaction. H^+-Y type zeolites possess a large-pore structure and contain high concentrations of active acid sites. This type of catalyst can depolymerise the biomass containing lignin into monomers and dimers by breaking the C–O–C bonds, and it could also link the monomers into dimers *via* alkylation or dimerization reactions, which is essential for the conversion of C_7–C_{18} to jet fuel range hydrocarbons. However, to increase the catalytic activity and acid strength, incorporation of non-sulfide metals in the zeolite results in an increased yield of jet fuel.

The diesel obtained directly by hydrodeoxygenation is mainly composed by C_{17} and C_{18} linear alkanes that have a high cetane number but poor cold flow properties as these hydrocarbons have a high melting (freezing) point between 20 and 28 °C. Therefore, a second step of selective hydroisomerization and cracking is necessary.[162] In this process, not only quality-improved diesel is obtained but also jet fuel, as the acid sites of the catalyst promote the conversion of larger alkyl chains into smaller chains.

Hydroisomerization is a key parameter to obtain methyl-branched alkanes. Shape-selective catalysts such as zeolites or other acidic supports are necessary. Normal alkanes in the boiling point range of diesel have generally a higher cetane number than that of their branched isomers. Conversely, iso-alkanes have lower freezing points than *n*-alkanes. Thus, an optimal intermediate situation exists. Hancsók *et al.*[163] and Simacek *et al.*[164] have studied the influence of the reaction temperature on the cetane number, which showed an inverse correlation. Hydroisomerization is desirable for green diesel to have a lower freezing point though with lower cetane numbers. The hydroisomerization of a green diesel, for instance, containing mainly *n*-C_{17} and *n*-C_{18} alkanes and with a cetane number close to 100 and a freezing point close to 20 °C may produce a green diesel with a cetane number close to 70 and a freezing point lower than −5 °C.

Once green diesel has an improved performance for cold weather, it may be used directly in vehicles or it can be mixed with petro-diesel. In the case of jet fuel, it has been mixed at 50% with petroleum-derived jet fuel or other percentages in one turbine and used in some documented commercial flights. The green jet fuel produced by the UOP/ENI process has been proved to meet all the main properties of petroleum-derived aviation fuels such as flash point, cold temperature performance, and good air oxidation stability.

The studies regarding one-step processes in which a catalyst is able to enhance both hydroisomerization and cracking are scarce. Ni/beta catalysts, particularly when a desilicated beta zeolite is utilized to introduce mesoporosity, allow a fine control of the hydroisomerization/hydrocracking properties in *n*-hexadecane conversion, used as a model compound.[165] Moreover the introduction of mesoporosity by controlled desilication improves activity, while maintaining a good selectivity, increasing the potential use in the production of biofuel from vegetable/algal oil, or the upgrading of biomass-derived FT-waxes. In addition, the desilication treatment improves the stability with respect to carbon formation. Chica *et al.*[166] showed that by varying the level of dealumination in zeolite (ITQ-6) materials, it is possible to change the hydroisomerization to hydrocracking ratio of long-chain alkanes. The ratio can be tuned by controlling the swelling time.

5.4 Biomass Thermochemical Conversion to Liquid Fuels

We may schematically differentiate between two primary biorefinery platforms: sugar and thermochemical. In the latter, biomass is converted into synthesis gas through gasification, or into bio-oils through (fast) pyrolysis and hydrothermal conversion (HTC). Bio-oils should be further upgraded to liquid fuels *via* catalytic treatment.[167] Other alternatives of thermochemical conversion processes include liquefaction and high-pressure hydroconversion.[168] Although the latter gives promising results, most of the studies are focused on pyrolysis bio-oil upgrading. Pyrolysis achieves thermal decomposition of biomass in the absence of oxygen to derive renewable oil, gas and char. Pyrolysis bio-oil is very acidic, viscous, thermally unstable and contains high amounts of oxygenated compounds, up to 40 wt%.[169,170] Attempts have been made to introduce catalysis during the pyrolysis step,[171,172] but they failed in general to reach practical implementation due to fast catalyst deactivation.

Pyrolysis is able to convert >65–70 wt% of the biomass into a crude bio-oil, which can be transported to centralized refineries for the production of transportation fuels and chemicals, although stability and corrosion of pyrolysis bio-oil remain an issue. The oil resulting from biomass pyrolysis

Table 5.1 Example of characteristics from typical wood-derived pyrolysis oil.

Characteristics	Value
C	46 wt%
H	7 wt%
N	<0.01 wt%
O (Balance)	47 wt%
Water content	25 wt%
Ash content	0.02 wt%
Solids content	0.04 wt%
Density	1.2 g l^{-1}
LHV MJ kg^{-1}	16 MJ kg^{-1}
pH	2.9
Kinematic viscosity (40 °C)	13 cSt

is a dark liquid with an elemental composition very similar to that of the biomass. These oils may contain up to 10–25 mass% water, together with small amounts of solid carbon and ashes. The bio-oil contains several types of species (such as acids, aldehydes, ketones, carbohydrates, furans, pyrans, aromatics, hydrocarbons). The density of the liquid is about 1170 kg m^{-3}, which is denser than fuel oil and significantly denser than the original biomass. The higher heating value (HHV) of pyrolysis oil is 16–23 MJ l^{-1} compared to fossil fuel, which is 37 MJ l^{-1}. Pyrolysis oil has a low pH value of around 3. Due to the large number of oxygenated components present, the oil has a polar nature and does not mix readily with hydrocarbons. Table 5.1 reports an example of properties from typical wood-derived pyrolysis oil.

The catalytic hydrotreating aims to improve product quality without appreciable alteration of the boiling range. It is a widely-used technology for converting oil heavy fractions. Bio-crude may in principle be processed by a conventional refinery (co-feeding) to increase yields, with the oxygen in bio-oils removed *via* hydrotreating. However, when using conventional hydrotreating catalysts, such as sulfide $CoMo/Al_2O_3$ or $NiMo/Al_2O_3$, there are severe problems of deactivation. Hydrotreating requires mild conditions, while the yield of bio-oil is relatively low. The process also produces a large amount of char, coke, and tar, which will also result in catalyst deactivation and reactor clogging. By introduction of acidic functions in the catalysts, together with hydrotreating, hydrocracking and hydroisomerization occur. These combined processes are thus enabled with dual-function catalysts, in which silica–alumina (or zeolite) catalysts provide the cracking/isomerization function, and platinum and tungsten oxide catalyze the reactions, or nickel provides the hydrogenation function. Alumina is by far the most widely-used support. Usually higher temperatures (>350 °C) and pressures (1000–2000 Psi) than for hydrotreatment are necessary, but larger amounts of light products are formed. H_2 consumption is also higher.

5.4.1 Catalysts for Pyrolysis Oil Hydroconversion

Due to the high oxygen content in pyrolysis bio-oil, oxygen removal through hydrodeoxygenation (HDO) is a primary objective of hydroconversion. Together with HDO, removal of N and S present in lower amounts together with oxygen also occurs. The removal of oxygen can occur *via* three main routes:[173] decarbonylation (removed as CO), decarboxylation (CO_2), and dehydration/condensation (H_2O). Other reactions will occur during the HDO process, for example: cracking or hydrocracking of long-chain molecules into shorter chain lengths, dealkylation, dealkoxylation, methyl transfer, and hydrogenation among others.[174,175] In aromatics, deoxygenation occurs through direct hydrogenolysis of the $C_{aromatic}$–O bond (indicated also as direct deoxygenation) or an initial ring hydrogenation followed by deoxygenation (faster process) to yield a cyclohexyl product.[174]

Removing oxygen as CO or CO_2 will result in a higher loss of carbon, reducing the yield, but reducing H_2 consumption. Removing oxygen as water will retain carbon, but at the expense of hydrogen, except when unsaturated products (alkenes and aromatics) are produced. The alkenes, however, induce instability in the biofuel, but can be eventually separated and utilized as chemicals.

In HDO, the relative strength of the C–O bonds is the main issue. The bond dissociation energies range from 80–90 kcal mol^{-1} for R–OR or –OH, to 100–110 kcal mol^{-1} for $C_{aromatic}$–OH or –OR and 180 kcal mol^{-1} for R=O.[173] Thus, aromatic C–O bonds and carbonyls are difficult to eliminate. If the aromatic ring is first saturated with hydrogen, then C–O bond cleavage becomes more facile. However, hydrogen consumption increases through saturation and the products obtained will be cyclohexyl species rather than aromatics. It results thus in a less convenient process. There is thus the challenge of a catalyst able to break carbonyl C=O bonds and aromatic C–O bonds without the need for ring saturation, for example through assisted hydrogenation mechanisms.

There are two possible strategies for catalytic upgrading of pyrolysis bio-oil through HDO. The pretreated biomass is fast pyrolyzed. In fast pyrolysis, the biomass is rapidly heated at >100 °C s^{-1} to a moderate temperature of 400–600 °C. The liquid yields in bio-oil are typically up to 70–80 wt% on a dry-feed basis with the remainder being gases (mainly CO and CO_2) and solid char. The biomass pretreatments are typically: (i) drying to reduce moisture content, (ii) milling or grinding to reduce particle size, (iii) water or acid washing to remove inorganic metals, or separating the biomass constituents (cellulose, hemicellulose, and lignin). The pretreated feedstock is then pyrolyzed in the presence of an inert heat-carrier gas, hydrogen, and/or eventually a catalyst. The solids and char are then removed through cyclones or hot gas filters with the vapours being swept to the next unit operation.

In the first upgrading strategy, the pyrolysis vapours are condensed into a bio-oil, which can be further processed as it is or preferably after phase separation into nonpolar organic and water-soluble polar fractions. The bio-oil is then catalytically hydroconverted. In the second option, the pyrolysis vapours before condensation are catalytically upgraded. During the pyrolysis step, the possibility of optimization of operative conditions and the absence of the solid components (carbon, tar, ashes) allow a longer life time of the catalysts. Differently from liquid-phase upgrading, the presence of tars in the vapour stream still creates relevant problems of deactivation.

5.4.1.1 Catalytic Upgrading in Liquid Phase

The initial studies on the catalytic upgrading in the liquid phase used a single-stage, higher-temperature (>300 °C) hydrotreatment, resulting in significant coking and heavy tar plugging in the reactor systems.[176] A two-stage hydrotreatment with a lower temperature first step (275 °C) and a second, higher-temperature step (350 °C) using a CoMo catalyst produces a more stable bio-oil with low oxygen content (about 2.3%). However, still, problems of coking and reactor plugging are observed.[177] A pretreatment at low temperatures (80–140 °C) with a Ru/C catalyst allows the time of operation (reduced plug formation) to be improved likely due to a reduction in carbonyls and sugars after the pretreatment. A room temperature and atmospheric pressure hydrogenation using Pd/C catalyst could also be beneficial,[178] because it reduces the pyrolytic phenolics (derived from lignin), which are also one of the causes for reactor plugging due to easy polymerization. The catalysts for the single high-temperature stage can be either a Pd–Re/C catalyst for operations at about 350 °C or a $CoMoS/Al_2O_3$ catalyst for operations at about 400 °C. In a two-stage hydrotreatment, Ru/C for the first stage at about 140 °C can be used and Pd/C can be used for the 2nd stage at about 370 °C. Catalyst fouling is less severe in the second case, especially at low liquid hourly space velocity (LHSV). At higher LHSV, a plugging of the feed line after 24 h TOS (time on stream) can be observed. As an indication of the performances, the red oak-derived phenolic oil feed containing initially about 22 wt% oxygen gave rise to a biofuel containing less than 5 wt% oxygen after the catalytic treatment. About 70–80% of the carbon is retained in the biofuel. A three-step hydroprocess to upgrade a crude white oak bio-oil and the water-soluble fraction from a pinewood bio-oil was also proposed.[179]

An initial two-stage hydrotreatment is introduced, the first at lower temperature (125 °C) over a Ru/C catalyst and the second at a higher temperature (250 °C) over a Pt/C catalyst. The third step uses a ZSM-5 zeolite catalyst. The hydrotreatment steps were carried out under high pressure (10 300 kPa H_2), while the last step is done at atmospheric pressure. The results for the water-soluble fraction were rather interesting: 76 C% hydrocarbon yield (15 C% of C_1–C_6 alkanes, 18 C% aromatics, and 43 C% alkenes), with 24% of the carbon converted into CO, CO_2, and coke. When the whole bio-oil is treated using the Ru/H_2^+zeolite process, the hydrocarbon yield halves, with a parallel increase in the CO_x and coke yields. High-pressure operations, however, increase the costs with respect to the milder conditions indicated above. Furthermore, a phase-separation step is necessary to obtain the water-soluble fraction. An alternative is to use a Ru/C catalyst in the first hydrotreatment stage (170 °C) and a CoMo/C catalyst for the second stage (405 °C), with both catalysts sulfided *in situ* before the reaction.[180] The operating pressure in both stages was 13 800 kPa. For a TOS of 11–14 h, the oxygen content in the product was very low (0.1–0.3 dry wt%), which increases to about 2.4 dry wt% after a TOS of 51 h, likely due to oxidation of the sulfided catalyst.

Further aspects of the process and catalysts for bio-oil hydrotreatments were reviewed by Nolte and Shanks,[173] Zhang *et al.*[181] and Zacher *et al.*,[182] but the above data evidence how the liquid-phase upgrading of bio-oil is still a significant challenge. Multiple stages are required, eventually with an additional phase-separation step. High pressures are needed and the loss of carbon as coke or CO/CO_2 may be relevant, resulting in low oil yields. In addition, some compounds in the bio-oil can polymerize, plugging the hydrotreatment reactors, causing pressure to drop and reactor downtime. The use of noble-metal catalysts may result in expensive catalysts and sulfided catalysts may require a co-feed with H_2S to maintain the sulfide phase and catalyst activity.

5.4.1.2 Catalytic Upgrading in the Vapour Phase

For these operations, conventional refinery catalysts, such as zeolites or other silica–alumina catalysts, could be used, or alternatively novel types of HDO catalysts.

$Na_{0.2}H_{0.8}$ faujasite (FAU) zeolites were found to give excellent HDO activity and hydrocarbon yields in the deoxygenation of pinewood pyrolysis vapours.[183] This catalyst shows an enhanced rate of decarbonylation and decarboxylation, with the higher-hydrocarbon yields being likely aromatics derived from the deoxygenation of the lignin-derived phenolic compounds. For a H-ZSM-5 zeolite, a maximum of 30 C% aromatic yield in cellulose catalytic fast pyrolysis was observed,[183] but this yield strongly depends on the lignin content of the starting biomass. Using switchgrass as the feed, low yields of hydrocarbons (26.8 C%) were produced along with noticeable yields of CO_x and coke (51 C% total). The aromatic yields depend on the pore size of the zeolite, being the highest for pore sizes in the range of 0.52–0.59 nm.[184] The smaller-pore zeolites do not produce aromatics, while larger-pore zeolites lead to a high degree of coking (up to 85 C%). ZSM-5 deactivation also occurred through dealumination of the zeolite framework.[185]

Bifunctional zeolites, containing Ni or Co as metals, were also utilized, for example for the catalytic upgrading of beech-wood pyrolysis oil.[186] Both zeolites reduce the oxygen content of the bio-oil from 40% down to 20–30%, but the total yield of the organic fraction depends on the type of catalyst. Various zeolites and Pt-loaded zeolites were used for HDO in cellulose pyrolysis oil.[187] There is, in general, a rapid deactivation due to coking, which could be removed by co-feeding O_2 and H_2 in the reaction gas mixture. However, the CO_x yield was nearly 50 C%.

Various alkali and alkaline-earth doped amorphous silica–alumina (ASA) catalysts were also investigated.[188] Na/ASA and K/ASA give the highest degree of decarboxylation to form CO_2, whereas Cs/ASA had the highest rate of decarbonylation and dehydration to form CO and H_2O, respectively. However, the degree of deoxygenation was modest in every case. About 8.5% hydrocarbons from the vapour products of pinewood pyrolysis was obtained using ASA-60 (60 wt% Al_2O_3).[189] $Na_2CO_3/\gamma\text{-}Al_2O_3$ produces a liquid with an organic fraction containing nearly 20% hydrocarbons in the HDO of the product vapours in pinewood pyrolysis.[190] However, the main products of the reactions are char and coke (34 wt% total), gas (23 wt%), and water (22 wt%). Therefore, zeolites and other types of conventional hydrotreatment catalysts produce low-to-moderate hydrocarbon yields, with a relatively high loss of carbon in the form of coke and CO_x. Deactivation due to coking or loss of acid sites is also relevant with poor long-term hydrothermal stability. It is thus not surprising that efforts are being made in the search for novel types of catalysts for the hydroconversion of pyrolysis oil, although the studies are mainly limited to model compounds. Ruddy *et al.*[174] summarized the recent activities on the vapour-phase model-compound HDO using transition-metal sulfides, carbides, nitrides, and phosphides, and noble-metal catalysts.

Large activities were found in the screening of HDO novel catalysts, including acids, bases, acid–base bifunctionals, noble metals, and transition-metal catalysts. Typical acid catalysts comprise zeolites or other silica–aluminas, but other acid catalysts were also tested, such as a SO_4^{2-}/ZrO_2 strong acid. In the deoxygenation of cellulose pyrolysis vapours, the strong acid promotes dehydration reactions. Basic catalysts like CaO or MgO promoted instead the formation of low-molecular-weight products. Mixed metal oxide minerals, such as serpentine exhibiting acid–base bifunctionality after treatment with H_2SO_4, give the highest HDO degree. Performances, however, are still relatively poor. Pt/SiO_2 or Pd/ZrO_2 catalysts produce toluene from *m*-cresol,[191,192] but Pd supported on SiO_2, on the contrary, hydrogenates the aromatic ring (3-methylcyclohexanone formation) and Ru/SiO_2 catalyzes the cleavage of C–C bonds, yielding methane. Guaiacol is another common model compound used in the literature. Pt/C gives about 90% conversion of guaiacol at 300 °C, with stable activity, but formation of a range of products was observed: phenol, catechol, and cyclopentanone.[193] The methoxy group of guaiacol undergoes hydrogenolysis to form methane and catechol. Phenol derives from the demethoxylation of guaiacol or dehydration of catechol. A variety of products were obtained also using Ru/TiO_2: dioxygenates (catechol, dimethoxybenzene) to mono-oxygenates (phenol, anisole, cresol) and aromatics (benzene, toluene, xylene).[194] A Ru/TiO_2 catalyst was also used to deoxygenate vapours from the pyrolysis of switchgrass and red oak.[195] A phase-separated liquid was obtained in which the organic fraction was 16.1 wt% of the feed, with relatively low oxygen content. Noble-metal catalysts are thus active in vapour-phase HDO, but performances are still not satisfactory, even if stability is not bad, although obtained mainly using model compounds. Transition-metal catalysts, particularly molybdenum-based catalysts,[196–199] also showed promising performances for biomass HDO. Under mild conditions, molybdenum carbide showed excellent HDO activity with model compounds, whereas molybdenum oxide has been demonstrated to fully deoxygenate the pyrolysis vapours of cellulose, lignin, and complete biomass. Both catalysts yielded alkenes and aromatics, but also hydrogenated products like alkanes and cycloalkanes can be observed. However, most of these studies can be considered to still be at the initial stage. A catalyst modification to reduce hydrogen consumption and saturation of alkenes and aromatics is also required. It may be thus concluded that optimal catalysts for the development of a process for upgrading of pyrolysis bio-oil, both in vapour and liquid phases, have still not been identified, and further effort is necessary.

5.5 Comparison of Technology: Potential of Raw Materials, Costs and Barriers for Large-scale Development

Food waste from the food industry has received the most attention due to its consistent yield, composition, and locality. Food waste from municipalities is from different sources and varies in composition and yield. Currently, both transesterification and hydrodeoxygenation of fats and oils have achieved industrial maturity as methods for production of diesel from biomass. Feedstock price of the oil or fat remain the largest cost in the production. In Europe, rapeseed oil is often used as a feedstock for FAME while soy bean oil is more common in North America and palm oil in the tropical regions of the world, for instance in South-East Asia or Latin America. The transesterification of fats and oils, and esterification of FFA are the most widespread methods. Benign production at ambient pressure at the boiling point of co-reactant methanol yields FAMEs. Industrially, the reaction is based on homogeneous mineral acids and bases as catalysts, which are used batch-wise and discarded after use, and heterogeneous analogues have been vastly investigated. These include solid oxides of basic or acidic character, meso- or microporous materials, organic functionalized resins or other solids. The application of reusable catalysts would cut costs and produce less waste, and allow for a simpler and more economical process design. The deoxygenation of fats and oils can be performed over three types of catalysts, namely supported sulfided transition metals, supported transition metals in metallic form, and porous acidic or basic materials for cracking. The sulfided metals require sulfur and hydrogen in the feed to remain active, so co-treating with hydrodesulfurization feedstock in refineries is an option. The metallic catalysts may formally run without hydrogen, but they require it to avoid deactivating side reactions. Hydrodeoxygenation of fats and oils is still in need of unsulfided catalysts to avoid adding sulfur to the feedstock, and of metal catalysts with a longer lifetime than those previously reported. Deactivation phenomena need to be studied and understood better to avoid them. Cracking of fats and oils generally still gives too poor yields of primary fuels like gasoline, jet fuel, and diesel due to high production of lighter gases and coke.

Fats and oils are sparse resources. The amount of waste oils (*e.g.*, from abattoirs or restaurants) is limited compared to the global diesel consumption, and the land use associated with the growth of vegetable oil resources cannot cope with demand for fuels either – only a minor percentage of the global diesel demand can be covered with the current two technologies based on fats and oils with the available resources in the near- and medium-term future.[200] The advent of algae-based triglyceride production on a large-scale could help to solve this feedstock issue and make it possible to produce triglycerides for biofuels on an industrial scale in an environmentally acceptable manner; however, this is yet to be industrially demonstrated.[201,202]

The growing of crops for biofuel production on productive soil has several drawbacks from an environmental point of view: Switching to growing agricultural crops for biofuels means that productivity of the remaining agricultural production must either be increased or forest or wilderness must be cleared to make room for new agricultural production with severe ecological implications for the invaded ecosystems.[202] Emissions of nitrous oxide (N_2O, a greenhouse gas several orders of magnitude stronger than CO_2)[203] from fertilization is one effect of intensive agriculture, and fertilizer production also emits CO_2. Farming in itself also requires fueling for tractors, *etc.*[203–205]

Thus, only lignocellulosic resources like straw, wood and waste appear to be sustainable feedstocks that are available in sufficient amounts to cover a large fraction of the global fuel demand. Until now, three catalytic strategies for conversion to fuels have been formulated for these feedstocks. Flash pyrolysis of biomass or near-critical hydrothermal upgrading of biomass can yield a type of bio-crude commonly referred to as bio-oil, containing up to 40 wt% oxygen. This oxygen may be removed by deoxygenation, which generally can use the same three types of catalysts that are used when upgrading fats and oils, namely supported metals, sulfided metals or porous materials. The hydrogen consumption and deactivation phenomena may, however, be even more severe and complex than for HDO of fats and oils. The cracking-type catalysts like zeolites appear to be most practical for bio-oils, requiring no hydrogen and allowing for regeneration of zeolites or metal oxides by oxidation by burning away coke formed as by-products.

Carbohydrates and hydrolyzed cellulose can be converted by condensations, reductions and reforming reactions in the aqueous phase.[21]

The reforming reactions can take place at quite moderate temperatures and usually over supported Pt or Pt–Re catalysts, and up to 95% energy efficiency contained in the fuel can be achieved for the integrated process. Furthermore, hydrogen is supplied internally by reforming. Similarly, bio-oils may be processed *via* condensation reactions, while possible alternative reaction paths from the conversion of bio-oil mixtures can be controlled.

The research into bio-oil production, upgrading and aqueous-phase-reforming may therefore play an increasingly prominent role in the future, despite their more challenging nature. The potential ability to use more parts of, for instance, wood or straw residues would make the aqueous phase upgrading even more promising, especially due to its high-energy efficiency and carbon containment in the product.

A final method, in principle possible with all types of biomass, is the complete gasification of the biomass into a producer gas, which can be cleaned into syngas (CO and H_2) and then liquefied as alkanes *via* FTS. Gasification requires high temperatures and expensive process equipment. The gasification products contain tars and impurities of sulfur and nitrogen from the biomass, which must be removed prior to FTS. This gas cleaning is complex to integrate. The consecutive FTS is usually catalyzed by Co and Fe metals, however, activity and selectivity can be vastly modified by varying, promoting or confining the active sites.

Major logistic challenges need to be resolved as well: Biomass is a local resource,[206] while gasifiers are usually huge equipment placed centrally. Regarding the logistics around a gasification fuel economy, some authors suggested that residues, like straw or wood, could be upgraded by flash pyrolysis first in small decentralized plants.[207] Then the bio-oil and char fraction, now with a much higher carbon- and energy density can be transported to and upgraded at central gasification-FTS plants.

Primary and easy-to-upgrade feedstocks like plant oils are not and cannot be available in quantities sufficient to replace the entire diesel oil demand, so waste products from food production, agriculture and forestry hold greater potential for use in the future. It is necessary that the processes for future biofuel production have sufficient tolerance to use diverse raw materials, primarily in the form of waste, as feedstocks.

5.6 Final Remarks

The utilization of biomass-derived heterogeneous catalysts for biodiesel production seems to be a promising choice as it eliminates the noxious problems faced by homogeneous operations. The exploration of biomass or waste as the source of a catalyst may reduce the associated costs for commercially available solid catalysts as well as provide new applications for the waste. However, further investigations and development of biomass-derived catalysts are necessary to improve the catalytic performance for biofuels production.

References

1. J. K. Nørskov, T. Bligaard, J. Rossmeisl and C. H. Christensen, *Nat. Chem.*, 2009, **1**, 37.
2. F. Ma and M. A. Hanna, *Bioresour. Technol.*, 1999, **70**, 1.
3. J. Van Gerpen, *Fuel Process. Technol.*, 2005, **86**, 1097.
4. G. Knothe, J. Van Gerpen and J. Krahl, *The Biodiesel Handbook*, AOCS Press, Urbana, Illinois, USA, 1st edn, 2005.
5. L. Meher, D. Vidya Sagar and S. Naik, *Renewable Sustainable Energy Rev.*, 2006, **10**, 248.
6. M. Mittelbach and C. Remschmidt, *Biodiesel: The Comprehensive Handbook*, Martin Mittelbach, Graz, Austria, 3rd edn, 2006.
7. G. Knothe, *Top. Catal.*, 2010, **53**, 714.
8. I. Kubičková, M. Snåre, K. Eränen, P. Mäki-Arvela and D. Y. Murzin, *Catal. Today*, 2005, **106**, 197.
9. M. Snåre, I. Kubičková, P. Mäki-Arvela, K. Eränen and D. Y. Murzin, *Ind. Eng. Chem. Res.*, 2006, **45**, 5708.
10. G. W. Huber, P. O′Connor and A. Corma, *Appl. Catal., A*, 2007, **329**, 120.
11. G. W. Huber and A. Corma, *Angew. Chem., Int. Ed.*, 2007, **46**, 7184.
12. B. Rozmysłowicz, P. Mäki-Arvela, S. Lestari, O. A. Simakova, K. Eränen, I. Simakova, D. Y. Murzin and T. Salmi, *Top. Catal.*, 2010, **53**, 1274.
13. A. Guzman, J. E. Torres, L. P. Prada and M. L. Nuñez, *Catal. Today*, 2010, **156**, 38.
14. P. Šimáček and D. Kubička, *Fuel*, 2010, **89**, 1508.
15. I. Simakova, O. Simakova, P. Mäki-Arvela and D. Y. Murzin, *Catal. Today*, 2010, **150**, 28.
16. O. V. Kikhtyanin, A. E. Rubanov, A. B. Ayupov and G. V. Echevsky, *Fuel*, 2010, **89**, 3085.
17. F. A. Twaiq, N. A. M. Zabidi, A. R. Mohameda and S. Bhatia, *Fuel Process. Technol.*, 2003, **84**, 105.
18. P. Tamunaidu and S. Bhatia, *Bioresour. Technol.*, 2007, **98**, 3593.
19. J.-G. Na, B. E. Yi, J. N. Kim, K. B. Yi, S.-Y. Park, J.-H. Park, J.-N. Kim and C. H. Ko, *Catal. Today*, 2009, **156**, 44.
20. D. Simonetti and J. Dumesic, *Catal. Rev.: Sci. Eng.*, 2009, **51**, 441.
21. G. W. Huber, J. N. Chheda, C. J. Barrett and J. A. Dumesic, *Science*, 2005, **308**, 1446.
22. J. N. Chheda, G. W. Huber and J. A. Dumesic, *Angew. Chem., Int. Ed.*, 2007, **46**, 7164.
23. J. C. Serrano-Ruiz, D. Wang and J. A. Dumesic, *Green Chem.*, 2010, **12**, 574.
24. J. R. Rostrup-Nielsen, *Catal. Rev.: Sci. Eng.*, 2004, **46**, 247.
25. J. R. Rostrup-Nielsen, *Science*, 2005, **308**, 1421.
26. W. Zhang, *Fuel Process. Technol.*, 2010, **91**, 866.
27. M. Stöcker, *Angew. Chem., Int. Ed.*, 2008, **47**, 9200.
28. T. R. Carlson, J. Jae, Y.-C. Lin, G. A. Tompsett and G. W. Huber, *J. Catal.*, 2010, **270**, 110.

29. J. Wildschut, F. H. Mahfud, R. H. Venderbosch and H. J. Heeres, *Ind. Eng. Chem. Res.*, 2009, **48**, 10324.
30. E. Laurent and B. Delmon, *J. Catal.*, 1994, **146**, 281.
31. S. Zhang, Y. Yan, T. Li and Z. Ren, *Bioresour. Technol.*, 2005, **96**, 545.
32. B. Donnis, R. G. Egeberg, P. Blom and K. G. Knudsen, *Top. Catal.*, 2009, **52**, 229.
33. D. Y. C. Leung, X. Wu and M. K. H. Leung, *Appl. Energy*, 2010, **87**, 1083.
34. M. Zabeti, W. A. W. Daud and M. K. Aroua, *Fuel Process. Technol.*, 2009, **90**, 770.
35. A. Macario and G. Giordano, *Catal. Lett.*, 2013, **143**, 159.
36. M. K. Lam, K. T. Lee and A. R. Mohamed, *Biotechnol. Adv.*, 2010, **28**, 500.
37. C. Cannilla, G. Bonura, E. Rombi, F. Arena and F. Frusteri, *Appl. Catal., A*, 2010, **382**, 158.
38. S. M. Smith, C. Oopathum, V. Weeramongkhonlert, C. B. Smith, S. Chaveanghong, P. Ketwong and S. Boonyuen, *Bioresour. Technol.*, 2013, **143**, 686.
39. E. Lotero, Y. J. Liu and D. E. Lopez, *Ind. Eng. Chem. Res.*, 2005, **44**, 5353.
40. D. E. López, J. G. Goodwin Jr and D. A. Bruce, *J. Catal.*, 2007, **245**, 381.
41. R. Tesser, M. Di Serio, M. Guida, M. Nastasi and E. Santacesaria, *Ind. Eng. Chem. Res.*, 2005, **44**, 7978.
42. N. Özbay, N. Oktar and N. A. Tapan, *Fuel*, 2008, **87**, 1789.
43. B. M. E. Russbueldt and W. F. Hoelderich, *Appl. Catal., A*, 2009, **362**, 47.
44. M. M. R. Talukder, J. C. Wu, S. K. Lau, L. C. Cui, G. Shimin and A. Lim, *Energy Fuels*, 2009, **23**, 1.
45. R. Tesser, M. Di Serio, L. Casale, L. Sannino, M. Ledda and E. Santacesaria, *Chem. Eng. J.*, 2010, **161**, 212.
46. N. Shibasaki-Kitakawa, H. Honda, H. Kuribayashi, T. Toda, T. Fukumura and T. Yonemoto, *Bioresour. Technol.*, 2007, **98**, 416.
47. M. Kouzu, A. Nakagaito and J. S. Hidaka, *Appl. Catal., A*, 2011, **405**, 36.
48. F. Frusteri, F. Arena, G. Bonura, C. Cannilla, L. Spadaro and O. Di Blasi, *Appl. Catal., A*, 2009, **367**, 77.
49. C. L. Bianchi, D. C. Boffito, C. Pirola and V. Ragaini, *Catal. Lett.*, 2009, **134**, 179.
50. J. Y. Park, D. G. Kim and J. S. Lee, *Bioresour. Technol.*, 2010, **101**, S62.
51. C. S. Caetano, L. Guerreiro, I. M. Fonseca, A. M. Ramos, J. Vital and J. E. Castanheiro, *Appl. Catal., A*, 2009, **359**, 41.
52. Y. H. Feng, B. Q. He, Y. H. Cao, J. X. Li, M. Liu, F. Yan and X. P. Liang, *Bioresour. Technol.*, 2010, **101**, 1518.
53. S. C. M. dos Reis, E. R. Lachter, R. S. V. Nascimento, J. A. Rodrigues Jr and M. G. Reid, *J. Am. Oil Chem. Soc.*, 2005, **82**, 661.
54. J. Fu, Z. Li, S. Xing, Z. Wang, C. Miao, P. Lv and Z. Yuan, *Fuel*, 2016, **181**, 1058.
55. R. Tesser, M. Di Serio, G. Carotenuto and E. Santacesaria, *Can. J. Chem. Eng.*, 2010, **88**, 1044.
56. J. Pérez-Pariente, I. Díaz, F. Mohino and E. Sastre, *Appl. Catal., A*, 2003, **254**, 173.

57. Q. Shu, B. Yang, H. Yuan, S. Qing and G. Zhu, *Catal. Commun.*, 2007, **8**, 2159.
58. K. H. Chung, D. R. Chang and B. G. Park, *Bioresour. Technol.*, 2008, **99**, 7438.
59. R. Jothiramalingam and M. K. Wang, *Ind. Eng. Chem. Res.*, 2009, **48**, 6162.
60. A. C. Carmo, L. K. C. de Souza, C. E. F. da Costa, E. Longo, J. R. Zamian and G. N. da Rocha Filho, *Fuel*, 2009, **88**, 461.
61. M. Sasidharan and R. Kumar, *J. Mol. Catal. A: Chem.*, 2004, **210**, 93.
62. K. S. Karmee and A. Chadha, *Bioresour. Technol.*, 2005, **96**, 1425.
63. M. Di Serio, R. Tesser, L. Pengmei and E. Santacesaria, *Energy Fuels*, 2008, **22**, 207.
64. I. V. Kozhevnikov, *Chem. Rev.*, 1998, **98**, 171.
65. T. Okuhara, *Chem. Rev.*, 2002, **102**, 3641.
66. N. Mizuno and M. Misono, *Chem. Rev.*, 1998, **98**, 199.
67. A. Alsalme, E. F. Kozhevnikova and I. V. Kozhevnikov, *Appl. Catal., A*, 2008, **349**, 170.
68. M. G. Kulkarni, R. Gopinath, L. C. Meher and A. K. Dalai, *Green Chem.*, 2006, **8**, 1056.
69. F. H. Cao, Y. Chen, F. Y. Zhai, J. Li, J. H. Wang, X. H. Wang, S. T. Wang and W. M. Zhu, *Biotechnol. Bioeng.*, 2008, **101**, 93.
70. W. Chu, X. Yang, X. Ye and Y. Wu, *Appl. Catal., A*, 1996, **145**, 125.
71. K. Nowinska and W. Kaleta, *Appl. Catal., A*, 2000, **203**, 91.
72. I. V. Kozhevnikov, A. Sinnema, R. J. J. Jansen, K. Pamin and H. Vanbekkum, *Catal. Lett.*, 1995, **30**, 241.
73. P. G. Vazquez, M. N. Blanco and C. V. Caceres, *Catal. Lett.*, 1999, **60**, 205.
74. K. Srilatha, N. Lingaiah, B. L. A. P. Devi, R. B. N. Prasad, S. Venkateswar and P. S. S. Prasad, *Appl. Catal., A*, 2009, **365**, 28.
75. G. D. Yadav and N. Kirthivasan, *Appl. Catal., A*, 1997, **154**, 29.
76. A. Concellón, P. Vázquez, M. Blanco and C. Cáceres, *J. Colloid Interface Sci.*, 1998, **204**, 256.
77. T. Nakato, M. Kimura, S. Nakata and T. Okuhara, *Langmuir*, 1998, **14**, 319.
78. T. Okuhara and T. Nakato, *Catal. Surv. Jpn.*, 1998, **2**, 31.
79. Y. Izumi, M. Ono, M. Kitagawa, M. Yoshida and K. Urabe, *Microporous Mater.*, 1995, **5**, 255.
80. A. Corma, A. Martínez and C. Martínez, *J. Catal.*, 1996, **164**, 422.
81. T. Okuhara, *Appl. Catal., A*, 2003, **256**, 213.
82. K. Narasimharao, D. R. Brown, A. F. Lee, A. D. Newman, P. F. Siril, S. J. Tavener and K. Wilsona, *J. Catal.*, 2007, **248**, 226.
83. F. Chai, F. Cao, F. Y. Zhai, Y. Chen, X. H. Wang and Z. M. Su, *Adv. Synth. Catal.*, 2007, **349**, 1057.
84. B. Hamad, R. O. Lopes de Souza, G. Sapaly, M. G. Carneiro Rocha, P. G. Pries de Oliveira, W. A. Gonzalez, E. Andrade Sales and N. Essayem, *Catal. Commun.*, 2008, **10**, 92.
85. X. Zhang, J. Li, Y. Chen, J. H. Wang, L. L. Feng, X. H. Wang and F. H. Cao, *Energy Fuels*, 2009, **23**, 4640.

86. A. Zięba, L. Matachowski, E. Lalik and A. Drelinkiewicz, *Catal. Lett.*, 2009, **127**, 183.
87. J. B. Moffat and S. Kasztelan, *J. Catal.*, 1988, **109**, 206.
88. C. F. Oliveira, L. M. Dezaneti, F. A. C. Garcia, J. L. de Macedo, J. A. Dias, S. C. L. Dias and K. Alvin, *Appl. Catal., A*, 2010, **372**, 153.
89. L. L. Xu, Y. H. Wang, X. Yang, X. D. Yu, Y. H. Guo and J. H. Clark, *Green Chem.*, 2008, **10**, 746.
90. C. M. Garcia, S. Teixeira, L. L. Marciniuk and U. Schuchardt, *Bioresour. Technol.*, 2008, **99**, 6608.
91. S. Furuta, H. Matsuhashi and K. Arata, *Catal. Commun.*, 2004, **5**, 721.
92. A. A. Kiss, A. C. Dimian and G. Rothenberg, *Adv. Synth. Catal.*, 2006, **348**, 75.
93. J. Jitputti, B. Kitiyanan, P. Rangsunvigit, K. Bunyakiat, L. Attanatho and P. Jenvanitpanjakul, *Chem. Eng. J.*, 2006, **116**, 61.
94. X. R. Chen, Y. H. Ju and C. Y. Mou, *J. Phys. Chem. C*, 2007, **111**, 18731.
95. D. E. López, J. G. Goodwin Jr, D. A. Bruce and S. Furuta, *Appl. Catal., A*, 2008, **339**, 76.
96. G. D. Yadav and A. D. Murkute, *J. Catal.*, 2004, **224**, 218.
97. B. S. Fu, L. J. Gao, L. Niu, R. P. Wei and G. M. Xiao, *Energy Fuels*, 2009, **23**, 569.
98. B. Peng, Q. Shu, J. Wang, G. Wang, D. Wang and M. Han, *Process Saf. Environ. Prot.*, 2008, **86**, 441.
99. J. A. Melero, J. Iglesias and G. Morales, *Green Chem.*, 2009, **11**, 1285.
100. M. K. Lam, K. T. Lee and A. R. Mohamed, *Appl. Catal., B*, 2009, **93**, 134.
101. S. R. Watkins, F. Lee and K. Wilson, *Green Chem.*, 2004, **6**, 335.
102. H. Kim, B. Kang, M. Kim, Y. M. Park, D. Kim, J. Lee and K. Lee, *Catal. Today*, 2004, **93**, 315.
103. C. R. V. Reddy, R. Oshel and J. G. Verkase, *Energy Fuels*, 2006, **20**, 1310.
104. M. L. Granados, M. D. Z. Poves, D. M. Alonso, R. Mariscal, F. C. Galisteo, R. Moreno-Tost, J. Santamaría and J. L. G. Fierro, *Appl. Catal., B*, 2007, **73**, 317.
105. C. R. V. Reddy, B. M. Fetterly and J. G. Verkade, *Energy Fuels*, 2007, **21**, 2466.
106. W. L. Xie, H. Peng and L. G. Chen, *J. Mol. Catal. A: Chem.*, 2006, **246**, 24.
107. Y. Liu, E. Lotero, J. G. Goodwin Jr and X. Mo, *Appl. Catal., A*, 2007, **331**, 138.
108. D. G. Cantrell, L. J. Gilie, A. F. Lee and K. Wilson, *Appl. Catal., A*, 2005, **287**, 183.
109. C. T. Fishel and R. J. Davis, *Langmuir*, 1994, **10**, 159.
110. W. Xie, H. Peng and L. Chen, *J. Mol. Catal. A: Chem.*, 2006, **246**, 24.
111. H. Zeng, Z. Feng, X. Deng and Y. Li, *Fuel*, 2008, **87**, 3071.
112. D. W. Lee, Y. M. Park and K. Y. Lee, *Catal. Surv. Asia*, 2009, **13**, 63.
113. J. I. Di Cosimo, V. K. Díez, M. Xu, E. Iglesia and C. R. Apesteguía, *J. Catal.*, 1998, **178**, 499.
114. N. Barakos, S. Pasias and N. Papayannakos, *Bioresour. Technol.*, 2008, **99**, 5037.

115. K. G. Georgogianni, A. P. Katsoulidis, P. J. Pomonis and M. G. Kontominas, *Fuel Process. Technol.*, 2009, **90**, 671.
116. G. S. Macala, A. W. Robertson, C. L. Johnson, Z. B. Day, R. S. Lewis, M. G. White, A. V. Iretskii and P. C. Ford, *Catal. Lett.*, 2008, **122**, 205.
117. X. Deng, Z. Fang, Y. H. Liu and C. L. Yu, *Energy*, 2011, **36**, 777.
118. Y. Z. Xi and R. J. Davis, *J. Catal.*, 2008, **254**, 190.
119. A. Kawashima, K. Matsubara and K. Honda, *Bioresour. Technol.*, 2008, **99**, 3439.
120. X. J. Liu, H. Y. He, Y. Wang and S. L. Zhu, *Catal. Commun.*, 2007, **8**, 1107.
121. X. J. Liu, H. Y. He, Y. J. Wang, S. L. Zhu and X. Piao, *Fuel*, 2008, **87**, 216.
122. J. M. Montero, P. Gai, K. Wilson and A. F. Lee, *Green Chem.*, 2009, **11**, 265.
123. S. Bancquart, C. Vanhove, Y. Pouilloux and J. Barrault, *Appl. Catal., A*, 2001, **218**, 1.
124. H. Zhu, Z. Wu, Y. Chen, P. Zhang, S. Duan, X. Liu and Z. Mao, *Chin. J. Catal.*, 2006, **27**, 391.
125. B. Yoosuk, P. Udomsap, B. Puttasawat and P. Krasae, *Chem. Eng. J.*, 2010, **162**, 135.
126. H.-X. Bai, X.-Z. Shen, X.-H. Liu and S.-Y. Liu, *Trans. Nonferrous Met. Soc. China*, 2009, **19**, S674.
127. V. B. Veljković, O. S. Stamenković, Z. B. Todorović, M. L. Lazić and D. U. Skala, *Fuel*, 2009, **88**, 1554.
128. L. Wang and J. Yang, *Fuel*, 2007, **86**, 328.
129. H. Mootabadi, B. Salamatinia, S. Bhatia and A. Z. Abdullah, *Fuel*, 2010, **89**, 188.
130. C. Cannilla, G. Bonura, F. Arena, E. Rombi and F. Frusteri, *Catal. Today*, 2012, **195**, 32.
131. S. L. Yan, H. F. Lu and B. Liang, *Energy Fuels*, 2008, **22**, 646.
132. C. C. S. Macedo, F. R. Abreu, A. P. Tavares, M. B. Alves, L. F. Zara, J. C. Rubim and P. A. Z. Suarez, *J. Braz. Chem. Soc.*, 2006, **17**, 1291.
133. F. Guo, Z. G. Peng, J. Y. Dai and Z. L. Xiu, *Fuel Process. Technol.*, 2010, **91**, 322.
134. M. Di Serio, M. Cozzolino, R. Tesser, P. Patrono, B. Pinzarij Bonelli and E. Santacesaria, *Appl. Catal., A*, 2007, **320**, 1.
135. J. M. Rubio-Caballero, J. Santamaría-González, R. Moreno-Tost, A. Jiménez-López and P. Maireles-Torres, *Appl. Catal., B*, 2009, **91**, 339.
136. X. J. Liu, X. G. Piao, Y. J. Wang, S. L. Zhu and H. Y. He, *Fuel*, 2008, **87**, 1076.
137. Y. D. Long, F. Guo, Z. Fang, X. F. Tian, L. Q. Jiang, X. Deng and F. Zhang, *Bioresour. Technol.*, 2011, **102**, 6884.
138. M. Di Serio, M. Ledda, M. Cozzolino, G. Minutillo, R. Tesser and E. Santacesaria, *Ind. Eng. Chem. Res.*, 2006, **45**, 3009.
139. H. Sun, K. Hu, H. Lou and X. M. Zheng, *Energy Fuels*, 2008, **22**, 2756.
140. A. P. Vyas, N. Subrahmanyam and P. A. Patel, *Fuel*, 2009, **88**, 625.
141. W. L. Xie, H. Peng and L. G. Chen, *Appl. Catal., A*, 2006, **300**, 67.
142. I. W. C. E. Aends and R. A. Sheldon, *Appl. Catal., A*, 2001, **212**, 175.

143. G. Arzamendi, I. Campo, E. Arguínarena, M. Sánchez, M. Montes and L. M. Gandá, *Chem. Eng. J.*, 2007, **134**, 123.
144. https://www.eni.com/docs/en_IT/enicom/publications-archive/company/operations-strategies/refining-marketing/eni_Green-Refinery_esecutivo.pdf, last accessed May 2017.
145. N. Arun, R. V. Sharma and A. K. Dalai, *Renewable Sustainable Energy Rev.*, 2015, **48**, 240.
146. A. Vonortas and N. Papayannakos, *Wiley Interdiscip. Rev.: Energy Environ.*, 2014, **3**, 3.
147. S. Bezergianni and A. Dimitriadis, *Renewable Sustainable Energy Rev.*, 2013, **21**, 110.
148. A. Demirbas and K. Dincer, *Energy Sources, Part A*, 2008, **30**, 1233.
149. J. Regalbuto, *Comput. Chem. Eng.*, 2010, **34**, 1393.
150. https://biorrefineria.blogspot.it/2015/09/hydrotreated-vegetable-oils-hvo.biorefineries.html, last accessed May 2017.
151. http://www.transdolomites.eu/wp-content/uploads/ENI-Ecofining-pdf.pdf, last accessed May 2017.
152. T. N. Kalnes, T. Marker, D. R. Shonnard and K. P. Koers, *Biofuel Technol.*, 7–11, available at: https://www.uop.com/?document=uop-hydrorefining-green-diesel-tech-paper&download=1, last accessed May 2017.
153. T. Kalnes, T. Marker and D. Shonnard, *Int. J. Chem. React. Eng.*, 2007, **5**, A48.
154. S. Bezergianni, A. Kalogianni and I. A. Vasalos, *Bioresour. Technol.*, 2009, **100**, 3036.
155. S. Abate, G. Giorgianni, P. Lanzafame, S. Perathoner and G. Centi, *Chem. Eng. Trans.*, 2016, **50**, 259.
156. W. Souverijns, J. A. Martens, G. F. Froment and P. A. Jacobs, *J. Catal.*, 1998, **174**, 177.
157. T. Ennaert, J. Van Aelst, J. Dijkmans, R. De Clercq, W. Schutyser, M. Dusselier, D. Verboekend and B. F. Sels, *Chem. Soc. Rev.*, 2016, **45**, 584.
158. B. Egia, J. F. Cambra, P. L. Arias, M. B. Guemez, J. A. Legarreta, B. Pawelec and J. L. G. Fierro, *Appl. Catal., A*, 1998, **169**, 37.
159. S. Gong, A. Shinozaki and E. W. Qian, *Ind. Eng. Chem. Res.*, 2012, **51**, 13953.
160. J. Zecevic, G. Vanbutsele, K. P. de Jong and J. A. Martens, *Nature*, 2015, **528**, 245.
161. M. Shahinuzzaman, Z. Yaakob and Y. Ahmed, *Renewable Sustainable Energy Rev.*, 2017, **77**, 1375.
162. L. Jeczmionek, Z. Burnus, G. Zak, L. Ziemianski, M. Wojtasik, W. Krasodomski, Z. Stepien, M. Rutkowska and A. Wegrzyn, *Energy Fuels*, 2014, **28**, 7569.
163. J. Hancsók, M. Krár, S. Magyar, L. Boda, A. Holló and D. Kalló, *Microporous Mesoporous Mater.*, 2007, **101**, 148.
164. P. Simacek, D. Kubicka, I. Kubicková, F. Homola, M. Pospisil and J. Chudoba, *Fuel*, 2011, **90**, 2473.
165. P. Lanzafame, S. Perathoner, G. Centi, E. Heracleous, E. F. Iliopoulou, K. S. Triantafyllidis and A. A. Lappas, *ChemCatChem*, 2017, **9**, 1632.

166. A. Chica, U. Diaz, V. Fornes and A. Corma, *Catal. Today*, 2009, **147**, 179.
167. T. Aysu and A. Sanna, *Bioresour. Technol.*, 2015, **194**, 108.
168. S. Li, S. Xu, S. Liu, C. Yang and Q. Lu, *Fuel Process. Technol.*, 2004, **85**, 1201.
169. M. Garcia-Perez, J. Shen, X. Shan Wang and C.-Z. Li, *Fuel Process. Technol.*, 2010, **91**, 296.
170. Q. Lu, W.-Z. Li and X.-F. Zhu, *Energy Convers. Manage.*, 2010, **50**, 1376.
171. T. Aysu, H. Durak, S. Güner, A. Şükrü Bengü and N. Esim, *Bioresour. Technol.*, 2016, **205**, 7.
172. S. Xiu, B. Zhang and A. Shahbazi, in *Biofuel's Engineering Process Technology*, ed. M. A. dos Santos Bernardes, InTech Pub., 2011, ch. 8, p. 167.
173. M. W. Nolte and B. H. Shanks, *Energy Technol.*, 2017, **5**, 7–18.
174. D. A. Ruddy, J. A. Schaidle, J. R. Ferrell III, J. Wang, L. Moens and J. E. Hensley, *Green Chem.*, 2014, **16**, 454.
175. P. M. Mortensen, J.-D. Grunwaldt, P. A. Jensen, K. G. Knudsen and A. D. Jensen, *Appl. Catal., A*, 2011, **407**, 1.
176. D. C. Elliot, *Energy Fuels*, 2007, **21**, 1792.
177. M. V. Olarte, A. H. Zacher, A. B. Padmaperuma, S. D. Burton, H. M. Job, T. L. Lemmon, M. S. Swita, L. J. Rotness, G. N. Neuenschwander, J. G. Frye and D. C. Elliot, *Top. Catal.*, 2016, **59**, 55.
178. M. R. Rover, P. H. Hall, P. A. Johnston, R. G. Smith and R. C. Brown, *Fuel*, 2015, **153**, 224.
179. T. P. Vispute, H. Zhang, A. Sanna, R. Xiao and G. W. Huber, *Science*, 2010, **330**, 1222.
180. A. H. Zacher, D. C. Elliot, M. V. Olarte, D. M. Santosa, F. Preto and K. Iisa, *Energy Fuels*, 2014, **28**, 7510.
181. J. Zhang, Y. S. Choi and B. H. Shanks, *ChemSusChem*, 2015, **8**, 4256.
182. A. H. Zacher, M. V. Olarte, D. M. Santosa, D. C. Elliot and S. B. Jones, *Green Chem.*, 2014, **16**, 491.
183. K. Wang, K. H. Kim and R. C. Brown, *Green Chem.*, 2014, **16**, 727.
184. J. Jae, G. A. Tompsett, A. J. Foster, K. D. Hammond, S. A. Auerbach, R. F. Lobo and G. W. Huber, *J. Catal.*, 2011, **279**, 257.
185. A. G. Gayubo, A. T. Aguayo, A. Atutxa, R. Prieto and J. Bilbao, *Ind. Eng. Chem. Res.*, 2004, **43**, 5042.
186. E. F. Iliopoulou, S. D. Stefanidis, K. G. Kalogiannis, A. Delimitis, A. A. Lappas and K. S. Triantafyllidis, *Appl. Catal., B*, 2012, **127**, 281.
187. N. E. Persson, S. D. Blass, C. Rosenthal, A. Bhan and L. D. Schmidt, *RSC Adv.*, 2013, **3**, 20163.
188. M. Zabeti, T. S. Nguyen, L. Lefferts, H. J. Heeres and K. Seshan, *Bioresour. Technol.*, 2012, **118**, 374.
189. M. Zabeti, K. B. S. S. Gupta, G. Raman, L. Lefferts, S. Schallmoser, J. A. Lercher and K. Seshan, *ChemCatChem*, 2015, **7**, 3386.
190. T. S. Nguyen, M. Zabeti, L. Lefferts, G. Brem and K. Seshan, *Bioresour. Technol.*, 2013, **142**, 353.
191. L. Nie and D. E. Resasco, *J. Catal.*, 2014, **317**, 22.

192. P. M. de Souza, L. Nie, L. E. P. Borges, F. B. Noronha and D. E. Resasco, *Catal. Lett.*, 2014, **144**, 2005.
193. D. Gao, C. Schweitzer, H. T. Hwang and A. Varma, *Ind. Eng. Chem. Res.*, 2014, **53**, 18658.
194. S. Boonyasuwat, T. Omotoso, D. E. Resasco and S. P. Crossley, *Catal. Lett.*, 2013, **143**, 783.
195. S. Wan, T. Pham, S. Zhang, L. Lobban, D. E. Resasco and R. Mallinson, *AIChE J.*, 2013, **59**, 2275.
196. S. Boullosa-Eiras, R. Lodeng, H. Bergem, M. Stocker, L. Hannevold and E. A. Blekkan, *Catal. Today*, 2014, **223**, 44.
197. M. M. Sullivan, J. T. Held and A. Bhan, *J. Catal.*, 2015, **326**, 82.
198. M. M. Sullivan and A. Bhan, *ACS Catal.*, 2016, **6**, 1145.
199. M. W. Nolte, J. Zhang and B. H. Shanks, *Green Chem.*, 2016, **18**, 134.
200. G. Knothe, *Prog. Energy Combust. Sci.*, 2010, **36**, 364.
201. L. Dinh, Y. Guo and M. Mannan, *Environ. Prog. Sustainable Energy*, 2009, **28**, 38.
202. P. Vasudevan and M. J. Briggs, *J. Ind. Microbiol. Biotechnol.*, 2008, **35**, 421.
203. P. Forster, V. Ramaswamy, P. Artaxo, T. Berntsen, R. Betts, D. Fahey, J. Haywood, J. Lean, D. Lowe, G. Myhre, J. Nganga, R. Prinn, G. Raga, M. Schulz and R. Van Dorland, Changes in Atmospheric Constituents and in Radiative Forcing, in *Climate Change 2007: The Physical Science Basis. Contribution of Working Group I to the Fourth Assessment Report of the Intergovernmental Panel on Climate Change*, ed. S. Solomon, D. Qin, M. Manning, Z. Chen, M. Marquis, K. B. Averyt, M. Tignor and H. L. Miller, 2007, Cambridge University Press, Cambridge, United Kingdom and New York, NY, USA.
204. M. S. de Paula Gomes and M. S. Muylaert de Araújo, *Renewable Sustainable Energy Rev.*, 2009, **13**, 2201.
205. C. B. Granda, L. Zhu and M. T. Holtzapple, *Environ. Prog. Sustainable Energy*, 2007, **26**, 233.
206. P. A. Willems, *Science*, 2009, **325**, 707.
207. E. Henrich, N. Dahmen and E. Dinjus, *Biofuels, Bioprod. Biorefin.*, 2009, **3**, 28.

CHAPTER 6

Catalytic Upgrading of Bio-oils

B. PAWELEC, R. M. NAVARRO AND J. L. G. FIERRO*

Institute of Catalysis and Petroleum Chemistry, CSIC, Cantoblanco, Marie Curie 2, 28049 Madrid, Spain
*E-mail: jlgfierro@icp.csic.es

6.1 Introduction

Current energy consumption is to a large extent based on fossil fuels: oil, natural gas and coal, whose combustion emits large amounts of CO_2 into the atmosphere thus accelerating global warming. This concern has boosted the importance of research into alternatives to petroleum fuel.[1] Consequently, the use of biomass for energy has recently received a great deal of attention due to environmental considerations and the increasing energy demands worldwide.[2] This is because the combustion of biomass-derived fuels produces less harmful gas emissions such as nitrogen oxides (NO_x), sulfur dioxide (SO_2) and soot compared to conventional fossil fuels. In addition, fewer or zero CO_2 emissions are possible from biomass-derived fuel combustion because the CO_2 released can be recycled by photosynthesis.[3] Thus, biomass can significantly participate in reducing worldwide CO_2 emissions.[4] Current technologies for producing liquid fuels from biomass are typically multistep and energy-intensive processes. These are depicted in Figure 6.1.

Among the various biomass conversion technologies, thermochemical conversion (*i.e.*, combustion, gasification and pyrolysis) of biomass is widely used for the production of renewable fuels. Compared with the gasification

Green Chemistry Series No. 56
Sustainable Catalysis for Biorefineries
Edited by Francesco Frusteri, Donato Aranda and Giuseppe Bonura

Published by the Royal Society of Chemistry, www.rsc.org

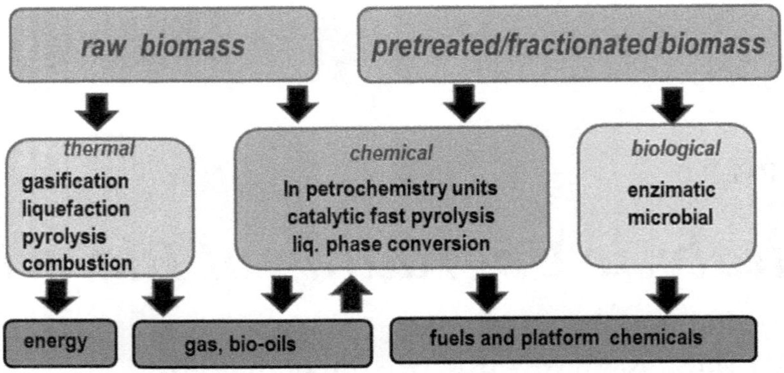

Figure 6.1 Schematic representation of the different modes of biomass conversion. Adapted from ref. 34 with permission from The Royal Society of Chemistry.

and combustion methods, the conversion of lignocellulosic biomass to bio-liquid by fast pyrolysis is a more effective method because a larger yield of liquid (up to 75 wt%) should be obtained.[5] Bio-oil is being considered as a promising C-containing source to replace petroleum fuels. However, bio-oil cannot be used directly in diesel and gasoline engines because its main component is water (10–30 wt%) and it possesses a high oxygen content (typically in the order of 35–40 atom%).[6] In addition, bio-oil exhibits high acidity, which can be corrosive for engines. The organic fraction contains aldehydes, ketones, sugars, carboxylic acids, esters, furans, guaiacols, and phenolics, which are present as oligomers.[7] Some of these, when oxygenated, are highly reactive, which makes them highly unstable, viscous and possess a low heating value. For instance, aldehydes, acids and ketones react to form acetals, hemiacetals and ethers, respectively.[8] Due to the high O-content, during ageing, bio-oil separates into two phases, *i.e.*, a non-polar viscous bottom phase, which consists of lignin extractives, and a more polar aqueous phase, which consists mainly of water-soluble compounds.[6] Therefore, bio-oils need to be upgraded to reduce their oxygen content in order to increase their stability, miscibility with conventional oil, and to increase their H/C ratio.

The production of renewable fuels, such as gasoline, diesel and jet fuel, using the Biomass to Liquid (BTL) *via* Fischer–Tropsch synthesis has been gaining increasing attention during recent years. These fuels are usually much cleaner and environmentally friendly and hence satisfy the upcoming stricter environmental regulations.[9] In the BTL process, woody biomass is firstly gasified with air or steam to produce raw bio-syngas, the cleaned bio-syngas is then directed into a catalytic reactor for FT synthesis to produce renewable liquid fuels.[10,11] Notwithstanding this, it has been shown that oxygen removal from bio-oils constitutes a feasible route for the production of synthetic fuels. The competitiveness of the O-removal processes is due to good economy when using bio-oil as a platform chemical (lower transport cost for large scale plants) and the flexibility with respect to the biomass feed.[12,13] Both cracking and HDO processes are referred to as catalytic bio-oil upgrading and these could become routes for the production of second generation bio-fuels in the future, even though both routes are still far from industrial application. Here, we present an overview of these two processes and discuss the influence of the operating conditions on catalyst performance together with mechanisms and catalyst deactivation during on-stream operation.

6.2 Bio-oil Upgrading: General Characteristics

Catalytic upgrading of bio-oil involves a complex reaction network due to the high number and diversity of oxygenate compounds present in the feed. Reactions such as hydrodeoxygenation, cracking/hydrocracking, decarbonylation, hydrogenation, decarboxylation and polymerization have been reported to take place in both acid and HDO catalysts.[14–16] Another important reaction is carbon deposition. Due to the large diversity of bio-oil sources, and hence the reaction network for each one, the evaluation of bio-oil upgrading is difficult. As a consequence, most of these evaluations are often restricted to model compounds. The chemistry of model compounds from pyrolysis bio-oils and the catalysts employed, including classical sulfided hydrotreating catalysts, noble metals, sulfides, phosphides, carbides, nitrides, non-precious metals, metal oxides, bimetallic amorphous boron-based catalysts, and reduced metal oxide bronzes, have recently been well-reviewed by He and Wang.[17]

A thermodynamic overview of the reactions can be obtained by looking at the thermodynamics of the hydrogenation/hydrogenolysis of model compounds such as phenol:

$$C_6H_5OH + H_2 \rightarrow C_6H_6 + H_2O \tag{6.1}$$

$$C_6H_5OH + 4H_2 \rightarrow C_6H_{12} + H_2O \tag{6.2}$$

This reaction path of phenol has been proposed by Massoth *et al.*[16] Calculating the thermodynamic equilibrium for the two reactions shows that complete conversion of phenol can be achieved at temperatures up to at least 600 °C under atmospheric pressure and stoichiometric conditions. Increasing either the pressure or the excess hydrogen will shift the thermodynamics even further toward complete conversion. Similar calculations have also been made with furfural, giving equivalent results. Thus, although the thermodynamics do not appear to constitute a constraint for the processes when evaluating model compounds (eqn (6.3)–(6.8)), in practice, it is difficult to evaluate the conversion of each individual component in the bio-oil.

$$\text{Cracking: } R_1\text{–}CH_2\text{–}CH_2\text{–}CH_2\text{–}CH_2\text{–}R_2 \rightarrow R_1\text{–}CH{=}CH_2 + CH_2{=}CH\text{–}R_2 \quad (6.3)$$

$$\text{Hydrocracking: } R_1\text{–}CH_2\text{–}CH_2\text{–}R_2 + H_2 \rightarrow CH_3\text{–}R_1 + CH_3\text{–}R_2 \quad (6.4)$$

$$\text{Decarboxylation: } R_1\text{–}COOH \rightarrow R_1\text{–}H + CO_2 \quad (6.5)$$

$$\text{Decarbonylation: } R_1\text{–}COH \rightarrow R_1\text{–}H + CO \quad (6.6)$$

$$\text{Hydrogenation: } R_1\text{–}CH{=}CH\text{–}R_2 + H_2 \rightarrow R_1\text{–}CH_2\text{–}CH_2\text{–}R_2 \quad (6.7)$$

$$\text{HDO: } R\text{–}OH + H_2 \rightarrow R\text{–}H + H_2O \quad (6.8)$$

Instead, two important parameters are the oil yield and the degree of deoxygenation:

$$Y_{\text{oil}} = (m_{\text{oil}}/m_{\text{feed}}) \times 100 \quad (6.9)$$

$$\text{DOD} = [1 - (O_{\text{product}}/O_{\text{feed}})] \quad (6.10)$$

where: Y_{oil} is the yield of oil, m_{oil} is the mass of oil produced, m_{feed} is the mass of the feed, DOD is the degree of oxygenation, and $O_{product}$ and O_{feed} are the wt% of oxygen in the product and in the feed, respectively. Both parameters (Y_{oil} and DOD) give a rough overview of the extent of the reaction, Y_{oil} provides the selectivity toward an oil product, and DOD describes the effectiveness of the oxygen removal and therefore indicates the quality of the oil produced. However, separately, the parameters are less descriptive, because a 100% yield can be achieved in the case of no reaction.

6.3 Catalytic Valorization of Bio-oils

The valorization of bio-oils can be achieved in different ways. One of these is a biological conversion such as a microbial or yeast-based fermentation process, *i.e.* the commercial conversion of carbohydrate biomass into ethanol or lactic acid.[18,19] Another process is thermal conversion, ranging from burning biomass for heat generation followed by gasification to biogas or liquefaction and pyrolysis to bio-oils.[20] Finally, bio-oils can be converted by chemical and catalytic processes to target products. There are a few excellent reviews dedicated to the catalytic upgrading of bio-oils to engine fuels.[8,21–25] Regarding the bio-oils produced *via* pyrolysis, there are two main strategies to enhance their quality for fuel applications: (i) bio-oil cracking–deoxygenation using an acid catalyst (currently zeolites), which can be achieved *in situ* in the pyrolysis reactor[26,27] or *ex situ* in a fixed bed reactor coupled on-line with the pyrolysis reactor,[28,29] and (ii) bio-oil catalytic hydrodeoxygenation (HDO) using bifunctional catalysts, which require a high H_2 consumption and rather severe reaction conditions.[21,30]

Both the catalytic cracking and hydrodeoxygenation methods are discussed in some detail below.

6.3.1 Cracking of Bio-oil Products

During fast pyrolysis, the biomass substrate is heated up rapidly to 400–600 °C for short residence times (1–2 s). By fast pyrolysis, 50–70% of liquid bio-oil is produced by rapid quenching and condensation of the vapors and coalescence of aerosols.[6,31] Recent detailed analysis has shown that pyrolysis bio-oil contains more than 400 different compounds,[32] most with a high molecular weight, resulting in an undesired high viscosity.[31] Therefore, a quantitative chemical characterization of bio-oils is necessary to evaluate their characteristics for downstream production.

The upgrading of the vapors of these compounds can be properly performed with acid zeolites.[33,34] Carlson *et al.*[35] have studied extensively the potential of Brønsted acidic zeolites in upgrading pyrolysis products. Basically, the pyrolysis is carried out in the presence of a zeolite catalyst, which leads to a more intense contact between the oil precursor and zeolite catalyst, which results in a much higher zeolite impact compared to the catalytic upgrading of pyrolysis vapors. A major disadvantage of this approach is the high quantity of coke deposited on the catalyst surface, leading to quick catalyst deactivation. Therefore, proper catalyst selection by optimal tuning of the catalyst pore structure and active sites is necessary to reduce the amount of coke and to increase the yield of aromatics. It was shown that both the Brønsted acidity and the crystalline pore structure are prerequisites for aromatic production from carbohydrate feedstocks.[36,37] Indeed, it was shown that zeolite ZSM-5 yields over 30% aromatics whereas zeolites USY and beta mainly produce coke, indicating the importance of the specific pore structure of ZSM-5.[35] These findings indicate that small pore zeolites produce mainly oxygenates and coke because the diffusion of reactants and products is strongly hindered. Despite the faster reactant diffusion in large pore zeolites, they are not well suited because the large pores allow the formation of polyaromatics due to the lack of reactant/transition state confinement.[34]

As the biomass-derived oils require not only the cracking of the C–C bond but also some of the C–O bonds, the Si/Al ratio of the zeolite has to be adjusted. C–O bonds have lower dissociation energies and hence do not require zeolites with high acid density or acid strength. This means that medium Si/Al zeolites show the best balance between the availability of Brønsted acid sites, on the one hand, and maintaining enough space between the Brønsted acid sites to prevent side reactions on the other hand.[38]

Because the zeolite catalyst becomes deactivated by coke deposition, which originates mainly from the lignin compounds present in the bio-oil feed,[39] it has been suggested that the removal of lignin prior to pyrolysis, or a more selective pyrolysis approach with lignin, will reduce the coke formation. As research into the conversion of lignin is very limited,[39] further investigation into the process of the production of benzene-rich gasoline or petrochemicals (toluene, phenol or benzene) will be of major importance for bio-oil upgrading in general. Furthermore, taking into account the advances achieved in the regeneration of zeolite catalysts used in commercial refinery plants, these ideas can be extrapolated to enhance the regeneration capacity of zeolites in the pyrolysis of bio-oils.

Coke deposition is the main source of catalyst deactivation in zeolite cracking. Coke is formed through the polymerization and polycondensation reactions, which lead to the blockage of the pores of the zeolites.[40] Gayubo *et al.*[41] studied the nature of coke residues formed on a HZSM-5 catalyst during operation with synthetic bio-oil in a fixed bed reactor at 400–450 °C with temperature programmed oxidation (TPO) and found two types of carbon: thermal carbon and catalytic carbon. The thermal carbon deposited only in the macropores of the catalyst. The catalytic carbon was found in the micropores of the zeolites and was ascribed to dehydrogenation, condensation, and hydrogen transfer reactions. This was found to have a lower hydrogen content compared to the thermal carbon. The thermal carbon was removed at lower temperatures by oxidation at (450–480 °C) compared to the catalytic carbon, which was removed at 520–550 °C. The higher temperature required to remove the catalytic carbon was associated with steric hindrance, as it is deposited in the micropores where it is strongly bound to the acidic sides of the zeolite, and it is less reactive due to the hydrogen deficient nature. This work nicely illustrates that the principal source of zeolite deactivation is the coke formed in the micropores, which blocks the internal acid sites of the catalyst, although thermal carbon also contributes to the deactivation.

A way of decreasing the extent of carbon formation on the catalyst was shown by Zhu *et al.*[42] by co-feeding H_2 and anisole over HZSM-5 in a fixed bed reactor at 400 °C. Their results indicated that the presence of H_2 diminishes the coke formation only slightly. It appears that H_2 has the affinity to react with adsorbed carbenium ions to form paraffins, but apparently this effect was not sufficient to increase the lifetime of the catalyst to any significant extent. A similar conclusion was derived during the deoxygenation of benzaldehyde over HZSM-5 in which it was reported that the presence of hydrogen did not influence the conversion, although a shift in selectivity occurred as an increase in toluene production was observed with H_2, which was ascribed to the hydrogenation/hydrogenolysis reactions taking place.[43]

In summary, all these results show that a hydrogen source in catalytic cracking has a positive effect on the stability of the zeolite catalyst. Therefore, it seems that the potential exists when using zeolites to incorporate a metal function and co-feed with hydrogen.

6.3.2 Cracking of Triglyceride-type Compounds

Triglyceride compounds and/or their derivatives can be converted into aviation fuels using bifunctional catalysts such as zeolites.[39,44,45] Hydrogenolysis and deoxygenation are carried out on the highly dispersed metal function, whereas the combined action with the acid sites of the acid substrate enables the hydro-isomerization and hydrocracking of the long-chain alkanes formed. For both diesel and aviation fuel, isomerization to branched alkanes is necessary to meet the cold-flow properties demanded. On the other hand, cracking to shorter-chain alkanes is only desired when aviation fuel is desired. The carbenium ion intermediate formed upon dehydrogenation (metal site) and protonation (acid site) can either be isomerized and again, hydrogenated, or cracked. Therefore, the metal/acid site balance is a crucial parameter of these bifunctional catalysts. In general, moderate acidic supports combined with an optimized, sufficiently high metal/acid balance favor isomerization, whereas more acidic supports or non-optimal metal loadings enhance the degree of cracking toward the aviation fuel range.

Heterogeneous catalysts have been used and compared for their activity in the cracking of triglycerides from various sources in a review by Taufiqurrahmi and Bhatia[46] that is summarized in Table 6.1. Most of these studies have been performed on FCC catalysts, zeolites, silica alumina and their hybrids at relatively high temperatures. The reaction is initiated by a thermal cracking by means of free radicals before the catalytic conversion to oxygenated compounds. In the presence of a catalyst with active sites accessible for triglyceride molecules, the ester group of the triglyceride decomposes on an acid site. Gusmao *et al.*[47] reported that the catalytic cracking of vegetable oil on a Ni/SiO_2 catalyst occurs in two steps: (i) the decomposition of triglycerides to carboxylic acids, acrolein and ketenes, which then recombine to form esters, carboxylic acids and hydrocarbons; and (ii) the carboxylic acids

Table 6.1 Feedstock, catalysts and reaction temperature of catalytic cracking of vegetable oil. Adapted from ref. 46 with permission from The Royal Society of Chemistry.

Feed	Catalyst	*T* (K)	References
Palm/soybean oil	FCC equilibrium catalyst	838	48
Cottonseed oil	FCC equilibrium catalyst	713–783	49
Rapeseed oil	FCC equilibrium catalyst	823	50
Woody oil	Al_2O_3/MCM-41/CaO	723, 773	51
Soybean oil	HZSM-5/MCM-41	693, 723	52
Soybean oil	Bauxite	653, 673	53
Canola oil	HZSM-5, H-mordernite, H-Y	540, 673	54
Olive oil	KOH, Na_2CO_3	540, 630	55
Sunflower oil	NaOH treated aluminium oxide	460, 630	56
Waste olive oil	Dolomite	773–1073	57
Used vegetable oil	HZSM-5/sulfated zirconia	653–703	58
Octanoid acid	Activated alumina	673, 648, 673	59
Used sunflower oil	Na_2CO_3	673, 693	60
Canola oil	HZSM-5/$SiAl_2O_3$/HS mix	673–773	61
Canola oil	Potassium-impregnated HZSM-5	673, 773	62
Canola oil	HZSM-5, H-mordernite, SiO_2Al2O_3	648–773	63

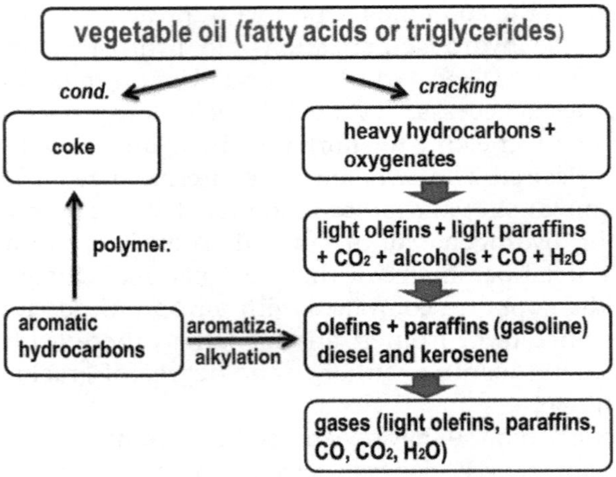

Figure 6.2 Proposed reaction pathway for the conversion of palm oil over a HZSM-5 catalyst to hydrocarbon products. Adapted from ref. 64.

are decarbonylated or decarboxylated yielding CO, olefins and H_2O or CO_2 and paraffins, respectively.

Maher and Bressler[64] reported carboxylic acids, ketones, esters, acrolein and hydrocarbons as the major reaction products of palm oil cracking (Figure 6.2). Idem *et al.*[65] reported that the organic liquid fraction of the reaction is a mixture of oxygenated and heavy hydrocarbons consisting of olefins, paraffins, alcohols, acids, ketones, aldehydes, C_6^+ aliphatic hydrocarbons, C_9 aromatic hydrocarbons, and gases. Other processes such as aromatization, alkylation, isomerization, oligomerization and condensation may also occur. A similar mechanism was reported by Melero *et al.*[48] for the catalytic cracking of mixtures of crude vegetable oils and non-edible animal fats with vacuum gasoil. They reported that once the triglyceride molecule had been primarily cracked to heavy oxygenated hydrocarbons such as fatty acids, ketones, esters, and aldehydes, further reactions yielding other products start by breaking the C–O and C–C bonds by β-scission reactions. These β-scission reactions are characterized by the scission of the beta bond (connected to an adjacent atom) to the atom bearing a radical. Katikaneni *et al.*[66] proposed another reaction scheme for the catalytic cracking of canola oil over zeolites in which the heavy oxygenated hydrocarbons and oxygenates undergo further catalytic cracking to produce esters.

6.3.3 Catalytic Deoxygenation (HDO)

Hydrodeoxygenation (HDO) is one of the most widespread processes used for upgrading bio-oils. Nevertheless, the harsh reaction conditions required to run the reactions means that the process is limited in its viability for commercial implementation.[67] Phenolic model compounds, such as guaiacol, anisole, cresols and phenol, all obtained from the depolymerization of the lignin fraction of woody biomass, have typically been used as model compounds to assess the reactivity of many catalysts for HDO reactions. This strategy involves hydrogenation–dehydration–hydrogenation reactions in which the dehydration step takes place on the acid sites while the hydrogenation step occurs on the metal sites. It is commonly assumed that the oxygen from oxy-compounds is absorbed on coordinatively unsaturated metal sites (oxygen vacancies) on metal oxide supports through a Lewis acid/base interaction, or on H in –OH groups, which are attached to non-metal oxides such as SiO_2, or even on metal sites such as noble metals. Thus, the donation of activated hydrogen species might be available directly from phosphides, carbides, nitrides, Brønsted acid –OH groups or –SH groups and from metals by H spillover. After the activated H species have reacted with the oxy-organics, hydrodeoxygenated products are obtained.[68]

The main challenge is to design and produce a high performance and selective heterogeneous catalyst, being capable of deoxygenating different types of oxygenate functional groups during the HDO process at mild conditions, while maintaining its hydrothermal stability during the long reaction time. As a consequence, several heterogeneous catalysts have been tested, such as hydrodesulfurization catalysts, acid catalysts, mainly zeolites, and supported noble or transition metal particles, either on reducible or non-reducible oxides or carbonaceous materials. Among the wide variety of heterogeneous catalysts tested, the bifunctional catalysts, in which two different types of active sites are exposed on the surface to facilitate the hydrogenation and dehydration reactions, are the most promising.[69] Bifunctional catalysts exhibit an active site on which the C–O scission can take place allowing the deoxygenation, and at the same time, a second type of active site to hydrogenate aromatic compounds. The reported bifunctional catalysts showed significant catalytic differences in terms of activity and selectivity depending on the chemical nature of the supported metal or the support, as well as on their structural properties.

In general, supported monometallic noble metals initially behave as the active centers for hydrogenating the double bonds.[70] Considering the bifunctional catalysts, either the acidic support or a second metal with oxophilic properties has the ability to act as the deoxygenation active center. When using metal oxides with acidic properties as supports, the oxophilic sites are strong enough to selectively adsorb and activate the C=O groups, thus increasing the scission of the C–O bond, and it appears that there is a correlation between the deoxygenation activity and the affinity of the metal to interact with oxygenated species. Some of the catalysts used for bio-oil HDO together with their reaction conditions and performance (DOD and Y) are summarized in Table 6.2.

Understanding the mechanism through which the HDO reaction takes place on different bifunctional catalytic systems can assist in the design of more selective and promising catalysts for the valorization of bio-oils into valuable hydrocarbons.

Table 6.2 Overview of catalysts used for catalytic upgrading of bio-oil. Adapted from *Applied Catalysis A: General*, **407**, P. M. Mortensen, J.-D. Grunwaldt, P. A. Jensena, K. G. Knudsen and A. D. Jensen, A review of catalytic upgrading of bio-oil to engine fuels, 1–19. Copyright 2011 with permission from Elsevier.[8]

Catalyst	Feed	Time (h)	P (bar)	T (K)	DOD (%)	Y (%)	References
Co–MoS_2/Al_2O_3	Bio-oil	4	300	370	100	33	71
Ni–MoS_2/Al_2O_3	Bio-oil	4	200	350	74	28	14
Ni–MoS_2/Al_2O_3	Bio-oil	0.5	85	400	28	84	72
Pd/C	Bio-oil	4	140	340	64	48	73
Pd/ZrO_2	Guaiacol	3	80	300	—	—	74
Ru/Al_2O_3	Bio-oil	0.2	230	350	73	38	75
Ru/C	Bio-oil	4	200	350	86	53	14
Ru/TiO_2	Bio-oil	4	200	350	77	57	14

6.3.3.1 Catalysts and Reaction Mechanisms

The reaction conditions and types of catalyst have significant effects on the quality of the liquid products obtained by the hydrodeoxygenation of bio-oils. Regarding the reaction conditions, a high H_2 consumption and rather severe reaction conditions are needed for the HDO reaction.[30] Regarding the catalysts, bifunctional heterogeneous systems having both metal and acid functions should be employed. Using optimized reaction conditions and an effective HDO catalyst, the product obtained can be enhanced for fuel applications. The recent advances in the HDO of pyrolysis bio-oils using many different types of catalysts and supports have been reviewed by He and Wang.[68] The work has concentrated on investigations into the reasons why the current catalysts have showed poor stability and have hindered the pyrolysis oil HDO process on an industrial scale.

6.3.3.1.1 Sulfide/Oxide Catalysts. Typical hydrotreating sulfide catalysts such Co–MoS_2 and Ni–MoS_2 are the most frequently used catalysts to perform HDO reactions, as these are also used in the traditional hydrotreating process.[22] However, in the absence of a sulfiding agent, the sulfided catalysts were deactivated quickly in the HDO reaction and the selectivity to different hydrocarbons changed with time on stream due to the oxidation of the active phases by the water produced during the HDO reaction.[22,76–82] This suggests that the addition of a sulfur agent during bio-oil processing could be beneficial for the HDO reaction. Accordingly, promising laboratory-scale tests have been conducted using sulfide NiMo/γ-Al_2O_3 catalysts in the simultaneous HDO of phenol and HDS of dibenzothiophene.[81] However, the refineries are not willing to adopt this co-processing approach due to the observed drop in the HDS activity of the industrial catalysts.[82]

The Mo-based hydrotreating catalysts consist of MoS_2 nanoslabs incorporating Co or Ni promoters on their edges, which act as electron donors to molybdenum atoms. As a consequence, the Mo–S bond is weakened and

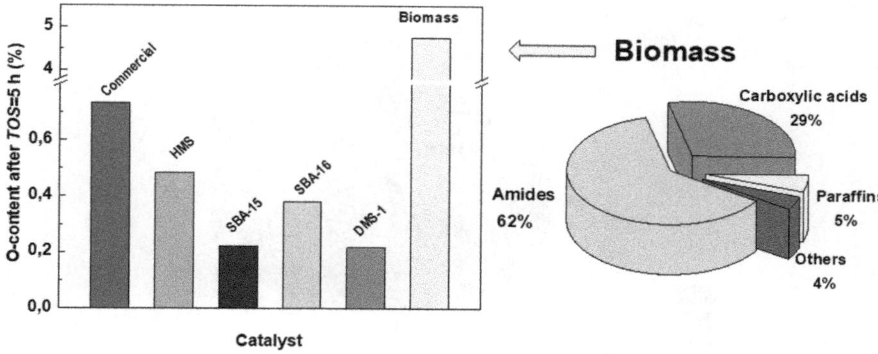

Figure 6.3 The comparison of the O-content after 5 h hydrotreating of olive oil waste over sulfide CoMo catalysts supported on different mesoporous silica materials (SBA-15, SBA-16, HMS and DMS-1). Adapted from *Applied Catalysis B: Environmental*, **92**, R. Nava, B. Pawelec, P. Castaño, M. C. Álvarez-Galván, C. V. Loricera, J. L. G. Fierro, Upgrading of bio-liquids on different mesoporous silica-supported CoMo catalysts, 154–167, Copyright 2009, with permission from Elsevier.[83]

thus generates a sulfur vacancy on the surface. The resulting S-vacancies are the active sites in both the HDS and HDO reactions.[16,83–85]

We have investigated the performance of CoMo catalysts supported on four different mesoporous silicates (DMS-1, SBA-15, SBA-16, HMA) for the HDO reaction of waste from olive oil processing.[83] The HDO reaction was performed in a flow reactor (552 K, 3 MPa, WHSV = 46.4 h^{-1}). The comparison of the O-content after 5 h reaction is shown in Figure 6.3. All synthesized catalysts produced higher activity at 5 h on stream than that of a conventional sulfide $CoMo/Al_2O_3$ catalyst.

The catalysts supported on SBA-15, SBA-16 and DMS-1 were more effective than that supported on HMS. The products formed after 5 h reaction time, mainly paraffins, esters and carboxylic acids, are shown in Figure 6.4. The enhancement of activity on the sulfided CoMo catalysts is linked with a higher active phase surface exposure and larger acidity than the CoMo/HMS.

Romero *et al.*[84] studied the HDO of 2-ethylphenol on MoS_2-based catalysts and proposed the reaction mechanism depicted in Figure 6.5. The oxygen of the molecule is believed to adsorb on an S-vacancy of a MoS_2 slab edge, activating the compound. As S–H species are also formed on the edge of the MoS_2 slabs, they donate H-donation to the attached molecule, which forms a carbocation. This can undergo a C–O bond cleavage, thus leading to the deoxygenated compound (ethylbenzene) while the O-atom forms water.

According to this mechanism, it is necessary for the O-group formed on the S-vacancy from the deoxygenation step be removed as water. After long operation times it has been observed that activity drops due to the conversion of the sulfide phase of the catalyst to the respective oxide form. To avoid the oxidation of the sulfide phase, it has been proved that co-feeding a very small

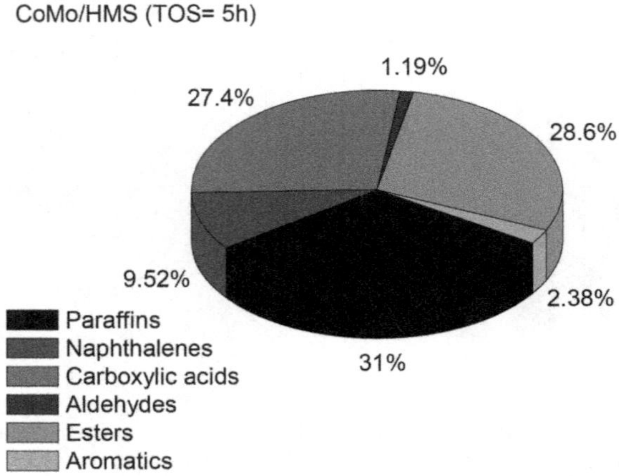

Figure 6.4 The products formed after 5 h HDO reaction of waste from olive oil production over sulfided CoMo/HMS catalyst.

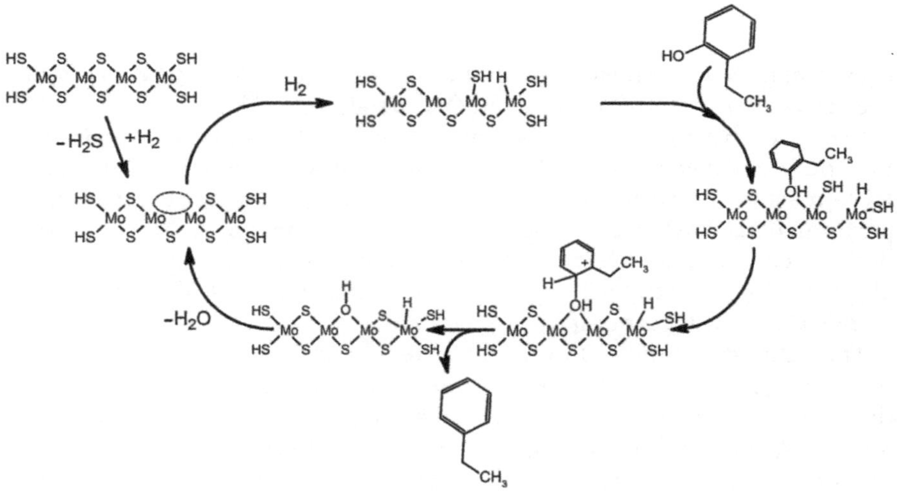

Figure 6.5 Proposed mechanism of HDO of 2-ethylphenol over a Co–MoS_2 catalyst. The dotted circle indicates the catalytically active S-vacancy site. Reprinted from *Applied Catalysis B: Environmental*, **98**, Y. Romero, F. Richard and S. Brunet, Hydrodeoxygenation of 2-ethylphenol as a model compound of bio-crude over sulfided Mo-based catalysts: promoting effect and reaction mechanism, 213–223, Copyright 2010, with permission from Elsevier.[84]

amount of H_2S into the reactor regenerates the MoS_2 phase and stabilizes the catalyst.[76] However, it was shown that trace amounts of thiols and sulfides are formed during the HDO reactions over sulfide catalysts when co-feeding up to 1000 ppm H_2S into the reactor.[76] This indicates that sulfur contaminates the deoxygenated product. Furthermore, theoretical calculations led to the conclusion that not only MoS_2 but also MoO_3 catalysts are active for HDO reactions.[86] These calculations assumed chemisorption of the O-containing molecules on the O-vacancies of the oxide, proton donation and desorption. Acidity is a key requirement for both sulfide and oxide catalysts. Thus, the first step of the HDO reaction is the adsorption of an O-containing molecule on a Lewis acid/base site, followed by proton donation. These protons originate from the Brønsted acid hydroxyl groups present on the catalyst surface. For WO_3, which is expected to have a high population of acid sites, we have shown[87] that carbon-supported W and Ni–W oxides are highly active for phenol HDO (Figure 6.6).

By feeding 1 wt% phenol in *n*-octane into a fixed bed reactor at 150–300 °C and 15 bar, it was proven that these catalysts were all active for HDO, and especially that the Ni–W system had the potential for the complete conversion of the model compound. Furthermore, a low affinity for carbon was observed during 6 h on stream. This low level of carbon deposition was ascribed to the beneficial effect of the non-acidic carbon support.

6.3.3.1.2 Noble Metal-based Catalysts. It is known that noble metal-based catalysts exhibit high intrinsic hydrogenation activity. However, they are expensive catalysts and display a low sulfur tolerance. Table 6.3 summarizes the type of catalysts, operating parameters, product yield, degree of deoxygenation, and product grade for some of the work conducted in the field of bio-oil upgrading.[72,88–96]

Taking into account that the catalysts needed for the HDO reaction should be bifunctional, the selection of the catalyst support is of great importance.[97] In the same way that conventional hydrotreating catalysts are supported on γ-Al_2O_3, the hydrodeoxygenation of bio-oils is generally carried out in the presence of alumina-supported catalysts. However, alumina does not appear to be particularly well suited because the bio-oil contains a number of oxy-molecules with two or more oxygen atoms and a large amount of water (up to ~30 wt%).[17] In addition, an alumina carrier tends to form large quantities of coke on the surface of Lewis type acid sites.[77] Furthermore, alumina is unstable in the presence of water. This difficulty can be overcome by replacing alumina with active carbon or titania.[98] Although these supports display lower acidity, the interaction of the oxide phases (Mo or Ni) on the surface is weaker than on alumina. Recently, encouraging results were reported by Cordero-Lanzac *et al.*[96] for the bifunctional Pt–Pd catalyst supported on active carbon modified with phosphorous (ACP). This catalyst exhibited outstanding activity and stability in the hydrodeoxygenation (HDO) of raw bio-oil.[96] Similarly, active carbon, CeO_2 and ZrO_2[91] were found to be effective in HDO reactions because of the additional activation of oxy-compounds on the support surface.[91]

6.3.3.1.3 Non-noble Metal Catalysts. Transition metal catalysts display good catalytic performance for bio-oil upgrading. They can be used as a potential alternative for noble metal and sulfided catalysts due to their relatively high activity and low cost. Zn-promoted nickel catalysts were found

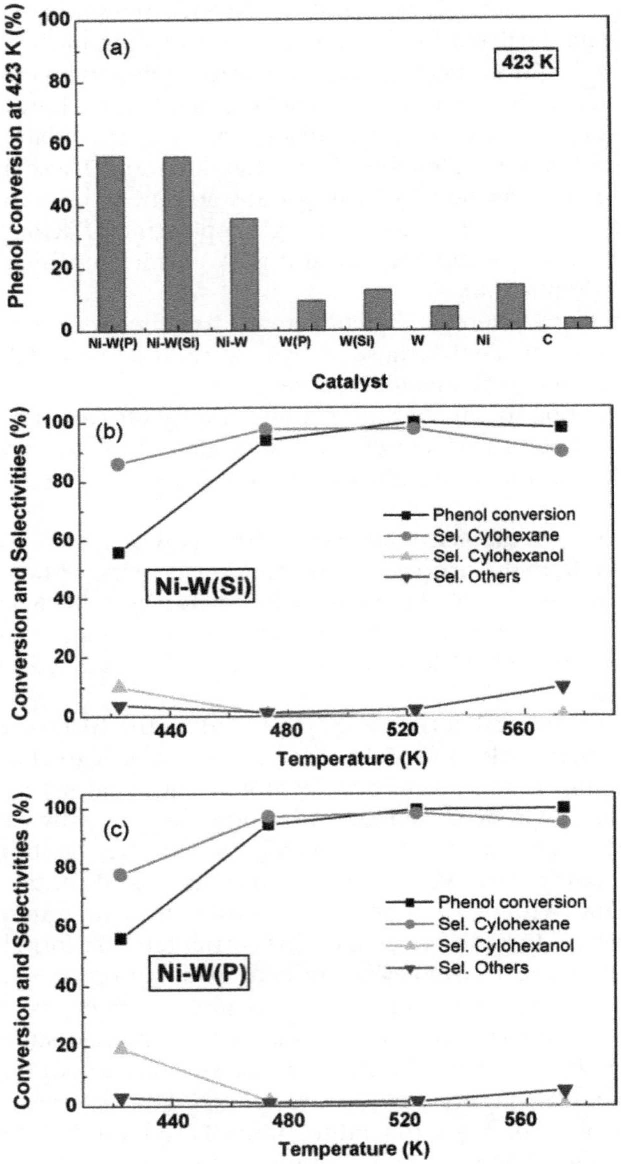

Figure 6.6 Comparison of the steady-state phenol conversions of different catalysts in HDO of phenol at $T = 423$ K, $P = 15$ bar and WHSV = 0.5 h^{-1} (a). The influence of temperature on the phenol conversion and selectivities in HDO of phenol (WHSV = 0.5 h^{-1}) over Ni–W(Si) (b) and Ni–W(P) catalysts (c). Reprinted from *Applied Catalysis B: Environmental*, **101**, S. Echeandia, P. L. Arias, V. L. Barrio, B. Pawelec, J. L. G. Fierro, Synergy effect in the HDO of phenol over Ni–W catalysts supported on active carbon: effect of tungsten precursors, 1–12, Copyright 2010, with permission from Elsevier.[87]

Table 6.3 Overview of noble metal-based catalysts used for HDO of model compounds.

Feed	Active phase	Support	Set-up	*T* (K)	*P* (MPa)	References
Soybean oil	Pt, Pd, Ru	Al_2O_3	Batch	673	9.2	88
Pyrolysis oil	Ru	Al_2O_3	Batch	598	12	89
Pine pyrolitic oil	Pt	SiO_2–Al_2O_3	Continuous trickle-bed	623–673	0.005	72
Pine pyrolitic oil	Pt	Al_2O_3/SiO_2	Continuous trickle-bed	623–673	0.01	90
Pyrolysis oil	Pt, Pd, Rh	ZrO_2	Batch	623	20	91
Crude and upgraded bio-oils	Pd	$SiO_4^{2}/ZrO_2/$ SBA-15	Batch	553	8.5–10.5	92
Lignin surrogates	Pd–Zn	C	Batch	423	2.0	93
Oleic & linoleic acids, methyl-oleate	Pd	C	Semi-batch	573–633	1.5–2.7	94
Pyrolysis oils	Ru, Pt, Pd	C	Batch	523, 623	10, 20	95
Pyrolysis oil	Ru	C	Batch	723	35	72
Raw bio-oil	Pt–Pd	P/AC	Continuous fixed-bed	673–728	6.5	96

to be effective for the liquid-phase hydrodearomatization (HDA) of synthetic diesel and for the gas-phase hydrodeoxygenation of phenol.[99] This is because the addition of Zn to Ni-based catalysts has a positive effect on Ni dispersion together with its ability to induce some electronic and geometric effects. In addition, Zn^{2+} ions have a more pronounced tendency toward tetrahedral coordination with oxygen ions at the support interface than Ni^{2+} ions. Thus, in the case of silica-supported catalysts, the incorporation of Zn might prevent the formation of undesired spinel $NiSiO_4$ phases, which are inactive in hydrotreating reactions.

Alumina-supported, non-sulfided MoNi, NiCu and NiFe catalysts have often been employed for bio-oil upgrading due to their good catalytic activity for the HDO of pyrolysis oil.[100–111] MoNi catalysts display good performance for hydrodeoxygenation of pine sawdust bio-oil by increasing its hydrogen content and lowering the acidity.[112] It has been shown that the addition of Mo to a nickel based catalyst resulted in an enhancement of nickel species dispersion and inhibited $NiAl_2O_4$ spinel formation at the alumina–MoNi interface. Furthermore, bimetallic Ni–Cu/Al_2O_3 catalysts are more active than their monometallic Ni/Al_2O_3 and Cu/Al_2O_3 counterparts for the HDO of fast pyrolysis oil.[91] This can be explained in terms of the smaller particle size and increased number of active NiCu clusters in the catalyst. Another non-noble metal catalyst, silica-supported iron (Fe/SiO_2), was found to be an active and selective catalyst for HDO upgrading of guaiacol to produce aromatic hydrocarbons, and it exhibited a good selectivity for BT (benzene, toluene) production.[113]

Inexpensive active carbon-supported transition metal catalysts such as Ni, Ni–Fe, Ni–Mo, and Ni–Cu were applied to the HDO of prairie cordgrass bio-oil.[114] Among these catalysts, Ni/AC produced upgraded bio-oil with the highest content of gasoline range hydrocarbons (32.63%). This was due to the ability of Ni to favor hydrogenation reactions. On the other hand, the Mo-promoted Ni catalyst (Ni–Mo/AC) produced the upgraded bio-oil with the highest content of gasoline blending alkyl-phenols at 38.41%. In addition, the Ni–Mo/AC catalyst inhibited C–C breaking and promoted C–O activation, which led to a higher degree of deoxygenation (66.02%) than Ni/AC (63.60%). Bifunctional metal/acid catalysts including HBeta zeolite-supported Ni, Fe and NiFe were used for the HDO of a simulated phenolic bio-oil consisting of phenol, *o*-cresol, and guaiacol.[115] The results indicated that the bimetallic NiFe catalyst showed higher HDO activity when compared to the monometallic Ni and Fe catalysts. This is due to the synergistic effect between the two metals. The bimetallic NiFe phases appear to convert phenolic compounds into oxygen-free products *via* hydrogenation and hydrogenolysis reactions.

Boride, carbide, nitride and phosphide catalysts have also been used for bio-oil upgrading. All these amorphous catalysts exhibit excellent activity and good selectivity during bio-oil HDO reactions because of their unique isotropic properties and high concentration of coordinative unsaturated sites. Boride amorphous catalysts such as Co–Mo–B and Co–Ni–Mo–B exhibited high catalytic activity in the HDO of bio-oil model compounds including phenol, benzaldehyde, and acetophenone.[116] This result was attributed to the smaller particle size and high content of Mo^{4+} Brønsted acid sites on the catalyst surface. Carbide, nitride, and phosphide catalysts have also drawn considerable attention due to their potential use for bio-oil HDO processes. For instance, silica-supported NiMoC phases were highly active and stable catalysts for the HDO of model bio-oil compounds including ethyl benzoate, acetone, and acetaldehyde. This reactivity was explained in terms of the acid sites on the surface of the NiMo carbide catalysts.[117] Similarly, nitride catalysts of the type Mo_2N were evaluated for the HDO of guaiacol in a batch autoclave at 300 °C and 5 MPa.[118] Active Mo_2N phases prepared with flowing ammonia showed the highest activity for guaiacol conversion due to the Mo_2N phase and high N/Mo atomic ratio present in the catalyst. In addition, phosphide catalysts such as nickel phosphide (Ni_2P/SiO_2) were compared with traditional catalysts (Ni/SiO_2, Pd/C, ZSM-5, SiO_2–Al_2O_3, and FCC) for direct catalytic upgrading of cedar chip pyrolysis bio-oil.[119] The results indicated that the deoxygenation activity of the Ni_2P/SiO_2 catalyst was higher than that of conventional catalysts. Activated carbon (AC) alone can be used as an effective catalyst for the deoxygenation and denitrogenation of bio-oil. This is attributed to the mesopore properties of AC including an affinity for heavy hydrocarbon compounds and an excellent ability to inhibit coke formation. Several commercial ACs were tested for catalytic upgrading of crude bio-oil produced from the hydrothermal liquefaction of duckweed.[120] The results indicated that ACs exhibited similar deoxygenation and denitrogenation activity when compared to Ru/C.

Kinetic Models. The kinetics of several model compounds has been reviewed in some detail by Furimsky.[22] However, due to the complexity of real bio-oil mixtures, information on the kinetics of the HDO of bio-oils is scarce and only mainly lumped kinetic expressions have been derived. The kinetics of the HDO of pine bio-oil has been investigated by Sheu *et al.*[72] over $Pt/Al_2O_3/SiO_2$, $Co\text{–}MoS_2/Al_2O_3$ and $Ni\text{–}MoS_2/Al_2O_3$ catalysts in a packed bed reactor. Kinetic data were evaluated in terms of a kinetic expression of the type:

$$-\mathrm{d}w_{\mathrm{oxy}}/\mathrm{d}Z = k \cdot w^{m} \cdot P^{n} \tag{6.11}$$

where w is the mass of oxygen in the product relative to the oxygen in the raw pyrolysis oil, Z is the axial position in the reactor, k is the rate constant given by an Arrhenius expression, P is the total pressure (mainly H_2), m is the reaction order for oxygen, and n is the reaction order for the total pressure. For the three types of catalysts, it was assumed that the rate could be described by a first order dependency with respect to the oxygen in the bio-oil ($m = 1$). Accordingly, the pressure dependency and activation energy were calculated and the values are shown in Table 6.4. A positive effect of an increase in pressure was reported when n was in the range of 0.3 to 1, and the activation energy was in the range of 45.5 to 71.4 kJ mol^{-1}, with $Pt/Al_2O_3/SiO_2$ having the lowest activation energy. This agrees with the higher degree of deoxygenation of this catalyst with respect to the two others. It is, however, emphasized that the rate term of eqn (6.11) has a non-fundamental form because the use of the axial position in the reactor is time dependent and therefore this term is very specific for the system used. Therefore, a correlation of these results with other systems would be difficult, if not impossible. Moreover, the assumption of a general first order dependency for w is a very rough assumption when developing a kinetic model.

A similar kinetic equation was proposed by Zhang *et al.*[121] for the HDO of bio-oil over $Co\text{–}MoS_2/Al_2O_3$ catalysts carried out in a batch reactor between

Table 6.4 Kinetic parameters for the kinetic model in eqn (6.11) of different catalysts. Experiments performed in a packed bed reactor between *ca.* 300–400 °C and 45–105 bar. Data taken from *Fuel Processing Technology*, **19**, Y. E. Sheu, R. G. Anthony and E. J. Soltes, Kinetic studies of upgrading pine pyrolytic oil by hydrotreatment, 31–50 Copyright 1988, with permission from Elsevier.[72]

Catalyst	m	N	E_a (kJ mol^{-1})
$Pt/Al_2O_3SiO_2$	1	1.0	45.5 ± 3.2
$Co\text{–}Mo/Al_2O_3$	1	0.3	71.4 ± 14.6
$Ni\text{–}Mo/Al_2O_3$	1	0.5	61.7 ± 7.1

360 and 390 °C. However, considering a general low dependency on the hydrogen partial pressure over a pressure interval from 15 to 30 bar, the H_2 pressure term was omitted. This led to the expression:

$$-dC_{oxy}/dt = k \cdot C^{2.3} \tag{6.12}$$

where C is the total concentration of all oxygenated compounds. A higher reaction order of 2.3 was found in this case, compared to the assumption of Sheu *et al.*[72] The apparent high reaction order may be correlated with the activity of the different oxygen-containing species; the very reactive species will entail a high reaction rate, but as these disappear quickly, a rapid decrease in the rate will be observed. The activation energy was found to be 91.4 kJ mol^{-1}, which is somewhat higher than that obtained by Sheu *et al.*[72]

Deactivation. One of the remaining problems of the technologies employed for bio-oil upgrading is the catalyst lifetime because carbon deposition occurs on the catalyst surface in a relatively short time on stream.[17] However, the extent of coking depends on the type of oxy-compounds, acidity of the catalyst, and operation conditions. The deactivated catalyst, if stable, can be regenerated easily without losing too much activity. This process is highly desirable for the hydroprocessing of pyrolysis bio-oils on an industrial scale.

All the catalysts are deactivated to some extent during the course of HDO reactions. Deactivation can occur through poisoning by nitrogen species or water, sintering of catalyst nanoparticles, metal deposition, or coking.[8] The extent of these phenomena depends on the catalyst type, but coke deposition has been proven to be the most common and main path of catalyst deactivation.[22] Carbon originates from the polymerization and polycondensation reactions on the catalytic surface, forming polyaromatic species and resulting in the coverage of the active sites of the catalysts.[122] Indeed, for the Co–MoS_2/Al_2O_3 catalyst, the fast growth of carbon within the substrate pores was observed as a consequence of the strong adsorption of polyaromatic species. The rates of the carbon forming reactions are, to a large extent, controlled not only by the feed, but also by the process conditions. With respect to the feed, alkenes and aromatics are more likely to form coke deposits than saturated hydrocarbons. For oxygen-containing compounds, it has been shown that those with more than one O-atom have a higher affinity for carbon formation by polymerization reactions on the catalyst surface.[123] Moreover, coking increases with increasing acidity of the catalyst. This is a problem because acid sites are also required for the HDO reactions.

In order to minimize carbon formation, operation parameters can be optimized. Hydrogen has been shown to decrease the carbon formation on Co–MoS_2/Al_2O_3 as it will convert carbon precursors into stable molecules by saturating adsorbed olefinic species on the catalyst surface.[17] In addition, temperature also affects the formation of carbon. At a high temperature, the rate of dehydrogenation increases, which gives an increase in the rate of polycondensation reactions.

6.3.4 Esterification

Bio-oil needs to be stabilized before further treatment under harsh conditions.[124] Esterification of alcohols is one method of stabilizing bio-oil through the removal of the reactive compounds such as sugars and furans *via* various acid-catalyzed reactions.[125] The acid catalysts that have been applied to the esterification of bio-oil include mineral acids, solid acidic resin, supported acid catalysts and sulfated metal oxides. The inorganic liquid acid catalysts can disperse homogeneously in bio-oil. However, the difficulty in separating them from the products and the high corrosiveness and low recyclability limit their application. Solid acid catalysts, solid base catalysts, ionic liquid catalysts,[126] HZSM-5, and aluminum silicate catalysts have been investigated for the esterification of bio-oils.[127] By comparison, solid acid catalysts can easily be recovered from the product mixture and be reused, and these have been widely employed as catalysts for the esterification of bio-oil.[128] The studies into bio-oil esterification have mainly focused on the transformation of bio-oil into the required products and the development of catalysts.[129] In these cases, coke formation occurs to some extent, which is a problem because it can deactivate solid acid catalysts, which in turn lowers the efficiency of the process, and increases the cost.

Polar solvents such as methanol, ethanol, and furfural have been used to homogenize and stabilize biomass oils.[130] The immediate effect of adding polar solvents not only decreases viscosity but also increases the heating value. Some recent studies have shown that reacting the oil with an alcohol (*e.g.*, ethanol) and an acid catalyst (*e.g.*, acetic acid), under mild conditions using reactive distillation, resulted in a better quality bio-oil.[131] The chemical reactions that can occur between the bio-oil and methanol or ethanol are esterification and acetalization (eqn (6.13) and (6.14)).

$$\text{R–COH} + 2\,\text{R}_1\text{–OH} \rightarrow \text{R–CH(OR}_1)_2 + \text{H}_2\text{O} \tag{6.13}$$

$$\text{R–COH} + \text{R}_1\text{–OH} \rightarrow \text{R–CO–R}_1 + \text{H}_2\text{O} \tag{6.14}$$

Here, the reactive molecules of bio-oil like organic acids and aldehydes are converted by reactions with alcohols to esters and acetals, respectively. Considering the simplicity and the low cost of some solvents such as methanol, and their beneficial effects on the oil, this method seems to be the most practical approach for bio-oil quality upgrading.

6.4 Conclusions

One of the problems remaining in the technologies employed for bio-oil upgrading is the catalyst lifetime because carbon deposition takes place on the catalyst surface in a relatively short time on stream. Another important restriction inherent to the most widely used HDO bio-oil upgrading technology is the high hydrogen consumption required to obtain a fuel with a relatively high H/C ratio. As HDO is being considered as a technology to compete with fossil fuels, several factors still have to be addressed. Those areas that still have to be elucidated before bio-oils can be used as renewable fuel precursors are: the design of more effective catalysts, an understanding of the mechanism of carbon deposition on the catalyst surface, an understanding of the kinetics, the elucidation of catalyst deactivation in the presence of S compounds, and the use of more sustainable sources of hydrogen.

Acknowledgements

This work was financially supported by the Spanish Ministry of Economy, Industry and Competitiveness (CTQ2016-76505-C3-1) and the Community of Madrid (S2013/MAE-2882) research projects.

References

1. A. V. Bridgwater, *Therm. Sci.*, 2004, **8**, 21.
2. O. Zhang, J. Chang, T. Wang and Y. Xu, *Energy Convers. Manage.*, 2007, **48**, 87.
3. S. Perathoner and G. Centi, *ChemSusChem*, 2014, **7**, 1274.
4. T. E. Bull, *Science*, 1999, **285**, 1209.
5. E. Buzetzki, J. Rimarčík, Z. Cvengrošová, A. Kleinová, I. Sarkozy and J. Cvengros, *Procedia Eng.*, 2012, **42**, 1133.
6. A. V. Bridgwater, D. Meier and D. Radlein, *Org. Geochem.*, 1999, **30**, 1479.
7. H. B. Goyal, D. Seal and R. C. Saxena, *Renewable Sustainable Energy Rev.*, 2008, **12**, 504.
8. P. M. Mortensen, J.-D. Grunwaldt, P. A. Jensena, K. G. Knudsen and A. D. Jensen, *Appl. Catal., A*, 2011, **407**, 1.
9. J. H. Yang, H. J. Kim, D. H. Chun, H. T. Lee, J. C. Hong, H. Jung and J. I. Yang, *Fuel Process. Technol.*, 2010, **91**, 285.
10. A. Demirbas, *Energy, Educ., Sci. Technol.*, 2006, **17**, 27.
11. M. J. A. Tijmensen, A. P. C. Faaij, C. N. Hamelinck and M. R. M. van Hardeveld, *Biomass Bioenergy*, 2002, **23**, 129.
12. D. C. Elliott, *Energy Fuels*, 2007, **21**, 1792.
13. J. G. Rogers and J. G. Brammer, *Biomass Bioenergy*, 2009, **33**, 1367.
14. J. Wildschut, F. H. Mahfud, R. H. Venderbosch and H. J. Heeres, *Ind. Eng. Chem. Res.*, 2009, **48**, 10324.
15. L. Wang, H. W. Lei, L. John, S. Chen, J. Tang and B. Ahring, *RSC Adv.*, 2013, **3**, 14609.
16. F. E. Massoth, P. Politzer, M. C. Concha, J. S. Murray, J. Jakowski and J. Simons, *J. Phys. Chem. B*, 2006, **110**, 14283.
17. Z. He and X. Wang, *Catal. Sustainable Energy*, 2012, **1**, 28.
18. P. V. Abdelnur, B. G. Vaz, J. D. Rocha, M. B. B. de Almeida, M. A. G. Teixeira and R. C. L. Pereira, *Fuels*, 2013, **27**, 6646.

19. S. Nanda, A. K. Dalai and J. A. Kozinski, *Energy Sci. Eng.*, 2014, **2**, 138.
20. A. V. Bridgwater, *Chem. Eng. J.*, 2003, **91**, 87.
21. Q. Bu, H. Lei, A. H. Zacher, L. Wang, S. Ren, J. Liang, Y. Wei, Y. Liu, J. Tang, Q. Zhang and R. Ruan, *Bioresour. Technol.*, 2012, **124**, 470.
22. E. Furimsky, *Appl. Catal., A*, 2000, **199**, 147.
23. D. C. Elliott, *Energy Fuels*, 2007, **21**, 1792.
24. E. Butler, G. Devlin, D. Meier and K. McDonnell, *Renewable Sustainable Energy Rev.*, 2011, **15**, 4171.
25. Q. Zhang, J. Chang, T. Wang and Y. Xu, *Energy Convers. Manage.*, 2007, **48**, 87.
26. R. Atutxa, A. G. Aguado, M. Gayubo, J. Olazar and J. Bilbao, *Energy Fuels*, 2005, **19**, 765.
27. N. Aho, K. Kumar, T. Eränen, M. Salmi, L. Hupa and D. Y. Murzin, *Fuel*, 2008, **87**, 2493.
28. E. F. Iliopoulou, S. D. Stefanidis, K. G. Kalogiannis, A. Delimitis, A. A. Lappas and K. S. Triantafyllidis, *Appl. Catal., B*, 2012, **127**, 281.
29. K. Iisa, R. J. French, K. A. Orton, M. M. Yung, D. K. Johnson, J. ten Dam, M. J. Watson and M. R. Nimlos, *Energy Fuels*, 2016, **30**, 2144.
30. M. Saidi, F. Samimi, D. Karimipourfard, T. Nimmanwudipong, B. C. Gates and M. R. Rahimpour, *Energy Environ. Sci.*, 2014, **7**, 103.
31. D. Mohan, C. U. Pittman Jr and P. H. Steele, *Energy Fuels*, 2006, **20**, 848.
32. G. Lyu, S. Wu and H. Zhang, *Front. Energy Res.*, 2015, **3**, 28.
33. P. T. Williams and P. A. Horne, *Fuel*, 1995, **74**, 1839.
34. T. Ennaert, J. van Aelst, J. Dijkmans, R. de Clercq, W. Schutyser, M. Dusselier, D. Verboekend and B. F. Sels, *Chem. Soc. Rev.*, 2016, **45**, 584.
35. T. R. Carlson, T. P. Vispute and G. W. Huber, *ChemSusChem*, 2008, **1**, 397.
36. K. Wang, K. H. Kim and R. C. Brown, *Green Chem.*, 2014, **16**, 727.
37. J. Q. Bond, A. A. Upadhye, H. Olcay, G. A. Tompsett, J. Jae, R. Xing, D. M. Alonso, D. Wang, T. Zhang, R. Kumar, A. Foster, S. M. Sen, C. T. Maravelias, R. Malina, S. R. H. Barrett, R. Lobo, C. E. Wyman, J. A. Dumesic and G. W. Huber, *Energy Environ. Sci.*, 2014, **7**, 1500.
38. A. J. Foster, J. Jae, Y.-T. Cheng, G. W. Huber and R. F. Lobo, *Appl. Catal., A*, 2012, **423**, 154.
39. C. Wang, Z. Tian, L. Wang, R. Xu, Q. Liu, W. Qu, H. Ma and B. Wang, *ChemSusChem*, 2012, **5**, 1974.
40. J. Huang, W. Long, P. K. Agrawal and C. W. Jones, *J. Phys. Chem. C*, 2009, **113**, 16702.
41. A. G. Gayubo, A. T. Aguayo, A. Atutxa, B. Valle and J. Bilbao, *J. Chem. Technol. Biotechnol.*, 2005, **80**, 12441.
42. X. X. Zhu, R. G. Mallinson and D. E. Resasco, *Appl. Catal., A*, 2010, **379**, 172.
43. A. Ausavasukhi, T. Sooknoi and D. E. Resasco, *J. Catal.*, 2009, **268**, 68.
44. H. Chen, Q. Wang, X. Zhang and L. Wang, *Appl. Catal., B*, 2015, **166–167**, 327.

45. B. Peng, Y. Yao, C. Zhao and J. A. Lercher, *Angew. Chem., Int. Ed.*, 2012, **51**, 2072.
46. N. Taufiqurrahmi and S. Bhatia, *Energy Environ. Sci.*, 2011, **4**, 1087.
47. J. Gusmao, D. Brodzki, G. Djega-Mariadassou and R. Frety, *Catal. Today*, 1989, **5**, 533.
48. J. A. Melero, M. M. Clavero, G. Calleja, A. Garcia, R. Miravalles and T. Galindo, *Energy Fuels*, 2010, **24**, 707.
49. L. Hong and S. Benxian, *Energy Sources, Part A*, 2010, **32**, 1159.
50. P. Bielansky, A. Reichhold and C. Schonberger, *Chem. Eng. Process.*, 2010, **49**, 873.
51. J. Xu, J. Jiang, J. Chen and Y. Sun, *Bioresour. Technol.*, 2010, **101**, 5586.
52. T. Ngo, J. Kim, S. K. Kim and S. Kim, *Energy*, 2010, **35**, 2723.
53. C. M. R. Prado and N. R. A. Filho, *J. Anal. Appl. Pyrolysis*, 2009, **86**, 338.
54. S. M. Sadrameli, A. E. S. Green and W. Seames, *J. Anal. Appl. Pyrolysis*, 2009, **86**, 1.
55. A. Demirbas, *Energy Sources, Part A*, 2008, **30**, 1060.
56. A. Demirbas, *Energy Sources, Part A*, 2009, **31**, 671.
57. J. M. Encinar, J. F. Gonzalez, G. Martiınez and J. M. Gonzalez, *Fuel Process. Technol.*, 2008, **89**, 1448.
58. W. Charusiri, W. Yongchareon and T. Vitidsant, *Korean J. Chem. Eng.*, 2006, **23**, 49.
59. F. Billaud, A. K. Tran Minh, P. Lozano and D. Pioch, *J. Anal. Appl. Pyrolysis*, 2001, **58–59**, 605.
60. L. Dandik and H. A. Aksoy, *Fuel Process. Technol.*, 1998, **57**, 81.
61. S. P. R. Katikaneni, J. D. Adjaye, R. O. Idem and N. N. Bakhshi, *J. Am. Oil Chem. Soc.*, 1998, **75**, 381.
62. S. P. R. Katikaneni, J. D. Adjaye, R. O. Idem and N. N. Bakhshi, *Ind. Eng. Chem. Res.*, 1996, **35**, 3332.
63. S. P. R. Katikaneni, J. D. Adjaye and N. N. Bakhshi, *Can. J. Chem. Eng.*, 1995, **73**, 484.
64. K. D. Maher and D. C. Bressler, *Bioresour. Technol.*, 2007, **98**(12), 2351.
65. R. O. Idem, S. P. R. Katikaneni and N. N. Bakhshi, *Fuel Process. Technol.*, 1997, **51**, 101.
66. S. P. R. Katikaneni, J. D. Adjaye and N. N. Bakhshi, *Energy Fuels*, 1995, **9**, 599.
67. A. H. Zacher, V. Mariefel, V. Olarte, D. M. Santosa, D. C. Elliott and S. B. Jones, *Green Chem.*, 2014, **16**, 491.
68. Z. He and X. Wang, *Catal. Sustainable Energy*, 2012, **1**, 28.
69. W. Song, Y. Liu, E. Baráth, C. Zhao and J. A. Lercher, *Green Chem.*, 2015, **17**, 1204.
70. A. M. Robinson, J. E. Hensley and J. W. Medlin, *ACS Catal.*, 2016, **6**, 4356.
71. W. Baldauf, U. Balfanz and M. Rupp, *Biomass Bioenergy*, 1994, **7**, 237.
72. Y. E. Sheu, R. G. Anthony and E. J. Soltes, *Fuel Process. Technol.*, 1988, **19**, 31.
73. D. C. Elliott, T. R. Hart, G. G. Neuenschwander, L. J. Rotness and A. H. Zacher, *Environ. Prog.*, 2009, **28**, 441.

74. A. Gutierrez, R. K. Kaila, M. L. Honkela, R. Slioor and A. O. I. Krause, *Catal. Today*, 2009, **147**, 239.
75. R. H. Venderbosch, A. R. Ardiyanti, J. Wildschut, A. Oasmaa and H. J. Heeres, *J. Chem. Technol. Biotechnol.*, 2010, **85**, 674.
76. O. I. Senol, T.-R. Viljava and A. O. I. Krause, *Catal. Today*, 2005, **106**, 186.
77. E. Laurent and B. Delmon, *J. Catal.*, 1994, **46**, 281.
78. E. Furimsky and F. E. Massoth, *Catal. Today*, 1999, **52**, 381.
79. V. La Vopa and C. N. Satterfield, *Energy Fuels*, 1987, **1**, 323.
80. M. W. Vogelzang, C.-L. Li, G. C. A. Schuit, B. C. Gates and L. Petrakis, *J. Catal.*, 1983, **84**, 170.
81. P. Platanitis, G. D. Panagiotou, K. Bourikas, C. Kordulis and A. Lycourghiotis, *Open Catal. J.*, 2014, **7**, 18.
82. V. N. Bui, G. Toussaint, D. Laurenti, C. Mirodatos and C. Geantet, *Catal. Today*, 2009, **143**, 172.
83. R. Nava, B. Pawelec, P. Castaño, M. C. Alvarez, C. V. Loricera and J. L. G. Fierro, *Appl. Catal., B*, 2009, **92**, 154.
84. Y. Romero, F. Richard and S. Brunet, *Appl. Catal., B*, 2010, **98**, 213.
85. H. Topsøe, B. S. Clausen and F. E. Massoth, in *Hydrotreating Catalysis*, Springer-Verlag, 1996.
86. D. R. Moberg, T. J. Thibodeau, F. G. Amar and B. G. Frederick, *J. Phys. Chem. C*, 2010, **114**, 13782.
87. S. Echeandia, P. L. Arias, V. L. Barrio, B. Pawelec and J. L. G. Fierro, *Appl. Catal., B*, 2010, **101**, 1.
88. B. Veriansyah, J. Y. Han, S. K. Kim, S.-A. Hong, Y. J. Kim, J. S. Lim, Y.-W. Shu, S.-G. Oh and J. Kim, *Fuel*, 2012, **94**, 578.
89. J. Gagnon and S. Kaliaguine, *Ind. Eng. Chem. Res.*, 1988, **27**, 1783.
90. Y. E. Hwa, E. Sheu, R. G. Anthony and E. J. Soltes, *Fuel Process. Technol.*, 1988, **19**, 31.
91. A. R. Ardiyanti, S. A. Khromova, R. H. Venderbosch, V. A. Yakovlev and H. J. Heeres, *Appl. Catal., B*, 2012, **117**, 105.
92. Y. Tang, S. Miao, B. H. Shanks and X. Zheng, *Appl. Catal., A*, 2010, **375**, 310.
93. T. H. Parsell, B. C. Owen, T. M. Jarrell, C. L. Marcum, L. J. Haupert, L. M. Amundson, H. I. Kenttämaa, F. Ribeiro, J. T. Miller and M. M. Abu-Omar, *Chem. Sci.*, 2013, **4**, 806.
94. M. Snåre, I. Kubičková, P. Mäki-Arvela, K. Eränen and D. Y. Murzin, *Ind. Eng. Chem. Res.*, 2006, **45**, 5708.
95. J. Wildschut, M. Iqbal, F. H. Mahfud, I. Melián, R. H. Venderbosch and H. J. Heeres, *Energy Environ. Sci.*, 2010, **3**, 962.
96. T. Cordero-Lanzac, R. Palos, J. M. Arandes, P. Castaño, J. Rodríguez-Mirasol, T. Cordero and J. Bilbao, *Appl. Catal., B*, 2017, **203**, 389.
97. V. A. Yakovlev, S. A. Khromova, O. V. Sherstyuk, V. O. Dundich, D. Y. Ermakov, V. M. Novopashina, M. Y. Lebedev, O. Bulavchenko and V. N. Parmon, *Catal. Today*, 2009, **144**, 362.
98. D. C. Elliott and T. R. Hart, *Energy Fuels*, 2009, **23**, 631.
99. C. V. Loricera, A. Infantes-Molina, I. Hita, P. Castaño, A. Gutiérrez, J. M. Arandes, J. L. G. Fierro and B. Pawelec, *Green Chem.*, 2012, **14**, 2759.

100. W. B. Widayatno, G. Q. Guan, J. Rizkiana, X. Du, X. G. Hao, Z. L. Zhang and A. Abudula, *Bioresour. Technol.*, 2015, **179**, 518.
101. Y. Xu, T. Wang, L. Ma, Q. Zhang and W. Liang, *Appl. Energy*, 2010, **87**, 2881.
102. N. Martínez, R. García, J. L. G. Fierro, C. Wheeler, R. N. Austin, J. R. Gallagher, J. T. Miller, T. R. Krause, N. Escalona and C. Sepúlveda, *Fuel*, 2016, **186**, 112.
103. N. T. Tran, Y. Uemura, S. Chowdhury and A. Ramli, *Appl. Catal., A*, 2016, **512**, 93.
104. J. Sun, A. M. Karim, H. Zhang, L. Kovarik, X. S. Li, A. J. Hensley, J. S. McEwen and Y. Wang, *J. Catal.*, 2013, **306**, 47.
105. Q. Lai, C. Zhang and J. H. Holles, *Appl. Catal., A*, 2016, **528**, 1.
106. T. A. Le, H. V. Ly, J. Kim, S. S. Kim, J. H. Choi, H. C. Woo and M. R. Othman, *Chem. Eng. J.*, 2014, **250**, 157.
107. S. H. Hakim, B. H. Shanks and J. A. Dumesic, *Appl. Catal., B*, 2013, **142**, 368.
108. M. Peroni, G. Mancino, E. Baráth, O. Y. Gutiérrez and J. A. Lercher, *Appl. Catal., B*, 2016, **180**, 301.
109. Y. Yang, A. Gilbert and C. C. Xu, *Appl. Catal., A*, 2009, **360**, 242.
110. A. Popov, E. Kondratieva, L. Mariey, J. M. Goupil, J. El Fallah, J. P. Gilson, A. Travert and F. Maugé, *J. Catal.*, 2013, **297**, 176.
111. I. Gandarias, V. L. Barrio, J. Requies, P. L. Arias, J. F. Cambra and M. B. Güemez, *Int. J. Hydrogen Energy*, 2008, **33**, 3485.
112. Y. Xu, T. Wang, L. Ma, Q. Zhang and W. Liang, *Appl. Energy*, 2010, **87**, 2886.
113. R. N. Olcese, M. Bettahar, D. Petitjean, B. Malaman, F. Giovanella and A. Dufour, *Appl. Catal., B*, 2012, **115**, 63.
114. C. S. Wei, L. Zhao, X. Kadis, E. Cao, Y. Julson and Z. Gu, *New Biotechnol.*, 2016, **33**, 440.
115. H. Shafaghat, P. S. Rezaei and W. M. A. W. Daud, *Ind. Eng. Chem. Res.*, 2016, **35**, 268.
116. W. Wang, Y. Yang, H. Luo, T. Hu and W. Liu, *Catal. Commun.*, 2011, **12**, 436.
117. W. Zhang, Y. Zhang, L. Zhao and W. Wei, *Energy Fuels*, 2010, **24**, 2052.
118. I. T. Ghampson, C. Sepúlveda, R. Garcia, B. G. Frederick, M. C. Wheeler, N. Escalona and W. J. DeSisto, *Appl. Catal., B*, 2012, **413**, 78.
119. N. Koike, S. Hosokai, A. Takagaki, S. Nishimura, R. Kikuchi, K. Ebitani, Y. Suzuki and S. T. Oyama, *J. Catal.*, 2016, **333**, 115.
120. P. Duan, C. Zhang, F. Wang, J. Fu, X. Lü, Y. Xu and X. Shi, *Catal. Today*, 2016, **274**, 73.
121. S. Zhang, Y. Yan and Z. Ren, *Energy Sources*, 2009, **31**, 639.
122. S. M. Richardson, H. Nagaishi and M. R. Gray, *Proc. 210th National Meeting of the American Chemical Society*, 1995, vol. 40, p. 455.
123. M. D. Argyle and C. H. Bartholomew, *Catalysts*, 2015, **5**, 145.
124. Q. Lu, W. Z. Li and X. F. Zhu, *Energy Convers. Manage.*, 2009, **50**, 1376.

125. X. Hu, R. Gunawan, D. Mourant, Y. Wang, C. Lievens, W. Chaiwat, L. P. Wu and C. Z. Li, *Bioresour. Technol.*, 2012, **123**, 249.
126. Y. Xiong, Z. H. Zhang, X. Wang, B. Liu and J. T. Lin, *Chem. Eng. J.*, 2014, **235**, 349.
127. J. Peng, P. Chen, H. Lou and X. Zheng, *Energy Fuels*, 2008, **22**, 3489.
128. M. Ahmad, S. Ahmed, F. U. Hassan, M. Arshad, M. A. Khan, M. Zafar and S. Sultana, *Afr. J. Biotechnol.*, 2010, **9**, 8630.
129. X. Hu, R. Gunawan, D. Mourant, Y. Wang, C. Lievens, W. Chaiwat, L. P. Wu and C. Z. Li, *Bioresour. Technol.*, 2012, **123**, 249.
130. A. Oasmaa, J. Korhonen and E. Kuoppala, *Energy Fuels*, 2011, **25**, 3307.
131. X. Jiang and N. Ellis, *Energy Fuels*, 2010, **24**, 1358.

CHAPTER 7

Noble Metal Based Bimetallic Catalysts for the Catalytic Hydrotreatment of Phenolic Model Components for (Pyrolytic) Lignins

W. YIN, A. KLOEKHORST, I. HITA AND H. J. HEERES*

Department of Chemical Engineering, University of Groningen, Nijenborgh 4, Groningen, 9747 AG, The Netherlands
*E-mail: h.j.heeres@rug.nl

7.1 Introduction

7.1.1 Lignin: General Features and Molecular Composition

The impending depletion of fossil resources together with environmental concerns have boosted research and development activities on renewable resources. An interesting alternative is (lignocellulosic) biomass, which is currently already being used for the production of carbon based transportation fuels (*e.g.* bioethanol, biodiesel) and has high potential for biobased chemicals.[1–4]

Lignin is the third largest bio-polymer found in lignocellulosic biomass. It has a rigid structure and provides the plant with tissue strength and

Green Chemistry Series No. 56
Sustainable Catalysis for Biorefineries
Edited by Francesco Frusteri, Donato Aranda and Giuseppe Bonura

Published by the Royal Society of Chemistry, www.rsc.org

hydrophobicity. The lignin content varies between plant species and, for instance, the lignin content of softwood (*e.g.* pine, spruce) is about 28%, whereas it is about 20% for hardwood (*e.g.* oak, birch).[5]

From a chemical perspective, lignin is an amorphous polyphenolic thermoset, which is made in the plant/tree by radical coupling of three aromatic monomers (coniferyl, sinapyl, and *p*-coumaryl alcohols).[6] A highly complex three-dimensional polymer is formed with different types of linkages between the aromatic nuclei. Among these, the β-O-4, α-O-4, β-5, and 5-5 are the most common, see Figure 7.1 for details. The main linkages (about 70%) in lignin are C–O–C ether bonds (β-O-4 and a smaller amount of α-O-4), while the remaining linkages consist primarily of C–C bonds (5-5 and β-5). The high amounts of aromatic fragments in lignin make it an excellent feedstock for the production of aromatic biobased chemicals, like benzene, toluene, xylene (BTX), and alkylphenolics. This is widely recognized now and the topic has attracted considerable interest from both academia and industry.[5,7–10]

The paper and pulp industry is a major producer of lignin with an estimated yearly production of 50 million tons world-wide, which is primarily used for steam and power production.[13,14] Two important processes are used in the paper industry, *viz.* Kraft and sulphite pulping.[15] During sulphite pulping, lignin is depolymerized and sulphonate groups are introduced

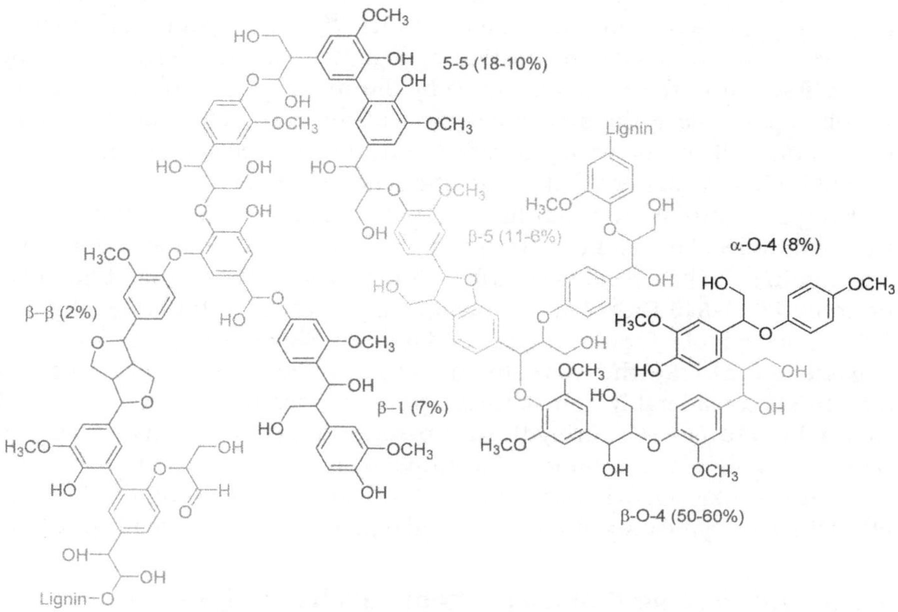

Figure 7.1 Structural model of spruce lignin (softwood) based on Adler *et al.*,[11] with typical amounts of the different types of linkages in soft- and hardwood.[12]

upon treatment with various salts of sulphurous acid. On the other hand, Kraft pulping uses sodium sulphite under alkaline conditions to solubilize the lignin.

Lignin may also be isolated by advanced lignocellulosic biomass pre-treatment techniques like steam explosion[16,17] or organosolv processes. In the latter, the biomass source is treated with a mixture of an organic solvent (*e.g.* ethanol) and water under pressure to separate the lignin from the other biopolymers.[18–20]

7.1.2 Pyrolytic Lignin

Another interesting form of lignin, also known as pyrolytic lignin, is produced from fast pyrolysis oils. Such pyrolysis oils can be obtained from (lignocellulosic) biomass using fast pyrolysis technology in yields of up to 70 wt% and are considered an attractive liquefied form of biomass that has considerable advantages compared to solid biomass. Examples are their ease of transportation, a higher energy density per volume, and less contaminants (minerals). In addition, fast pyrolysis technology has been shown to be economically viable in small scale units allowing decentralized biomass processing.

The pyrolysis oil may be separated into two main fractions (an aqueous and an organic phase) with distinct differences in chemical composition. The aqueous phase typically contains high amounts of polar organics like sugars, organic acids, furanics, and phenolics.[21,22] The organic phase mainly consists of the so called pyrolytic lignin, oligomeric lignins derived from the lignin fraction in the biomass source by thermal depolymerization. Valorization of particularly the pyrolytic lignin fraction is still in a state of infancy even though it is potentially a very interesting source for low molecular weight biobased chemicals like aromatics and phenolics.[23]

Pyrolytic lignin and natural lignin differ considerably in chemical structure and molecular weight distribution. Bayerbach *et al.* characterized pyrolytic lignin from beech using several techniques and showed that it has a M_w of around 560–840 Da.[24] This is considerably lower than for typical lignins (1600–2000 g mol^{-1} for Alcell hardwood lignin,[23] 4800–5200 g mol^{-1} for soda bagasse lignin).[25] Furthermore, the molecular structure of pyrolytic lignins also differs considerably from natural lignins as it has already been processed thermally (450–600 °C) during the fast pyrolysis process. The main bonds in pyrolytic lignin are C–C bonds, like stilbene and diphenyl (5-5) bonds, and C–O bonds, like phenyl coumaran (β-1) and resinol (β-β) bonds while the ether linkages typically present in natural lignin are absent[26] (see Figure 7.2).

7.1.3 Interesting Biobased Chemicals from Lignin

Lignins and pyrolytic lignins are interesting renewable sources for aromatic chemicals like benzene, toluene, xylene (BTX), and phenolic compounds (see Figure 7.3). These are important base chemicals that are commercially

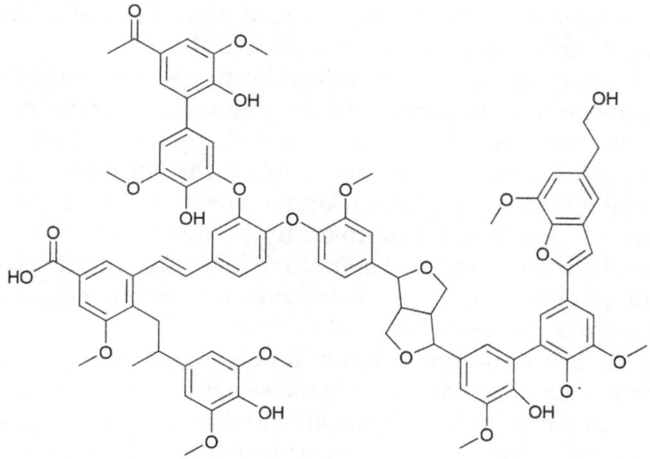

Figure 7.2 Structural model for pyrolytic lignin.[26]

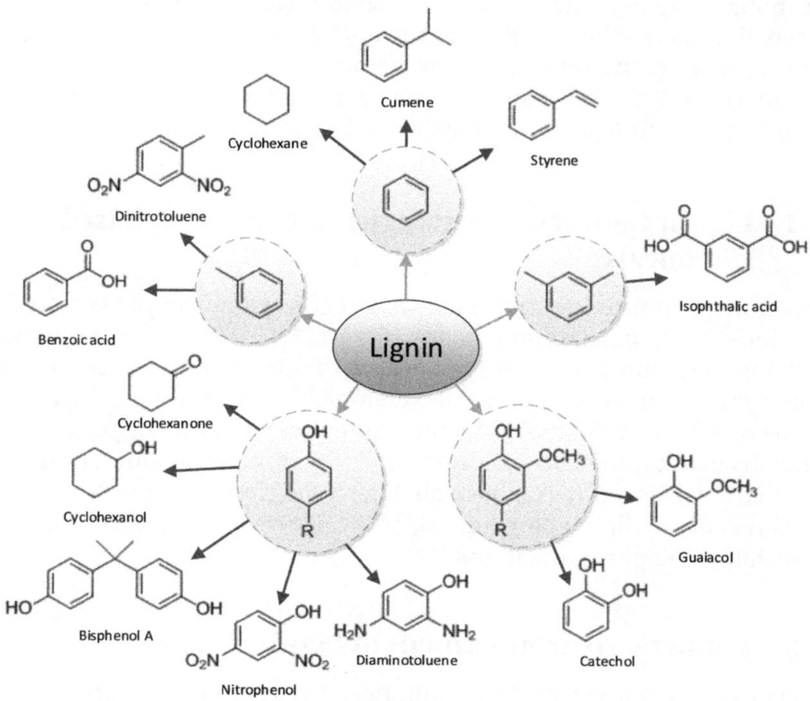

Figure 7.3 Examples of chemicals from lignin.[23]

applied as intermediates for a wide range of other derivatives like styrene, benzoic acid, cyclohexane, and isophthalic acid.[23]

Another interesting class of chemicals from (pyrolytic) lignin are alkyl substituted phenolics. The parent phenol is globally produced at about 8 Mt per year (2008) and is widely used as a base chemical for the production of bisphenol-A, an important monomer applied for polycarbonate synthesis (48%), phenolic resins, *e.g.* phenol formaldehyde resins (25%), cyclohexanone production (the feed for caprolactam production) (11%), and conversion to other base chemicals.[27] The production of renewable aromatic and phenolic compounds from lignin would thus have a beneficial effect on the carbon footprint of the petrochemical industry.

(Pyrolytic) lignin valorization to low molecular weight aromatics and phenolics requires depolymerization of the lignin structure. However, this is not sufficient, as the aromatic nuclei in lignin are heavily substituted, *e.g.* with at least a three-carbon side chain with multiple substituents and additional methoxy and hydroxyl groups. While for the production of aromatics, full deoxygenation is required, for alkylphenolics, removal of the methoxy group(s) is essential. As previously mentioned and exemplified in Figure 7.1, lignins contain significant though variable amounts of ether bonds. Thermodynamic calculations indicate that these ether bonds have the lowest bond dissociation energy and as such, these are in principle the most reactive of all bonds within the lignin structure (Table 7.1). The main challenge is therefore to cleave the remaining C–C linkages (up to 30%), which have a much higher bond dissociation energy.

7.1.4 Lignin Conversion Strategies to Obtain Biobased Chemicals

Lignin conversions to low molecular weight compounds have been achieved by various (catalytic) methodologies (Figure 7.4). These can be classified according to the process severity (pressure and temperature). Examples are lignin pyrolysis including catalytic versions,[7,31] hydrothermal liquefaction in sub- or super critical water[32] and lignin solvolysis in organic solvents in the absence or presence of catalysts.[33,34] Lignin conversion strategies at milder conditions (<300 °C) have also been developed, examples are enzymatic breakdown (lignin peroxidase),[35,36] oxidative depolymerization,[9,37] and base-catalysed depolymerizations.[38,39]

7.1.5 Catalytic Hydrotreatment of Lignin

The catalytic hydrotreatment of lignin has proven to be an interesting technology for the synthesis of (oxygenated) aromatics, and has recently been reviewed by Zakzeski *et al.*[5] For instance, Oasmaa *et al.*[41] and Meier *et al.*[42] showed that lignin may be converted to monomeric alkylphenolics like phenol, and *o*/*p*/*m*-cresol, albeit in relatively low yield (5–11 wt% of the lignin feed). In addition, aromatics (BTX) may also be obtained.[5,7] Analogous to

Table 7.1 Bond dissociation enthalpies (BDEs) for lignin linkages.

Lignin compounds	BDE (kJ mol^{-1})	Lignin compounds	BDE (kJ mol^{-1})
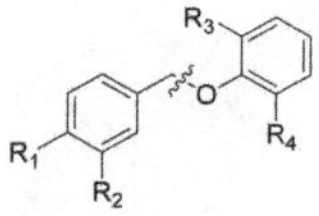 α-O-4	220 ± 16[b]	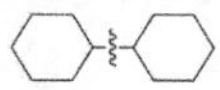	369 ± 4[a]
		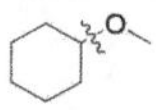	376 ± 4[a]
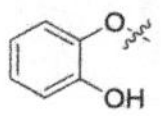	235 ± 4[a]	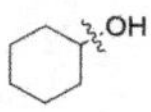	398 ± 4[a]
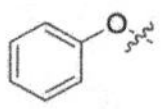	263 ± 4[a]	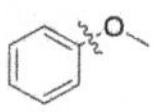	419 ± 4[a]
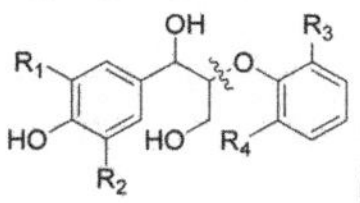 β-O-4	292 ± 5[b]	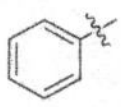	427 ± 4[a]
	281 ± 1[c]	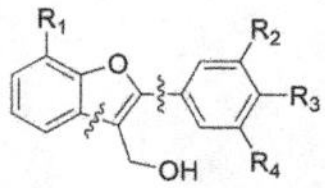 β-5	430 ± 8[b]
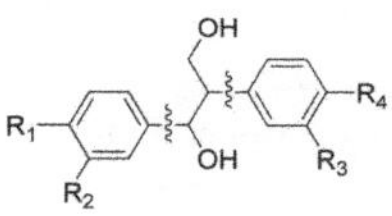 β-1	289 ± 2[b]	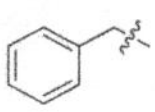	435 ± 4[a]
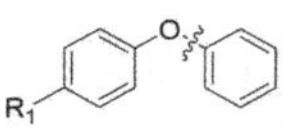 4-O-5	332 ± 7[c]	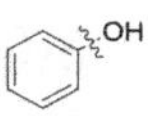	463 ± 4[a]
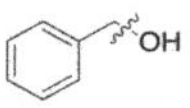	334 ± 4[a]	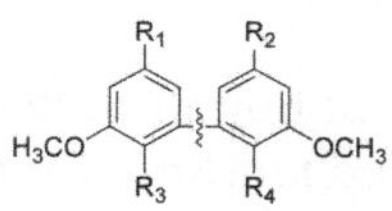 5-5	482 ± 12[b]
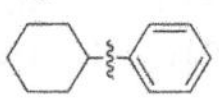	356 ± 4[a]		488 ± 12[c]
	358 ± 4[a]		

[a]BDE from Benson.[28]
[b]Calculated BDE values from Parthasarathi *et al.*[29]
[c]Calculated BDE values from Kim *et al.*[30]

lignin, the catalytic hydrotreatment of pyrolytic lignin may also be a very interesting valorization approach to obtain monomeric biobased building blocks like alkylphenolics and aromatics. Compared to lignin, the use of pyrolytic lignin may have an additional advantage as the molecular weight is considerably lower, which may have a positive effect on the monomer yields. However, very limited research has been reported on the catalytic hydrotreatment of pyrolytic lignins.[43–45]

Catalytic hydrotreatment is a process where a (pyrolytic) lignin reacts with hydrogen in the presence of a catalyst at elevated temperatures and pressures. Typical reaction temperatures are between 200 and 450 °C and hydrogen pressures between 50 and 250 bar. Hydrogen can be supplied *ex situ* by the addition of molecular hydrogen or formed *in situ* by the use of hydrogen-donating compounds (*e.g.* isopropanol or formic acid). Catalytic hydrotreatment of lignin aims for (i) depolymerization of the lignin to lower molecular weight fragments by cleavage of C–O–C and C–C bonds, and

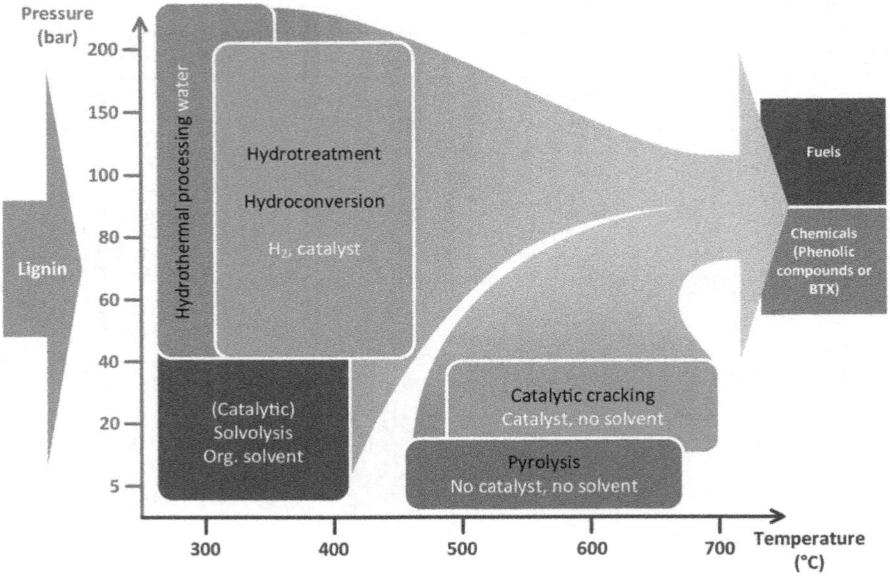

Figure 7.4 Various thermochemical pathways and process parameters (*P*, *T*) for the depolymerisation of lignin.[40]

(ii) (partial) removal of bound oxygen for the production of phenolics and aromatics. A simplified reaction for the catalytic hydrotreatment of lignin to phenol is given in eqn (7.1).

$$C_6H_{6.36}O_2 + 0.82H_2 \rightarrow C_6H_6O + H_2O \quad (7.1)$$

Various types of heterogeneous catalysts have been applied for the catalytic hydrotreatment of (pyrolytic) lignins into both biobased chemicals and biofuels (for an overview, the reader is referred to ref. 5, 46 and 47). Well known examples are noble metals on various supports (*e.g.* Ru, Pd) and cheaper Ni and Co based catalysts, often in combination with a promoter.

There has been an increased interest in the use of bimetallic catalysts for various catalytic transformations. In a number of cases, these have shown superior electronic and chemical properties that are distinctly different from their parent metals, resulting in enhanced selectivity, activity, and stability.[48] Bimetallic catalysts have also been tested for the catalytic hydrotreatment of lignins, even though the reported studies are mainly limited to NiMo and CoMo catalysts and using particularly alumina as the support.[49,50]

7.1.6 Scope of the Current Review

As discussed above, lignin is a highly cross-linked polymer with a large number of linkages between the aromatic nuclei and, as such, it is very challenging to obtain information about the reaction pathways occurring during its catalytic hydrotreatment. Therefore, the use of model components is highly

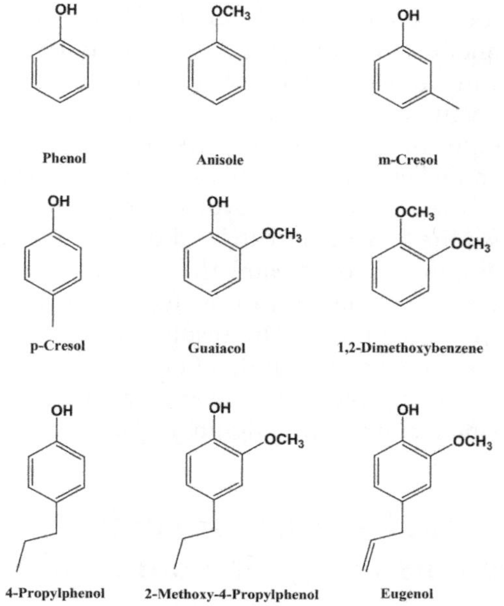

Scheme 7.1 Examples of phenolic model components.

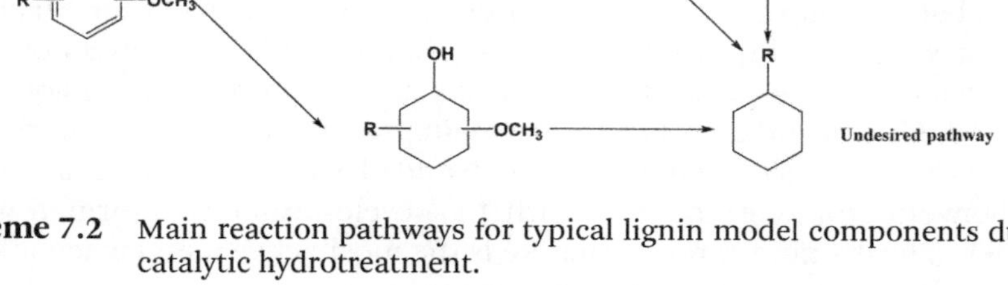

Scheme 7.2 Main reaction pathways for typical lignin model components during catalytic hydrotreatment.

advantageous and can provide valuable insights into the reactivity of the different bonds in the lignin structure. With this knowledge, the selectivity towards, for instance, aromatics or alkylphenolics can be optimized through proper catalyst design and selection. Typical model components used to determine reactivity patterns of catalysts are given in Scheme 7.1.

For such model components, a number of reactions can be envisaged, *viz.*, (i) hydrogenation of C=C bonds, (ii) hydrogenolysis reactions of C–O bonds, and (iii) hydrogenolysis of C–C bonds, see Scheme 7.2 for details. The reactivity of the model components will be determined by the catalyst properties and process conditions. When aiming for aromatics, the formation of saturated alkanes by hydrogenation of C=C bonds must be prevented as the economic value of such alkanes is much lower than for aromatics.

The primary objective of the current review is to provide an overview of the use of bimetallic noble metal based catalysts for the catalytic hydrotreatment of lignin-derived model components. Bimetallic catalysts provide an opportunity to enhance the efficiency of hydrodeoxygenation catalysts, as the interaction between the two metals can modify the geometric and electronic properties, resulting in improved catalytic activity, selectivity and stability. The bimetallic catalysts discussed in the chapter are limited to those that contain at least one noble metal. The results in this review are structured based on the type of model compound, followed by a critical discussion section to identify the most promising options for future catalytic hydrotreatment reactions using real (pyrolytic) lignin feeds.

7.2 Experimental Studies for Lignin Model Components Using Bimetallic Noble Metal Catalysts

7.2.1 Overview for Anisole

An overview of reported studies on the use of anisole is given in Table 7.2. Reactions were either carried out in an apolar solvent or in solvent-free conditions using either continuous or batch set-ups. Typically, temperatures between 300–450 °C and hydrogen pressures between 1 and 30 bar were applied.

Ghampson *et al.*[51] reported the catalytic hydrotreatment of anisole in dodecane using bimetallic Re–MoO_x/TiO_2 and Re–VO_x/TiO_2 catalysts in a stirred-batch set-up at 300 °C at a hydrogen pressure of 30 bar for 4 h (Table 7.2, Entry 1). Initially, monometallic Re, Ga, Ni, and Co catalysts on TiO_2 were tested and the product distributions at 10% anisole conversion were reported. Phenol and cresols were the main products (up to 98%) for the Ga, Ni, and Co catalysts (Scheme 7.3, route b and c), whereas up to 19% aromatics (benzene, toluene) together with 18% cyclohexane was formed using Re/TiO_2. These findings reveal that Re is the most active metal when aiming for fully deoxygenated products like aromatics.

Bimetallic Mo and V promoted Re, Ga, Ni, and Co on a TiO_2 support were also tested and the product distributions at 10% anisole conversion were reported. For the bimetallic Re–Mo and Re–V/TiO_2 catalysts, the selectivity to aromatics increased to values of up to 60% and 48%, respectively (compared to 19% for the monometallic Re catalyst).

It was observed that performance of the bimetallic Re–Mo and Re–V catalysts is a function of the Re/M ratio and catalysts with equimolar amounts led to the highest yield of aromatics. For instance, a 35% yield of benzene and toluene at 48% conversion of anisole was achieved for Re–MoO_x/TiO_2,

Table 7.2 Summary of experimental studies on the catalytic hydrotreatment of anisole using bimetallic noble metal based catalysts.

Entry	Catalyst	Solvent	Reactor	Conditions	Conversion/selectivity/yield	Reference
1	$ReMoO_x/TiO_2$ $Re–VO_x/TiO_2$	Dodecane	Batch	300 °C, 30 bar H_2, 4 h	Conversion 32% for Mo/TiO_2 17% for V/TiO_2 30% for Re/TiO_2 48% for $Re–MoO_x/TiO_2$ (Re : Mo = 50 : 50) 30% for $Re–VO_x/TiO_2$ (Re : V = 50 : 50) Yield Mo/TiO_2 27% phenol 5% cresols V/TiO_2 14.5% phenol 2.5% cresols Re/TiO_2 9% phenol 7% cresols 6.5% benzene 5.5% cyclohexane 2% toluene $Re–MoO_x/TiO_2$(Re : Mo = 50 : 50) 25% benzene 7% toluene 5% phenol 5% cresol 4% cyclohexane $Re–VO_x/TiO_2$(Re : V = 50 : 50) 13% benzene 2% toluene 5% phenol 3% cresol 5% cyclohexane	51

(continued)

Table 7.2 (*continued*)

Entry	Catalyst	Solvent	Reactor	Conditions	Conversion/selectivity/yield	Reference
2	PtZn/C	*n*-Heptane	Flow reactor	325 °C and 27.5 bar, W/F = 0.5 g min ml^{-1} for 10 wt% Pt/C, W/F = 0.1667 gmin ml^{-1} for 1 wt% Pt/C, W/F = 0.5 gmin ml^{-1} for 10 wt% PtZn/C	Conversion 75% for 10 wt% Pt/C 31% for 1 wt% Pt/C 27% for 10 wt% PtZn/C Selectivity 74% methoxycyclohexane for 10 wt% Pt/C 88% methoxycyclohexane for 1 wt% Pt/C 55% phenol for 10 wt% PtZn/C	52
3	Ru–Ni/AlSBA-15	Solvent free	Flow reactor	350–450 °C, WHSV = 4.1771 h^{-1}; atmosphere pressure; H_2 = 50 ml min^{-1}, anisole flow rate = 1.0 ml h^{-1}.	Ru–Ni/AlSBA-15 (0.4 wt% Ru, 7 wt% Ni) 400 °C, TOS = 15 h 15% conversion Selectivity 61.6% benzene 33% phenol 3% toluene Ru/AlSBA-15 450 °C, TOS = 6h 3% conversion Selectivity 62% phenol 38% benzene Ni/AlSBA-15 450 °C, TOS = 21 h 18% conversion Selectivity 35% benzene 63% phenol 2% toluene	53

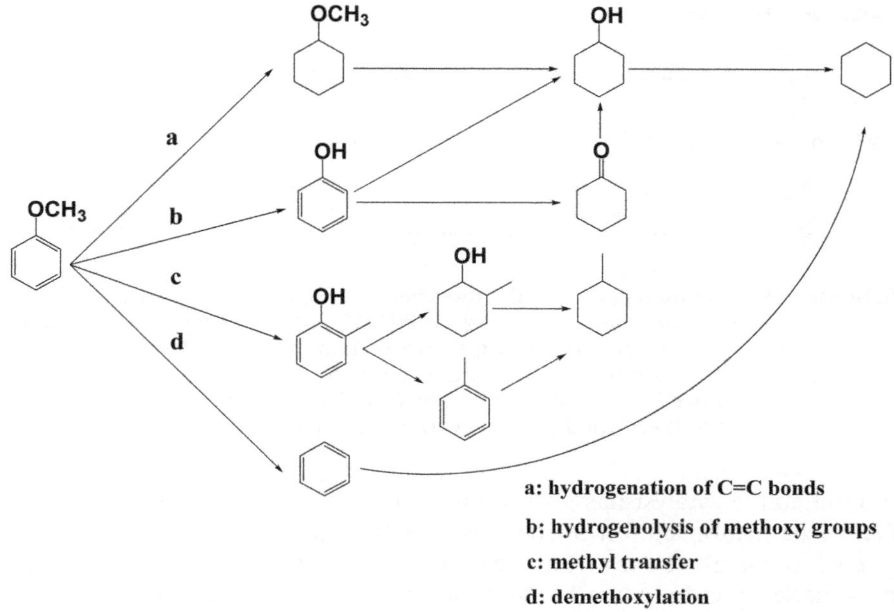

Scheme 7.3 Reaction network for anisole during catalytic hydrotreatment.

compared to 8.5% aromatics at around 30% conversion of anisole using the monometallic Re/TiO_2 catalyst. Catalyst characterization revealed that the introduction of Mo and V cations dramatically increases the extent of adsorption of anisole on the catalyst surface by interactions with MoO_x or VO_x, which promotes the demethoxylation of anisole to aromatics.

Shi *et al.*[52] reported the catalytic hydrotreatment of anisole using a bimetallic PtZn/C catalyst in a continuous flow reactor and using *n*-heptane as the solvent (Table 7.2, Entry 2) and compared its performance with monometallic Pt/C catalysts (10 and 1 wt%). For the latter, the anisole conversion was 75% with a high selectivity to methoxycyclohexane (74%), indicating substantial hydrogenation of the C=C bonds (Scheme 7.3, route a). Minor amounts of cyclohexane and cyclohexanol were also formed.

Product selectivity at lower conversions was also determined using the 1 wt% Pt/C catalyst. At the selected conditions (325 °C, 27.5 bar, $W/F = 0.5$ gmin ml^{-1}), the anisole conversion was reduced to 31%, with an increase in selectivity to methoxycyclohexane (88%).

Experiments with the bimetallic PtZn/C catalyst showed that this catalyst is less active than Pt/C under similar reaction conditions, but it is much more selective for the formation of phenol (selectivity of 66% at 27% conversion). This indicates that the introduction of Zn leads to a reduction of the rate of hydrogenation of C=C bonds (Scheme 7.3, route b) and allows for the removal of the methoxy group by C–O hydrogenolysis (Scheme 7.3, route b). A reaction mechanism was proposed suggesting that the Pt(111) plane in the

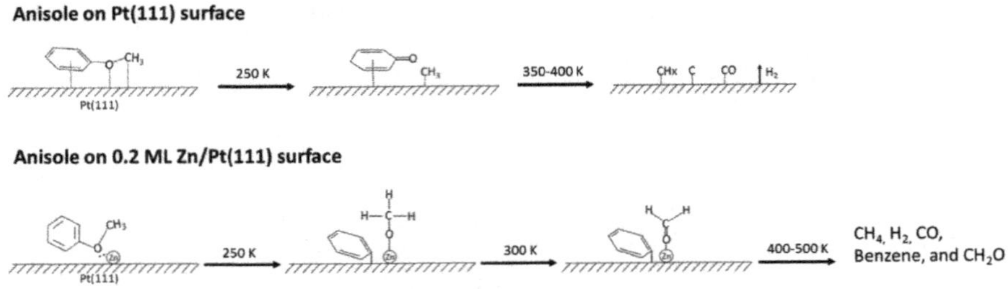

Scheme 7.4 Proposed pathways and intermediates for the adsorption and reaction of anisole on Pt(111) and Zn/Pt(111) surfaces. Adapted from *Journal of Catalysis*, **340**, D. Shi, L. Arroyo-Ramirez and J. M. Vohs, The use of bimetallics to control the selectivity for the upgrading of lignin-derived oxygenates: Reaction of anisole on Pt and PtZn catalysts, 219–226, Copyright 2016, with permission from Elsevier.[52]

monometallic catalyst has a strong interaction with the phenyl ring, which facilitates hydrogenation of the ring. For the bimetallic PtZn catalyst, the phenyl ring is tilted away from the surface and bonding primarily occurs *via* oxygen. Selective C–O bond scission at (or near) Zn sites takes place, which explains why PtZn catalysts are more selective towards phenol formation (see Scheme 7.4 for details).

Pichaikaran *et al.*[53] studied the vapor phase hydrotreatment of anisole using both monometallic Ni and Ru catalysts (metal contents of 7 wt% and 1.0 wt%, respectively) and their bimetallic versions (Ru contents between 0.2 and 1.0 wt% at a fixed Ni content of 7 wt%) on a mesoporous aluminosilicate in a downflow fixed-bed reactor operated between 300–450 °C at atmospheric hydrogen pressure (Table 7.2, Entry 3). The monometallic Ni and Ru catalysts showed the highest activity at 450 °C, in contrast with the bimetallic Ru–Ni catalysts, which performed better at 400 °C. Phenol was detected as an intermediate compound for all catalysts (Scheme 7.3, route b), together with benzene and small amounts of toluene. Among the bimetallic catalysts, Ru–Ni (0.4 wt% Ru) showed the highest selectivity towards benzene (61.6% at 15% anisole conversion, TOS of 15 h, 400 °C), together with 33% phenol and 3% toluene. For the monometallic Ni and Ru catalysts at similar conversion levels of anisole, the selectivity towards aromatics (benzene and toluene) was lower (38%) in comparison with their bimetallic counterpart.

The highest conversion of anisole at the start of the reaction (TOS < 12 h) was achieved for the bimetallic Ru–Ni catalyst with 1.0 wt% Ru content. Rapid catalyst deactivation was observed, as evidenced from a drop in the anisole conversion from 47% (3 h TOS) to 7% (12 h TOS). Similar observations were reported for the monometallic Ni catalyst, while a slight increase in activity was obtained using the monometallic Ru catalyst. Characterization of the spent catalysts using TGA revealed that coke formation on the bimetallic catalysts was much lower than for the monometallic catalysts, suggesting that coke formation on the Ni catalysts can be suppressed by the addition of Ru.

7.2.2 Overview for *m*-cresol

An overview of experimental studies on the use of bimetallic catalysts for the hydrotreatment of *m*-cresol is given in Table 7.3. All reactions were carried out in continuous set-ups and in solvent-free conditions. Temperatures were between 250 and 300 °C and pressures between 0.5 and 5 bar were applied. A reaction network with possible products is given in Scheme 7.5.

Do *et al.*[54] investigated the catalytic hydrotreatment of *m*-cresol in the vapor phase in a packed bed flow reactor on γ-Al_2O_3-supported Pt catalysts at near atmospheric hydrogen pressure (Table 7.3, Entry 1). In the first phase of experimentation, monometallic Pt/Al_2O_3 was tested, giving an *m*-cresol conversion of 38% (W/F of 63.0 g of cat.h mol^{-1} at 30 min TOS). A high selectivity towards deoxygenated products like toluene (65%) and methylcyclohexane (11%) was obtained (Scheme 7.5). The reactivity of the Pt catalyst was enhanced by the incorporation of Co and Ni, achieving *m*-cresol conversions of 56.7 and 62.7% using the Pt–Co/Al_2O_3 and Pt–Ni/Al_2O_3 catalysts, respectively. Furthermore, the selectivity of the bimetallic catalysts was also improved, obtaining higher amounts of methylcyclohexane. The authors conclude that these improvements are due to a higher hydrogenation activity combined with the presence of sites for alcohol (methylcyclohexanol) dehydration followed by hydrogenation to methylcyclohexane.

Robinson *et al.*[55] investigated the catalytic hydrotreatment of *m*-cresol using two bimetallic Pt–Mo/γ-Al_2O_3 catalysts with different Pt to Mo ratios (50 bar, 250 °C). The conversion of *m*-cresol was tuned by varying the weight hourly space velocity (WHSV, 100–1000 h^{-1}) (Table 7.3, Entry 2). The reactivity and product selectivity of the two bimetallic catalysts (Pt10Mo1/γ-Al_2O_3, Pt1Mo1/γ-Al_2O_3) were compared with the monometallic Pt/γ-Al_2O_3 catalyst. At conversion levels higher than 90%, 42% methylcyclohexanol, 29% methylcyclohexanone and 29% methylcyclohexane were obtained using the monometallic Pt/γ-Al_2O_3catalyst. The product composition shifted to methylcyclohexane when using the bimetallic Pt–Mo/γ-Al_2O_3 catalysts. The highest selectivity to methylcyclohexane (95%) was obtained using Pt1Mo1/γ-Al_2O_3 at 95% *m*-cresol conversion, indicating that the bimetallic catalysts yield higher amounts of saturated, fully deoxygenated compounds. Two main pathways for the hydrodeoxygenation of *m*-cresol were proposed, *i.e.* (i) hydrogenation to methylcyclohexanol, followed by further dehydration–hydrogenation to methylcyclohexane and (ii) tautomerization of *m*-cresol to an unsaturated cyclohexanone followed by deoxygenation to methylcyclohexane (Scheme 7.6). DFT calculations indicate that the Mo sites on the PtMo bimetallic surface not only dramatically lower the barrier for *m*-cresol tautomerization and subsequent deoxygenation but also enhance the adsorption of *m*-cresol.

Yang *et al.*[56] investigated the catalytic hydrotreatment of *m*-cresol using mono- and bimetallic Ni–Re catalysts in the vapor phase at atmospheric pressure (300 °C, Table 7.3, Entry 3). The bimetallic Ni–Re/SiO_2 catalyst showed superior performance in terms of *m*-cresol conversion and toluene yield

Table 7.3 Summary of experimental studies on the catalytic hydrotreatment of *m*-cresol using bimetallic noble metal based catalysts.

Entry	Catalyst	Solvent	Reactor	Conditions	Conversion/selectivity/yield	Reference
1	$PtCo/Al_2O_3$, $Pt–Ni/Al_2O_3$	Solvent free	Packed-bed reactor	$T = 260$ °C, TOS = 30 min $P_{H2} = 0.505$ bar, $P_{N2} = 0.505$ bar $W/F = 63$ g of cat h mol^{-1} Hydrogen/*m*-cresol ratio was 70	Conversion 38% for Pt/Al_2O_3 56.7% for $Pt–Co/Al_2O_3$ 62.7% for $Pt–Ni/Al_2O_3$ Selectivity Pt/Al_2O_3 65% toluene 20% 3-methylcyclohexanol 11% cyclohexane $Pt–Co/Al_2O_3$ 44.7% toluene 52.4% methylcyclohexane 1.2% methylcyclohexanol $Pt–Ni/Al_2O_3$ 51.0% toluene 43.2% methylcyclohexane 3.1% methylcyclohexanol	54
2	$Pt–Mo/\gamma$-Al_2O_3	Solvent free	Packed-bed reactor	5 bar and 250 °C With 12 : 1 molar ratio of H_2 : oxygenate	Conversion > 90% Selectivity Pt/γ-Al_2O_3 29% cyclohexane 29% methylcyclohexanone 42% methylcyclohexanol $Pt10Mo1/\gamma$-Al_2O_3 65% cyclohexane 6% methylcyclohexanone 29% methylcyclohexanol $Pt1Mo1/\gamma$-Al_2O_3 95% cyclohexane 2.5% methylcyclohexanone 2.5% methylcyclohexanol	55

3	Ni–Re/SiO_2	Solvent free	Packed-bed reactor	T = 300 °C, P = 1 bar H_2/*m*-cresol = 60 W/F = 1 h, TOS = 30 min	Ni/SiO_2 10.3% conversion Re/SiO_2 3.0% conversion Ni–Re/SiO_2 47.6% conversion Yield Ni/SiO_2 1.9% toluene 0.1% benzene 2.4% phenol 2.2% methylcyclohexanone 0.5% methylcyclohexanol Re/SiO_2 2.9% toluene Ni–Re/SiO_2 23.7% toluene 2.6% benzene 5.9% phenol 3.0% methylcyclohexanone 0.7% methylcyclohexanol 0.04% methylcyclohexane	56

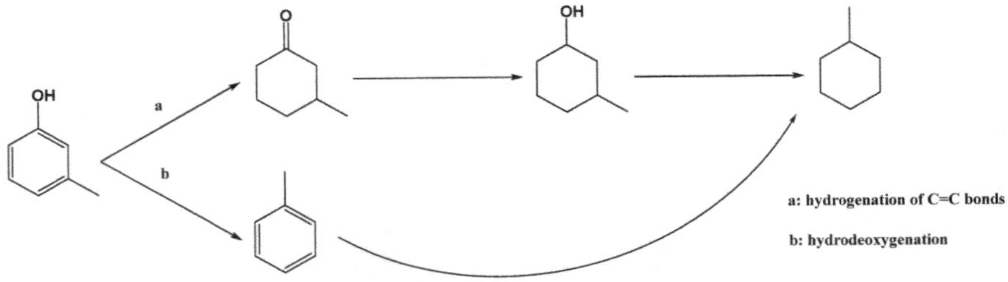

Scheme 7.5 Possible routes for the catalytic hydrotreatment of *m*-cresol.

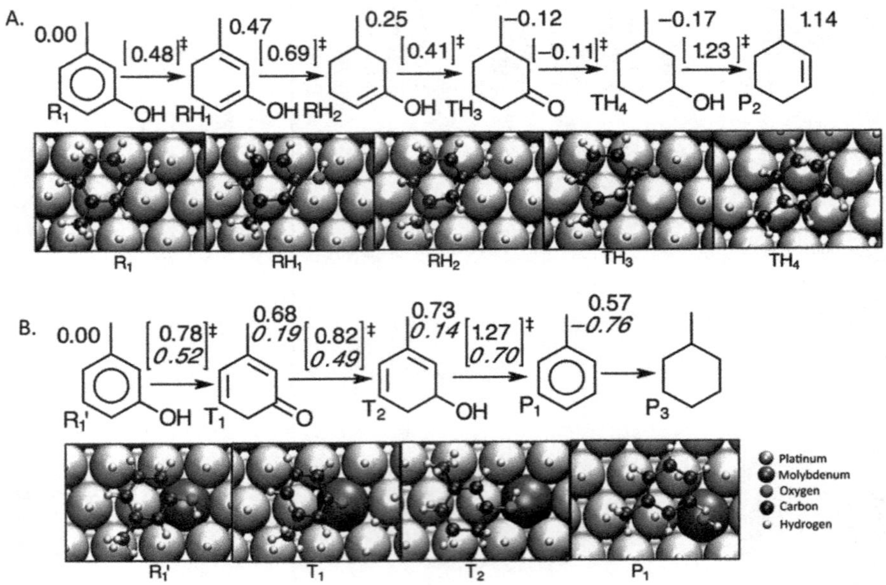

Scheme 7.6 Reaction network calculated for direct ring hydrogenation on H/Pt(111) shown in pathway A and tautomerization shown over both monometallic and bimetallic surfaces in pathway B. All values are in eV. In pathway B, the values in italics are reaction energies and barriers over the Mo site in the H/Pt(111) surface, and the values in plain text are the corresponding value on an unmodified H/Pt(111) surface. Important minima are represented by ball and stick models with the designation corresponding to the reaction networks. Reaction barriers are denoted in brackets with a ‡, and other reported values are reaction energies. Adapted with permission from A. Robinson, G. A. Ferguson, J. R. Gallagher, S. Cheah, G. T. Beckham, J. A. Schaidle, J. E. Hensley, J. W. Medlin, *ACS Catal.*, 2016, **6**, 4356. Copyright 2016 American Chemical Society.[55]

compared to the monometallic Ni and Re catalysts. For instance, at similar conditions, the conversion for the bimetallic catalyst was 47.6%, compared to 10.3% for Ni/SiO_2 and 3% for Re/SiO_2. Further studies were performed using different metal ratios in the bimetallic catalysts (Ni: 5 wt%, Re: 2.5 to 10 wt%). The reaction rate of *m*-cresol increased from 0.536 μmol g_{cat}^{-1} s^{-1} for the monometallic catalyst (5% Ni) to 3.367 and 6.764 μmol g_{cat}^{-1} s^{-1} for 5Ni2.5Re and 5Ni5Re, respectively. Catalyst characterization revealed that the formation of a Ni–Re alloy on the catalytic surface of the Ni–Re catalysts occurs after reduction and that addition of Re increases the Ni dispersion. DFT calculations indicate that the Ni–Re sites are the active site for C–O bond breakage by coordination of the O atom to Re and the phenyl ring on the neighboring Ni atom, which facilitates deoxygenation to toluene.

7.2.3 Overview for Guaiacol

An overview of experimental studies on the use of guaiacol as a model component for the catalytic hydrotreatment is given in Table 7.4. Reactions were either carried out in water or saturated hydrocarbons or solvent free. Mainly batch set-ups were used, though also a number of experiments in continuous set-ups were reported. Temperatures were between 100 and 450 °C and hydrogen pressures were between 0.4 and 80 bar. The proposed reaction pathways including intermediates and products are given in Scheme 7.7.

7.2.3.1 Experiments in Batch Set-ups

Two experimental studies in batch set-ups have been reported using water as solvent. Ishikawa *et al.*[57] studied the catalytic hydrotreatment of guaiacol using bimetallic carbon-supported Ru-based catalysts at 160 °C and 15 bar H_2 (Table 7.4, Entry 1). Results were compared with monometallic Ru catalysts on various supports, including activated carbon, carbon black, SiO_2, TiO_2, ZrO_2 and AlOOH. The main products for the monometallic catalysts were demethoxylated compounds like cyclohexanol and phenol (together with methanol, see Scheme 7.7). Ru on carbon black showed the highest selectivity to cyclohexanol (72% selectivity at >99% conversion) after 2 h of reaction. The bimetallic catalysts with Ru/C as the basis and Fe, Co, Ni, Cu, Zn, Rh, Pd, and Pt as the promoters gave nearly full conversion of guaiacol (except those with Fe and Zn). The highest cyclohexanol selectivity (79%) was achieved using Mn as the promoter, which is slightly higher than those found for the monometallic catalysts. Further optimization studies were performed with bimetallic Ru–Mo catalysts with varying Ru to Mn ratios (Ru/Mn ratio of 0.25–2) and the best results were obtained using Ru–MnO_x/C with a Ru to Mn ratio of 1. The sum of the selectivity to cyclohexanol and phenol for this catalyst was 79% at 69% guaiacol conversion. By tuning the reaction conditions, 82% selectivity to cyclohexanol with full conversion of guaiacol could be achieved in 16 h. The presence of methanol (around 70% selectivity) combined with the high selectivity of phenol at the start of the reaction

Table 7.4 Summary of experimental studies on the catalytic hydrotreatment of guaiacol using bimetallic noble metal based catalysts.

Entry	Catalyst	Solvent	Reactor	Conditions	Conversion/selectivity/yield	Reference
1	Ru–M(MO_x)/C (M = Mn, Fe, Co, Ni, Cu, Zn, Rh, Pd, Pt)	Water	Batch	15 bar H_2, 160 °C, 2 h	>98% conversion for monometallic Ru/C, SiO_2, TiO_2, ZrO_2 and AlOOH >99% conversion for bimetallic Ru–M catalysts (95% for Fe, 78% for Zn) Selectivity Monometallic Ru on various supports 64–72% cyclohexanol 4–12% cyclohexane 14–28% methoxycyclohexanol <1% phenol Bimetallic Ru–M/C catalysts 64–79% for cyclohexanol 4–13% for cyclohexane 15–27% for methoxycyclohexanol	57
2	Ru–M/HY, M = Fe, Ni, Cu, Zn	Water	Batch	40 bar H_2, 250 °C, 2 h	91% conversion for Ru/HY >99% conversion for Ru–Cu/HY Yield Ru/HY 30.0% hydrocarbon 57.0% mono-oxygenated 10.0% gas phase product Ru–Cu/HY 62.4% hydrocarbon 33.2% mono-oxygenated 4.4% gas phase product	58

3	RhPt, RhPd, PdPt/ZrO_2	*n*-Hexadecane	Batch	100 °C, 80 bar, 5 h	Conversion ~10% for Pt/ZrO_2 ~10% for Pd/ZrO_2 ~100% for Rh/ZrO_2 ~5% for PdPt/ZrO_2 ~30% for RhPd/ZrO_2 ~100% for RhPt/ZrO_2 Selectivity Pt/ZrO_2 63% cyclohexanol 16% 1,2-dimethoxybenzene 15% 1-methyl-1,2-cyclohexanediol Pd/ZrO_2 40% cyclohexanol 19% 1,2-dimethoxybenzene 37% 1-methyl-1,2-cyclohexanediol Rh/ZrO_2 12% cyclohexanol 75% 1-methyl-1,2-cyclohexanediol PdPt/ZrO_2 73% cyclohexanol 18% 1,2-dimethoxybenzene 3% 1-methyl-1,2-cyclohexanediol RhPd/ZrO_2 32% cyclohexanol 10% 1,2-dimethoxybenzene 45% 1-methyl-1,2-cyclohexanediol 10% other ring hydrogenated products RhPt/ZrO_2 6% cyclohexanol 92% 1-methyl-1,2-cyclohexanediol	59

(continued)

Table 7.4 (*continued*)

Entry	Catalyst	Solvent	Reactor	Conditions	Conversion/selectivity/yield	Reference
4	Pd–WO_x/γ-Al_2O_3	*n*-Decane	Batch	300 °C, 70 bar H_2, 2.5 h	Pd–WO_x/γ-Al_2O_3, Pd/γ-Al_2O_3 Full conversion of guaiacol Pd/γ-Al_2O_3 62% yield of cyclohexane Pd–WO_x/γ-Al_2O_3 88% yield of cyclohexane	60
5	Re–Cu/SiO_2	*n*-Dodecane	Batch	300 °C, 50 bar H_2	Re/SiO_2 22% conversion 15% yield of phenol 4% yield of BTX 4% yield of cyclohexane Results for highest activity Cu(1.58)–Re/SiO_2 Full conversion at 3 h 75% selectivity to cyclohexane in 3h for Cu(1.58)–Re/SiO_2	61
6	Pt–Sn/Inconel, Pt–Sn/CNF/Inconel	Solvent free	Continuous fixed-bed reactor	400 °C, 1 bar, a gas mixture (50 ml min^{-1}) stream of N_2 (83%) and H_2 (17%)	Guaiacol conversion 10–25% for Sn/Inconel 5–33% for Pt/Inconel 80–100% for PtSn/Inconel 100% for PtSn/CNF/Inconel Yield 10–15% phenol, 0–8% *O*-cresol for Sn/Inconel 4–21% phenol, 2–3% *O*-cresol for Pt/Inconel 55–80% phenol, 0–38% benzene for Pt–Sn/Inconel 30–70% phenol, 5–60% benzene for Pt–Sn/CNF/Inconel TOS = 0–180 min, W/F is 3.2 g of catalyst (g of reactant)$^{-1}$ h^{-1}	62

7	Ni@Pd/ SiO_2–Al_2O_3	Heptane	Continuous fixed-bed reactor	450 °C, TOS = 30 min, W/F = 4.0 g of catalyst/ (g of reactant h^{-1}), the molar ratio of H_2 to guaiacol was about 25	83% conversion for Pd/SiO_2–Al_2O_3 97% conversion for Ni@Pd/SiO_2–Al_2O_3 Selectivity Pd/SiO_2–Al_2O_3 19% benzene 38% toluene 18% xylene 4% cresol 21% phenol 1% anisole Ni@Pd/SiO_2–Al_2O_3 36% benzene 32% toluene 15% xylene 8% cresol 4% phenol 2% anisole	63
8	PdFe/C	Solvent free	Continuous fixed-bed reactor	$P_{guaiacol}$ = 0.004 bar; P_{H2} = 0.4 bar, W/F = 0.15 s.g. STP ml^{-1}, 450 °C	Full conversion for PdFe/C 96% conversion for Fe/C 98% conversion for Pd/C Yield Fe/C 36% benzene 38% phenol 5% toluene/trimethylbenzene Pd/C 79% phenol 3% benzene PdFe/C 75% benzene 8% toluene/trimethylbenzene	64

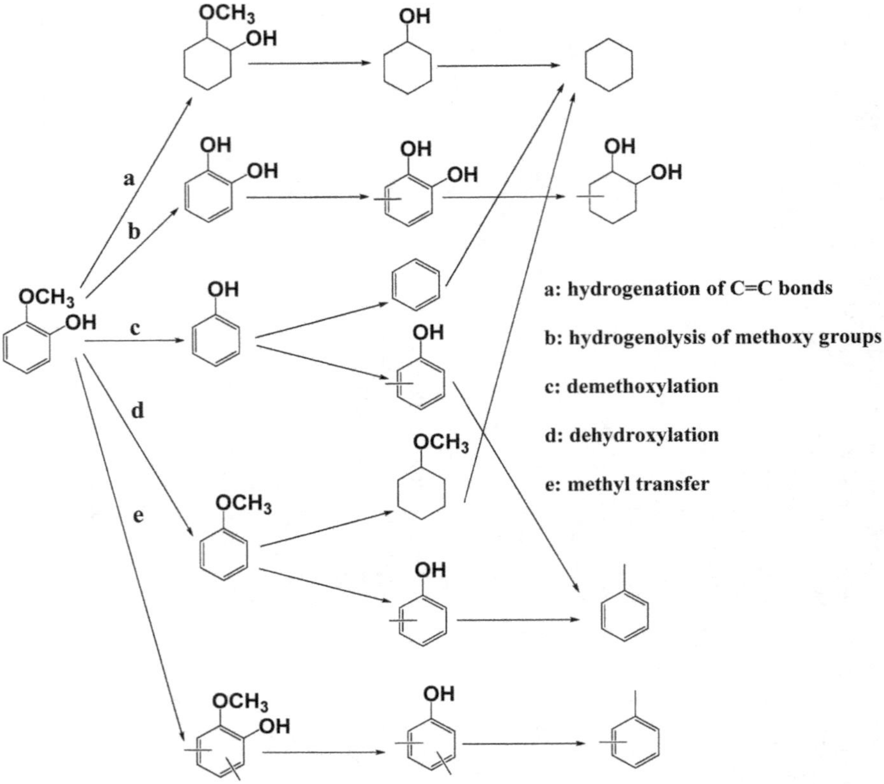

Scheme 7.7 Reaction network for the catalytic hydrotreatment of guaiacol.

implies that the main reaction pathways involves demethoxylation at the initial stage of the reaction, followed by further C=C bond hydrogenation to cyclohexanol.

Catalyst characterization studies indicate that the MnO species in the reduced Ru–MnO_x/C catalyst are located nearby the highly dispersed Ru nanoparticles. Catalyst recycling tests showed that the catalysts can be reused without significant loss of activity.

Wang *et al.*[58] investigated the catalytic hydrotreatment of guaiacol using Ru based bimetallic catalysts promoted with Ni, Fe, Cu and Zn supported on zeolite Y in a batch autoclave in water (Table 7.4, Entry 2). The bimetallic Ru based catalysts showed higher activity than the monometallic Ru one (91% conversion of guaiacol for Ru/HY to >95% for bimetallic catalysts). The best catalyst in terms of conversion (>99%) was Ru–Cu/HY. Main products were cyclohexanone, cyclohexanol and hydrocarbons (cyclohexane, dodecane).

The authors proposed that hydrogenation of the aromatic ring of guaiacol to cyclohexanols occurs on the metal sites, whereas acid sites on the catalysts play an important role in the dehydration reactions. Catalyst characterization studies (XRD and EDX) revealed that the bimetallic catalysts contain bimetallic phases and metal oxides. The average Ru nanoparticle sizes in the bimetallic catalysts were smaller (2–5 nm for Ru–Cu/HY) than for the monometallic Ru catalysts (10–15 nm), which explains the higher activity of the bimetallic catalysts. The Ru–Cu/HY catalyst showed the highest number of total acid sites, especially strong acid sites (0.123 ml NH_3/g_{cat} for Ru–Cu/HY *versus* 0.089 ml NH_3/g_{cat} for Ru/HY, NH_3-TPD). These acid sites are expected to favor dehydration reactions, leading to higher amounts of hydrocarbons, in line with the experimental data.

Besides water, also a number of hydrocarbons have been used as the solvent. Gutierrez *et al.*[59] reported the hydrotreatment of guaiacol in *n*-hexadecane using monometallic Pt, Pd, and Rh catalysts supported on ZrO_2 and bimetallic catalysts thereof in a batch reactor at 100 °C under 80 bar hydrogen pressure (Table 7.4, Entry 3). Around 10% conversion of guaiacol was achieved using the monometallic Pt/ZrO_2 and Pd/ZrO_2catalysts while the Rh/ZrO_2 catalyst showed superior activity and full conversion of guaiacol after 5 h of reaction. The selectivity was strongly affected by the catalyst, with 65% selectivity to cyclohexanol using both Pt and Pd/ZrO_2, and 75% selectivity to 1-methyl-1,2-cyclohexanediol with Rh/ZrO_2 (Scheme 7.7). The results indicate that the Rh based catalyst has a much higher activity for methyl transfer reactions compared to the Pt and Pd counterparts. The bimetallic PdPt/ZrO_2 catalyst was less active than the monometallic Pd and Pt counterparts and only 5% conversion of guaiacol was achieved in 5 h compared to 10% using monometallic Pd and Pt/ZrO_2. The bimetallic Rh based catalysts (RhPt and RhPd/ZrO_2) were as active as the parent monometallic Rh/ZrO_2 catalyst and full conversions of guaiacol were observed. The selectivity heavily depends on the metal promoter, and the addition of Pt to Rh/ZrO_2 resulted in a further improvement of the selectivity to 1-methyl-1,2-cyclohexanediol to 92%, while it decreased considerably to 45% by adding Pd.

Hong *et al.*[60] reported the hydrodeoxygenation of guaiacol in *n*-decane using a bimetallic Pd–WO_x/γ-Al_2O_3 catalyst in a batch reactor (300 °C, 70 bar H_2, Table 7.4, Entry 4). The highest yield of cyclohexane (88%) at full conversion of guaiacol was obtained using the bimetallic catalyst, compared to a 62% yield using monometallic Pd/γ-Al_2O_3. Reactions with the bimetallic catalyst at 200 °C instead of 300 °C gave 2-methoxycyclohexanol as the main product with a yield of around 85%. These findings indicate that ring hydrogenation is dominant at low temperatures and that deoxygenation to cyclohexane requires more severe reaction conditions. Catalyst characterization studies revealed that the presence of tungsten affects the surface acidity of the catalysts.

Martínez *et al.*[61] studied the catalytic hydrotreatment of guaiacol using a series of Cu promoted Re/SiO_2 catalysts in a batch autoclave using *n*-dodecane as the solvent (Table 7.4, Entry 5). The activity of the bimetallic catalyst (full guaiacol conversion after 3 h) was higher than for the monometallic Re catalyst (22% conversion after 3 h). For the monometallic catalyst, a product mixture consisting of phenol, and aromatic and aliphatic hydrocarbons was observed. The selectivity for the bimetallic catalyst was mainly to cyclohexane (75%). Upon prolonged reaction times, the amount of cyclohexane decreased, likely due to ring opening reactions. Catalyst characterization studies revealed that the Re dispersion increased with the amount of Cu until about 1.58% Cu, and then dropped due to the formation of Re aggregates. The amounts of Ru^0 in the bimetallic catalysts were higher than for the monometallic one, indicating that the presence of Cu enhances the reducibility of ReO_x in the Cu(x)ReO_x/SiO_2 catalysts.

7.2.3.2 Experiments in Continuous Set-ups

Borja *et al.*[62] studied the catalytic hydrotreatment of guaiacol using Pt–Sn catalysts supported on Inconel monoliths and coated monoliths with *in situ*-grown carbon nanofibers (CNFs) in a fixed-bed reactor (400 °C, 1 bar, Table 7.4, Entry 6). The reactivity was compared with monometallic catalysts (Sn/Inconel, Pt/Inconel). A maximum guaiacol conversion of 33% was achieved using the monometallic Sn and Pt catalysts at a TOS of 30 min, while for a longer TOS, the catalyst activity dropped to less than 10%. Phenol and *o*-cresol were detected as the main products for both monometallic catalysts.

The activity of the bimetallic Pt–Sn catalysts on Inconel was found to be higher, and essentially quantitative conversion was obtained at the start of the reaction. However, also for these catalysts, deactivation was evident and the conversion slowly dropped to 80% at a TOS of 180 min. Phenol and benzene were the main products, even though the amount of phenol increased at lower conversions, suggesting that phenol is one of the primary products in the reaction network. The stability of the bimetallic Pt–Sn catalyst was improved by coating the catalyst support (Inconel) with CNFs, which provide increased surface area and anchoring sites for the active metals. Full conversion of guaiacol was achieved and no clear drop in conversion was observed for a TOS between 0 and 180 min. However, the product selectivity shifted from benzene (60 to 5%) to phenol (30 to 70%) at prolonged TOS. This indicates that the reactions were performed at too high conversions of guaiacol at the start of the reaction, which obscures determination of the extent of catalyst deactivation.

The catalytic hydrotreatment of guaiacol in a fixed-bed tubular quartz reactor using bimetallic Ni@Pd and Ni@Pt (core@shell) overlayer catalysts was reported by Lai *et al.*[63] (Table 7.4, Entry 7). Such catalysts with a monolayer of a metal on top of a base metal were prepared by direct deposition of Pd and Pt on the parent Ni surface.[65,66] The performance of the bimetallic Ni@Pd and Ni@Pt overlayer catalysts was compared with monometallic Ni, Pt, Pd and bimetallic Ni–Pt, Ni–Pd/SiO_2–Al_2O_3 catalysts. The highest TOF was obtained using the overlayer Ni@Pd and Ni@Pt catalysts (*e.g.* 0.07 sec^{-1} for Ni@Pd compared to 0.02 sec^{-1} for Ni–Pd, 0.175 sec^{-1} for Ni@Pt compared to 0.03 sec^{-1} for Ni–Pt). Further studies using the monometallic Pd and overlayer Ni@Pd catalyst showed that the main products at low guaiacol conversion are phenol, cresol, non-deoxygenation products and anisole. At higher guaiacol conversions, xylenols and completely deoxygenated products like benzene, toluene, and xylene (BTX) were observed. This implies that phenol, *o*-cresol, anisole, xylenols and non-deoxygenation products are reaction intermediates. At 97% guaiacol conversion, BTX products were the main products (selectivity: 36% benzene, 32% toluene, 15% xylene) with small amounts of mono-oxygenated phenolics (8% cresol, 4% phenol, 2% anisole).

Sun *et al.*[64] investigated the vapor-phase hydrotreatment of guaiacol in a micro-tubular fixed-bed reactor using mono and bimetallic Fe and Pd based carbon-supported catalysts with the prime objective of obtaining fully hydrodeoxygenated products (*e.g.* benzene, Table 7.4, Entry 8). The reaction was carried out at a temperature between 250 and 450 °C under ambient pressure and guaiacol was fed in combination with a H_2/N_2 gas mixture ($P_{guaiacol}$ = 0.004 bar; P_{H2} = 0.4 bar with a *W*/*F* of 0.15 s.g. STP ml^{-1}). The bimetallic PdFe/C catalyst proved to be more active than Pd/C and Fe/C alone. Phenol was found to be the major intermediate for the monometallic Pd and Fe catalysts, and the highest yield of phenol was 79% for Pd/C at 350 °C at almost full conversion of guaiacol. At higher guaiacol conversions, saturation of the aromatic rings occurred to mainly cyclohexanone and cyclohexanol using the noble metal based catalysts, followed by ring opening to form gaseous products (mainly C_1 products including methane, CO, and CO_2). Ru was the most active catalyst for gasification, as evidenced by the 100% yield to gas-phase products for 5Ru/C as opposed to 16.2% and 38.8% for 5Pd/C and 5Pt/C, respectively.

7.2.4 Overview for 4-propylphenol

An overview of experimental studies on the use of 4-propylphenol is given in Table 7.5. Only reactions in water have been reported using batch set-ups, at 300 °C and hydrogen pressures in the 20–40 bar range.

Ohta *et al.*[67] investigated the selective hydrodeoxygenation of 4-propylphenol to *n*-propylbenzene in water in a batch reactor using mono- and bimetallic Pt catalysts supported on zirconia (Table 7.5, Entry 1). The conversion of 4-propylphenol by the monometallic Pt/ZrO_2 was 62%, increasing to 67% upon the addition of Re. In addition, the yield of *n*-propylbenzene was also higher for the bimetallic catalyst (57% compared to 40%, Scheme 7.8, route b). Promotion with Mo and Bi resulted in a clear reduction of the activity to less than 10% after 1 h. Subsequently, the best bimetallic combination (Pt–Re) was also tested on other supports, including TiO_2, Al_2O_3, CeO_2, SiO_2 and activated carbon. The use of TiO_2, Al_2O_3, CeO_2, and SiO_2 resulted in less active catalysts compared to ZrO_2, while activated carbon showed the highest activity (90% conversion of 4-propylphenol by Pt–Re/AC, compared to 67% by Pt–Re/ZrO_2). However, the main product for Pt–Re on carbon was

Table 7.5 Summary of experimental studies on the catalytic hydrotreatment of 4-propylphenol using bimetallic noble metal based catalysts.

Entry	Catalyst	Solvent	Reactor	Conditions	Conversion/selectivity/yield	Reference
1	Pt–M/ZrO_2 (M = Re, Sn, Ir, Ga, Fe, Au, Pd, W, In, Mo, Bi)	Water	Batch reactor	Initial H_2 pressure at RT = 20 bar, 300 °C, 1 h	Pt/ZrO_2 62% conversion 40% yield of propylbenzene	67
	Pt–Re/(TiO_2, Al_2O_3, CeO_2, SiO_2, AC)				Pt–Re/ZrO_2 67% conversion 57% yield of propylbenzene	
2	M–Ni/ZrO_2, M = Re, Ga, Cu, Zn, Sn, Bi, Mo, Re–Ni supported on TiO_2, Al_2O_3, SiO_2, CeO_2, Nb_2O_5 and activated carbon	Water	Batch reactor	300 °C, initial H_2 pressure at RT 40 bar, 1 h	Ni/ZrO_2 41% conversion Yield 6% *n*-propylbenzene 6% *n*-propylcyclohexane 1% *n*-propylcyclohexene 11% 4-propylcyclohexanol No activity for Re/ZrO_2 Re–Ni/ZrO_2 100% conversion Yield 54% *n*-propylbenzene 17% *n*-propylcyclohexane 1% *n*-propylcyclohexene 5% 4-propylcyclohexanol	68

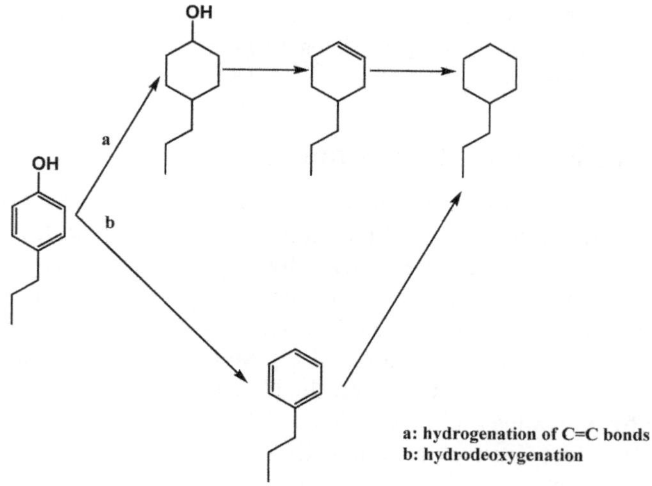

Scheme 7.8 Possible reaction network for the catalytic hydrotreatment of 4-propylphenol.

4-propylcyclohexanol (62% yield, Scheme 7.8, route a) instead of *n*-propylbenzene. Thus, there was a significant change in chemoselectivity based on support selection.

Feng *et al.*[68] reported the hydrotreatment of 4-propylphenol in water, in this case using a bimetallic Re–Ni/ZrO_2 catalyst (Table 7.5, Entry 2). The results were compared with monometallic Ni and Re catalysts supported on ZrO_2 (300 °C, initial H_2 pressure at RT of 30 bar and for 1 h reaction time). The monometallic Ni/ZrO_2 catalyst gave a mixture of *n*-propylbenzene (6% yield), aliphatic hydrocarbons (7% yield), oxygenates (4-propylcyclohexanol, 11% yield) and some unidentified products at 41% 4-propylphenol conversion. In contrast, the monometallic Re/ZrO_2 was inactive under the selected reaction conditions. Promotion of the Ni catalyst with Re, Ga, Cu, Zn, Sn, Bi or Mo had significant effects on catalyst activity. Adding Ga, Cu or Zn to Ni/ZrO_2 resulted in moderate activities with 21–57% conversions of 4-propylphenol, whereas promotion with Sn, Bi or Mo resulted in lower catalytic activities (4-propylphenol conversions between 1–9%). The addition of Re, however, resulted in a great increase in conversion (96%). The main products were *n*-propylbenzene (26% yield), aliphatic hydrocarbons (22% yield) and 4-propylcyclohexanol (27% yield).

Bimetallic Ni–Re catalysts on other supports were also tested, namely TiO_2, Al_2O_3, SiO_2, CeO_2, Nb_2O_5 and activated carbon, and important support effects were found on both selectivity and activity. Among the catalysts tested, Re–Ni/ZrO_2 and Re–Ni/TiO_2 were the most effective for the conversion of 4-propylphenol into *n*-propylbenzene in water. Further optimization of reaction conditions using Re–Ni/ZrO_2 showed that full conversion of 4-propylphenol with a 54% yield of *n*-propylbenzene was possible. Based on catalyst characterization studies, the authors postulate that the enhanced catalytic activity of the bimetallic Re–Ni/ZrO_2catalyst is due to the presence of Re oxides that act as anchor points for Ni nanoparticles and lead to a reduction in the average Ni nanoparticle size.

7.2.5 Miscellaneous Model Component

Two other model components have been tested, *viz.* phenol and 2-methoxy-4-propylphenol, and the results are given in Table 7.6.

Ghampson *et al.*[69] studied the catalytic hydrotreatment of phenol in dodecane using Re promoted Co catalysts on various supports in a batch reactor at 300 °C under a hydrogen pressure of 30 bar for 4 h (Table 7.6, Entry 1). The effects of catalyst support including Al_2O_3, SiO_2–Al_2O_3, ZrO_2 and TiO_2 were also examined. Among the monometallic Co catalysts, the highest activities were found for the silica and mixed SiO_2–Al_2O_3 supports.

The main product was cyclohexane with an 82% yield for the zirconia based catalyst. Three reaction mechanisms were proposed: (i) ring hydrogenation of phenol to cyclohexanone/cyclohexanol, followed by dehydration to cyclohexene and hydrogenation to cyclohexane, (ii) direct deoxygenation of the $C_{aromatic}$–O bond to give benzene, which is converted to cyclohexane by hydrogenation and (iii) an initial keto–enol tautomerization (TAU) of phenol to form a cyclohexadienone intermediate, followed by hydrogenation of the carbonyl group to form 2,4-cyclohexadienol and subsequent rapid dehydration to benzene or hydrogenation to cyclohexanol, which is then dehydrated to form cyclohexene, and finally hydrogenated to cyclohexane (Scheme 7.9). Based on the concentration *versus* batch time profiles of the intermediates and products, it was concluded that the Co/Al_2O_3 and Co/SiO_2–Al_2O_3 catalysts mainly follow route (i) for cyclohexane formation, whereas route (iii) is dominant for Co/ZrO_2.

The performance of Re promoted Co catalysts was determined and compared with the monometallic Co catalysts, and higher activity was observed for the former, achieving 70% phenol conversion using the ReCo/TiO_2 catalyst compared to 11% conversion for Co/TiO_2. The main product for the bimetallic catalyst was cyclohexane (80% yield at full conversion of phenol). Catalyst characterization studies indicated that catalyst promotion by Re enhanced catalytic activity by lowering the reduction temperature for Co, improving the Co dispersions, and forming Re–Co alloys.

Yohe *et al.*[70] studied the catalytic hydrotreatment of 2-methoxy-4-propylphenol (dihydroeugenol, DHE) to hydrocarbons using mono- and bimetallic Pt based catalysts supported on multi-walled carbon nanotubes (WCNTs) in a fixed-bed reactor at 300 °C and 240 bar H_2 (Table 7.6, Entry 2).

The monometallic Pt catalyst provided quantitative DHE conversion with propylcyclohexanol as the main product (47% selectivity), while the selectivity changed to mainly propylcyclohexane (87%) when using a bimetallic Mo promoted Pt catalyst (Scheme 7.10). Through catalyst characterization studies, it was postulated that the addition of Mo to the Pt catalysts promotes dehydration reactions and thus the formation of hydrocarbons as it renders the catalysts more acidic.

Table 7.6 Summary of experimental studies on the catalytic hydrotreatment of phenol and 2-methoxy-4-propylphenol using bimetallic noble metal based catalysts.

Entry	Catalyst	Substrate	Solvent	Reactor	Conditions	Conversion/selectivity/yield	Reference
1	ReCo supported on Al_2O_3, SiO_2–Al_2O_3, ZrO_2, TiO_2	Phenol	Dodecane	Batch	300 °C, 30 bar H_2, 4 h	Conversion 100% for Co/Al_2O_3 100% for Co/SiO_2–Al_2O_3 98% for Co/ZrO_2 11% for Co/TiO_2 100% for ReCo/Al_2O_3 100% for ReCo/SiO_2–Al_2O_3 100% for ReCo/ZrO_2 70% for ReCo/TiO_2 Yield Co/Al_2O_3 38% cyclohexane Co/SiO_2–Al_2O_3 61% cyclohexane 30% methylcyclopentane Co/ZrO_2 82% cyclohexane 9% benzene Co/TiO_2 7.5% cyclohexane 3% benzene ReCo/Al_2O_3 60% cyclohexane ReCo/SiO_2–Al_2O_3 80% cyclohexane 17.5% methylcyclopentane ReCo/ZrO_2 80% cyclohexane, 10% benzene ReCo/TiO_2 61% cyclohexane 8% benzene	69

(continued)

Table 7.6 (*continued*)

Entry	Catalyst	Substrate	Solvent	Reactor	Conditions	Conversion/selectivity/yield	Reference
2	PtMo supported on multi-walled carbon nanotubes	2-Methoxy-4-propylphenol (DHE)		Fixed-bed reactor	300 °C, 240 bar H_2, and varying WHSV (PtMo: 5–660; Pt: 5–4290 g DHE h^{-1} g cat^{-1})	Pt/WCNT 100% conversion of DHE, 47% carbon selectivity of propylcyclohexanol PtMo/WCNT 100% conversion of DHE 87% carbon selectivity of propylcyclohexane	70

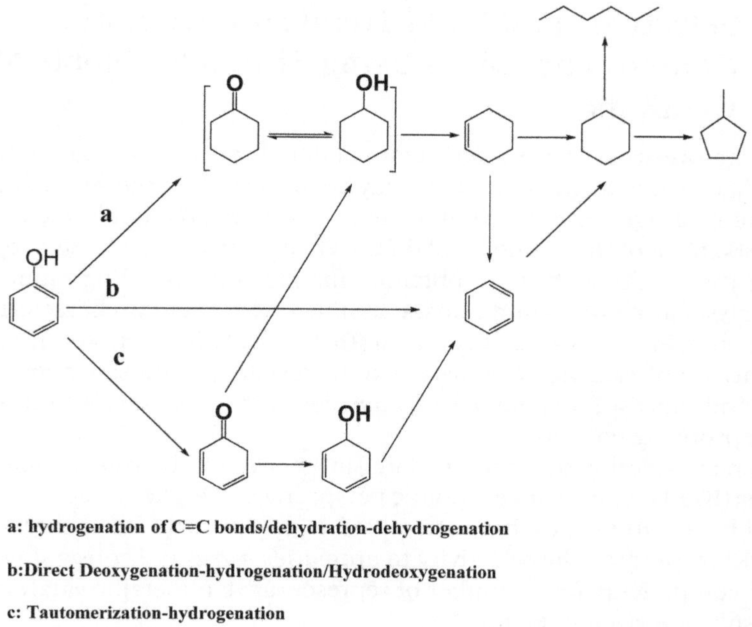

Scheme 7.9 Possible routes for catalytic hydrotreatment of phenol. Adapted from ref. 69 with permission from The Royal Society of Chemistry.

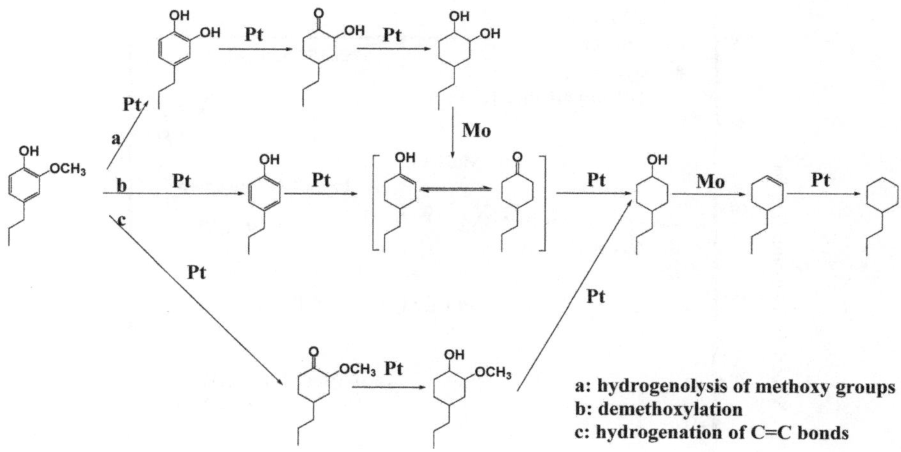

Scheme 7.10 Proposed reaction network for the catalytic hydrotreatment of dihydroeugenol (DHE) over a bimetallic 5%Pt2.5%Mo/MWCNT catalyst. Adapted from *Journal of Catalysis*, **344**, S. L. Yohe, H. J. Choudhair, D. D. Mehta, P. J. Dietrich, M. D. Detwiler, C. M. Akatay, E. A. Stach, J. T. Miller, W. N. Delgass, R. Agrawal and F. H. Ribeiro, High-pressure vapor-phase hydrodeoxygenation of lignin-derived oxygenates to hydrocarbons by a PtMo bimetallic catalyst: Product selectivity, reaction pathway, and structural characterization, 535–552, Copyright 2016, with permission from Elsevier.[70]

7.3 Selectivity and Yield Trends for Aromatic Components when Using Bimetallic Noble Metal Catalysts

From the overview of the model component studies discussed above, it is clear that a wide range of products may be formed upon the catalytic hydrotreatment of lignin model components with bimetallic noble metal based catalysts. Most of the studies explain the effects of metal promoters by comparing the results with those obtained for the corresponding monometallic catalysts at similar conditions, combined with catalyst characterization studies and, in some cases, theoretical (DFT) calculations. However, a proper comparison of catalyst selectivity is only possible at the same conversion levels and this aspect is not always considered in the studies and this leads to inappropriate conclusions.

When considering that deoxygenated aromatics like benzene, toluene and xylenes (BTX) are, from an economic perspective, of higher interest than saturated hydrocarbons, we have attempted to compensate for conversion differences by plotting the selectivity to aromatics *versus* the conversion of the model components for a number of representative bimetallic catalysts, and the results are given in Figure 7.5.

The aromatics are intermediates in the reaction network and are prone to further hydrogenation to aliphatic hydrocrabons. As such, catalysts with

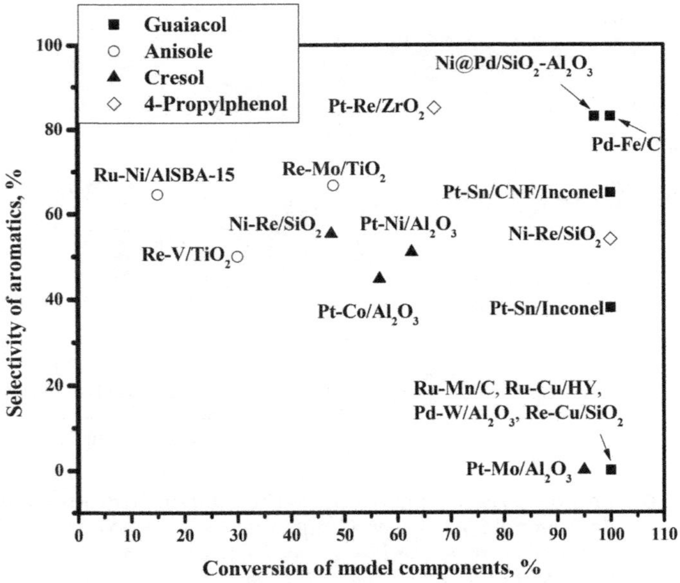

Figure 7.5 Selectivity to deoxygenated aromatics (*e.g.* benzene, toluene, *etc.*) *versus* the conversion of the phenolic model compound using noble metal based bimetallic catalysts.

a reduced tendency for hydrogenation of aromatics are preferred. It is clear that the best results, *i.e.* high aromatic selectivity at high conversion of the phenolic compound, are obtained for the overlayer bimetallic Ni@Pd/SiO_2–Al_2O_3catalyst and the Pd–Fe/C catalyst, which allowed for aromatic yields over 80%. As such, it appears that the addition of particularly Pd to Ni or Fe based catalysts leads to high aromatic yields. Both studies were performed in continuous flow reactors at relatively low hydrogen pressures.

7.4 Conclusions

The catalytic hydrotreatment of lignin and pyrolytic lignins is a promising technology for the production of (oxygenated) aromatics. Due to the high complexity of the lignin structure, it is very difficult to obtain information about the reaction pathways occurring during the catalytic hydrotreatment of lignins. As such, studies with lignin model components like phenolics have been shown to be able to provide valuable information regarding the reactivity of the various bonds within the lignin structure.

It is evident that the noble metal based bimetallic catalysts show profoundly different performances than their monometallic counterparts. In most of the cases, the activity is enhanced and the stability is improved, though the effect on selectivity is difficult to assess as catalyst performance is often not compared at the same conversion levels. Reaction pathways have been identified by profiling the product distributions at various conversion levels and rationalized using catalyst characterization studies in combination with theoretical calculations. When aiming for non-oxygenated aromatics, bimetallic Fe and Ni based catalysts promoted by Pd seem to be a good choice and aromatics yields of up to 80% have been reported.

References

1. A. E. Atabani, A. S. Silitonga, I. A. Badruddin, T. M. I. Mahlia, H. H. Masjuki and S. Mekhilef, *Energy Rev.*, 2012, **16**, 2070.
2. M. Balat, H. Balat and C. Öz, Progress in bioethanol processing, *Prog. Energy Combust. Sci.*, 2008, **34**, 551.
3. J. J. Bozell, *Clean: Soil, Air, Water*, 2008, **36**, 641.
4. T. Dickerson and J. Soria, *Energies*, 2013, **6**, 514.
5. J. Zakzeski, P. C. A. Bruijnincx, A. L. Jongerius and B. M. Weckhuysen, *Chem. Rev.*, 2010, **110**, 3552.
6. C. W. Dence and S. Y. Lin, *Methods in Lignin Chemistry*, Springer, Berlin, Heidelberg, 1992, pp. 3–19.
7. C. Amen-Chen, H. Pakdel and C. Roy, *Bioresour. Technol.*, 2001, **79**, 277.
8. H. Kobayashi, H. Ohta and A. Fukuoka, *Catal. Sci. Technol.*, 2012, **2**, 869.
9. M. P. Pandey and C. S. Kim, *Chem. Eng. Technol.*, 2011, **34**, 29.
10. M. Saidi, F. Samimi, D. Karimipourfard, T. Nimmanwudipong, B. C. Gates and M. R. Rahimpour, *Energy Environ. Sci.*, 2014, **7**, 103.

11. E. Adler, *Wood Sci. Technol.*, 1977, **11**, 169.
12. C. Heitner, D. Dimmel and J. A. Schmidt, *Lignin and Lignans: Advances in Chemistry*, Taylor & Francis, Boca Raton, FL, 2010.
13. A. Higson and C. Smith, *NNFCC Renewable Chemicals Factsheet: Lignin*, 2014.
14. R. J. A. Gosselink, E. de Jong, B. Guran and A. Abächerli, *Ind. Crops Prod.*, 2004, **20**, 121.
15. R. G. Macdonald and J. N. Franklin, *Joint Textbook Committee of the Paper Industry, the Pulping of Wood*, McGraw-Hill, New York, 2nd edn, 1969.
16. B. K. Avellar and W. G. Glasser, *Biomass Bioenergy*, 1998, **14**, 205.
17. J. Li, G. Gellerstedt and K. Toven, *Bioresour. Technol.*, 2009, **100**, 2556–2561.
18. W. J. J. Huijgen, J. H. Reith and H. den Uil, *Ind. Eng. Chem. Res.*, 2010, **49**, 10132.
19. E. Pye and J. Lora, *Tappi J.*, 1991, **74**, 113.
20. D. T. A. Huibers and M. W. Jones, *Can. J. Chem. Eng.*, 1980, **58**, 718.
21. R. H. Venderbosch and W. Prins, *Biofuels, Bioprod. Biorefin.*, 2010, **4**, 178.
22. C. R. Vitasari, G. W. Meindersma and A. B. de Haan, *Bioresour. Technol.*, 2011, **102**, 7204.
23. J. E. Holladay, J. F. White, J. J. Bozel and D. Johnson, *Top Value-added Chemicals from Biomass: Volume II. Resultsof Screening for Potential Candidates from Biorefinery Lignin, Technical Report PNNL-16983, Visited 2013*, 2007.
24. R. Bayerbach, V. D. Nguyen, U. Schurr and D. Meier, *J. Anal. Appl. Pyrolysis*, 2006, **77**, 95.
25. S. Baumberger, A. Abaecherli, M. Fasching, G. Gellerstedt, R. Gosselink, B. Hortling, J. Li, B. Saake and E. de Jong, *Holzforschung*, 2007, 459.
26. R. Bayerbach and D. Meier, *J. Anal. Appl. Pyrolysis*, 2009, **85**, 98.
27. R. J. A. Gosselink, *Lignin as a Renewable Aromatic Resource for the Chemical Industry*, Wageningen University, 2011.
28. S. W. Benson, *Thermochemical Kinetics Methods for the Estimation of Thermochemical Data Andrate Parameters*, Wiley, New York, 1968.
29. R. Parthasarathi, R. A. Romero, A. Redondo and S. Gnanakaran, *J. Phys. Chem. Lett.*, 2011, **2**, 2660.
30. S. Kim, S. C. Chmely, M. R. Nimlos, Y. J. Bomble, T. D. Foust, R. S. Paton and G. T. Beckham, *J. Phys. Chem. Lett.*, 2011, **2**, 2846.
31. P. de Wild, R. Van der Laan, A. Kloekhorst and E. Heeres, *Environ. Prog. Sustainable Energy*, 2009, **28**, 461.
32. S. Kang, X. Li, J. Fan and J. Chang, *Renewable Sustainable Energy Rev.*, 2013, **27**, 546.
33. M. Kleinert and T. Barth, *Chem. Eng. Technol.*, 2008, **31**, 736.
34. M. Kleinert and T. Barth, *Energy Fuels*, 2008, **22**, 1371.
35. H. E. Schoemaker, *Recl. Trav. Chim. Pays-Bas*, 1990, **109**, 255.
36. H. E. Schoemaker and K. Piontek, *Pure Appl. Chem.*, 1996, 2089.
37. H. Lange, S. Decina and C. Crestini, *Eur. Polym. J.*, 2013, **49**, 1151.
38. V. M. Roberts, V. Stein, T. Reiner, A. Lemonidou, X. Li and J. A. Lercher, *Chem.–Eur. J.*, 2011, **17**, 5939.
39. R. W. Thring, *Biomass Bioenergy*, 1994, **7**, 125.

40. B. Joffres, D. Laurenti, N. Charon, A. Daudin, A. Quignard and C. Geantet, *Oil Gas Sci. Technol.*, 2013, **68**, 753.
41. A. Oasmaa, R. Alén and D. Meier, *Bioresour. Technol.*, 1993, **45**, 189.
42. D. Meier, R. Ante and O. Faix, *Bioresour. Technol.*, 1992, **40**, 171.
43. F. de Miguel Mercader, M. J. Groeneveld, S. R. A. Kersten, C. Geantet, G. Toussaint, N. W. J. Way, C. J. Schaverien and K. J. A. Hogendoorn, *Energy Environ. Sci.*, 2011, **4**, 985.
44. J. Piskorz, P. Majerski, D. Radlein and D. S. Scott, *Energy Fuels*, 1989, **3**, 723.
45. Z. Tang, Y. Zhang and Q. Guo, *Ind. Eng. Chem. Res.*, 2010, **49**, 2040.
46. M. Besson, P. Gallezot and C. Pinel, *Chem. Rev.*, 2014, **114**, 1827.
47. C. Li, X. Zhao, A. Wang, G. W. Huber and T. Zhang, *Chem. Rev.*, 2015, **115**, 11559.
48. W. Yu, M. D. Porosoff and J. G. Chen, *Chem. Rev.*, 2012, **112**, 5780.
49. C. R. Kumar, N. Anand, A. Kloekhorst, C. Cannilla, G. Bonura, F. Frusteri, K. Barta and H. J. Heeres, *Green Chem.*, 2015, **17**, 4921.
50. A. Narani, R. K. Chowdari, C. Cannilla, G. Bonura, F. Frusteri, H. J. Heeres and K. Barta, *Green Chem.*, 2015, **17**, 5046.
51. I. T. Ghampson, G. Pecchi, J. L. G. Fierro, A. Videla and N. Escalona, *Appl. Catal., B*, 2017, **208**, 60.
52. D. Shi, L. Arroyo-Ramírez and J. M. Vohs, *J. Catal.*, 2016, **340**, 219.
53. S. Pichaikaran and P. Arumugam, *Green Chem.*, 2016, **18**, 2888.
54. P. T. M. Do, A. J. Foster, J. Chen and R. F. Lobo, *Green Chem.*, 2012, **14**, 1388.
55. A. Robinson, G. A. Ferguson, J. R. Gallagher, S. Cheah, G. T. Beckham, J. A. Schaidle, J. E. Hensley and J. W. Medlin, *ACS Catal.*, 2016, **6**, 4356.
56. F. Yang, D. Liu, H. Wang, X. Liu, J. Han, Q. Ge and X. Zhu, *J. Catal.*, 2017, **349**, 84.
57. M. Ishikawa, M. Tamura, Y. Nakagawa and K. Tomishige, *Appl. Catal., B*, 2016, **182**, 193.
58. H. Wang, H. Ruan, M. Feng, Y. Qin, H. Job, L. Luo, C. Wang, M. H. Engelhard, E. Kuhn, X. Chen, M. P. Tucker and B. Yang, *ChemSusChem*, 2017, **10**, 1846.
59. A. Gutierrez, R. K. Kaila, M. L. Honkela, R. Slioor and A. O. I. Krause, *Catal. Today*, 2009, **147**, 239.
60. Y.-K. Hong, D.-W. Lee, H.-J. Eom and K.-Y. Lee, *Appl. Catal., B*, 2014, **150–151**, 438.
61. N. Martínez, R. García, J. L. G. Fierro, C. Wheeler, R. N. Austin, J. R. Gallagher, J. T. Miller, T. R. Krause, N. Escalona and C. Sepúlveda, *Fuel*, 2016, **186**, 112.
62. M. Á. González-Borja and D. E. Resasco, *Energy Fuels*, 2011, **25**, 4155.
63. Q. Lai, C. Zhang and J. H. Holles, *Appl. Catal., A*, 2016, **528**, 1.
64. J. Sun, A. M. Karim, H. Zhang, L. Kovarik, X. S. Li, A. J. Hensley, J.-S. McEwen and Y. Wang, *J. Catal.*, 2013, **306**, 47.
65. M. P. Latusek, B. P. Spigarelli, R. M. Heimerl and J. H. Holles, *J. Catal.*, 2009, **263**, 306.

66. M. D. Skoglund, C. L. Jackson, K. J. McKim, H. J. Olson, S. Sabirzyanov and J. H. Holles, *Appl. Catal., A*, 2013, **467**, 355.
67. H. Ohta, B. Feng, H. Kobayashi, K. Hara and A. Fukuoka, *Catal. Today*, 2014, **234**, 139.
68. B. Feng, H. Kobayashi, H. Ohta and A. Fukuoka, *J. Mol. Catal. A: Chem.*, 2014, **388–389**, 41.
69. I. T. Ghampson, C. Sepulveda, A. B. Dongil, G. Pecchi, R. Garcia, J. L. G. Fierro and N. Escalona, *Catal. Sci. Technol.*, 2016, **6**, 7289.
70. S. L. Yohe, H. J. Choudhari, D. D. Mehta, P. J. Dietrich, M. D. Detwiler, C. M. Akatay, E. A. Stach, J. T. Miller, W. N. Delgass, R. Agrawal and F. H. Ribeiro, *J. Catal.*, 2016, **344**, 535.

CHAPTER 8

Microwaves in the Catalytic Valorisation of Biomass Derivatives

PETER PRIECEL[a], JAVIER EDUARDO PEREZ MEJIA[a], PIERA DEMMA CARÀ[a,b] AND JOSE A. LOPEZ-SANCHEZ*[a,b]

[a]Stephenson Institute for Renewable Energy, Department of Chemistry, University of Liverpool, Liverpool, L69 7ZD, UK; [b]MicroBioRefinery Facility, Department of Chemistry, University of Liverpool, Liverpool, L69 7ZD, UK
*E-mail: jals@liverpool.ac.uk

8.1 Introduction

Valuable feedstock materials for industrial applications can be obtained from various types of biomass and be converted into a wide range of derivatives *via* chemical processes, for example hydrolysis, reduction, oxidation, dehydration or esterification. Such transformations typically require catalysts to ensure the high yields required to make such processes economically viable. Among the wide variety of chemicals that can be obtained from biomass, sugars and polyols (*e.g.* glucose, sorbitol, glycerol), organic acids (*e.g.* succinic, levulinic, itaconic, lactic acids) and furans (*e.g.* furfural and 5-hydroxymethylfurfural (HMF)) deserve particular mention because they are some of the key biomass-derived platform molecules with the highest commercial potential.[1] The second characteristic of these molecules is that

Green Chemistry Series No. 56
Sustainable Catalysis for Biorefineries
Edited by Francesco Frusteri, Donato Aranda and Giuseppe Bonura

Published by the Royal Society of Chemistry, www.rsc.org

they are typically soluble in polar media and obtained in aqueous media, which makes the application of microwaves in their production and/or upgrading particularly promising.

In 2016, Jérôme and co-workers[2] discussed the most important non-thermal technologies (no external source of heating involved) successfully used for the selective depolymerisation of cellulose to processable glucans. These technologies are mechanocatalysis, non-thermal atmospheric plasma, ultrasound (US) and microwaves (MWs). The advantage of using non-thermal technologies is mainly the high selectivity reached at a relatively low temperature (<100 °C). Among these techniques, microwaves and ultrasound have emerged as the most environmentally friendly procedures for converting biomass into highly valuable fine chemicals. The combination of both microwaves and ultrasound produces a synergic effect that can boost yields and reduce reaction times. In fact, the highest reducing sugar yield (57%) from complex dry potato sludge was obtained under US/MW irradiation with a 77% depolymerisation of potato starch.[1]

Furthermore, microwaves represent an efficient energy source to obtain value-added chemicals from biomass.[1,3–7] MWs can be used at different temperatures in different types of processes. For example, biomass treated at temperatures above 400 °C undergoes thermochemical conversions such as pyrolysis and gasification, whereas at temperatures below 400 °C and in the presence of water, biomass can undergo hydrolytic conversions to liberate the main structural components (*e.g.* monosaccharides from cellulosic biomass, which can be used, for example, for bioethanol generation).[3] There has been, therefore, an extensive amount of work dedicated to the conversion of biomass with microwaves for a range of processes and technologies that, due to its relevance and extension, deserves a whole book by itself. Indeed, the direct conversion of biomass with microwaves has extensively been reviewed recently by others.[3,7,8] Instead, we will dedicate this chapter to the emerging area of combining microwave reactors with catalysis, with particular emphasis on heterogeneous catalysis for the conversion of biomass derivatives, rather than ion pretreatment or conversion of "real" biomass.

Heterogeneous catalysts can synergise with microwave irradiation to develop eco-friendly and efficient processes for the chemocatalytic conversion of lignocellulosic biomass and derived streams. The creation of hot spots is typically used for energy saving, and specifically in carbohydrate chemistry where high yields and selectivities can be obtained after very short reaction times at relatively low temperatures.[6] In the following sections, we will summarise the state of the art in the combination of catalysis and microwave irradiation in biomass transformations, such as in the hydrolysis of cellulose, hemicellulose and lignin, hydrogenation under hydrogen pressure and with hydrogen transfer agents, pressurised selective oxidation and oxidative degradation, dehydration of sugars to heterocycles (such as HMF or furfural) and platform acids (*e.g.* levulinic acid), esterification of cellulose and its derivatives or transesterification of vegetable oils to produce biodiesel.

8.2 Hydrolysis of Biopolymers

8.2.1 The Importance of Lignocellulosic Biomass as a Source of Cellulose, Hemicellulose and Lignin

Yearly, 200 billion metric tons of lignocellulosic biomass are produced in nature, positioning it as the most abundant biomass source on Earth.[9] Since lignocellulosic biomass does not necessarily compete with food sources and represents an environmentally friendly feedstock, lignocellulosic biomass has emerged as a promising alternative platform to produce green fuels and value-added chemicals with the potential to replace non-renewable fossil feedstocks.[10,11] This biomass is mainly constituted by cellulose (35–50%), hemicellulose (20–35%) and lignin (10–25%).[12] In Scheme 8.1, the utilisation of these biopolymers to obtain chemical products or fuels is represented. Even though most of the research done on biomass conversion has focused on cellulose, there has been an extensive amount of work dedicated to the valorisation of hemicellulose[13,14] and lignin conversion.[13]

Cellulose, hemicellulose and lignin are biopolymers that can contain thousands of monomer units depending on their nature.[15] Glucose is the monomeric unit present in cellulose, whereas xylose is the predominant repeated molecule in hemicellulose, but it also contains glucose and other sugars. On the other hand, lignin is formed by phenolic units.[14] Usually, the initial stage to produce value-added chemicals or fuels from lignocellulosic biomass, after the separation of its three main components, is its depolymerisation or hydrolysis. Once the depolymerisation has been achieved, the monomeric units contained in the biopolymer can be transformed further by chemo- and biological routes, such as hydrogenation, oxidation or dehydration.[13–15]

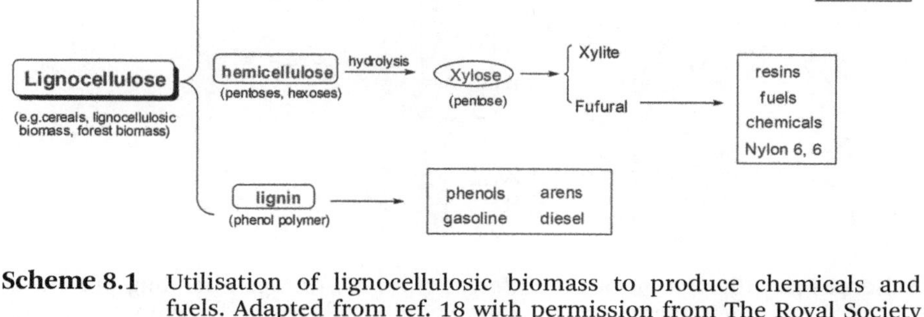

Scheme 8.1 Utilisation of lignocellulosic biomass to produce chemicals and fuels. Adapted from ref. 18 with permission from The Royal Society of Chemistry.

8.2.2 Hydrolysis of Cellulose

Cellulose is a biopolymeric compound present in the cell wall of green plants. This polysaccharide, represented in Scheme 8.2, consists of glucose units linked by β-1,4-glycosidic bonds forming at first a complex linear molecule. The degree of polymerisation (DP) of cellulose varies depending on the source. For instance, the DP in wood pulps ranges from 300 to 1700, and from 800 to 10 000 in plant fibres.[16]

Since glucose units present in cellulose are entirely surrounded by hydroxyl groups, strong hydrogen bonds are generated within the linear polymer chain, but also across, reaching adjacent linear polymeric units. In Scheme 8.3, a representation of the intra-, inter-molecular and inter-sheet hydrogen bonding in cellulose is shown. The presence of a vast quantity of hydrogen bonds in cellulose generates a robust crystalline structure, which is insoluble

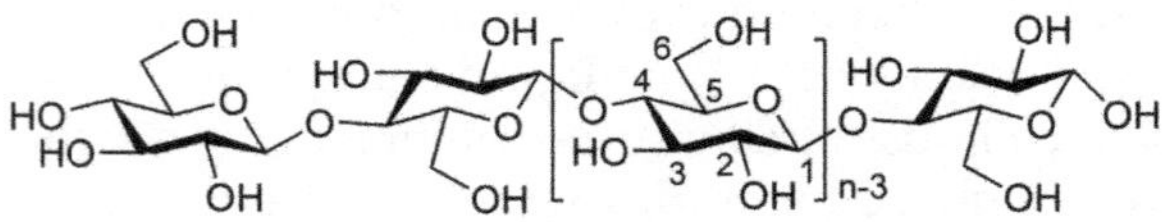

Scheme 8.2 Structure of cellulose (*n* stands for degree of polymerisation). Adapted with permission from *Applied Catalysis B: Environmental*, **174**, L. Hu, L. Lin, Z. Wu, S. Zhou and S. Liu, Chemocatalytic hydrolysis of cellulose into glucose over solid acid catalysts, 225–243, Copyright 2015, with permission from Elsevier.[9]

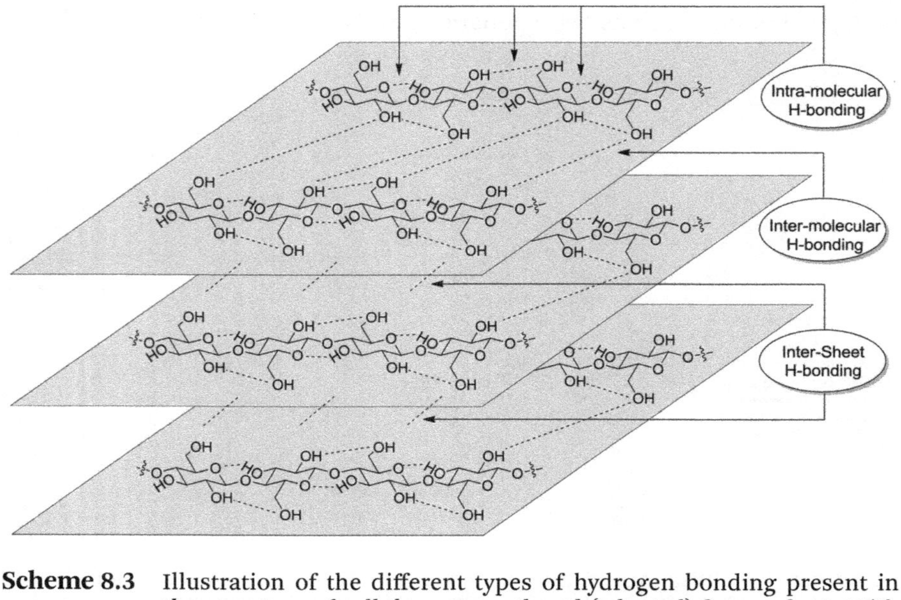

Scheme 8.3 Illustration of the different types of hydrogen bonding present in the structure of cellulose. Reproduced (adapted) from ref. 16. With permission from the Royal Society of Chemistry.

in water and most organic solvents and therefore its depolymerisation still remains a challenge.[12] The presence of β-type glycosidic bonds also makes the breaking down of cellulose much harder in comparison to polysaccharides containing α-type glycosidic bonds such as in starch.

Nowadays, processes developed for the hydrolysis of cellulose utilise a conventional source of heating. During the hydrolysis, the biopolymers are converted to fermentable sugars.[17] This is the first step to generate chemical products and fuels from lignocellulosic biomass. The hydrolysis of cellulose is the cleavage of the β-1,4-glycosidic bonds between two anhydroglucose units accelerated by the presence of the catalyst.[5] The reaction products and by-products depend on the features of the catalyst and reaction conditions. The cellulose hydrolysis is especially challenging because of the recalcitrant crystalline structure of the biopolymeric molecule and the inter/intra-molecular bonds involved.

Most of the hydrolysis processes are carried out by either dilute acids or enzymes.[12] The most common liquid acids utilised for the hydrolysis of cellulose are: sulphuric acid, hydrochloric acid, nitric acid, perchloric acid, phosphoric acid, hydrofluoric acid and formic acid.[9] This reaction system is simple and can provide good cellulose conversion. However, there are disadvantages such as reactor corrosion, additional post-hydrolysis separation and a poor recyclability of the catalyst. Enzymes, on the other hand, can overcome some of the issues found in the homogeneous acid hydrolysis, and their high selectivity is a marked advantage. However, enzymes can be very expensive, and their hydrolytic process requires long reaction times.

Various types of solid acids have been developed for biomass hydrolysis because of the ease of catalyst/product separation, recyclability or better reactor-friendliness.[12,18] These materials also have the potential to be applied in flow fixed-bed reactors, which is a priority for industrial applications to reduce process time and operation expenses. On the other hand, solid acid catalysts typically present low reaction rates in cellulose hydrolysis because of the little contact between the reactants and catalyst, both being solids. The demanding reaction conditions required for cellulose hydrolysis also typically result in some leaching of the active phase immobilised on the support, which results in catalyst deactivation. Recently, several authors have published excellent reviews with a comprehensive analysis of the catalytic activity of acid catalysts for this reaction.[5,9,12,18]

Hu *et al.* have classified solid acids in ten groups: acid resins, metal oxides, H-form zeolites, heteropoly acids, functionalised silicas, supported metals, immobilised acidic ionic liquids, carbonaceous acids, magnetic acids or others, according to their nature and properties.[9] Table 8.1 summarises some of the most representative investigations where the hydrolysis of cellulose has been catalysed by solid acids using conventional heating, and high levels of cellulose conversion and glucose yields have been achieved.

Out of the investigations shown in Table 8.1, the highest glucose yield was reported by Shuai *et al.*[19] They have demonstrated that –Cl groups anchored

Table 8.1 Selected works on hydrolysis of cellulose catalysed by solid acids under conventional heating with glucose as a main product. AC stands for activated carbon, BMIM for 1-butyl-3-methylimidazolium, C for carbon, CMK for type of ordered mesoporous carbon, CP is chloromethyl polystyrene resin, IL is ionic liquid, SBA is Santa Barbara, which is a name for a specific type of hexagonal mesoporous silica, ZSM-5 is a type of zeolite.

Entry	Catalyst	Pretreatment method	Solvent	*T* (°C)	Time (h)	Yield (%)	References
1	AC-*N*-SO_3H-250	Ball-milling	H_2O	150	24	62.6	22
2	CMK-3-SO_3H	Ball-milling	H_2O	150	24	74.5	22
3	$Si_{33}C_{66}$-823-SO_3H	Ball-milling	H_2O	150	24	50.4	361
4	AC + HCl (0.012%)	Mix-milling	H_2O	180	0.5	88	362
5	CP-SO_3H	—	H_2O	120	10	93	19
6	SBA-15-SO_3H	IL treatment	H_2O	150	3	52	363
7	$H_3BW_{12}O_{40}$	—	H_2O	60	48	77	364
8	C-SO_3H	—	[BMIm] Cl/H_2O	110	4	72.7	365
9	H-Y zeolite	—	[BMIm] Cl/H_2O	130	2	50	366
10	H-ZSM-5	—	[BMIm] Cl/H_2O	130	3.5	42.3	366

on polystyrene-based acid catalysts improved the activity of the acid catalyst in the hydrolysis of cellulose. After 10 h at 120 °C, their catalyst was able to produce a 93% glucose yield from microcrystalline cellulose (Avicel), which is one of the highest glucose yields obtained so far.

Sulfonated carbons are a group of catalysts with exceptionally high performance in the cellulose hydrolysis reaction because of the diversity of functional groups available on their surface.[9,20] This type of catalyst has also shown an admirable performance in the production of biodiesel, as shown by Toda *et al.*[21]

The highest cellulose conversion and glucose yield over carbon-based catalysts were reached by Pang *et al.* in 2010 over a structured mesoporous carbon catalyst (CMK-3).[22] Before the development of the CMK-3 catalyst, Pang *et al.*[22] evaluated the reaction with various types of sulfonated carbons, which were synthesised by modifying the sulfonation temperature and nitric acid pretreatment, using a commercial activated carbon (AC) as a support. In this group of experiments, the catalyst with the best performance was the one that was sulfonated at 250 °C and pretreated with nitric acid. This resulted in 74.3% cellulose conversion and 62.6% glucose yield. Finally, the catalyst was optimised using CMK-3 as the carbonaceous source and the ball-milled cellulose conversion and glucose yield increased to 94.4% and 74.5% at 150 °C for 24 hours, respectively. The performance of this catalyst was attributed to the high-temperature sulfonation and the well-ordered mesoporous structure of the carbon.

8.2.2.1 Microwave Hydrolysis of Cellulose

8.2.2.1.1 Microwaves and Cellulose. Recently, microwave irradiation has emerged as a promising technique applied to the hydrolysis of cellulose since it exhibits multiple advantages in comparison to conventional heating. Even though high yields of reducing sugars are afforded under mild conditions, the hydrolysis of cellulose under microwave irradiation still presents several unexplored areas.

According to Wu *et al.*,[23] a key factor that promotes cellulose hydrolysis is the destruction of cellulose crystals by the microwave irradiation, as was confirmed by XRD. The estimated crystallinity of the cellulose was about 75% before the treatment and decreased to 58% after the exposure to microwave irradiation. Their explanation is based on the fact that microwaves can be absorbed deeply in the folding layers of cellulose. Also, they proposed that the microwave irradiation can enhance the number of collisions between the cellulose particles and the catalyst.

Nevertheless, most of the investigations based on the application of microwave irradiation as a source of heating have been done adding liquid acids as the catalyst.[24–39] Sulphuric acid,[27,31,32,35,36] phosphoric acid,[28,38] and hydrochloric acid[24,30,34] are the most popular liquid acids for this reaction due to their high availability and low cost. However, their use has a considerable environmental cost, and solid acids may be preferred if their performance and cost are satisfactory. Additionally, the use of ionic liquids in the reaction system enhances the depolymerisation of the biopolymer.[25,28,29,33] Although the majority of the investigations have been focused on the production of reducing sugars, higher value-added products such as levulinic acid, HMF and furfural can be obtained with good yields.[24,28,33,35] In these cases, the hydrolysis step is followed by a dehydration reaction, as will be discussed in a later part (8.3.3).

To overcome the limitations of using homogeneous catalysts in the hydrolysis reaction, a few papers have been published applying solid acid catalysts to the hydrolysis of cellulose under microwave irradiation.[23,40–45] Three main types of solid acid catalysts have been reported in this reaction: functionalised carbonaceous materials,[23,40–43] acidified polymers,[44] and zeolites.[45] When ionic liquids are added to the reaction mixture, the interaction between the cellulose and the catalyst is enhanced, promoting higher yields of products. Solid acid catalysts have shown recyclability over a number of cycles (typically below five), which represents an advantage over the soluble acids. Among the available catalysts, activated carbon (AC) is a great microwave susceptor, and it became the most common support for heterogeneous catalysts used in lignocellulosic conversions. Polyoxometalates (POMs) are super-strong acids but less corrosive than mineral acids, hence they can be used as an alternative to them for the hydrolysis of cellulose as well as for others transformations, such as esterification[46] or transesterification.[47] When POMs are supported on activated carbon *via* the wet impregnation method, the catalytic system obtained is heterogeneous and, due to this, the catalyst can be easily separated from the mixture and recycled. Moreover, the AC-POMs exhibited a higher saccharification of carbohydrates under microwaves as compared to induction heating.[48]

In Table 8.2, the most relevant results of the hydrolysis of cellulose over solid and liquid acid catalysts under microwave irradiation are summarised.

Acidified polymers have been explored and can offer tuneable properties due to their variable composition. Li *et al.* developed a random copolymer poly(acrylic acid)-random-poly(styrene sulfonic acid) (PAA-*r*-PSSH) synthesised by the direct thermolysis of the precursor poly(*tert*-butyl acrylate)-random-poly(neopentyl styrenesulfonate) (PtBA-*r*-PNSS). It was compared with a block copolymer possessing the same acid concentration, but PAA-*r*-PSSH displayed higher activity in the conversion of cellulose in aqueous media under microwave irradiation.[44] They attributed the enhanced activity to the synergy between the sulfonic and carboxylic groups on the surface of the catalyst, which have also been proposed to produce superior activity in the sulfonated carbon catalyst.[9] At a mild temperature of 120 °C, the ball-milled cellulose was successfully hydrolysed after 120 min achieving a glucose yield of 35%.

Several investigations have been published on the use of sulfonated carbons as a solid acid catalyst for the hydrolysis of cellulose.[9,12] Even though most of these investigations have been carried out in conventional heating, some works have been performed in microwave-assisted systems.[23,40,41,43] Wu *et al.* studied the hydrolysis of cellulose over various biomass char sulfonic acids (BC–SO_3H) under different temperatures and reaction times, obtaining a glucose yield of 19.8% at the optimised conditions (1 h at 90 °C). The catalyst showed a turn-over number (TON) of 1.73. The microwave irradiation produced activities up to 5 times higher than conventional heating under the same conditions. The promoter effect of the microwave irradiation was attributed to the fact that the microwaves can be deeply absorbed into the molecular layers of cellulose, which facilitated its decomposition. Cellulose was then exposed to microwave irradiation for 3 h and the crystallinity degree of cellulose decreased to 17% of the original value (100%).

The incorporation of ionic liquids into the hydrolysis of cellulose has been proved to be an efficient way to enhance the cellulose conversion under microwave irradiation. Zhang *et al.* tested several H-form zeolites using [C_4mim]Cl as a solvent obtaining the best results with the lower Si/Al ratio zeolites and larger surface areas.[45] They achieved a glucose yield of 36.9% and a TRS yield of 47.5% at 100 °C for 8 min displaying a better activity than a commercial ion-exchange resin. When the same zeolite was used under the same conditions, but in an oil bath instead of the microwave irradiation, only a 2.1% glucose yield and a 7.1% TRS yield were obtained.

More solid acids have been evaluated in the presence of ionic liquids as a solvent by Ishida *et al.*[42] Ion-exchange resins, acidified carbons and zeolites were tested to hydrolyse untreated cellulose in the [BMIM]Cl solvent at 120 °C for 20 min. The catalyst with the highest activity was the commercial

Table 8.2 Hydrolysis of cellulose catalysed by liquid or solid acids under microwave irradiation.

Entry	Catalyst	Catalyst type	Solvent	*T* (°C)	Time (min)	Main product	Yield (%)	References
1	PAA-*r*-PSSH	Solid	H_2O	120	120	Glucose	35	44
2	BC-SO_3H	Solid	H_2O	90	60	Glucose	19.8	23
3	BC-SO_3H–IL	Solid	H_2O	90	120	Reducing sugars	33.4	43
4	BCSA-IL-Cl/BCSA-IL-F_1	Solid	H_2O	80	180	HMF	27.94	40
5	BC-SO_3H-IL–Zn	Solid	[BMIM]Cl	90	120	HMF	22.7	41
6	Amberlyst-15	Solid	[BMIM]Cl	120	20	Glucose	65	42
7	Zeolites	Solid	[C_4mim]Cl	100	8	Glucose	36.9	45
8	H_2SO_4	Liquid	H_2O	160	15	Reducing sugars	59.2	31
9	H_3PO_4	Liquid	H_2O	160	5	Reducing sugars	60	38
10	H_3PO_4	Liquid	H_2O	175	5	Glucose	90	49
11	H_2SO_4	Liquid	H_2O	170	50	Levulinic acid	34.2	35
12	H_2SO_4	Liquid	H_2O	160	5	Glucose	73.3	36
13	H_2SO_4	Liquid	H_2O	160	10	Glucose	59.1	11
14	HCl	Liquid	H_2O	-	7	Glucose	67	34
15	HCl	Liquid	H_2O	200	3.32	Levulinic acid	42.7	24
16	HCl	Liquid	NaCl	160	10	Glucose	36.7	30
17	H_2SO_4/Oxalic acid	Liquid	H_2O	120	120	Glucose	29	32
18	Phosphotungstic acid	Liquid	H_2O	120–220	5	Glucose	37	39
19	$H_3PW_{12}O_{40}$	Solid	H_2O	90	3	Glucose	75.6	37
20	H_2SO_4	Liquid	H_2O/isopropanol	180	30	Glucose	50	50
21	TiO_2	Solid	ChCl/oxalic acid	140–200	5–60	HMF and furfural	53.4	33
22	Methyltrioxorhenium	Liquid	[AMIM]Cl	150	30	Reducing sugars	51.2	25
23	—	Liquid	$MgCl_2$–NaCl	230	10	Reducing sugars	13.8	26
24	H_2SO_4	Liquid	Sulfolane	160	5	Reducing sugars	71.9	27
25	H_3PO_4	Liquid	[BMIM]Cl	170	60	Levulinic acid	67.3	28
26	$[(CH_3)_4N]ReO_4$	Liquid	[AMIM]Cl	150	30	Reducing sugars	97	29

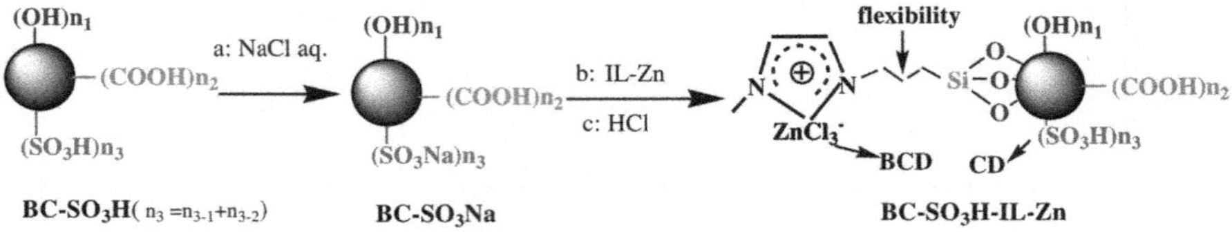

Scheme 8.4 Synthesis process for the biomimetic catalyst BC-SO_3H-IL-Zn. Adapted from *Cellulose*, Biochar sulfonic acid immobilized chlorozincate ionic liquid: an efficiently biomimetic and reusable catalyst for hydrolysis of cellulose and bamboo under microwave irradiation, **21**, 2014, 1227–1237, C. Zhang, Z. Fu, B. Dai, S. Zen, Y. Liu, Q. Xu, S. R. Kirk and D. Yin, © Springer Science + Business Media Dordrecht 2014, with permission of Springer.[41]

ion-exchange resin Amberlyst-15, producing a 65% yield of glucose at the mentioned conditions. When sequential additions of water were done, the glucose yield was enhanced since the formation of humins was observed when the reaction mixture had low concentrations of water.

Aiming towards a system free of liquid catalysts and non-aqueous solvents, Zhang *et al.* have developed some sulfonated biochar solid catalysts modified with fluorine- or chlorine-containing ionic liquids obtained by grafting the ionic liquids onto the sulfonated biochar in a multi-step process.[40,41,43] The schematic representation of the synthesis is presented in Scheme 8.4. Using the fluorine-containing catalyst and water as solvent, the cellulose can be hydrolysed to produce a 15.56% yield of reducing sugars and a 27.94% yield of HMF at a mild temperature (80 °C) for 3 h. After the 3 h of reaction time, the HMF yield did not display a significant increase and the reducing sugars yield started to decrease. A high TON of 4.89 was obtained in the reaction and the catalyst showed the same level of activity even after 6 runs. When the chlorine-containing catalyst was utilised, a 36.2% yield of reducing sugars and a 22.7% yield of HMF were obtained at 90 °C after 2h. The effect of the solvents (water, [BMIM]Cl) was evaluated as well as the functionalisation of the sulfonated char. Under the same conditions, a 22% yield of TRS was obtained with the bare sulfonated char whereas 33.4% was achieved utilising the ionic liquid containing sulfonated char. Furthermore, the TRS yield was improved from 33.4 to 64.6% by the addition of [BMIM]Cl as a solvent. The explanation of the high performance of these catalysts is based on the synergistic effect on the catalyst surface since the grafted ionic liquids possess a high affinity to cellulose and they also improve the acidity of the sulfonic groups of the biochar, which enhances the cleavage of the β-1,4-glycosidic bonds.

Among the studies carried out with homogeneous catalysts, Orozco *et al.* studied the hydrolysis of cellulose with H_3PO_4 and evaluated the effect of the catalyst concentration, temperature and reaction time in a microwave reactor.[49] The optimum conditions were at 175 °C with a catalyst concentration of 2.5% for 3–5 min where a 90% yield of glucose was obtained. Through a kinetic study, they determined that the reaction followed a pseudo-homogeneous consecutive first order model with reaction rate constants that were higher than those reported for conventional heating. By means of the kinetic study, it was concluded that the microwave irradiation causes a high reaction rate constant at mild temperatures and it also prevented the formation of hot spots in the reactor, which would be responsible for selectivity loss.

In their research, Ching *et al.* performed the hydrolysis of untreated microcrystalline cellulose through several cycles of reaction with sulphuric acid as catalyst and found an enhancement with the addition of isopropanol as co-solvent to water.[50] Each cycle consisted of 10 min of reaction at 180 °C and at the end of each cycle, the liquid phase was removed and fresh sulphuric acid solution was added to proceed with the second run. After the first 10 min cycle, more than 60% of the cellulose had been converted with a selectivity above 50% towards glucose, HMF, furfural and levulinic acid. After 30 min of total reaction time, an overall cellulose conversion of 76% was achieved keeping the high selectivity towards the desired products of 84%. Their results were explained on the basis that the microwave irradiation is a dielectric heating process that enhances the reaction by disrupting hydrogen bonds within the cellulose molecules. At temperatures above 180 C, cellulose becomes soft and the microwaves can rotate the CH_2OH groups provoking the breakdown of the glycosidic bonds.

In an investigation conducted by Yuan *et al.*, it was demonstrated that the hydrolysis of cellulose in 1-allyl-3-methylimidazolium chloride ([AMIM]Cl) was promoted by using a methyltrioxorhenium (MTO) or quaternary ammonium perrhenate catalyst.[25,29] When the reaction was carried out over the MTO catalyst, a 51.2% yield of TRS was accomplished after 30 min at 150 °C. However, when the reaction time was extended or the temperature increased, the yield of degradation products was higher due the decrease in the viscosity of the reaction mixture, which causes a complete interaction of the catalyst and the cellulose, leading to the formation of degradation products. It was claimed that the nucleophilic attack of the electron-rich O atom in the cellulose glycosidic bond to the electron-poor atom of the catalyst played a key role in the reaction. On the other hand, the application of the quaternary ammonium perrhenate catalyst produced a 97% yield of TRS and 42% glucose yield under the same reaction conditions. In the latter case, the hydrogen bonding between the cellulose hydroxyl groups and the perrhenate is a fundamental step in the reaction.

Sun *et al.*[11] reported a pretreatment method with α-cellulose to increase its conversion during the hydrolysis reaction. A cellulose hydrogel was obtained by treating the cellulose with an aqueous solution of NaOH/urea followed by a regeneration in water, achieving a significant increase from 0.42 to 44.6% glucose yield in the hydrolysis of cellulose at 160 °C for 10 min with diluted sulphuric acid as catalyst. Furthermore, when cellulose was exposed to ozone before the regeneration step, the final glucose yield was 59.1% with only a 0.28% yield of degradation products such as levulinic acid, HMF and furfural detected.

The role of electrolytes in the hydrolysis of cellulose under microwave irradiation was recently studied by Tsubaki *et al.*[26] When the hydrolysis of cellulose was conducted at 230 °C for 10 min without catalyst, the TRS yield was 3.1%. However, when NaCl or $MgCl_2$ was added as the electrolyte into the solution, the TRS yield increased to 6.9 and 13.8%, respectively. They explained that this positive effect was achieved by the electrolyte solution, which effectively absorbs the microwave irradiation. The dielectric loss of water caused by the high temperature was compensated by the addition of the electrolytes. Moreover, the addition of these electrolytes also contributes to reducing the energy consumption of the microwave source since the provided irradiation is used more effectively. A similar investigation about the effect of ions in solutions was executed by Qin *et al.*, where a 67.3% yield of levulinic acid was reached from cellulose after 60 min at 170 °C with phosphoric acid as catalyst in the presence of alkali metal halides.[28] The concentration of the ions accelerated the rate of the reaction and it was further promoted by the microwave irradiation. To demonstrate the promoting effects of the microwave irradiation, the reaction was carried out with the same conditions under conventional heating and their results are presented in Figure 8.1. They found that whereas a 57.4% yield of levulinic acid was achieved in 30 min of reaction under microwave irradiation with 1.5 M H_3PO_4, only a 6.0% yield was achieved at the same conditions under conventional heating. The reaction rate enhancement was attributed to the nature of the microwave heating model and the solution being rich in ions.

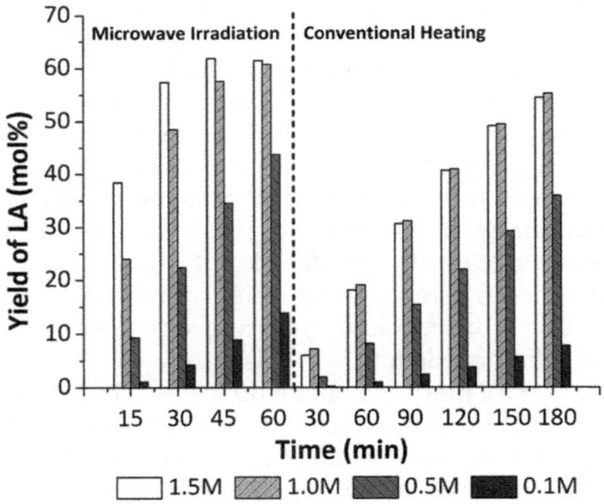

Figure 8.1 Yield of levulinic acid from α-cellulose in a high salt acidic catalytic system consisting of KCl and H_3PO_4 with different concentrations from 0.1 M to 1.5 M at 170 °C under microwave irradiation and conventional heating, respectively. Adapted from ref. 28 with permission from The Royal Society of Chemistry.

Another key factor to achieve an efficient hydrolysis of cellulose is the crystal structure of the cellulose during the hydrolysis reaction.[27] In the research developed by Teng *et al.*, regenerated cellulose was hydrolysed by diluted sulphuric acid with the addition of sulfolane at various concentrations. They found that the higher the concentration of sulfolane, the higher the TRS yield obtained. Subsequently, the crystalline structure of the residual cellulose was monitored. When a 40% content of sulfolane was used, the percentage of crystalline cellulose was still 20–30%, but this percentage was reduced to 2% when 80% sulfolane was utilised, facilitating the hydrolysis of cellulose. On the other hand, when no sulfolane was used, the change in the crystalline structure was almost negligible. Finally, the optimum concentration of sulfolane was 80%, producing a 98% cellulose conversion with a 71.9% yield of TRS.

8.2.2.1.2 How the Degree of Polymerisation and Crystallinity of Cellulose Affect Hydrolysis. When comparing reaction rates and yields in the hydrolysis of cellulose across the literature, it is very important first to establish whether the cellulose used presents the same level of crystallinity and degree of polymerisation, as these characteristics determine how recalcitrant is the substrate. Boissou *et al.* demonstrated that both degree of polymerisation (DP) and crystallinity of cellulose play a crucial role in the hydrolysis reaction of cellulose under microwave irradiation (Scheme 8.5).[51] The main aim of their study was to control the depolymerisation of cellulose with a [TBMA]Cl/BHC pretreatment (TBMA, tributylmethylammonium; BHC, betaine hydrochloride). This pretreatment decreased the DP from 200, corresponding to the original microcrystalline cellulose, to 65, and the crystallinity was reduced as well. They tested the pretreated cellulose in the microwave-assisted cellulose hydrolysis with water as solvent at 150 °C for 1 h utilising the commercial catalyst Amberlyst 35. The conversion of cellulose pretreated with [TBMA]Cl/BHC was compared with cellulose pretreated by ball-milling and untreated cellulose. The highest glucose yield was 43%, obtained with the cellulose pretreated with [TBMA]Cl/BHC (9/1), while the ball-milled cellulose and untreated cellulose produced only 15 and <5% glucose yields, respectively. Although the ball-milling pretreatment reduced the crystallinity of cellulose, the DP remained the same. So, the crucial difference between [TBMA]Cl/BHC pretreated cellulose and the ball-milled pretreated cellulose

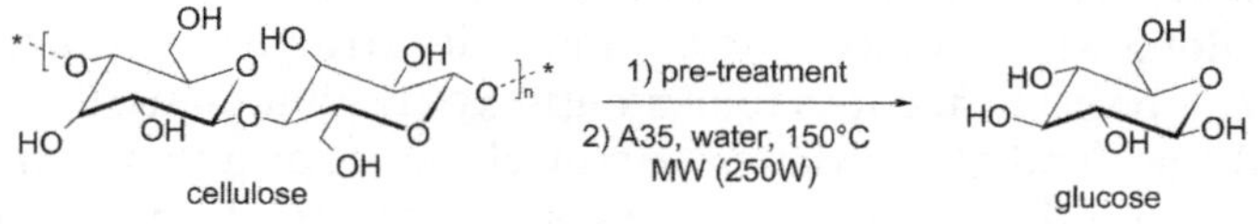

Scheme 8.5 Glucose yield in the microwave-assisted cellulose hydrolysis with different pretreatments. Adapted with permission from F. Boissou, K. De Oliveira Vigier, B. Estrine, S. Marinkovic and F. Jérôme, *ACS Sus. Chem. Eng.*, 2014, 2, 2683. Copyright 2014 American Chemical Society.[51]

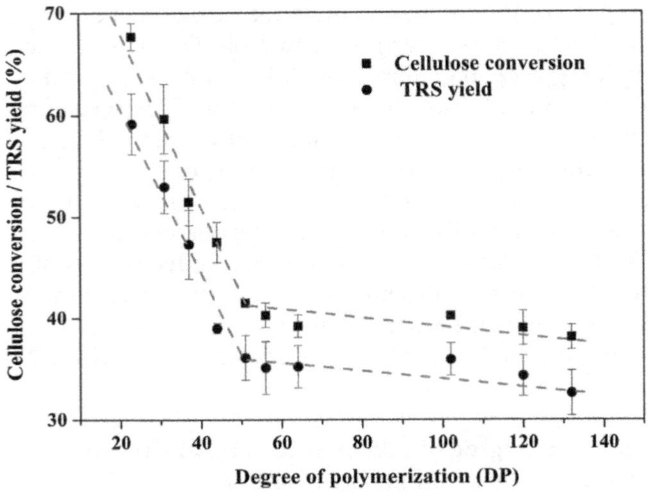

Figure 8.2 Cellulose conversion and total reducing sugars (TRSs) yield from regenerated celluloses with various degrees of polymerisation. Adapted from *Bioresource Technology*, **191**, J. Ni, N. Teng, H. Chen, J. Wang, J. Chu and H. Na, Hydrolysis behavior of regenerated celluloses with different degree of polymerization under microwave radiation, 229–233, Copyright 2015, with permission from Elsevier.[31]

was the DP. Therefore, it can be concluded that a low DP was the key element asociated with cellulose conversion rather than crystallinity.

Ni *et al.* evaluated the behaviour of regenerated celluloses with different degrees of polymerisation in the microwave-assisted cellulose hydrolysis using a dilute acid as catalyst.[31] The regeneration process consisted of treating the microcrystalline cellulose in 85% H_3PO_4 at different temperatures and durations followed by regeneration in water. They achieved a 59.2% yield of total reducing sugars (TRSs) with a DP of 23, this was the lowest DP evaluated in the study (Figure 8.2) and a steeply rising trend in the cellulose conversion and TRS yield is clearly presented as the DP decreases below 51. The reaction was carried out for 15 min at 160 °C.

8.2.3 Hydrolysis of Hemicellulose

Hemicellulose is a heterogeneously branched polymer present in the cell wall of plants along with cellulose. According to its structural properties, hemicellulose can be classified into four groups: xylans, mannans, β-glucans with mixed linkages, and glucoxylans.[52] Hemicellulose has an amorphous structure soluble in most common solvents, such as water. Due to these properties, the hydrolysis of hemicellulose is a lot easier than cellulose hydrolysis and it is usually carried out under milder conditions.[53] An example of a xylan structure is represented in Scheme 8.6.

The diverse polymer structures present in hemicellulose make this polysaccharide an attractive feedstock to produce biopolymers that may be applied

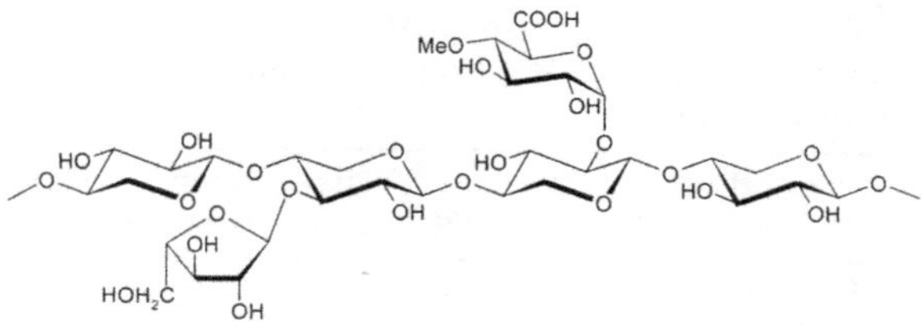

Scheme 8.6 Structure of (L-arabino)-4-*O*-methyl-D-glucurono-D-xylan. Adapted from *Polysaccharides I*, Hemicellulose, 2005, pp. 1–67, A. Ebringerová, (© Springer-Verlag Berlin Heidelberg 2005) With permission of Springer.[52]

in the food and non-food industries.[52] Although a range of value-added products can be obtained from hemicellulose, it is the most wasted polysaccharide fraction in ethanol plants.[13] The monomers in hemicellulose are linked by β-1,4-glycosidic bonds, thus the hydrolysis of hemicellulose consists of the selective cleavage of the C–O bonds.[54] Unlike cellulose, hemicellulose is a biopolymer soluble in water, so the conditions required for the hydrolysis of hemicellulose are milder than the ones for the hydrolysis of cellulose, and reaction rates are faster.[55]

The information regarding the hydrolysis of hemicellulose under microwave irradiation is still limited, although it might be becoming more popular due to the solubility of the substrate.[24,54,56–59] The hydrolysis of hemicellulose catalysed by acids is usually carried out with hydrochloric, sulphuric or organic acids.[24,57–59] Regarding the application of solid acid catalysts in this reaction, only a couple of investigations have been published.[54,56] Xylose is the main product from the hydrolysis step, however sequential reactions of dehydration will produce furfural.[24,59] Corncobs are an abundant agricultural waste with a high hemicellulose content (about 30–40 wt% of the dry material). Typically, oligomers contained in the corncob hydrolysates can rapidly be converted into monomers under MW irradiation using HCl as the catalyst, in reactions that yield susbtantial yields of furfural with respect to the total hemicellulose content in the raw material.[1] Table 8.3 summarises the available literature where the hydrolysis of hemicellulose has been performed with homogeneous or heterogeneous catalysts under microwave irradiation.

In 2015, Bai and co-workers[54] reported the microwave-assisted conversion of biomass-derived hemicelluloses into xylo-oligosaccharides in the presence of bamboo-based catalysts. This catalyst has $-SO_3H$, –COOH and phenolic –OH groups, similar to active carbon and proved to be active in the hydrolysis of hemicellulose yielding above 40% xylo-oligosaccharides at temperatures between 140 and 160 °C. Unfortunately, recycling studies showed deactivation of the catalyst primarily due to the leaching of sulfonic groups.

Sweygers *et al.*[24] found that the production of furfural from xylan does not follow the same optimised conditions as those for cellulose due to the differences in the biopolymers. The best conditions for xylan conversion

Table 8.3 Hydrolysis of hemicellulose catalysed by solid and liquid acids under microwave irradiation. AC is activated carbon, N/A is not available, THF is tetrahydrofuran.

Entry	Catalyst	Catalyst type	Solvent	*T* (°C)	Time (min)	Main product	Yield (%)	References
1	PAA-*r*-PSSH	Solid	H_2O	100	60	Glucose	94.5	44
2	AC-SO_3H	Solid	H_2O	130	60	Xylose	77.3	367
3	$NbOPO_4$	Solid	H_2O	170	30	Furfural	23	56
4	AC-SO_3H	Solid	H_2O	150	45	Xylooligosaccharides	>40	54
5	$AlCl_3$–$6H_2O$/NaCl	Liquid	THF	160	60	Furfural	64	59
6	H_2SO_4	Liquid	H_2O	180	30	Xylose + furfural	45.4	368
7	HCl	Liquid	H_2O	200	1	Furfural	31.2	24
8	Organic acids	Liquid	H_2O	126	8	Xylooligosaccharides	39.4	57
9	H_2SO_4	Liquid	H_2O	200	5	Monosaccharides	N/A	58

were 0.36 M HCl at 200 °C for 1 min and they gave a 31.2% yield of furfural. The high solubility of xylan in water also played an important role in this optimisation.

8.2.4 Hydrolysis of Starch

Due to its solubility in water, lack of crystallinity and weaker bonding between sugar units, the hydrolysis of starch proceeds readily. Li *et al.*[44] used their random copolymer PAA-*r*-PSSH in the hydrolysis of starch and studied the effect of microwave irradiation as compared to conventional heating. From Figure 8.3, it is evident that the microwave irradiation accelerated the hydrolysis of starch, where a 94.5% glucose yield was reached after 60 minutes at 100 °C under microwave irradiation, which is 5 times higher than the result obtained using conventional heating (Figure 8.3d points).

8.2.5 Microwave-assisted Lignin Conversion

Unlike cellulose and hemicellulose, lignin is a complex three-dimensional amorphous polymer containing methoxylated phenylpropanoid units that provide strength and rigidity to the cell walls in plants.[11,60] Lignin is the largest renewable source of aromatic-containing compounds, since it is formed by three phenylpropanoid monomers: coniferyl, sinapyl and *p*-coumaryl alcohols (Scheme 8.7), interlinked by C–C and C–O–C bonds.[61] Owing to its limited reactivity and complexity, lignin is considered as a low-value resource

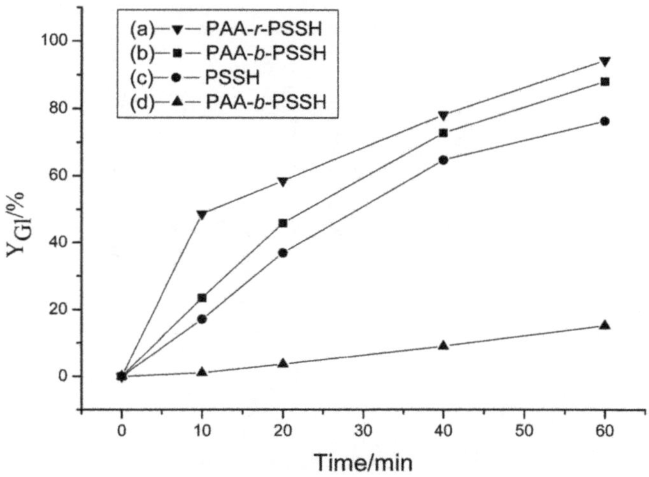

Figure 8.3 Hydrolysis of starch over various acid polymers under (a–c) microwave irradiation and (d) conventional heating (oil bath). Adapted from ref. 44 with permission from The Royal Society of Chemistry.

and therefore its depolymerisation to produce a natural source of aromatics is highly sought after. Several authors have produced excellent reviews on the catalytic valorisation of lignin[14,60,62] although some other alternative catalytic routes to aromatics exist with cellulose, alcohols and sugars as feedstocks.[63] As of 2004, 50 million tons of lignin were extracted in the pulp and paper industry, however only 2% of this lignin was utilised to obtain value-added products and the rest was used as low-value fuel to burn.[60,64]

As opposed to the typical hydrolysis pathway for cellulose and hemicellulose, lignin is not usually hydrolysed. However, there are many studies in which microwaves are utilised in the pretreatment of lignocellulosic biomass, often prior to enzymatic depolymerisation.[65–70] There are two major pathways for microwave-assisted lignin conversion: pyrolysis and solvolysis. Pyrolysis is a temperature treatment in the absence of air or oxidising agent, while solvolysis utilises solvent to help with the dissolution/depolymerisation steps (hydrolysis when the solvent is water). Both processes can occur with or without a catalyst but its presence can enhance the efficiency of the process significantly. Pyrolysis usually needs a higher temperature (300–900 °C) than solvolysis, which can occur under milder conditions (*ca.* 100–400 °C).[71] Both pyrolysis and solvolysis of different types of lignin performed under both conventional and microwave heating were reviewed recently by Yunpu and co-workers.[71] Biomass pyrolysis deserves more than a chapter on its own and this is out of the scope of this chapter. The reader is here directed to some recent reviews and book chapters dealing with this subject in detail and the use of microwaves in pyrolysis[72–75] and the book chapter by J. Clark *et al.* on low temperature microwave pyrolysis also published by the RSC Green Chemistry series.[76]

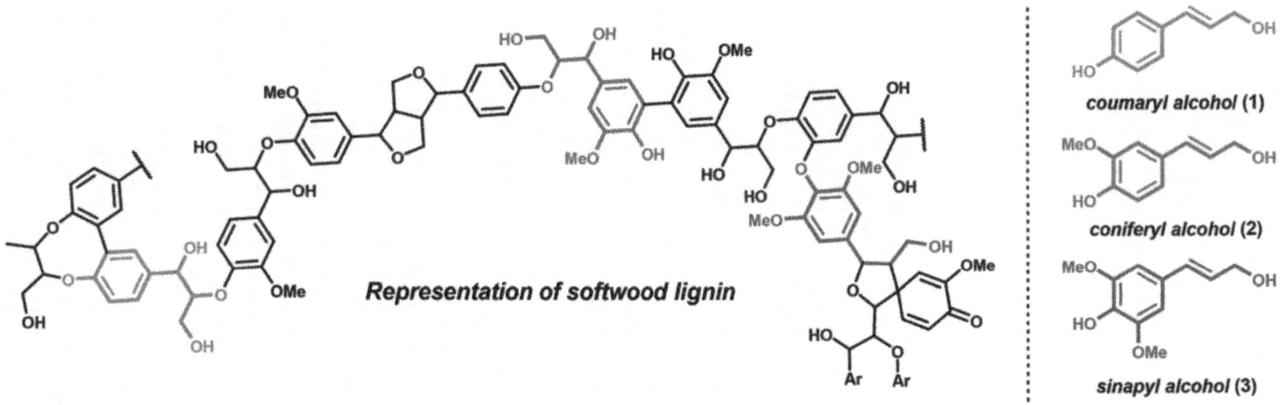

Scheme 8.7 Structure of lignin and the three main monomeric molecules. Adapted from ref. 359 with permission from The Royal Society of Chemistry.

Some of the early work in 2002 by Nagpurkar, Chaudhari and Ekhe[77] attempted the selective oxidation of purified alkali lignin assisted by nitrobenzene. Although non-catalytic, the reported domestic microwave system delivering full power to the solution of lignin and nitrobenzene for 50 minutes gave 12% conversion as compared to 8% achieved after 4 hours with conventional heating. The two main identified products were *p*-hydroxybenzaldehyde and vanillin although several other unidentified products were observed. Based on the available data, the authors suggested that these could be less polar than the *p*-hydroxybenzaldehyde and vanillin compounds. The results of this work suggest that the use of a catalytic system in the selective conversion of lignin could provide further needed improvement of lignin conversion. Since then, much work has been carried out in the microwave-assisted solvolysis of lignin, and ultrastable faujasite zeolite,[78] sulphuric acid,[79,80] formic acid,[81,82] KOH[78,83] or glycerol/methanol[79] have been used as catalysts and/or solvents to produce monomeric[79–83] or oligomeric[78] lignin units (*i.e.* guaiacols, syringin, phenolics or mesitol) with different degrees of success. The maximum reported yield of 63.27% monomeric and oligomeric products was produced from black liquor lignin by Dong and co-workers[81] when ultrastable zeolite Y was used together with formic acid as a solvent (160 °C, 30 min). Whereas for oxidative routes, J. Clark and co-workers demonstrated the efficiency of using hydrogen peroxide with microwave irradiation to cleave a number of lignin model compounds, using a range of mesoporous silicas and Co(salen) using hydrogen peroxide as oxidant.[84–87]

8.2.5.1 Use of Hydrogen-donating Agents

R. Luque and co-workers[82] investigated a range of hydrogen-donating solvents using Ni 10%Al-SBA-15 as a heterogeneous catalyst with lignin. The phenolic monomeric products obtained depended on the hydrogen-donating agent, with formic acid giving better results than tetralin, isopropanol and glycerol. In a subsequent work, they incorporated several metals on the Al–SBA-15 support, including nickel, palladium, platinum and ruthenium. It was found that the 10% Ni/Al–SBA-15 material was the best for the depolymerisation of lignin, giving the highest yield of >30% bio-oil composed mainly of mono- and dimeric compounds after 30 minutes at *ca.* 140 °C at a fixed power of 400 W.

8.2.5.2 Oxidative Lignin Conversion

The group of James Clark[84–87] has shown that lignin model compounds can be selectively oxidised in microwaves using hydrogen peroxide. In the oxidation of the lignin model compound apocynol, the authors first showed that a mixture of acetovanillone, vanillin and 2-methoxybenzoquinone (Scheme 8.8) can be produced over an SBA-15 catalyst when oxidised by H_2O_2.[84]

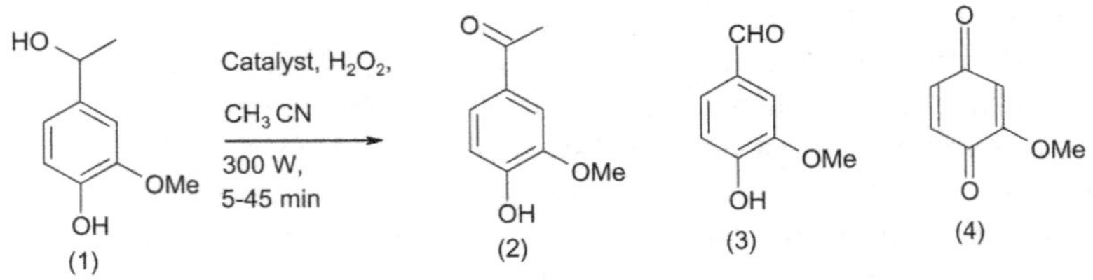

Scheme 8.8 Reaction scheme of the oxidation of 1,2-(4-hydroxy-3methoxy-phenoxy) ethanol (apocynol) (1) to acetovanillone (2), vanillin (3) and 2-methoxybenzoquinone (4) using hydrogen peroxide as an oxidant in acetonitrile and under microwave irradiation. Adapted from *Catalysis Communications*, **9**, S. K. Badamali, J. H. Clark and S. W. Breeden, Microwave assisted selective oxidation of lignin model phenolic monomer over SBA15, 2168–2170, Copyright 2008 with permission from Elsevier.[84]

Generally, yields lower than 51% were achieved after 30–45 min reaction times. This could be due to the low oxidative ability of the metal-free SBA-15. The activity of the catalyst decreased only by 5% in the second run and produced a 14% yield of acetovanillone as compared to 17.7% in the first run. Subsequent publications showed that the conversion of the substrate could be further increased by choosing the right silica catalyst, and commercial mesoporous silica-5 gave 95% conversion and MCM-41 gave 72% conversion as compared to the 36% conversion obtained over SBA-15. However, the yields of the desired products were not improved. At best, the same 18% yield of acetovanillone as with SBA-15 previously was obtained over MCM-41 and 5 and 23% yields of vanillin and 2-methoxybenzoquinone were observed.[85] Addition of Co as an active metal by supporting [*N*,*N*′-bis(salicylidene)ethane-1,2-diaminato]Cobalt(II) (Co(salen) complex) on SBA-15 led to an increase in activity. However, this did not help the yields of the desired products. In fact, the highest yield obtained was 15% acetovanillone at 63% conversion after 15 minutes of reaction. Moreover, when almost complete conversion of apocynol after 40 minutes was achieved, none of the three above-mentioned products could be detected, and authors attributed this to the complete oxidative degradation of the substrate and formation of high molecular weight compounds including phenolic dimers and quinines.[86]

When the oxidation with the same catalyst Co(salen)/SBA-15 was performed on the lignin model phenolic dimer [1-(4-hydroxy-3-methoxyphenyl)-2-(2-methoxyphenoxy)propane-1,3-diol, 2-methoxyphenol was produced (Scheme 8.9). This could be due to either the oxidative or non-oxidative cleavage of the C–O bond at the ether part of the molecule. Again, only high molecular weight products were obtained during the reaction under conventional heating the while activity of the homogeneous Co(salen) complex was very low compared to the supported catalyst.[87]

Recently, Pan *et al.* studied the degradation of model compounds, such as 2-phenoxy-1-phenylethanol, vanillyl alcohol, and 4-hydroxybenzyl alcohol, with 14 types of metal salts using hydrogen peroxide and found that

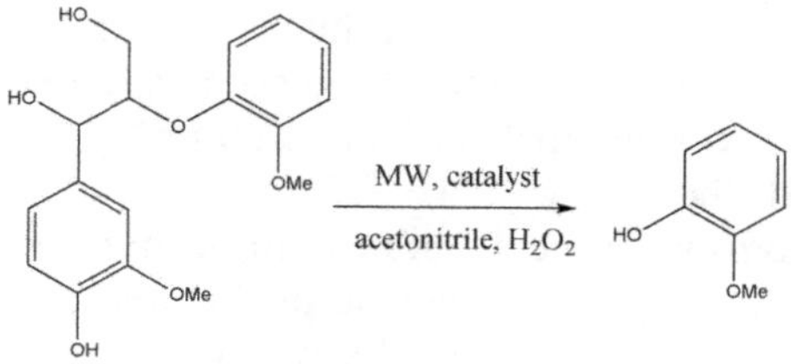

Scheme 8.9 Transformation of [1-(4-hydroxy-3-methoxyphenoxy)-2-(2-methoxyphenoxy)-propane-1,3-diol (1) to 2-methoxyphenol (2) over Co(salen)/SBA-15 under microwave irradiation and H_2O_2 as an oxidation agent. Adapted from *Catalysis Communications*, **12**, S. K. Badamali, R. Luque, James H. Clark, S. W. Breeden, Co(salen)/SBA15 catalysed oxidation of a βO4 phenolic dimer under microwave irradiation, 993–995, Copyright 2011, with permission from Elsevier.[87]

microwave irradiation accelerated the reaction as compared to conventional heating.[88] The best results were obtained with $CrCl_3$ and $MnCl_2$, and a mechanism based on the activation of the dimers with hydroxy radicals was discussed; the acidity of the metal salt promoted the oxidative degradation.

Gu and co-workers[89] used La-based SBA-15 catalysts in the microwave-assisted oxidation with hydrogen peroxide, but instead of model compounds used in the prior work, they reacted organosolv beech wood lignin with hydrogen peroxide. The main compounds identified in the study were vanillin, syringaldehyde, vanillic acid, syringic acid, acetovanillone and acetosyringone, and the highest yields reported were 9.56, 15.66, 5.19, 1.27, 0.35 and 0.1 mol%, respectively, after a 10 minutes reaction time (15 g of lignin, 5 g of catalyst, 20 g of NaOH, 10 ml of 35% H_2O_2 in a total of 250 ml mixture).

8.3 Catalytic Valorisation of Bioderived Compounds in Microwave Reactors

8.3.1 Catalytic Hydrogenation in Microwave Reactors

Although microwave technology has been used in kitchens for more than 40 years and it is widely used in some industries, this is not the case in the chemical industry sector.[90] Although catalytic hydrogenation has played a substantial role in the chemical industry since the beginning of the 20th century,[91] microwave-assisted hydrogenation was first reported only *ca.* 25 years ago and it is still in its infancy in the valorisation of biomass.[92–95] Indeed, there are only a few examples of the catalytic hydrogenation of bioderived compounds using microwave reactors and particularly in the application of heterogeneous catalysis and the use of H_2 as the reducing agent. We believe that this is an area of great academic and industrial interest and in order to highlight possible new opportunities in the valorisation of biomass, we will first review some of the most critical work that utilises microwave reactors in catalytic hydrogenations. For a detailed analysis on microwave-assisted reductions in general, we recommend the comprehensive review published by Schmöger and co-workers in 2011[96] or Chapter 2.8.5 in the book "Aqueous Microwave Assisted Chemistry".[97]

Interestingly, progress in the development of microwave reactors has dictated the type of chemistry that can be performed and that is the reason why the vast majority of studies are based on transfer hydrogenation, *i.e.* using a chemical agent that is generally a liquid (*e.g.* alcohols[98–101]) or a solid (such as formate salts[92,102–104]). Recently, the development of microwave reactors able to dispense and hold pressure of reactive gas in both academia[105–107] and industry[108–114] has enabled studies of reactions using pressurised gases as reactants such as hydrogen.[96,98,114–124] Although this is true, pressurised microwave reactors are still not widely utilised and most hydrogenation studies are still performed as catalytic transfer hydrogenations (CTH)[92–94,98–104,125–141] or with molecular hydrogen at atmospheric pressure.[142,143]

In the last ten years, catalytic hydrogenations performed under microwave irradiation under atmospheric[144–146] or pressurised hydrogen gas[98,114–116] have started to attract interest. Some of this research combined the development of pressure flow[115,116,146] or batch reactors,[114] which in the last few years have become available commercially and can operate under a pressure of up to 14 bar.[109] Furthermore, more recently, high pressure (up to 199 bar), parallel testing (up to 22 at the same time) and possible scale up (up to 1 litre) have become available commercially from Milestone S.r.l.[110] and commercial high pressure (up to 30 bar) flow MW reactors have also been commercialised.[108,111]

One of the first microwave-assisted reactions using molecular hydrogen was published by Wan and co-workers in 1984[147] and 1990[148] and used in a patent in 1986.[149] Researchers used a tailor-made reactor (Figure 8.4) of brass able to withstand pressure and used an oscilloscope to monitor the microwaves produced by a 3 kW magnetron. This showed that gas–solid reactions can be performed under microwave irradiation. The authors tested mixed nickel hydride $CaNi_5H_5$ to hydrogenate various alkenes (ethylene, propylene, bytene, 1,3-butadiene, cyclohexene) but also performed the water-gas shift reaction.

Although they did not compare their system to conventional heating, they showed the potential of the use of microwave irradiation in this catalysis. They also studied the conversion of 0.1 bar methane[148,149] over microwave-reduced nickel catalysts and produced 85–97% C_2 and the rest being C_3 hydrocarbons when using 2 kW power pulses during a total of 5 s irradiation time. It was found that the increasing number of pulse packets dramatically enhanced the yields of the products.

Other early hydrogenation MW studies focused on dechlorination of aromatics[144,146] and trichloroethelene,[142,143] model studies such as nitrobenzene reduction to aniline,[115] ethyl pyruvate to ethyl lactate,[98,122] and on screening a variety of substrates such as the dearomatisation of pyridine-2-carboxylic

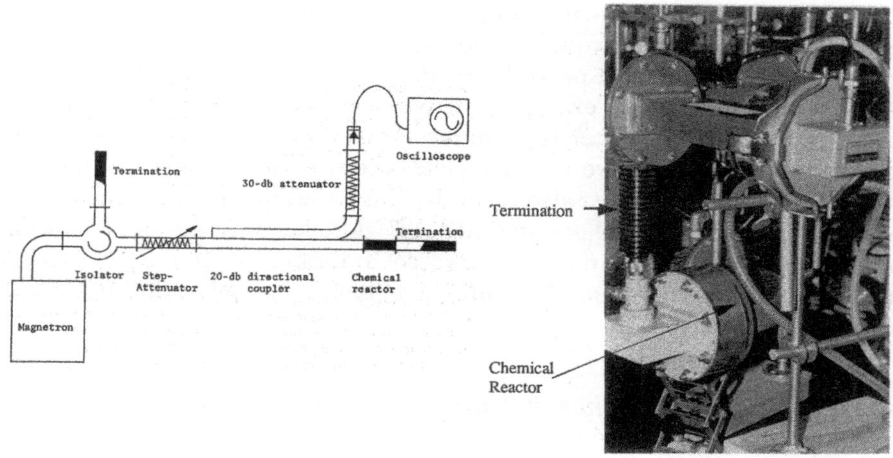

Figure 8.4 A schematic diagram (left) and a photograph (right) of an early design of a custom-made microwave reactor system utilised for gas–solid phase chemical reactions. Adapted with permission from ref. 148. Copyright 1990 Taylor & Francis.

acid, (*S*)-5-hydroxy-2-piperidone and strychnine reduction[116] or differently substituted nitrobenzenes, cholesterol, phenylacetylene, conjugated dibenzodienes, aromatic ketones or 2-octene.[114]

Within these studies, Heller and co-workers[116] demonstrated the hydrogenation of a variety of substrates, performing, for example, the dearomatisation of pyridine-2-carboxylic acid, *N*-debenzylation of AFDX 384 derivative (a drug that acts as a selective antagonist of the muscarinic acetylcholine receptors) but also the reduction of naturally occurring strychnine. Although they acknowledged that it is very hard to directly compare microwave and conventional heating, they found that all of these transformations could be done with significantly shorter reaction times under microwave irradiation. Also, the authors suggested that this difference could not be explained only by the higher temperatures they employed and agreed with the literature that the dipole moment could be induced in hydrogen bound to the catalyst surface.[150–152]

In another study, Vanier[114] demonstrated one of the first commercial microwave reactors able to operate at pressures of up to 14 bar and dose reactive gases (CEM Corporation, model Discover S/SP with gas addition accessory).[109] He compared the reactivity of more than 15 different organic substrates such as differently substituted nitrobenzenes, cholesterol, phenylacetylene, conjugated dibenzodienes, aromatic ketones or 2-octene in the hydrogenation reaction over Pd/C or PtO_2 at 3.45 bar H_2. Generally, short reaction times of around 5 minutes were reported while achieving >98% yields of the corresponding hydrogenated products. Furthermore, specific construction of the utilised microwave reactor allowed for simultaneous cooling by air/nitrogen while heating, which enabled the use of a higher microwave power than would be necessary at the same temperature and this technique further improved the yield of the desired product after the same reaction time. For example, nitrobenzene was hydrogenated to give a 70% yield of aniline after 10 minutes at 80 °C (<10 W power used) while the same conditions gave an 85% yield under simultaneous cooling (*ca.* 100 W power used). Vanier also carefully compared the reaction under conventional and microwave-assisted heating while keeping the same heating rate and hold time and in the same vessel and obtained 3 and 99% yields of the reduced product (80 °C, 10 min, 3.45 bar H_2, 1 mol% Pd/C, 0.5 mmol cholesterol).

8.3.1.1 *The Importance of Catalytic Hydrogenation in Biomass Conversion*

As described earlier, biomass is composed of highly oxygenated molecules whose valorisation towards platform molecules, bioderived monomers or biofuels typically requires the reduction and/or elimination of such functionalities to the required level to suit the application. Catalytic dehydration, hydrogenolysis and hydrogenation are therefore fundamental tools within the biorefining toolkit. In particular, catalytic hydrogenation using hydrogen for upgrading biomass has received extensive attention by researchers and industry alike, and the area has been extensively reviewed.[153–161]

8.3.1.2 *Transfer Hydrogenations in the Microwave Reactor*

Catalytic transfer hydrogenation in the microwave reactor has been studied on a wide range of non-bioderived substrates. One of the earliest transfer hydrogenations was published in 1993 by Gordon *et al.*,[93] who used a household microwave oven and formic acid as a hydrogen transfer source to reduce benzaldehyde to benzyl alcohol over a $RuHCl(CO)(PPh_3)_3$ catalyst. The authors reported a nearly 24-fold increase in turn-over frequency (TOF) when microwave and conventional heating were compared. Several authors used transfer hydrogenation to reduce different side groups attached mostly to the aromatic ring, such as the nitro or imine group to amine,[92,101,103,126,127,136,138] ketones to alcohols or reduction of other oxygen containing groups[100,129,131,135,139,140] or saturation of double/triple bonds.[102,126,132,137,141,162,163] In the above-mentioned studies, authors generally report shorter reaction times[101,162] and increased turn-over frequency (TOF) values.[129] In some cases, improved selectivities[103] and milder reaction conditions (*e.g.* temperature, catalyst, amount of hydrogen donor)[101,132] are reported for the microwave-assisted processes. The use of microwave irradiation was found to play a beneficial role in these transformations. Compounds with similar structure or functionalities can be derived from biomass, although they generally contain more oxygen in their structures,[164] and their transformations might also be susceptible to being accelerated by microwave irradiation.

8.3.1.2.1 Vegetable Oils. Transfer hydrogenations related to vegetable oils were studied by Leskovsek and co-workers[94] and Prabhavathi Devi *et al.*,[134] who reacted soybean oil using aqueous sodium formate and safflower oil with ammonium formate as the hydrogen transfer agent over Pd/C. The former found up to an 8-fold increase in reaction rates when compared to conventional heating and suggested this could be explained by an increase of the formation of the oil–water interface at the catalyst surface although authors acknowledged that they did not have direct proof. The latter group focused on the optimisation of parameters of CTH to produce monounsaturated fatty acids from diunsaturated ones. CTH could be performed successfully with no emulsifier and at low water content (only 20%) as compared to the literature (100–500%).[165,166]

8.3.1.2.2 Citronellal to Menthol. Menthol is in high demand in multiple industries[167] and its production at mild temperatures of 100–130 °C in an open vessel system could be one step closer to the commercialisation of microwave technology in industry. The group of Rafael Luque[168] showed the feasibility of the microwave-assisted one-pot conversion of citronellal to menthol (Scheme 8.10) with 2-propanol and a base over Pt/Ga–MCM-41 catalyst and obtained menthol selectivities >85% in 15 minutes. The effects of microwave power, catalyst quantity and time of irradiation were studied for the best catalyst. Importantly, the catalyst integrity was examined after reaction and Pt showed no sintering or leaching during reaction. Finally, when enantiomerically pure (+)-citronellal was employed instead of the racemic compound, a 55% maximum selectivity to (−)-menthol was obtained. Importantly, the addition of small quantities of cinchonidine as chiral modifier increased the selectivity to (−)-menthol up to 75% and this highlights that the utilisation of microwaves under these conditions does not impede the action of chiral modifiers, which find applicability in a wide range of applications.

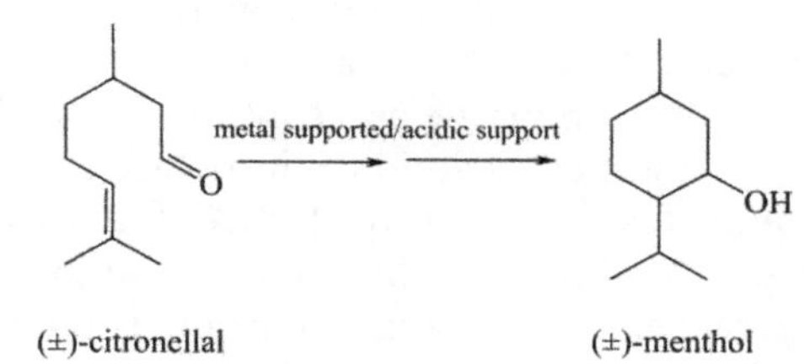

Scheme 8.10 Scheme of hydrogenation of citronellal to menthol. Adapted from ref. 168 with permission from The Royal Society of Chemistry.

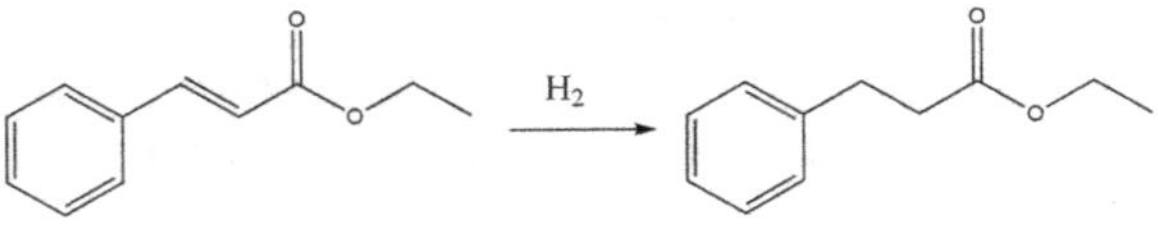

Scheme 8.11 Scheme of hydrogenation of ethyl cinnamate to ethyl hydrocinnamate over chitosan based Pd catalyst at 50 °C and 10 bar hydrogen pressure in microwave reactor.[169]

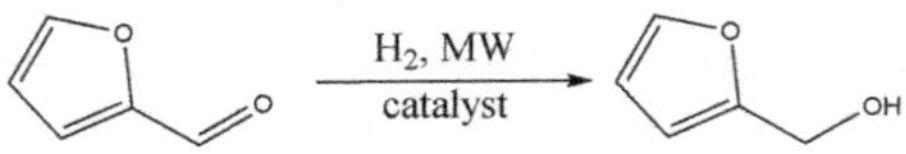

Scheme 8.12 Scheme of selective hydrogenation of furfural to furfuryl alcohol under hydrogen pressure and microwave irradiation.[119]

8.3.1.3 Catalytic Microwave Hydrogenations Under Pressurised Hydrogen

Hydrogenations with H_2 can offer several advantages in the use of sacrificial hydrogen donors. Atom efficiency/economy with hydrogen is inherently superior, but also, hydrogen is a gas and will not require additional separation and purification steps as compared to hydrogen transfer.

8.3.1.3.1 Ethyl Cinnamate to Ethyl Hydrocinnamate. Raspolli Galletti *et al.*[169] utilised microwaves for the synthesis of green catalysts based on biopolymer chitosan and as a heating source in the hydrogenation of ethyl cinnamate to ethyl hydrocinnamate (Scheme 8.11). They performed the synthesis of the catalyst and the test reaction in one pot. The authors showed that chitosan can act as a suitable and reusable biopolymeric support for metal nanoparticles, such as Pd. Moreover, they achieved 100% conversion of ethyl cinnamate within 3 minutes at 50 °C and 10 bar H_2 in the microwave reactor as compared to 30 minutes in an autoclave.

8.3.1.3.2 Citral (3,7-dimethylocta-2,6-dienal) to Citronellal. Citronellal, or 3,7-dimethyloct-6-en-1-al, is a bioderived compound with application as an insect repellent and aroma. Gallert and co-workers[170] carried out the mild hydrogenation of citral (3,7-dimethylocta-2,6-dienal) to citronellal at 70 °C, 4 bar H_2, 90 min and using *n*-heptane as a solvent by applying a porous glass supported Pd catalyst modified by imidazolium-based ionic liquids. Among the several ionic liquids tested, $[C_2mim]^+$ (1-ethyl-3-methylimidazolium) cation and $[OAc]^-$ (acetate) and $[DCA]^-$ (dicyanamide) anions were found to be the most suitable to prevent further reductions of citronellal and gave >95% selectivities to this desired product with nearly quantitative yields. Furthermore, $[C_2mim][OAc]$ was stable for five consecutive runs without the loss of conversion or selectivity.

8.3.1.3.3 Hydrogenation of Furfural to Furfuryl Alcohol. Among other catalytic hydrogenations, our research group has studied the hydrogenation of furfural to furfuryl alcohol over a wet-impregnated and reduced Cu/TiO_2 catalyst under 10 bar hydrogen in a commercial microwave reactor[119] (Scheme 8.12).

We found that the use of microwave reactors significantly increased the rate of the reaction while also improving the selectivity to furfuryl alcohol, which increased the product yield to 99+% yield of furfuryl alcohol at 125 °C in less than 3 hours as compared to <30% yield under conventional heating (Figure 8.5 left). The local heating of the catalyst could be responsible for the increase in reaction rate[169,171,172] and this effect was temperature dependent and diminished with increasing temperature (Figure 8.5 right). The positive effect on hydrogenation is however very tantalising and further research is required to elucidate the underlying phenomena responsible. Even though some ceramic materials such as SiO_2 or Al_2O_3 do not couple well with microwaves, metal particles do, thanks to their high electrical conductivity.[173] Xia and co-workers have shown that while stoichiometric TiO_2 has a low response to the microwave irradiation, hydrogenation turns it into a good microwave

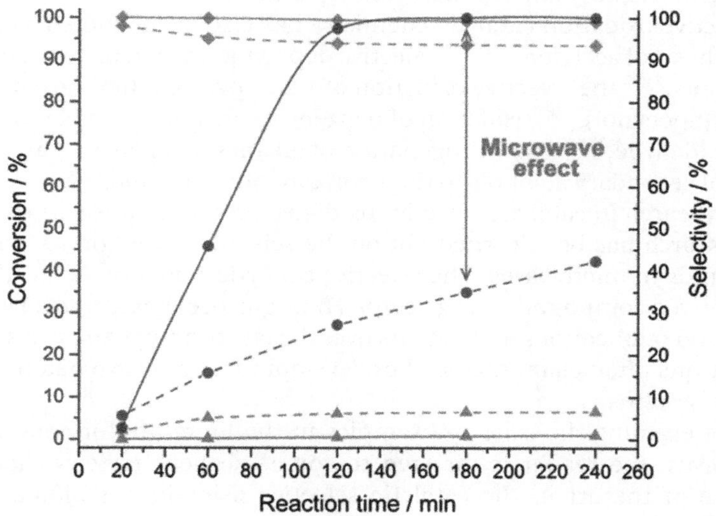

Figure 8.5 Hydrogenation of furfural under microwave (solid line) and conventional heating (dashed line) over Cu/TiO_2. Legend: conversion (circle symbol); furfuryl alcohol selectivity (diamond symbol); 2-methylfuran selectivity (triangle symbol). Reaction conditions (conventional/MW): 125 °C (left), 175 °C (right); 10 bar H_2; 60/10 mg of 10% Cu/TiO_2; 30/5 mL of a 40 mM furfural solution in cyclopentylmethyl ether. Reproduced from ref. 119, http://dx.doi.org/10.1002/cssc.201601398,

absorbing material (MAM).[174–176] X-ray diffractograms of both studies of Xia *et al.*[175] and our own work on Cu/TiO_2[119] suggest that this fact could be a contributing factor to the improved catalyst activity. Additionally, Heller and co-workers[116] suggested that a dipole moment could be induced in hydrogen adsorbed on the catalyst surface during the hydrogenation of strychnine enhancing hydrogenation rate and this could also enhance the reactivity or affect the selectivity in the furfural hydrogenation system.

8.3.2 Catalytic Oxidation

The use of microwaves in catalytic oxidative routes has often been associated with the degradation or decomposition of organic pollutants and substrates (*e.g.* phenolics, bacteria, pesticides, sludges),[86,87,177–220] decomposition of harmful gases such as NO,[221–226] SO_2,[227,228] ammonia,[229–233] ethylene,[234] and CO,[235] degradation of oleo-chemicals,[236–239] oxidative transformation of methane,[229,240–244] and treatment of dyes in wastewater.[212,245–269] Many of these examples combine the use of microwaves with a degradation element, such as Fenton chemistry or UV irradiation. Firstly, it is important to note that many of these technologies can have application in the treatment of waste streams from the biorefinery and this deserves separate attention. More importantly, the selective oxidation catalytic chemistry has also been studied, such as in the synthesis of acetylene,[270,271] selective depolymerisation of lignin to monomeric units,[77,89] the selective oxidation of 1,2-(4-hydroxy-3methoxy-phenoxy) ethanol (apocynol),[84,85] oxidation of benzene to phenol[272] and benzyl alcohol to benzaldehyde,[273,274] dehydrogenation of alkanes to alkenes,[275] and the oxidation of secondary alcohols to their corresponding ketones.[276] In view of all of this research in catalytic selective oxidative oxidation, it is surprising that little research has been carried out on the selective oxidation of bioderived compounds in microwaves. The selective catalytic oxidation of a wide range of bioderived compounds (*i.e.* glycerol, HMF, glucose) has attracted well over a thousand publications and patents over the last twenty years, and scientific publications often claim the need of developing more active heterogeneous catalytic systems.

Before examing the existing examples in the literature for conversion in microwaves, the reader is directed to some excellent reviews that report the state of the art in the catalytic selective oxidation of biomass derivatives;[158,159,161] some reviews put particular emphasis on metal catalysis,[157,277–279] the review by Collinson *et al.* focuses on the oxidation of starch, cellulose and lignin[280] and Villa *et al.* examine glycerol conversion in particular.[281]

Rautiainen and co-workers[282] first oxidised glucose to gluconic acid in the microwave reactor over Au loaded hydrotalcite ($MgAl_2O_4$) and alumina (Scheme 8.13). A 72% yield of gluconic acid (83% conversion of glucose and 87% selectivity to gluconic acid) was obtained after 10 minutes at 120 °C over 0.09 mol% Au/Al_2O_3 under base-free conditions with 2.2 molar equivalent of H_2O_2.

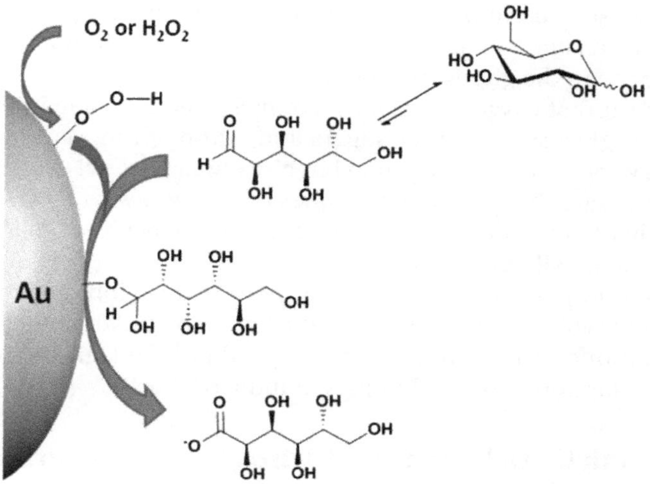

Scheme 8.13 Scheme of oxidation of glucose catalysed by gold nanoparticle as proposed by Omri *et al.*[283] Adapted with permission from M. Omri, G. Pourceau, M. Becuwe and A. Wadouachi, *ACS Sus. Chem. Eng.*, 2016, **4**, 2432. Copyright 2016 American Chemical Society.

This produced a TOF of 12 900 $mol_{glucose}$ per mol_{Au} per hour. The alumina-based catalyst was also stable for four consecutive runs and maintaned both activity and selectivity. Unfortunately, the hydrotalcite supported catalyst was neither as active nor as stable, as it was demonstrated that the activity of $Au/MgAl_2O_4$ decreased in the second run from 54 to 36% conversion. It was shown that the particle size of gold increased from 3.8 to 4.4 nm between the first and second run and the presence of large *ca.* 100 nm particles was mentioned. The sintering effect needs to be fully addressed and understood in the future due to potential safety concerns in the microwave reactor.

Omri *et al.* extended the selection of substrates to other monosaccharides such as galactose, mannose, methylglucoside or oligosaccharides such as trehalose, maltose, lactose or cellobiose also using hydrogen peroxide as oxidant.[283] The temperature was decreased to 60 °C and NaOH and K_2CO_3 were used as a base and salt, respectively, to improve the reactivities as this is well-known in the oxidations of alcohols.[278,284,285] The authors managed to obtain a 95% yield of sodium gluconate (the rest being glucarate) in 10 minutes over only 0.004 mol% Au-supported alumina catalyst, which gave a TOF of 312 000 $mol_{glucose}\ mol_{Au}^{-1}\ h^{-1}$. An even higher TOF of 438 000 $mol_{glucose}\ mol_{Au}^{-1}\ h^{-1}$ was achieved on a ceria-based catalyst although this only produced an 85.6% yield of gluconate (the rest was still glucarate). Similarly high activities in the oxidation of other shown mono- or oligosaccharides were reported after 20 minutes of reaction, generally >95% (at least 71% for mannose and lactose) conversions and >95% selectivities to the corresponding aldonates (at least

74% in the case of mannose). The only two exceptions were methylglucoside and related trehalose, which were unreacted due to their anomeric carbons being engaged in the glycosidic bond.

It is evident that oxygen is preferred to hydrogen peroxide in the selective oxidation of glucose and other sugars and, although we have not yet published this work, our research group has already carried out extensive studies that demonstrate that also superior rates of reaction with >99% conversion and selectivity to the desired gluconic acid can be achieved with oxygen as the only oxidant with gold catalysts.[286]

The high selectivities and conversions, reduced reaction times and impressive TOF values obtained in the literature above demonstrate that microwave reactors can offer extraordinary advantages that might, in the near future, justify the adoption of the technology at industrial scale.

8.3.3 Catalytic Dehydration: Hydroxymethylfurfural and Furfural Production

This section focuses on the conversion of carbohydrates (monosaccharides, dimers or polysaccharides) or their immediate products (*i.e.* HMF (5-(hydroxymethyl)-2-furaldehyde or 5-(hydroxymethyl)furfural)) into products such as HMF[287,288] but also levulinic acid[35,289,290] or furfural.[291,292] These conversions generally occur *via* hydrolysis, dehydration and isomerisation (Scheme 8.14) and are activated by Lewis or Brønsted acids and often in the absence of water.[293] Many studies were carried out with mineral or organic acids or salts as catalysts[35,289–292,294–311] although we aim to highlight mainly heterogeneously catalysed conversions here.[287,288,312–314] Many researchers have used ionic liquids either to facilitate the dissolution of less soluble biopolymer substrates such as cellulose[315,316] or as catalysts.[289,305,309,316] Alternatively to ionic liquids, biphasic systems were used[291,306,307,311,317] to enhance the yield of the product by separating it from the substrate but also to minimise by-product formation.[307,311,317] Carbohydrate conversions in microwave reactors were reviewed by Pistarà *et al.*,[318] who also included a small section dedicated explicitly to the dehydration of carbohydrates (synthesis of furfural derivatives).

Table 8.4 highlights the most notable literature that uses catalysis in the dehydration of bioderivatives. The papers that are the most relevant to this chapter will be discussed in the text.

Hydroxymethylfurfural is probably the most promising and versatile platform chemical derived from biomass, and developments into more efficient production routes are highly desirable.[319] HMF is produced by the dehydration of hexoses *via* catalytic dehydration. A positive effect of microwave irradiation has been reported in the production of HMF for a number of catalysed reactions using inorganic acids,[292] salts,[292,297,303] ionic liquids[289,305] and solid catalysts.[312,313]

The group of Raspolli Galletti[314] found that even though NbPO had a lower total amount and density of acid sites than ZrPO, it was more active in fructose dehydration. This was explained to be thanks to the conversion of

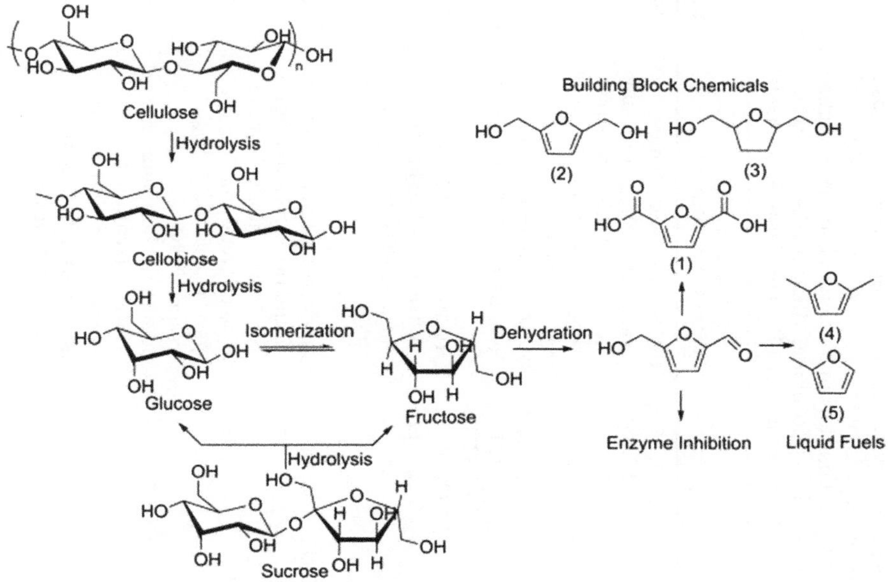

Scheme 8.14 Scheme of the pathway of the conversion of cellulose to hydroxymethylfurfural. Adapted from *Biomass and Bioenergy*, **55**, S. Dutta, S. De and B. Saha, Advances in biomass transformation to 5-hydroxymethylfurfural and mechanistic aspects, 355–369, Copyright 2013, with permission from Elsevier.[293]

Lewis-acid sites of NbPO in water and formation of stronger Brønsted acid sites than initially in ZrPO, which were unchanged under hydrous conditions. On the other hand, ZrPO was found to be more selective towards HMF due to the prevalence of Brønsted sites of medium strength as compared to the strong acid sites formed *in situ*, which promoted the formation of polymerised heavy compounds.

Wu and co-workers[313] established that the fast *in situ* exfoliation of layered niobic acid (Figure 8.6) was critical for its performance in the fructose and inulin conversion to HMF. *In situ* exfoliation of HNb_3O_8 was found to be crucial as it created fewer defects in comparison to pre-exfoliation. Microwave irradiation visibly promoted fast exfoliation in the reaction mixture and increased the reaction rate of dehydration, at the same time possibly avoiding the tendency to form unwanted by-products.

Dutta and co-workers[287,288] used titanium-based catalysts in the production of HMF from mono-, di- and polysaccharides, such as glucose, fructose, sucrose, maltose, cellobiose, cellulose or sugarcane bagasse. Although no conclusions on the microwave effect in these cases were provided, both mesoporous TiO_2 and hierarchical macro/mesoporous titanium phosphate were shown to be suitable and reusable materials for dehydration of biomass-derived carbohydrates. Both studied catalytic systems contained Lewis-acid sites and had high surface areas (326 $m^2\ g^{-1}$ for mesoporous TiO_2 and

Table 8.4 Summary of the selected literature on catalysed biomass-related dehydration with microwave heating. If more than one substrate is mentioned, conversions or yields are reported in the same order. Ref., reference; X, conversion; Y, yield of the main product or products. Abbreviations: AcN, acetonitrile; Bmim, 1-butyl-3-methylimidazolium; $[C_3SO_3Hmim]HSO_4$, 1-methyl-3-(3-sulfopropyl)imidazolium hydrogen sulphate; DMA, dimethylacetamide; DMSO, dimethyl sulfoxide; MTiP-1, hierarchical macro/mesoporous titanium phosphate; MIBK, methylisobutyl ketone; NKC-9, macroporous sulfonated polystyrene ion-exchange resin; NMP, *N*-methylpyrollidone; US, ultrasound; MEK, methylethyl ketone; EtOAc, ethyl acetate.

Substrates Studied	Main products	Catalyst system	*T* °C	Time, min	*X*, %	*Y*, %	MW impact summary	References
Glucose, fructose, sucrose	HMF	$ZnCl_2$, DMSO solvent, reflux	189	8	—	54.6, 55.1, 80.6	Higher yields were achieved *vs* longer time of conventional heating (60 min); possibly "specific" or "non-thermal" effect (cited)	297
Glucose, fructose, sucrose, maltose, cellobiose	HMF	Mesoporous TiO_2, H_2O, MIBK, AcN, NMP, DMSO	120–140	2–10	—	37.2, 54.1, 21, 14.1, 18.7	(DMSO better than water; commercial TiO_2 inactive)	287
Glucose, fructose, sucrose, mannose, cellulose, sugarcane bagasse	HMF	MTiP-1, solvent: H_2O, MIBK, DMA–LiCl	140	5	—	22, 44, 27, 23, 17, 26	(First use of MTiP-1 for conversion of cellulose/bagasse to HMF)	288
Glucose, fructose	HMF	Lignin-derived solid acid catalyst, DMSO, [Bmim]Cl	160, 110	50, 10	99, 98	68, 84	Y_{HMF} decreased in order: Ionic liquid (MW) ≫ water (MW) ⋙ conventional heating	312
Glucose, fructose, cellulose	HMF	$ZrCl_4$, [Bmim]Cl	220	3.5	72.3, 98.9, -	47.8, 92.9, 51.4	MW *vs* conventional heating gave 47.8 *vs* 12.4% Y_{HMF}, respectively.	303
Cellulose	Levulinic acid (glucose)	$[C_3SO_3Hmim]HSO_4$, water	160	30	—	55	One step further: Substrate → HMF → levulinic acid; combination of IL and its sulfonation to obtain IL-based catalyst	289

Fructose, inulin	HMF	Exfoliated HNb_3O_8	155	18	85.1, 100	55.9, 43.3	Very slow reaction under conventional heating *vs* MW; MW effect on *in situ* exfoliation of niobic acid; MW to also help by keeping short reaction times	313
Cellulose	HMF	Various ionic liquids	132	48	—	27.7	MW all higher yields than conventional.	305
Xylose	furfural	NaCl/HCl, water	200	7.3	100	76	MW similar to "conventional" (SiC) but MW 30% lower power consumption.	292
Glucose	HMF	NKC-9 + Al_2O_3, TiO_2, ZrO_2, [Bmim]Cl	140	20	—	62.09	Best combination NKC-9 and Al_2O_3 calcined at 550 °C. More catalyst than substrate (7 : 1).	310
Fructose	HMF	HCl, solvent: H_2O, MIBK, MEK, EtOAc	70, - (70 W US), - (800 W MW)	150, 15 (US), 1.3 (MW)	—	87, 62 (US), 47 (MW)	Homogeneous but comparison of MW, US and conventional heating. Reaction time: conv. ≫ US ≫ MW. Yields not based on optimised conditions for each method.	308
Fructose, inulin	HMF	Nb/Zr-based phosphate	180, 190	10, 8	84.2, 98.1	39.5 (+13.3 furfural), 42.1	Comparison of two phosphates with different acidities only in MW	314

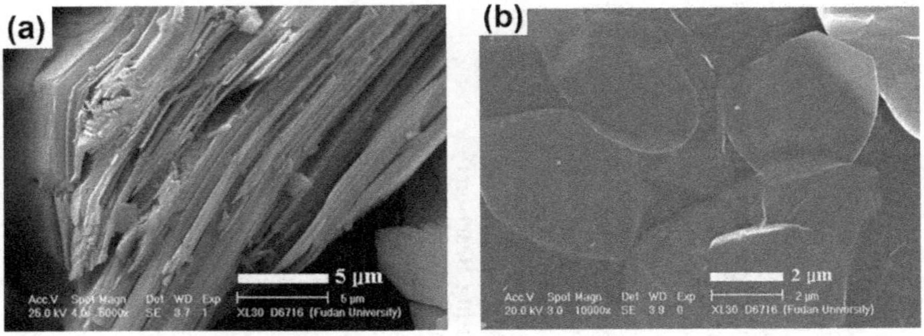

Figure 8.6 Example of exfoliation of niobic acid catalyst as evidenced by SEM: (a) layered sheet structure of niobic acid; (b) individual nanosheets of exfoliated niobic acid. Adapted from *Journal of Catalysis*, **280**, Z.-J. Yang, Y.-F. Li, Q.-B. Wu, N. Ren, Y.-H. Zhang, Z.-P. Liu and Y. Tang, Layered niobic acid with selfexfoliatable nanosheets and adjustable acidity for catalytic hydration of ethylene oxide, 247–254, Copyright 2011, with permission from Elsevier.[360]

193 m^2 g^{-1} for hierarchical titanium phosphate), which were identified as advantageous for the dehydration activity.

Guo, Fang and Zhou[312] used biomass-derived catalysts made from glucose, fructose, cellulose, lignin, bamboo and *Jatropha* hulls by carbonisation and sulfonation with sulphuric acid. They produced materials with varying composition and carbonisation degree, which was also reflected in their acidities. By absolute numbers, a lignin-derived catalyst was the most active one, both in an aqueous system and in DMSO with [BMIM]Cl giving as much as an 84% yield of HMF after 10 minutes at 110 °C. Furthermore, 81.7% Y_{HMF} was obtained (after 10 min at 110 °C) at as high as 10 wt% fructose concentration that was tested as necessary for economic industrial production of hydroxymethylfurfural. However, when the catalytic activities of the tested catalysts are compared by normalising per number of acid sites, the order of the HMF yields changes. While the lignin-based catalyst is still the most active in the water/no IL system and the catalysts made from cellulose and *Jatropha* hulls are the second and third most active, respectively, in DMSO with [BMIM]Cl, the order changes to *Jatropha* hulls > cellulose ≫ lignin. Although this difference could come from the change in the substrate to catalyst ratio and reaction conditions (*e.g.* temperature), it was shown that the catalyst with the highest acid density is not necessarily the most active one.

Also assisted by [BMIM]Cl ionic liquid, Wang and co-workers[310] utilised a commercial macroporous sulfonated polystyrene ion-exchange resin (NKC-9) combined with metal oxides such as alumina, zirconia or titania. The combination that gave the highest HMF yield of 62.1% was NKC-9 with alumina calcined at 550 °C (140 °C, 20 min).

When considering the application of homogeneous catalysts in the dehydration of fructose to HMF, Esmaeili and co-workers[308] went one step further and compared conventional heating with both microwave and ultrasound using HCl as catalyst. Their main finding was that ultrasound decreased the reaction time from 2.5 hours of conventional heating (Y_{HMF} = 87%) to only 15 minutes (Y_{HMF} = 62%) and the reaction time was further reduced to 80 seconds in a microwave reactor (Y_{HMF} = 47%). The amount of HMF produced increased with reaction time using both ultrasound and microwaves.

Furfural is another important biomass-derived compound that can be obtained from the xylose present in hemicellulose *via* dehydration. A very recent work from Xiouras *et al.*[292] compares microwave and conventional heating methods in the dehydration of xylose to furfural. They obtained similar reaction rates to those reported by De Jong in a previous study when using conventional heating.[320] To investigate the existence of non-thermal microwave effects potentially induced by the microwave irradiation, they carried out xylose dehydration to furfural by HCl with the addition of sodium chloride with microwave irradiation and simulated conventional heating in the same microwave reactor by employing a SiC vessel. SiC strongly absorbs microwaves but doesn't allow them to pass through the material, essentially simulating fast and relatively uniform "conventional" heating but induced by microwaves. The same heating speed and also yields of furfural were obtained using both glass (MW) and SiC ("conventional") vessels. However, the MW process used as much as 30% less power to get the same results, which shows the economic value of the microwave-based processes even when reaction rates appear unchanged.

8.3.4 Esterification and Transesterification

Esterification is a chemical reaction in which an alcohol is reacted with a carboxylic acid to form an ester eliminating a water molecule. Transesterification is a similar reaction but an ester is the substrate, and it reacts with an alcohol forming a new ester and eliminating an alcohol. An example of an important transesterification reaction in biomass conversion is biodiesel production, as shown in Scheme 8.15. It is a critical reaction that can

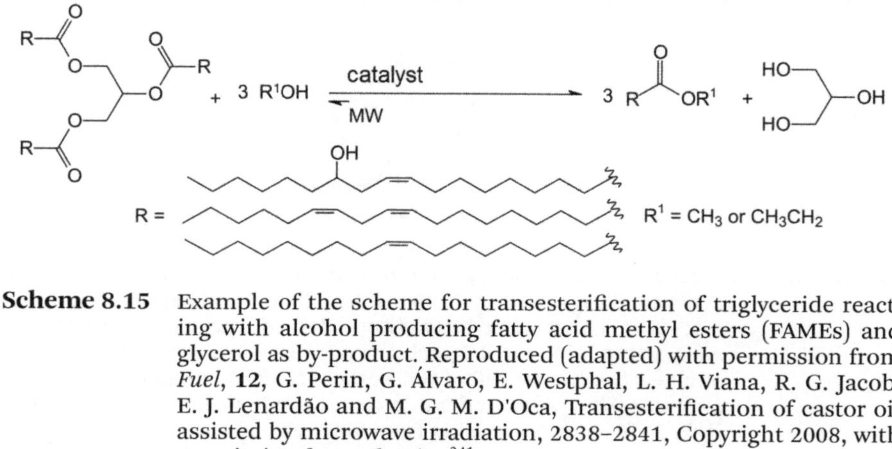

Scheme 8.15 Example of the scheme for transesterification of triglyceride reacting with alcohol producing fatty acid methyl esters (FAMEs) and glycerol as by-product. Reproduced (adapted) with permission from *Fuel*, **12**, G. Perin, G. Álvaro, E. Westphal, L. H. Viana, R. G. Jacob, E. J. Lenardão and M. G. M. D'Oca, Transesterification of castor oil assisted by microwave irradiation, 2838–2841, Copyright 2008, with permission from Elsevier.[341]

transform triglycerides (vegetable oils) into fuels (*i.e.* biodiesel) also called FAMEs (fatty acid methyl esters), which generally uses base catalysts (*e.g.* KOH, NaOH, sodium methoxide).[321–326] In this particular transesterification reaction, glycerol is obtained as the eliminated alcohol by-product, and it is now considered a crucial renewable platform chemical for the production of fuels, fuel additives, chemicals or even hydrogen.[327–330]

In this section, we provide examples of the esterification of biomass-derived compounds such as cinnamaldehydes and cinnamyl alcohols,[331] glucuronic acid[46,332] or glucuronan trisaccharide,[333] cellulose[334] or free fatty acids[335–337] and transesterification of oils[47,321–324,326,338–347] or original biomass sources.[348–354] Although effectively all industrial transesterification processes utilise homogeneous alkali-catalysed technology,[355–357] we will particularly focus on heterogeneously catalysed microwave-assisted processes.[47,339,341,354,358] Microwave-assisted biodiesel production was recently reviewed by Motasemi and Ani[325] and their excellent review covers many topics from feedstocks, policies and standards to chemistry, production technologies including the use of microwave reactors and sustainability aspects of biodiesel production by microwaves.

8.3.4.1 *Esterification*

Most of the aforementioned esterification reactions aim to produce esters that can serve as industrially relevant membranes (cellulose acetate),[334] sunscreen agents, plasticisers, perfumes, lubricants (cinnamic esters),[331] detoxification or anticancer drugs (*e.g.* glycosides or esters of glucuronic acid)[46,332,333] or biofuels (fatty acid methyl esters).[335–337]

Among these reports, El Nemr, Ragab and El Sikaily[334] combined commercial cotton cellulose and acetic anhydride over $ZnCl_2$ as a catalyst to produce cellulose di- and triacetate (Scheme 8.16) under solvent-free conditions in a microwave reactor in just 3 minutes.

The authors highlighted the ability to use a solvent-free environment thanks to the use of acetic anhydride as a source of acetate, which is a liquid

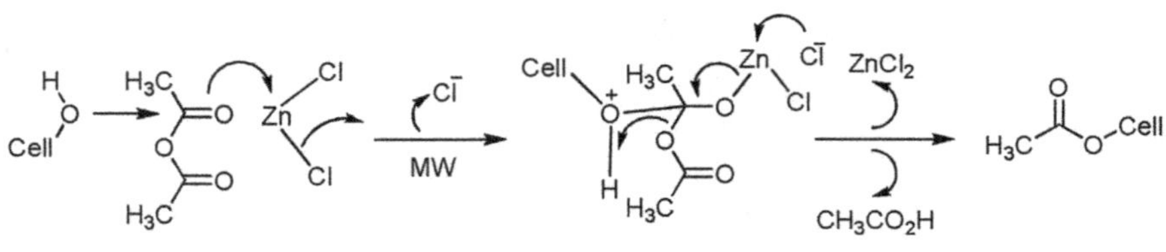

Scheme 8.16 Mechanism of the acetylation of cellulose over $ZnCl_2$ as Lewis acid catalyst as suggested by El Nemr, Ragab and El Sikaily. Cell is cellulose; MW is microwave irradiation. Adapted from *Carbohydrate Polymers*, **151**, A. El Nemr, S. Ragab and A. El Sikaily, Testing zinc chloride as a new catalyst for direct synthesis of cellulose di and tri acetate in a solvent free system under microwave irradiation, 1058–1067, Copyright 2016, with permission from Elsevier.[334]

and facilitates the reaction. Also, as compared to many esterification reactions, the acetate source was used in a small excess of 3.5 $g_{anhydride}/g_{cellulose}$ together with a high substrate to catalyst ratio of 25 $g_{cellulose}/g_{catalyst}$. The reported yield of the combination of di- and triacetate was 95.83% together with as high as a 2.87 degree of acetate substitution, which means that most of the product was cellulose triacetate. Although a domestic microwave was employed in this study and temperature during the reaction was not reported, the authors confirmed that no degradation of the original cellulose occurred. Moreover, it was demonstrated that microwave irradiation gave a higher yield of cellulose acetate in 3 minutes as compared to 72 hours under conventional heating and using 5 times more catalyst.

For the synthesis of industrially important (*E*)-cinnamic esters, Sinha and co-workers[331] used a combination of Amberlyst-15 and DDQ (2,3-dichloro-5,6-dicyanobenzoquinone) as an oxidation agent to perform an oxidative esterification of substituted cinnamaldehydes and cinnamyl alcohols (Scheme 8.17). Microwave irradiation helped to increase the yield of the desired products from 92 to 96% and shorten the reaction time from 6 hours to 30 minutes as compared to the conventional heating. The authors also showed similarly high yields of differently substituted esters, which can be produced using a variety of alcohols. Furthermore, catalysts such as alumina or silica can be used for the same conversion successfully although Amberlyst-15 proved to be the most active.

Anne Wadouachi and José Kovensky utilised microwaves to produce glycosides or esters of D-glucuronic acid (Scheme 8.18), and the triacetate of its 6,1-lactone ((2S,3S,5S)-7-oxo-6,8-dioxabicyclo[3.2.1]octane-2,3,4-triyl triacetate)[46,332] or its trimer glucuronan trisaccharide,[333] which could serve as detoxification or anticancer drugs. The authors used a plethora of acid catalysts such as $ZnCl_2$, $SnCl_4$, $FeCl_3$, H_2SO_4, *p*-TsOH (*p*-toluenesulfonic or *p*-tosylic acid), $SOCl_2$, $BF_3{\cdot}OEt_2$ (Et = ethyl), and $Yb(OTf)_3$ (Tf = triflate or trifluoromethanesulfonate).

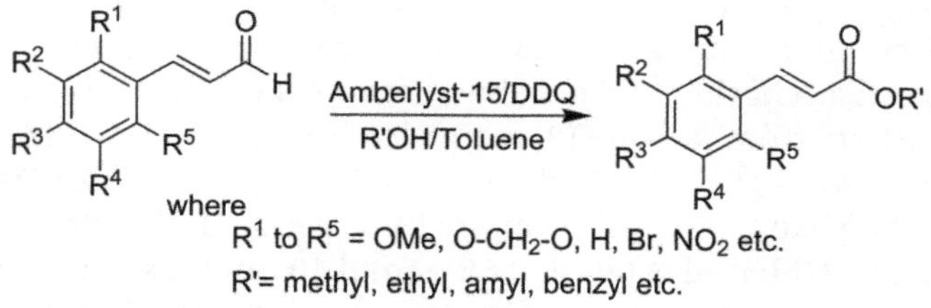

Scheme 8.17 Scheme of esterification of different substituted (E)-cinnamic esters from cinnamaldehydes or cinnamyl alcohols. DDQ is 2,3-dichloro-5,6-dicyanobenzoquinone used as oxidation agent. Adapted from *Tetrahedron*, **63**, A. K. Sinha, A. Sharma, A. Swaroop and V. Kumar, An efficient chemoselective strategy for the preparation of (*E*)-cinnamic esters from cinnamaldehydes using heterogeneous catalyst and DDQ, 1000–1007, Copyright 2007, with permission from Elsevier.[331]

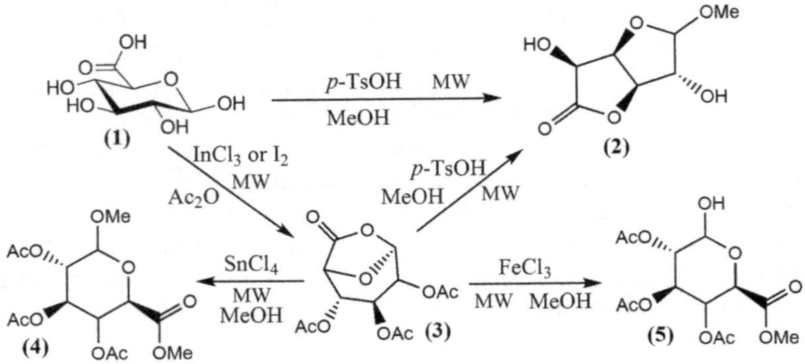

Scheme 8.18 Example of the scheme of catalysed esterification and lactonisation of D-glucuronic acid (**1**) to (3*S*,3 a*R*,6*R*,6 a*R*)-3,6-dihydroxy-5-methoxy-tetrahydrofuro[3,2-*b*]furan-2(3*H*)-one (D-glucofuranosidurono-6,3-lactone) (**2**) or to (2*S*,3*S*,5*S*)-7-oxo-6,8-dioxabicyclo[3.2.1]octane-2,3,4-triyl triacetate (**3**) and (**3**) to (3*S*,4*R*,6*R*)-2-methoxy-6-(methoxycarbonyl)tetrahydro-2*H*-pyran-3,4,5-triyl triacetate (**4**) or (3*S*,4*R*,6*R*)-2-hydroxy-6-(methoxycarbonyl)tetrahydro-2*H*-pyran-3,4,5-triyl triacetate (**5**) under microwave irradiation.[332]

A maximum yield of methyl ester (**2**) of 72% was obtained over *p*-TsOH (0.3 substrate equivalent) when starting from glucuronic acid (**1**) while the same catalyst gave a 93% yield of ester (**2**) when starting from the triacetate of the lactone form (**3**) and the same gave a 98% yield of the product (**5**) with $FeCl_3$ (1 equiv.).[332] Surprisingly, heterogeneous montmorillonite clay catalysts (K10, KSF) gave no reaction (K10) and only a product (**5**) yield of 37% (KSF) from lactone (**3**). Moreover, $SnCl_4$ led to a 93% yield of product (**4**) from (**3**). All these yields were obtained in the temperature range of 60–115 °C and relatively short reaction times between 2 and 10 minutes in the microwave reactor, while comparable reactions took *ca.* 15 hours under conventional heating. The authors also showed that the same substitution could be achieved with higher alcohols[332] and in their later paper, they demonstrated successful simultaneous glycosylation, esterification and etherification of glucuronan trisaccharide.[333] Their investigations on the synthesis of glucuronic acid esters from lactone (**3**) were also expanded to include the use of $[\alpha_1\text{-M}(H_2O)_4P_2W_{17}O_{61}]^{n-}$ (M = Yb^{III}, n = 7; Hf^{IV} and Zr^{IV}, n = 6) Dawson-type Lewis-acid polyoxometalates.[46] Up to 81% yield of the product (**5**) was obtained over a Hf-based catalyst at 65 °C and 10 minutes reaction (MeOH in 15 equiv.) and the yield was further increased to 94% over Zr polyoxometalate when the MeOH : substrate ratio was increased to 150.

Finally, esterification reactions have also been carried out to produce biodiesel when the starting feedstock is free fatty acids.[335–337] Such esterifications need to be performed using acid catalysts to avoid saponification even at the cost of slower reaction rates.

Lieu, Yusup and Moniruzzaman[337] catalysed the reaction of free fatty acids present in *Ceiba pentandra* seed oil with methanol to reduce the acid value in the crude substrate. They used sulphuric acid as catalyst and found that 94.43% conversion could be achieved at 60 °C and 12 minutes (MeOH : substrate = 10 mol mol^{-1}, substrate : catalyst = 50 g g^{-1}) as compared to 20.83% conversion for the uncatalysed reaction.

On the other hand, the group of Jung[335,336] used heterogeneous acid catalysts to improve the conversion of free fatty acids to their methyl esters. In their first communication,[336] the authors compared microwave and conventional heating in the esterification of oleic acid over sulfonated zirconia and found a significant increase in reaction rate for the microwave-assisted reaction. They achieved 91.9% conversion to FAMEs in 20 minutes under MWs as compared to 77.3% in 60 minutes when heated conventionally (MeOH : acid = 20 mol mol^{-1}, substrate : catalyst = 20 g g^{-1}, 60 °C). Maximum conversions reported at the same conditions were 95.2 *vs* 96.3% after 90 and 180 min for MW and conventional heating, respectively. Although this MW-assisted conversion is slower than in the previous report,[337] one has to consider that the same mass of sulfonated zirconia contains fewer acid sites than sulphuric acid. Lieu *et al.*[337] also used oil, which contains only a small amount of free fatty acids as compared to the neat oleic acid used by Kim *et al.*[336] Nevertheless, the latter study showed that microwaves enhance esterification of fatty acids with methanol and the authors also calculated energy consumption, which decreased to 67% using microwave irradiation. A second study from the same group[335] compared continuous and pulsed microwaves of different pulse duration to the period between cycles ratios (Figure 8.7) for the esterification of oleic acid over Amberlyst 15 at 60 °C (MeOH : acid = 20 mol mol^{-1}, substrate : catalyst = 10 g g^{-1}). In this case, pulses of microwave energy further increased the activity of the reaction heated by continuous microwave irradiation. The total amount of dispensed power in this comparison was kept

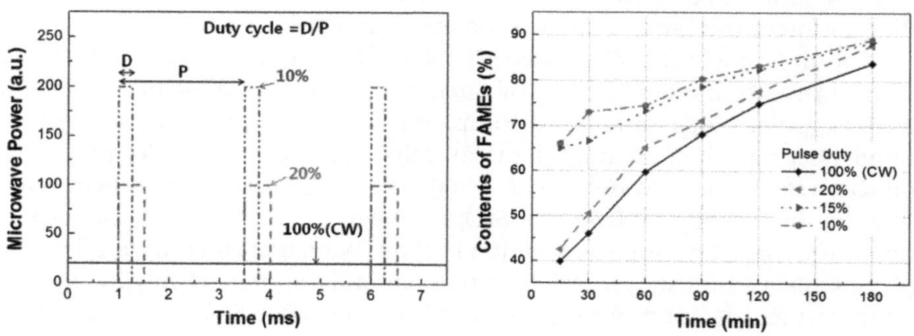

Figure 8.7 Plot of microwave pulse cycle ratio (duration (D) to period (P) between cycles) (left) and yields of FAME (fatty acid methyl ester) for different D/P ratios (right). CW stands for continuous microwave irradiation. Adapted from *Bioresource Technology*, **102**, D. Kim, J. Choi, G.-J. Kim, S. K. Seol and S. Jung, Accelerated esterification of free fatty acid using pulsed microwaves, 7229–7231, Copyright 2011, with permission from Elsevier.[335]

constant, and it was found that the higher the ratio between the pulse duration (D) to period (P) between cycles, the higher the yield of FAMEs. As can be seen from the right side plot in Figure 8.7, FAME yield after 15 minutes of reaction increased from 39.9 to 66.1% for continuous MW irradiation and 10% D/P ratio, respectively. It was suggested that probably due to the reversible nature of the esterification reaction and dependence on the methanol to substrate ratio, both pulse and continuous MWs give similar yields of ester, 88 and 83.9% after 3 hours, respectively. This work suggests that using high-power pulsed microwaves is preferable to low-power continuous microwave irradiation.

8.3.4.2 *Transesterification for the Production of Biodiesel*

As was explained at the beginning of this section and shown in Scheme 8.15, both biodiesel (FAMEs) and glycerol can be produced by transesterification most commonly *via* alkali-catalysed processes. However, in this section, we will highlight heterogeneously catalysed microwave-assisted processes[47,339,341,354,358] for transesterification of oils to FAMEs over acid[47] or base catalysts.[339,341,354,358]

Among the acid catalysts, only Zhang and co-workers[47] used heteropoly acid catalysts with composition $Cs_{2.5}H_{0.5}PW_{12}O_{40}$ to obtain a 96.2% yield of FAMEs in 10 minutes (60 °C, molar ratio of methanol/oil 12 : 1, 1 wt% catalyst) in MW. The majority of the reported heteropoly acid catalysts (based on the combination of P, W, Mo, Ag, Si) in this work gave the same FAME yield at the same conditions. When sulphuric acid was used as a catalyst, 30 minutes were needed to achieve the same yield in the MW and with the Cs-based catalyst under conventional heating, 60 minutes were necessary to obtain a comparable yield. The catalyst was shown to be reusable up to 9 times despite some weight loss and colour change, which were most probably due to the dissolution of the catalyst and composition changes.

Patil and co-workers[339] showed that both BaO and SrO are much more active in the transesterification of *Camelina sativa* oil and achieved 83 and 80% Y_{FAME} with BaO and SrO as compared to 30% with CaO under conventional heating (3 h, methanol/oil molar ratio 9, 12 and 15; 100, 60 and 100 °C, oil/catalyst weight ratio 100, 200 and 200 for BaO, SrO and CaO, respectively). The two best catalysts, BaO and SrO, then gave 94 and 80% Y_{FAME}, respectively, after only 4 minutes of reaction in the MW reactor (methanol/oil molar ratio 9, 65 °C, oil/catalyst weight ratio 66.6 and 20, respectively). However, temperature was not reported and could be unexpectedly higher when considering the 800 W output power and that a domestic microwave was used. Nevertheless, this provided a close to 100 times increase in reaction rate for the MW-assisted transesterification.

Koberg and co-workers[344] decreased the time of the reaction even further and obtained 99.8% conversion using waste cooking oil in only 10 seconds with a SrO catalyst utilising a 1100 W domestic microwave, which included magnetic stirring. The authors claim that the reaction mixture temperature was kept at 60 °C (measured after the end of reaction) (methanol/oil molar ratio 6, oil/catalyst weight ratio 54.4). Apart from the very short reaction time, which would significantly improve the economics of the biodiesel production, the authors also highlighted the fact that SrO is recovered from the biodiesel phase and this results in a higher glycerol purity compared to when homogeneous catalysts are used. A $ZnO/La_2O_2CO_3$ layered composite catalyst employed by Jin and co-workers[358] produced >95% yield of methyl esters after only 5 minutes reaction at 85 °C (methanol/oil molar ratio 20, oil/catalyst weight ratio 20) and showed that the catalyst was reusable up to three times before losing activity.

Summarising the aforementioned studies, the alcohol to oil molar ratios of 6–20 seem to be the most often used for heterogeneously catalysed transesterification to FAMEs under microwave irradiation. Although most of the studies used a temperature below the boiling point of alcohol (*ca.* 60 °C for methanol), higher temperatures such as 85 °C were also utilised in the case where it was possible to keep the microwave reactor pressurised.[358] Although it was not fully justified, microwave-induced heating is able to significantly shorten the reaction time required to produce biodiesel down to 4–5 minutes[339,358] or even lower.[344] Although these results are very encouraging, one can expect that the utilisation of microwave reactors in the real industrial production of biodiesel will find a few obstacles, mainly due to the particularly large scale at which biodiesel production is carried out, the limited scalability of microwave technology and the relatively low value of the reaction product. Potentially, continuous flow microwave reactors can offer a solution to scalability issues in microwave technology, but such systems will require a substantial investment and higher risks as compared to alternative commercial processes using conventional heating.

8.4 Conclusions

The most important conclusion is that the substantial amount of literature examined in this chapter points to a definitive positive effect of microwaves on reaction rates and selectivities in both hetero- and homogeneously catalysed reactions for a range of biomass-derived reaction systems. In view of this, it is necessary to acknowledge the increasing potential role of microwaves within biorefining, not only regarding biomass pretreatment or depolymerisation but also as a useful tool for the catalytic valorisation of biomass derivatives. On the other hand, it is particularly noticeable that explanations of the promotional effects of microwaves are still poorly understood and rarely studied by researchers. In many instances, a decrease of activation energy barriers and the effect of local hot spots are acknowledged, whereas the potential effect of specific microwave interaction effects is particularly poorly understood and often only speculative.

Some heterogeneous catalytic systems offer particularly good promise, and the limitation of MWs due to the loss tangent and dielectric constants of the solvent/gas can partly be overcome with heterogeneous catalysts, especially carbon-based catalysts, which are suitable MW absorbers and could counter the solvent MW inactivity for some reactions. Combining the use of ionic liquids with microwaves in biomass conversion also appears a successful strategy, particularly in the dehydration of cellulose/glucose/fructose due to the ionic character, which particularly suits MW activation.

Nevertheless, rather than limiting ourselves to what has been explored to date, this chapter aims to prompt the researcher to further exploration in the use of microwaves in catalytic conversion of biomass in reactions where this has not yet been studied. There is particular promise in gas-assisted reactions in microwaves, which have been demonstrated to accelerate reactivity by several fold (*e.g.* oxidation, hydrogenation). It is particularly relevant that most catalytic selective oxidations of biomass derivatives are carried out in water or other polar solvents, well-suited for microwave irradiation, but still the use of microwaves is neglected. Researchers still carry out the catalytic oxidations in stainless steel reactors, which take long times to heat up and cool down and for much longer reaction times as compared to microwave reactors. Catalytic selective oxidation and hydrogenation offer a myriad of opportunities to develop catalytic systems in combination with microwave technology, which can potentially offer reaction rates orders of magnitude faster than those currently reported that might indeed result in new continuous processes in the future.

Additionally, in some cases, the benefits of microwave technology could be exploited by coupling two or more in a one-pot system such as: (i) hydrolysis and hydrogenation to give polyols (*e.g.* sorbitol and xylitol from cellulose and hemicellulose, respectively), (ii) hydrolysis and dehydration to give heterocycles (*e.g.* HMF and furfural) or levulinic acid, and (iii) hydrolysis and oxidation to provide gluconic or even glucaric acid from cellulose (glucose). These combined processes could offer advantages in microwave-based biorefineries because the same reactors can be used in a multi-step process and the manufacture of the desired product can be achieved by merely adapting the reaction conditions and the catalyst used. Finally, it is crucial to state that some of the benefits of the use of microwaves can be compromised when scaling-up processes. To address this limitation, the study and development of flow microwave reactors appears to be the most promising way forward for a technology that, due to superior reaction rates, might be better suited for new processes or process intensification as compared to well-established large-scale batch processes.

Abbreviations

AC	Activated carbon
ACN	Acetonitrile
BHC	Betaine hydrochloride
BMIM	1-Butyl-3-methylimidazolium
C_2MIM	1-Ethyl-3-methylimidazolium
C_3SO_3HMIM	1-Methyl-3-(3-sulfopropyl)imidazolium

CMK-3	Type of an ordered mesoporous carbon
CTH	Catalytic transfer hydrogenation
DCA	Dicyanamide
DMA	Dimethylacetamide
DMSO	Dimethyl sulfoxide
DP	Degree of polymerisation
Et	ethyl
EtOAc	Ethyl acetate
HMF	(5-(hydroxymethyl)-2-furaldehyde or 5-(hydroxymethyl)furfural)
MAM	Microwave absorbing material
MEK	Methylethyl ketone
MIBK	Methylisobutyl ketone
MTiP-1	Hierarchical macro/mesoporous titanium phosphate
MW	microwave
NKC-9	Macroporous sulfonated polystyrene ion-exchange resin
NMP	*N*-Methylpyrollidone
OAc	acetate
POM	polyoxometalate
p-TsOH	*p*-toluenesulfonic or *p*-tosylic acid
salen	[*N*,*N*′-bis(salicylidene)ethane-1,2-diaminato] complex
SBA	Santa Barbara; city of university where this type of hexagonal mesoporous silica was first synthesised; the most known type of SBA is SBA-15 and SBA-16
TBMA	Tributylmethylammonium
Tf	Triflate or trifluoromethanesulfonate
TOF	Turn-over frequency
US	Ultrasound, ultrasonic
XRD	X-ray diffraction

Acknowledgements

The authors thank the EPSRC (grant EP/K014773/1) and the UK Department of Business Skills and Innovation (Regional Growth Fund, MicroBioRefinery). Javier Eduardo Perez Mejia also thanks Consejo Nacional de Ciencia y Tecnologia de Mexico (CONACYT) for the scholarship 409325.

References

1. S. Tabasso, D. Carnaroglio, E. Calcio Gaudino and G. Cravotto, *Green Chem.*, 2015, **17**, 684.
2. F. Jérôme, G. Chatel and K. De Oliveira Vigier, *Green Chem.*, 2016, **18**, 3903.
3. A. Richel and N. Jacquet, *Biomass Convers. Biorefin.*, 2015, **5**, 115.
4. A. Richel, P. Laurent, B. Wathelet, J. P. Wathelet and M. Paquot, *C. R. Chim.*, 2011, **14**, 224.

5. S. Van de Vyver, J. Geboers, P. A. Jacobs and B. F. Sels, *ChemCatChem*, 2011, **3**, 82.
6. A. Richel, P. Laurent, B. Wathelet, J. P. Wathelet and M. Paquot, *Catal. Today*, 2011, **167**, 141.
7. S. Tabasso and G. Cravotto, in *Production of Biofuels and Chemicals with Microwave*, ed. Z. Fang, J. L. R. Smith and X. Qi, Springer Netherlands, Dordrecht, 2015, pp. 129–144, DOI: 10.1007/978-94-017-9612-5_7.
8. J. Xu, in *Pretreatment of Biomass: Processes and Technologies*, 2014, pp. 157–172, DOI: 10.1016/B978-0-12-800080-9.00009-8.
9. L. Hu, L. Lin, Z. Wu, S. Zhou and S. Liu, *Appl. Catal., B*, 2015, **174–175**, 225.
10. T. Mitani and T. Watanabe, in *Microwaves in Catalysis*, Wiley-VCH Verlag GmbH & Co. KGaA, 2015, pp. 323–336, DOI: 10.1002/9783527688111.ch16.
11. B. Sun, L. Duan, G. Peng, X. Li and A. Xu, *Bioresour. Technol.*, 2015, **192**, 253.
12. J. Wang, J. Xi and Y. Wang, *Green Chem.*, 2015, **17**, 737.
13. F. M. Gírio, C. Fonseca, F. Carvalheiro, L. C. Duarte, S. Marques and R. Bogel-Łukasik, *Bioresour. Technol.*, 2010, **101**, 4775.
14. M. P. Pandey and C. S. Kim, *Chem. Eng. Technol.*, 2011, **34**, 29.
15. D. Klemm, B. Heublein, H.-P. Fink and A. Bohn, *Angew. Chem., Int. Ed.*, 2005, **44**, 3358.
16. P. Bhaumik and P. L. Dhepe, in *Biomass Sugars for Non-fuel Applications*, The Royal Society of Chemistry, 2016, pp. 1–53, DOI: 10.1039/9781782622079-00001.
17. A. Verardi, I. De Bari, E. Ricca and V. Calabro, *Bioethanol*, 2012, 95.
18. Y.-B. Huang and Y. Fu, *Green Chem.*, 2013, **15**, 1095.
19. L. Shuai and X. Pan, *Energy Environ. Sci.*, 2012, **5**, 6889.
20. S. Suganuma, K. Nakajima, M. Kitano, D. Yamaguchi, H. Kato, S. Hayashi and M. Hara, *J. Am. Chem. Soc.*, 2008, **130**, 12787.
21. M. Toda, A. Takagaki, M. Okamura, J. N. Kondo, S. Hayashi, K. Domen and M. Hara, *Nature*, 2005, **438**, 178.
22. J. Pang, A. Wang, M. Zheng and T. Zhang, *Chem. Commun.*, 2010, **46**, 6935.
23. Y. Wu, Z. Fu, D. Yin, Q. Xu, F. Liu, C. Lu and L. Mao, *Green Chem.*, 2010, **12**, 696.
24. N. Sweygers, R. Dewil and L. Appels, *Waste Biomass Valorization*, 2017, 1–13.
25. Y. Yuan, J. Wang, N. Fu and S. Zang, *Catal. Commun.*, 2016, **76**, 46.
26. S. Tsubaki, K. Oono, A. Onda, K. Yanagisawa, T. Mitani and J. I. Azuma, *Carbohydr. Polym.*, 2016, **137**, 594.
27. N. Teng, J. Ni, H. Chen, Q. Ren, H. Na, X. Liu, R. Zhang and J. Zhu, *ACS Sustainable Chem. Eng.*, 2016, **4**, 1507.
28. K. Qin, Y. Yan, Y. Zhang and Y. Tang, *RSC Adv.*, 2016, **6**, 39131.
29. J. Wang, M. Zhou, Y. Yuan, Q. Zhang, X. Fang and S. Zang, *Bioresour. Technol.*, 2015, **197**, 42.

30. B. Sun, G. Peng, L. Duan, A. Xu and X. Li, *Bioresour. Technol.*, 2015, **196**, 454.
31. J. Ni, N. Teng, H. Chen, J. Wang, J. Zhu and H. Na, *Bioresour. Technol.*, 2015, **191**, 229.
32. L. Gavilà, M. Constantí and F. Medina, *Cellulose*, 2015, **22**, 3089.
33. V. da Silva Lacerda, J. B. López-Sotelo, A. Correa-Guimarães, S. Hernández-Navarro, M. Sánchez-Bascones, L. M. Navas-Gracia, P. Martín-Ramos, E. Pérez-Lebeña and J. Martín-Gil, *Bioresour. Technol.*, 2015, **180**, 88.
34. I. N. Pulidindi, B. B. Kimchi and A. Gedanken, *Renewable Energy*, 2014, **71**, 77.
35. Á. Szabolcs, M. Molnár, G. Dibó and L. T. Mika, *Green Chem.*, 2013, **15**, 439.
36. J. Ni, H. Wang, Y. Chen, Z. She, H. Na, and J. Zhu, *Bioresour. Technol.*, 2013, **137**, 106.
37. X. Li, Y. Jiang, L. Wang, L. Meng, W. Wang and X. Mu, *RSC Adv.*, 2012, **2**, 6921.
38. A. M. Orozco, A. H. Al-Muhtaseb, A. B. Albadarin, D. Rooney, G. M. Walker and M. N. M. Ahmad, *RSC Adv.*, 2011, **1**, 839.
39. S. Tsubaki, K. Oono, T. Ueda, A. Onda, K. Yanagisawa, T. Mitani and J. I. Azuma, *Bioresour. Technol.*, 2013, **144**, 67.
40. C. Zhang, Z. Cheng, Z. Fu, Y. Liu, X. Yi, A. Zheng, S. R. Kirk and D. Yin, *Cellulose*, 2017, **24**, 95.
41. C. Zhang, Z. Fu, B. Dai, S. Zen, Y. Liu, Q. Xu, S. R. Kirk and D. Yin, *Cellulose*, 2014, **21**, 1227.
42. K. Ishida, S. Matsuda, M. Watanabe, H. Kitajima, A. Kato, M. Iguchi, T. M. Aida, R. L. Smith Jr, X. Qi, T. Tago and T. Masuda, *Biomass Convers. Biorefin.*, 2014, **4**, 323.
43. C. Zhang, Z. Fu, Y. C. Liu, B. Dai, Y. Zou, X. Gong, Y. Wang, X. Deng, H. Wu, Q. Xu, K. R. Steven and D. Yin, *Green Chem.*, 2012, **14**, 1928.
44. X. Li, Y. Jiang, L. Shuai, L. Wang, L. Meng and X. Mu, *J. Mater. Chem.*, 2012, **22**, 1283.
45. Z. Zhang and Z. K. Zhao, *Carbohydr. Res.*, 2009, **344**, 2069.
46. M. Bosco, S. Rat, N. Dupré, B. Hasenknopf, E. Lacôte, M. Malacria, P. Rémy, J. Kovensky, S. Thorimbert and A. Wadouachi, *ChemSusChem*, 2010, **3**, 1249.
47. S. Zhang, Y.-G. Zu, Y.-J. Fu, M. Luo, D.-Y. Zhang and T. Efferth, *Bioresour. Technol.*, 2010, **101**, 931.
48. S. Tsubaki, K. Oono, A. Onda, T. Ueda, T. Mitani and M. Hiraoka, *RSC Adv.*, 2017, **7**, 12346.
49. A. Orozco, M. Ahmad, D. Rooney and G. Walker, *Process Saf. Environ. Prot.*, 2007, **85**, 446.
50. T. W. Ching, V. Haritos and A. Tanksale, *Carbohydr. Polym.*, 2017, **157**, 1794.
51. F. Boissou, K. De Oliveira Vigier, B. Estrine, S. Marinkovic and F. Jérôme, *ACS Sustainable Chem. Eng.*, 2014, **2**, 2683.

52. A. Ebringerová, Z. Hromádková and T. Heinze, in *Polysaccharides I: Structure, Characterization and Use*, ed. T. Heinze, Springer Berlin Heidelberg, Berlin, Heidelberg, 2005, pp. 1–67, DOI: 10.1007/b136816.
53. P. Kumar, D. M. Barrett, M. J. Delwiche and P. Stroeve, *Ind. Eng. Chem. Res.*, 2009, **48**, 3713.
54. Y. Y. Bai, L. P. Xiao and R. C. Sun, *Biomass Bioenergy*, 2015, **75**, 245.
55. P. Mäki-Arvela, T. Salmi, B. Holmbom, S. Willför and D. Y. Murzin, *Chem. Rev.*, 2011, **111**, 5638.
56. H. Gómez Bernal, L. Bernazzani and A. M. Raspolli Galletti, *Green Chem.*, 2014, **16**, 3734.
57. Q. Lin, H. Li, J. Ren, A. Deng, W. Li, C. Liu and R. Sun, *Carbohydr. Polym.*, 2017, **157**, 214.
58. S. Kuittinen, Y. P. Rodriguez, M. Yang, M. Keinänen, O. Pastinen, M. Siika-aho, O. Turunen and A. Pappinen, *BioEnergy Res.*, 2016, **9**, 344.
59. Y. Yang, C. W. Hu and M. M. Abu-Omar, *ChemSusChem*, 2012, **5**, 405.
60. J. Zakzeski, P. C. A. Bruijnincx, A. L. Jongerius and B. M. Weckhuysen, *Chem. Rev.*, 2010, **110**, 3552.
61. P. Chen, Q. Zhang, R. Shu, Y. Xu, L. Ma and T. Wang, *Bioresour. Technol.*, 2017, **226**, 125.
62. C. Li, X. Zhao, A. Wang, G. W. Huber and T. Zhang, *Chem. Rev.*, 2015, **115**, 11559.
63. A. Maneffa, P. Priecel and J. A. Lopez-Sanchez, *ChemSusChem*, 2016, **9**, 2736.
64. H. Lihong, P. Hui, Z. Yonghong and Z. Meng, *BioResources*, 2011, **6**, 1.
65. P. Intanakul, M. Krairiksh and P. Kitchaiya, *J. Wood Chem. Technol.*, 2003, **23**, 217.
66. S. Zhu, Y. Wu, Z. Yu, C. Wang, F. Yu, S. Jin, Y. Ding, R. a. Chi, J. Liao and Y. Zhang, *Biosyst. Eng.*, 2006, **93**, 279.
67. H. Ma, W.-W. Liu, X. Chen, Y.-J. Wu and Z.-L. Yu, *Bioresour. Technol.*, 2009, **100**, 1279.
68. S. Zhu, Y. Wu, Z. Yu, J. Liao and Y. Zhang, *Process Biochem.*, 2005, **40**, 3082.
69. J. Cheng, H. Su, J. Zhou, W. Song and K. Cen, *Int. J. Hydrogen Energy*, 2011, **36**, 2093.
70. S. Zhu, Y. Wu, Z. Yu, X. Zhang, H. Li and M. Gao, *Bioresour. Technol.*, 2006, **97**, 1964.
71. W. Yunpu, D. Leilei, F. Liangliang, S. Shaoqi, L. Yuhuan and R. Roger, *J. Anal. Appl. Pyrolysis*, 2016, **119**, 104.
72. N. M. Mokhtar, R. Omar and A. Idris, *Energy Sources, Part A*, 2012, **34**, 2104.
73. B. Krieger-Brockett, *Res. Chem. Intermed.*, 1994, **20**, 39.
74. S. S. Lam and H. A. Chase, *Energies*, 2012, **5**, 4209.
75. F. Motasemi and M. T. Afzal, *Renewable Sustainable Energy Rev.*, 2013, **28**, 317.
76. J. Fan, V. Budarin, M. J. Gronnow and J. H. Clark, in *Alternative Energy Sources for Green Chemistry*, The Royal Society of Chemistry, 2016, pp. 64–92, DOI: 10.1039/9781782623632-00064.

77. L. Nagpurkar, A. Chaudhari and J. Ekhe, *Asian J. Chem.*, 2002, **14**, 1387.
78. D. Shen, N. Liu, C. Dong, R. Xiao and S. Gu, *Chem. Eng. J.*, 2015, **270**, 641.
79. J. Xie, J. Qi, C. Hse and T. F. Shupe, *J. For. Res.*, 2015, **26**, 261.
80. X. Ouyang, G. Zhu, X. Huang and X. Qiu, *J. Energy Chem.*, 2015, **24**, 72.
81. C. Dong, C. Feng, Q. Liu, D. Shen and R. Xiao, *Bioresour. Technol.*, 2014, **162**, 136.
82. A. Toledano, L. Serrano, A. Pineda, A. A. Romero, R. Luque and J. Labidi, *Appl. Catal., B*, 2014, **145**, 43.
83. H. G. Kim and Y. Park, *Ind. Eng. Chem. Res.*, 2013, **52**, 10059.
84. S. K. Badamali, J. H. Clark and S. W. Breeden, *Catal. Commun.*, 2008, **9**, 2168.
85. S. K. Badamali, R. Luque, J. H. Clark and S. W. Breeden, *Catal. Commun.*, 2013, **31**, 1.
86. S. K. Badamali, R. Luque, J. H. Clark and S. W. Breeden, *Catal. Commun.*, 2009, **10**, 1010.
87. S. K. Badamali, R. Luque, J. H. Clark and S. W. Breeden, *Catal. Commun.*, 2011, **12**, 993.
88. J. Pan, J. Fu and X. Lu, *Energy Fuels*, 2015, **29**, 4503.
89. X. Gu, C. Kanghua, H. Ming, Y. Shi and Z. Li, *Maderas: Cienc. Tecnol.*, 2012, **14**, 31.
90. E. T. Thostenson and T. W. Chou, *Composites, Part A*, 1999, **30**, 1055.
91. P. Sabatier and J. B. Senderens, *C. R. Hebd. Seances Acad. Sci.*, 1903, **137**, 301.
92. A. K. Bose, B. K. Banik, K. J. Barakat and M. S. Manhas, *Synlett*, 1993, **1993**, 575.
93. E. M. Gordon, D. C. Gaba, K. A. Jebber and D. M. Zacharias, *Organometallics*, 1993, **12**, 5020.
94. S. Leskovsek, A. Smidovnik and T. Koloini, *J. Org. Chem.*, 1994, **59**, 7433.
95. P. S. S. Prasad, N. Lingaiah, P. K. Rao, F. J. Berry and L. E. Smart, *Catal. Lett.*, 1995, **35**, 345.
96. C. Schmöger, A. Stolle, W. Bonrath and B. Ondruschka, *Curr. Org. Chem.*, 2011, **15**, 151.
97. V. Polshettiwar and R. S. Varma, *Aqueous Microwave Assisted Chemistry: Synthesis and Catalysis*, Royal Society of Chemistry, 2010.
98. B. Toukoniitty, O. Roche, J.-P. Mikkola, E. Toukoniitty, F. Klingstedt, K. Eränen, T. Salmi and D. Y. Murzin, *Chem. Eng. J.*, 2007, **126**, 103.
99. K. Yoshida, C. Gonzalez-Arellano, R. Luque and P. L. Gai, *Appl. Catal., A*, 2010, **379**, 38.
100. T. Marimuthu and H. B. Friedrich, *ChemCatChem*, 2012, **4**, 2090.
101. Ó. Pablo, D. Guijarro and M. Yus, *Eur. J. Org. Chem.*, 2014, **2014**, 7034.
102. T. N. Danks and B. Desai, *Green Chem.*, 2002, **4**, 179.
103. G. R. Srinivasa, S. N. Narendra Babu, C. Lakshmi and D. C. Gowda, *Synth. Commun.*, 2004, **34**, 1831.
104. B. Dayal, N. H. Ertel, K. R. Rapole, A. Asgaonkar and G. Salen, *Steroids*, 1997, **62**, 451.

105. M. Nishioka, M. Miyakawa, Y. Daino, H. Kataoka, H. Koda, K. Sato and T. M. Suzuki, *Ind. Eng. Chem. Res.*, 2013, **52**, 4683.
106. M. C. Bagley, R. L. Jenkins, M. C. Lubinu, C. Mason and R. Wood, *J. Org. Chem.*, 2005, **70**, 7003.
107. J. M. Sauks, D. Mallik, Y. Lawryshyn, T. Bender and M. Organ, *Org. Process Res. Dev.*, 2014, **18**, 1310.
108. Pacific Microwave Technologies, *Microwave Flow Reactor System*, http://www.pacificmicrowave.com/products.html, accessed 23/3/2017.
109. CEM Corporation, *10 mL Gas Addition Kit for Discover SP and Discover S-class Systems*, http://cem.com/en/10-ml-gas-addition-kit-for-discover-sp-and-discover-s-class-systems.html, accessed 23/3/2017.
110. Milestone S.r.l., *SynthWAVE*, http://www.milestonesci.com/synthwave, accessed 23/3/2017.
111. Milestone S.r.l., *FlowSYNTH*, http://milestonescicom.powweb.com/flowsynth.html, accessed 23/3/2017.
112. T. N. Glasnov and C. O. Kappe, *Macromol. Rapid Commun.*, 2007, **28**, 395.
113. E. Gjuraj, R. Kongoli and G. Shore, *Chem. Biochem. Eng. Q.*, 2012, **26**, 285.
114. G. S. Vanier, *Synlett*, 2007, **2007**, 0131.
115. C. Bonnet, L. Estel, A. Ledoux, B. Mazari and A. Louis, *Chem. Eng. Process.*, 2004, **43**, 1435.
116. E. Heller, W. Lautenschläger and U. Holzgrabe, *Tetrahedron Lett.*, 2005, **46**, 1247.
117. M. Irfan, M. Fuchs, T. N. Glasnov and C. O. Kappe, *Chem.–Eur. J.*, 2009, **15**, 11608.
118. M. Irfan, E. Petricci, T. N. Glasnov, M. Taddei and C. O. Kappe, *Eur. J. Org. Chem.*, 2009, **2009**, 1327.
119. P. N. Romano, J. M. A. R. de Almeida, Y. Carvalho, P. Priecel, E. Falabella Sousa-Aguiar and J. A. Lopez-Sanchez, *ChemSusChem*, 2016, **9**, 3387.
120. J.-H. Yang and D. Ma, *RSC Adv.*, 2013, **3**, 10131.
121. L. Piras, E. Genesio, C. Ghiron and M. Taddei, *Synlett*, 2008, **2008**, 1125.
122. B. Toukoniitty, J. Wärnå, J. P. Mikkola, M. Helle, H. Saxén, D. Y. Murzin and T. Salmi, *Chem. Eng. Process.*, 2009, **48**, 837.
123. S. Tussing and J. Paradies, *Dalton Trans.*, 2016, **45**, 6124.
124. Z. Wu, E. Calcio Gaudino, L. Rotolo, J. Medlock, W. Bonrath and G. Cravotto, *Chem. Eng. Process.*, 2016, **110**, 220.
125. B. K. Banik, K. J. Barakat, D. R. Wagle, M. S. Manhas and A. K. Bose, *J. Org. Chem.*, 1999, **64**, 5746.
126. H. Berthold, T. Schotten and H. Honig, *Synthesis*, 2002, 1607.
127. N. Chapman, B. Conway, F. O'Grady and M. D. Wall, *Synlett*, 2006, **2006**, 1043.
128. P. Dutta, C. D. Pegu, M. L. Deb and P. K. Baruah, *Tetrahedron Lett.*, 2015, **56**, 4115.
129. M. J. Gracia, J. M. Campelo, E. Losada, R. Luque, J. M. Marinas and A. A. Romero, *Org. Biomol. Chem.*, 2009, **7**, 4821.

130. M. J. Jin, M. S. Sarkar and J. Y. Jung, in *Recent Progress in Mesostructured Materials, Proceedings of the 5th International Mesostructured Materials Symposium - Imms2006*, ed. D. Zhao, S. Qiu, Y. Tang and C. Yu, 2007, vol. 165, pp. 455–458.
131. K. Leijondahl, A.-B. L. Fransson and J.-E. Bäckvall, *J. Org. Chem.*, 2006, **71**, 8622.
132. Y.-M. Ma, X.-Y. Wei, X. Zhou, K.-Y. Cai, Y.-L. Peng, R.-L. Xie, Y. Zong, Y.-B. Wei and Z.-M. Zong, *Energy Fuels*, 2009, **23**, 638.
133. P. K. Mandal and J. S. McMurray, *J. Org. Chem.*, 2007, **72**, 6599.
134. B. L. A. Prabhavathi Devi, M. S. L. Karuna, K. Narasimha Rao, P. S. Saiprasad and R. B. N. Prasad, *J. Am. Oil Chem. Soc.*, 2003, **80**, 1003.
135. M. Quai, C. Repetto, W. Barbaglia and E. Cereda, *Tetrahedron Lett.*, 2007, **48**, 1241.
136. J. F. Quinn, C. E. Bryant, K. C. Golden and B. T. Gregg, *Tetrahedron Lett.*, 2010, **51**, 786.
137. J. F. Quinn, D. A. Razzano, K. C. Golden and B. T. Gregg, *Tetrahedron Lett.*, 2008, **49**, 6137.
138. J. S. M. Samec, L. Mony and J.-E. Bäckvall, *Can. J. Chem.*, 2005, **83**, 909.
139. M. S. Sarkar and M. J. Jin, in *Advances in Nanomaterials and Processing, Pts 1 and 2*, ed. B. T. Ahn, H. Jeon, B. Y. Hur, K. Kim and J. W. Park, 2007, vol. 124–126, pp. 1785–1788.
140. S. M. Sarkar, M. E. Ali, M. S. Islam, M. L. Rahman, S. S. Rashid, M. R. Karim and M. Mohd Yusoff, *J. Nanomater.*, 2015, **2015**, 6.
141. A. Sharma, V. Kumar and A. K. Sinha, *Adv. Synth. Catal.*, 2006, **348**, 354.
142. H. Takashima, L. Ren and Y. Kanno, *Catal. Commun.*, 2004, **5**, 317.
143. L. Ren and X. Pan, *Catal. Commun.*, 2011, **12**, 1366.
144. W. Yuji, Y. Hengbo, K. Takayuki and Y. Shozo, *Chem. Lett.*, 2000, **29**, 632.
145. T. Gustafsson, M. Hedenström and J. Kihlberg, *J. Org. Chem.*, 2006, **71**, 1911.
146. U. R. Pillai, E. Sahle-Demessie and R. S. Varma, *Green Chem.*, 2004, **6**, 295.
147. J. K. S. Wan, K. Wolf and R. D. Heyding, *Stud. Surf. Sci. Catal.*, 1984, **19**, 561.
148. J. Wan, M. Tse, H. Husby and M. Depew, *J. Microwave Power Electromagn. Energy*, 1990, **25**, 32.
149. J. K. S. Wan, US4574038 A, 1986.
150. P. R. Antoniewicz, *Phys. Rev. Lett.*, 1974, **32**, 1424.
151. S. Bordiga, L. Regli, F. Bonino, E. Groppo, C. Lamberti, B. Xiao, P. S. Wheatley, R. E. Morris and A. Zecchina, *Phys. Chem. Chem. Phys.*, 2007, **9**, 2676.
152. D. Scarano, S. Bertarione, G. Spoto, A. Zecchina and C. Otero Areán, *Thin Solid Films*, 2001, **400**, 50.
153. D. Sun, S. Sato, W. Ueda, A. Primo, H. Garcia and A. Corma, *Green Chem.*, 2016, **18**, 2579.
154. K. A. Rogers and Y. Zheng, *ChemSusChem*, 2016, **9**, 1750–1772.

155. S. De, S. Dutta and B. Saha, *Catal. Sci. Technol.*, 2016, **6**, 7364.
156. M. Bartók and Á. Molnár, in *PATAI'S Chemistry of Functional Groups*, John Wiley & Sons, Ltd, 2009, DOI: 10.1002/9780470682531.pat0133.
157. D. M. Alonso, S. G. Wettstein and J. A. Dumesic, *Chem. Soc. Rev.*, 2012, **41**, 8075.
158. J. N. Chheda, G. W. Huber and J. A. Dumesic, *Angew. Chem., Int. Ed.*, 2007, **46**, 7164.
159. A. Corma, S. Iborra and A. Velty, *Chem. Rev.*, 2007, **107**, 2411.
160. G. W. Huber, S. Iborra and A. Corma, *Chem. Rev.*, 2006, **106**, 4044.
161. C.-H. Zhou, X. Xia, C.-X. Lin, D.-S. Tong and J. Beltramini, *Chem. Soc. Rev.*, 2011, **40**, 5588.
162. H. Cheng, R. Liu, Q. Wang, C. Wu, Y. Yu and F. Zhao, *New J. Chem.*, 2012, **36**, 1085.
163. Z. Zhang, K.-Z. Fan, Q.-P. Song and G.-W. Wang, *Synth. Commun.*, 2010, **40**, 1052.
164. P. McKendry, *Bioresour. Technol.*, 2002, **83**, 37.
165. O. Arkad, H. Wiener, N. Garti and Y. Sasson, *J. Am. Oil Chem. Soc.*, 1987, **64**, 1529.
166. A. Smidovnik, J. Kobe, S. Leskovsek and T. Koloini, *J. Am. Oil Chem. Soc.*, 1994, **71**, 507.
167. T. Patel, Y. Ishiuji and G. Yosipovitch, *J. Am. Acad. Dermatol.*, 2007, **57**, 873.
168. A. M. Balu, J. M. Campelo, R. Luque and A. A. Romero, *Org. Biomol. Chem.*, 2010, **8**, 2845.
169. A. M. Raspolli Galletti, C. Antonetti, M. Bertoldo and F. Piccinelli, *Appl. Catal., A*, 2013, **468**, 95.
170. T. Gallert, M. Hahn, M. Sellin, C. Schmöger, A. Stolle, B. Ondruschka, T. F. Keller and K. D. Jandt, *ChemSusChem*, 2011, **4**, 1654.
171. E. A. Anumol, P. Kundu, P. A. Deshpande, G. Madras and N. Ravishankar, *ACS Nano*, 2011, **5**, 8049.
172. X. Zhang, D. O. Hayward, D. Michael and P. Mingos, *Chem. Commun.*, 1999, 975.
173. J. R. Thomas, *Catal. Lett.*, 1997, **49**, 137.
174. T. Xia, Y. Cao, N. A. Oyler, J. Murowchick, L. Liu and X. Chen, *ACS Appl. Mater. Interfaces*, 2015, **7**, 10407.
175. T. Xia, C. Zhang, N. A. Oyler and X. Chen, *Adv. Mater.*, 2013, **25**, 6905.
176. T. Xia, C. Zhang, N. A. Oyler and X. Chen, *J. Mater. Res.*, 2014, **29**, 2198.
177. J. K. S. Wan, US4345983 A, 1982.
178. T. R. J. Dinesen, M. Y. Tse, M. C. Depew and J. K. S. Wan, *Res. Chem. Intermed.*, 1991, **15**, 113.
179. X. Quan, Y. Zhang, S. Chen, Y. Zhao and F. Yang, *J. Mol. Catal. A: Chem.*, 2007, **263**, 216.
180. S. Horikoshi, F. Hojo, H. Hidaka and N. Serpone, *Environ. Sci. Technol.*, 2004, **38**, 2198.
181. A. B. Ahmed, B. Jibril, S. Danwittayakul and J. Dutta, *Appl. Catal., B*, 2014, **156–157**, 456.

182. Y.-Y. Peng, J. Tao, J.-W. Feng, X.-Y. Wang, L.-B. Liao and G.-C. Lv, *J. Synth. Cryst.*, 2014, 7, 012.
183. J. Sun, L. Bo, L. Yang, X. Liang and X. Hu, *RSC Adv.*, 2014, **4**, 14385.
184. B. Y. Jibril, A. Y. Atta, Y. M. Al-Waheibi and T. K. Al-Waheibi, *J. Ind. Eng. Chem.*, 2013, **19**, 1800.
185. R. Carta and F. Desogus, *J. Environ. Chem. Eng.*, 2013, **1**, 1292.
186. P. Li, L. Zhang, W. Wang, J. Su and L. Feng, *Environ. Sci. Technol.*, 2011, **45**, 4521.
187. W. Liao, T. Zheng, P. Wang, S. Tu and W. Pan, *J. Environ. Sci.*, 2010, **22**, 1800.
188. L. Bo, X. Quan, S. Chen, H. Zhao and Y. Zhao, *Water Res.*, 2006, **40**, 3061.
189. X.-y. Bi, P. Wang, H. Jiang, H.-y. Xu, S.-j. Shi and J.-l. Huang, *J. Environ. Sci.*, 2007, **19**, 1510.
190. X. Bi, P. Wang and H. Jiang, *J. Hazard. Mater.*, 2008, **154**, 543.
191. N. Li, P. Wang, C. Zuo, H. Cao and Q. Liu, *Environ. Eng. Sci.*, 2010, **27**, 271.
192. X. Li, F. Xu, J. Wang, C. Zhang, Y. Chen, S. Zhu and S. Shen, *Environ. Technol.*, 2010, **31**, 433.
193. H.-y. Yang, X.-y. Xu, X.-y. Bi and P.-s. Sun, *Environ. Sci. Survey*, 2012, **2**, 016.
194. X. Y. Bi, H. Y. Yang, P. S. Sun and X. Y. Xu, Advanced Materials Research Conference, 2013, vol. 610, pp. 2023–2027.
195. H. Iboukhoulef, A. Amrane and H. Kadi, *Environ. Technol.*, 2013, **34**, 853.
196. J. Zhang, W. Q. Pan, T. Zheng and P. Wang, Advanced Materials Research Conference, 2013, vol. 777, pp. 101–105.
197. X. Qi and Z. Li, *Methods*, 2016, **10**, 12.
198. X. Qi and Z. Li, *Desalin. Water Treat.*, 2016, **57**, 11756.
199. U. M. Nascimento and E. B. Azevedo, *J. Environ. Sci. Health, Part A*, 2013, **48**, 1056.
200. A. A. Kenge, P. H. Liao and K. V. Lo, *J. Environ. Sci. Health, Part A*, 2008, **43**, 1431.
201. W. T. Wong, W. I. Chan, P. H. Liao and K. V. Lo, *J. Environ. Sci. Health, Part A*, 2006, **41**, 2623.
202. Y. Pang and H. Lei, *Chem. Eng. J.*, 2016, **287**, 585.
203. J. Gao, G. Zhao, W. Shi and D. Li, *Chemosphere*, 2009, **75**, 519.
204. M. Ravera, A. Buico, F. Gosetti, C. Cassino, D. Musso and D. Osella, *Chemosphere*, 2009, **74**, 1309.
205. Z. Zhang, Y. Xu, M. Shen, D. D. Dionysiou, D. Wu, Z. Chen, F. Li, D. Liu and F. Zhang, *Environ. Prog. Sustainable Energy*, 2013, **32**, 181.
206. H.-S. Tai and C.-J. G. Jou, *Chemosphere*, 1999, **38**, 2667.
207. X. Liu, X. Quan, L. Bo, S. Chen and Y. Zhao, *Carbon*, 2004, **42**, 415.
208. X. Liu and G. Yu, *Chemosphere*, 2006, **63**, 228.
209. X. Liu, Q. Zhang, G. Zhang and R. Wang, *Chemosphere*, 2008, **72**, 1655.
210. S. Yuan, M. Tian and X. Lu, *J. Hazard. Mater.*, 2006, **137**, 878.
211. L. Lin, S. Yuan, J. Chen, L. Wang, J. Wan and X. Lu, *Chemosphere*, 2010, **78**, 66.
212. N. Remya and J.-G. Lin, *Chem. Eng. J.*, 2011, **166**, 797.

213. N. Serpone, S. Horikoshi and A. V. Emeline, *J. Photochem. Photobiol., C*, 2010, **11**, 114.
214. C. Eskicioglu, N. Terzian, K. J. Kennedy, R. L. Droste and M. Hamoda, *Water Res.*, 2007, **41**, 2457.
215. C. Eskicioglu, A. Prorot, J. Marin, R. L. Droste and K. J. Kennedy, *Water Res.*, 2008, **42**, 4674.
216. T.-L. Lai, C.-C. Lee, K.-S. Wu, Y.-Y. Shu and C.-B. Wang, *Appl. Catal., B*, 2006, **68**, 147.
217. T.-L. Lai, J.-Y. Liu, K.-F. Yong, Y.-Y. Shu and C.-B. Wang, *J. Hazard. Mater.*, 2008, **157**, 496.
218. N. Wang, T. Zheng, J. Jiang, W.-s. Lung, X. Miao and P. Wang, *Chem. Eng. J.*, 2014, **239**, 351.
219. V. Homem, A. Alves and L. Santos, *Chem. Eng. J.*, 2013, **220**, 35.
220. X. Ouyang, Z. Lin, Y. Deng, D. Yang and X. Qiu, *Chin. J. Chem. Eng.*, 2010, **18**, 695.
221. H.-S. Roh, Y.-k. Park and S.-E. Park, *J. Ind. Eng. Chem.*, 2003, **9**, 341.
222. R. Hyun-Seog, P. Yong-Ki and P. Sang-Eon, *Chem. Lett.*, 2000, **29**, 578.
223. H.-S. Roh, Y.-K. Park and S.-E. Park, *J. Ind. Eng. Chem.*, 2001, **7**, 326.
224. W. Xu, J. Zhou, Z. You, Y. Luo and Y. Ou, *ChemCatChem*, 2015, **7**, 450.
225. W. Xu, J. Cai, J. Zhou, Y. Ou, W. Long, Z. You and Y. Luo, *ChemCatChem*, 2016, **8**, 417.
226. W. Xu, J. Zhou, Z. Su, Y. Ou and Z. You, *Catal. Sci. Technol.*, 2016, **6**, 698.
227. M. Yat Tse, M. C. Depew and J. K. S. Wan, *Res. Chem. Intermed.*, 1990, **13**, 221.
228. X. Zhang, D. O. Hayward, C. Lee and D. M. P. Mingos, *Appl. Catal., B*, 2001, **33**, 137.
229. I. Hartmann, W. D. Einicke, H. Mäurer and S. Mrokwa, *Chem. Ing. Tech.*, 2005, **77**, 1609.
230. D. Varisli, C. Korkusuz and T. Dogu, *Appl. Catal., B*, 2017, **201**, 370.
231. F. K. Rabah and M. S. Darwish, *Environ. Nat. Resour. Res.*, 2013, **3**, 24.
232. L. Lin, J. Chen, Z. Xu, S. Yuan, M. Cao, H. Liu and X. Lu, *J. Hazard. Mater.*, 2009, **168**, 862.
233. L. Lin, S. Yuan, J. Chen, Z. Xu and X. Lu, *J. Hazard. Mater.*, 2009, **161**, 1063.
234. M. Nishioka, K. Sato, A. Onodera, M. Miyakawa, D. A. P. Tanaka, M. Kasai, A. Miyazawa and T. M. Suzuki, *Ind. Eng. Chem. Res.*, 2014, **53**, 1073.
235. H. Einaga, Y. Nasu, M. Oda and H. Saito, *Chem. Eng. J.*, 2016, **283**, 97.
236. K. L. Cameron, M. C. Depew and J. K. S. Wan, *Res. Chem. Intermed.*, 1991, **16**, 57.
237. M. C. Depew, S. Lem and J. K. S. Wan, *Res. Chem. Intermed.*, 1991, **16**, 213.
238. T. Sumi and S. Horikoshi, *J. Oleo Sci.*, 2013, **62**, 443.
239. Y. Sun, Y. Zhang and X. Quan, *Sep. Purif. Technol.*, 2008, **62**, 565.
240. G. Roussy, J. M. Thiebaut, M. Souiri, E. Marchal, A. Kiennemann and G. Maire, *Catal. Today*, 1994, **21**, 349.

241. G. Roussy, E. Marchal, J. M. Thiebaut, A. Kiennemann and G. Maire, *Fuel Process. Technol.*, 1997, **50**, 261.
242. X. Zhang, C. S. M. Lee, D. M. P. Mingos and D. O. Hayward, *Appl. Catal., A*, 2003, **249**, 151.
243. A. Domínguez, B. Fidalgo, Y. Fernández, J. J. Pis and J. A. Menéndez, *Int. J. Hydrogen Energy*, 2007, **32**, 4792.
244. B. Ni, C. Lee, R.-C. Sun and X. Zhang, *React. Kinet. Catal. Lett.*, 2009, **98**, 287.
245. S.-T. Liu, A.-B. Zhang, K.-K. Yan, Y. Ye and X.-G. Chen, *Sep. Purif. Technol.*, 2014, **135**, 35.
246. X. Liu, S. An, W. Shi, Q. Yang and L. Zhang, *J. Mol. Catal. A: Chem.*, 2014, **395**, 243.
247. S.-p. Jiang, P. Wang, G. Zhang, G. Hong and H. Ma, *China Water Wastewater*, 2004, **20**, 13.
248. G. Zhang, P. Wang, S. Jiang, G. Hong and H.-j. Ma, *Environ. Sci.*, 2004, **25**, 52.
249. G. Hong, P. Wang, G.-y. Zhang, H.-j. Ma and S.-p. Jiang, *J. Microwaves*, 2005, **1**, http://en.cnki.com.cn/Article_en/CJFDTOTAL-WBXB20050100F.htm.
250. S. B.-p. F. Jian-min and D. Y. LI Lan-qing-zi, *J. Anyang Univ.*, 2005, **3**, 001.
251. L. Yilu, L. Weiqi and L. Jichang, *Guangdong Chem. Ind.*, 2006, **33**, 63.
252. J. L. Wang and L. J. Xu, *Crit. Rev. Environ. Sci. Technol.*, 2012, **42**, 251.
253. B. Yu, G. Cheng, P. Tong and B. Dong, *Ind. Wastewater*, 2010, **41**, 9.
254. W.-H. Kuan, C.-Y. Chen, C.-Y. Hu and Y.-M. Tzou, *Int. J. Photoenergy*, 2013, **2013**, 9.
255. X. Wang, L. Mei, X. Xing, L. Liao, G. Lv, Z. Li and L. Wu, *Appl. Catal., B*, 2014, **160–161**, 211.
256. D. Xu, F. Cheng, Q. Lu and P. Dai, *Ind. Eng. Chem. Res.*, 2014, **53**, 2625.
257. U. Riaz, S. M. Ashraf and A. Ruhela, *J. Environ. Chem. Eng.*, 2015, **3**, 20.
258. X. Wang, G. Lv, L. Liao and G. Wang, *RSC Adv.*, 2015, **5**, 55595.
259. X. Liu, D. Xu, D. Zhang, G. Zhang and L. Zhang, *Appl. Catal., B*, 2016, **186**, 193.
260. Y. Ju, R. Liu, X. Tan, Y. Yu, D. D. Dionysiou and J. Feng, *Sep. Purif. Technol.*, 2016, **170**, 57.
261. J. Yin, J. Cai, C. Yin, L. Gao and J. Zhou, *J. Environ. Chem. Eng.*, 2016, **4**, 958.
262. D. Xu, X. Lai, W. Guo and P. Dai, *J. Water Process Eng.*, 2017, **16**, 270.
263. N. Wang and P. Wang, *Chem. Eng. J.*, 2016, **283**, 193.
264. F. Parolin, U. M. Nascimento and E. B. Azevedo, *Environ. Technol.*, 2013, **34**, 1247.
265. Z. Ai, Y. Wang, M. Xiao, L. Zhang and J. Qiu, *J. Phys. Chem. C*, 2008, **112**, 9847.
266. J. Hong, N. Yuan, Y. Wang and S. Qi, *Chem. Eng. J.*, 2012, **191**, 364.
267. X. Zhang, Y. Wang, G. Li and J. Qu, *J. Hazard. Mater.*, 2006, **134**, 183.
268. S. Horikoshi, A. Saitou, H. Hidaka and N. Serpone, *Environ. Sci. Technol.*, 2003, **37**, 5813.
269. S.-T. Liu, J. Huang, Y. Ye, A.-B. Zhang, L. Pan and X.-G. Chen, *Chem. Eng. J.*, 2013, **215–216**, 586.

270. G. Bamwenda, E. Moore and J. K. S. Wan, *Res. Chem. Intermed.*, 1992, **17**, 243.
271. G. Bamwenda, M. C. Depew and J. K. S. Wan, *Res. Chem. Intermed.*, 1993, **19**, 553.
272. S. Gopalakrishnan, J. Münch, R. Herrmann and W. Schwieger, *Chem. Eng. J.*, 2006, **120**, 99.
273. Y. Su, L.-C. Wang, Y.-M. Liu, Y. Cao, H.-Y. He and K.-N. Fan, *Catal. Commun.*, 2007, **8**, 2181.
274. Y. Liu, H. Tsunoyama, T. Akita and T. Tsukuda, *J. Phys. Chem. C*, 2009, **113**, 13457.
275. L. M. Kustov, A. V. Kucherov and E. D. Finashina, *Russ. J. Phys. Chem. A*, 2013, **87**, 345.
276. M. Sutradhar, L. M. D. R. S. Martins, M. F. C. Guedes da Silva, E. C. B. A. Alegria, C.-M. Liu and A. J. L. Pombeiro, *Dalton Trans.*, 2014, **43**, 39667.
277. N. Dimitratos, J. A. Lopez-Sanchez and G. J. Hutchings, Supported Metal Nanoparticles in Liquid-Phase Oxidation Reactions, in *Handbook of Advanced Methods and Processes in Oxidation Catalysis*, Imperial College Press, 2014, pp. 631–678.
278. N. Dimitratos, J. A. Lopez-Sanchez and G. J. Hutchings, *Chem. Sci.*, 2012, **3**, 20.
279. M. Besson, P. Gallezot and C. Pinel, *Chem. Rev.*, 2014, **114**, 1827.
280. S. R. Collinson and W. Thielemans, *Coord. Chem. Rev.*, 2010, **254**, 1854.
281. A. Villa, N. Dimitratos, C. E. Chan-Thaw, C. Hammond, L. Prati and G. J. Hutchings, *Acc. Chem. Res.*, 2015, **48**, 1403.
282. S. Rautiainen, P. Lehtinen, M. Vehkamäki, K. Niemelä, M. Kemell, M. Heikkilä and T. Repo, *Catal. Commun.*, 2016, **74**, 115.
283. M. Omri, G. Pourceau, M. Becuwe and A. Wadouachi, *ACS Sustainable Chem. Eng.*, 2016, **4**, 2432.
284. B. N. Zope, D. D. Hibbitts, M. Neurock and R. J. Davis, *Science*, 2010, **330**, 74.
285. C. D. Pina, E. Falletta and M. Rossi, *Chem. Soc. Rev.*, 2012, **41**, 350.
286. J. A. Lopez-Sanchez, 254th ACS National Meeting & Exposition, 2017, CATL-51, http://washingtondc2017.acs.org/i/860887-254th-american-chemical-society-national-meeting-expo/375.
287. S. Dutta, S. De, A. K. Patra, M. Sasidharan, A. Bhaumik and B. Saha, *Appl. Catal., A*, 2011, **409–410**, 133.
288. A. Dutta, A. K. Patra, S. Dutta, B. Saha and A. Bhaumik, *J. Mater. Chem.*, 2012, **22**, 14094.
289. H. Ren, Y. Zhou and L. Liu, *Bioresour. Technol.*, 2013, **129**, 616.
290. V. B. Kumar, I. N. Pulidindi and A. Gedanken, *RSC Adv.*, 2015, **5**, 11043.
291. F. Delbecq, Y. Wang and C. Len, *J. Mol. Catal. A: Chem.*, 2016, **423**, 520.
292. C. Xiouras, N. Radacsi, G. Sturm and G. D. Stefanidis, *ChemSusChem*, 2016, **9**, 2159.
293. S. Dutta, S. De and B. Saha, *Biomass Bioenergy*, 2013, **55**, 355.
294. R. Kumar, A. Sharma, N. Sharma, V. Kumar and A. K. Sinha, *Eur. J. Org. Chem.*, 2008, **2008**, 5577.

295. T. S. Hansen, J. M. Woodley and A. Riisager, *Carbohydr. Res.*, 2009, **344**, 2568.
296. C. Li, Z. Zhang and Z. K. Zhao, *Tetrahedron Lett.*, 2009, **50**, 5403.
297. Z. Baohui, Z. Fang, J. Cheng and Y. Jiang, *J. Chem. Sci.*, 2010, **65**, 168.
298. X. Qi, M. Watanabe, T. M. Aida and R. L. Smith Jr, *ChemSusChem*, 2010, **3**, 1071.
299. S. Wu, C. Wang, Y. Gao, S. Zhang, D. Ma and Z. Zhao, *Chin. J. Catal.*, 2010, **31**, 1157.
300. Z. Zhang and Z. K. Zhao, *Bioresour. Technol.*, 2010, **101**, 1111.
301. S. De, S. Dutta and B. Saha, *Green Chem.*, 2011, **13**, 2859.
302. J. Lu, Y. Yan, Y. Zhang and Y. Tang, *RSC Adv.*, 2012, **2**, 7652.
303. B. Liu, Z. Zhang and Z. K. Zhao, *Chem. Eng. J.*, 2013, **215–216**, 517.
304. X. Zhou, Z. Zhang, B. Liu, Z. Xu and K. Deng, *Carbohydr. Res.*, 2013, **375**, 68.
305. Q. Dong and Y. Xiong, *Bioresour. Technol.*, 2014, **171**, 127.
306. S. L. Guenic, F. Delbecq, C. Ceballos and C. Len, *J. Mol. Catal. A: Chem.*, 2015, **410**, 1.
307. J. Sun, X. Yuan, Y. Shen, Y. Yi, B. Wang, F. Xu and R. Sun, *Ind. Crops Prod.*, 2015, **70**, 266.
308. N. Esmaeili, M. J. Zohuriaan-Mehr, H. Bouhendi and G. Bagheri-Marandi, *Korean J. Chem. Eng.*, 2016, **33**, 1964.
309. D. Mathur and S. Chakraborty, *Chem. Eng. Trans.*, 2016, **52**, 967.
310. P. Wang, L. Ren, Q. Lu and Y. Huang, *Chem. Eng. Commun.*, 2016, **203**, 1507.
311. P. Wrigstedt, J. Keskiväli and T. Repo, *RSC Adv.*, 2016, **6**, 18973.
312. F. Guo, Z. Fang and T.-J. Zhou, *Bioresour. Technol.*, 2012, **112**, 313.
313. Q. Wu, Y. Yan, Q. Zhang, J. Lu, Z. Yang, Y. Zhang and Y. Tang, *ChemSusChem*, 2013, **6**, 820.
314. C. Antonetti, M. Melloni, D. Licursi, S. Fulignati, E. Ribechini, S. Rivas, J. C. Parajó, F. Cavani and A. M. Raspolli Galletti, *Appl. Catal., B*, 2017, **206**, 364.
315. Y.-Y. Dong, J. He, S.-L. Sun, M.-G. Ma, L.-H. Fu and R.-C. Sun, *Carbohydr. Polym.*, 2013, **98**, 168.
316. L. Zhang and Z. Le, *Chem. Bull.*, 2011, **74**, 917.
317. R. Weingarten, J. Cho, W. C. Conner Jr and G. W. Huber, *Green Chem.*, 2010, **12**, 1423.
318. V. Pistarà, A. Rescifina, M. A. Chiacchio and A. Corsaro, *Curr. Org. Chem.*, 2014, **18**, 417.
319. R.-J. van Putten, J. C. van der Waal, E. de Jong, C. B. Rasrendra, H. J. Heeres and J. G. de Vries, *Chem. Rev.*, 2013, **113**, 1499.
320. G. Marcotullio and W. De Jong, *Green Chem.*, 2010, **12**, 1739.
321. N. Azcan and A. Danisman, *Fuel*, 2007, **86**, 2639.
322. K.-S. Chen, Y.-C. Lin, K.-H. Hsu and H.-K. Wang, *Energy*, 2012, **38**, 151.
323. R. Kumar, G. Ravi Kumar and N. Chandrashekar, *Bioresour. Technol.*, 2011, **102**, 6617.
324. N. Azcan and A. Danisman, *Fuel*, 2008, **87**, 1781.

325. F. Motasemi and F. N. Ani, *Renewable Sustainable Energy Rev.*, 2012, **16**, 4719.
326. E. Martinez-Guerra and V. G. Gude, *Fuel*, 2014, **137**, 100.
327. P. D. Vaidya and A. E. Rodrigues, *Chem. Eng. Technol.*, 2009, **32**, 1463.
328. X. Fan, R. Burton and Y. Zhou, *Open Fuels Energy Sci. J.*, 2010, **3**, 17.
329. M. Ayoub and A. Z. Abdullah, *Renewable Sustainable Energy Rev.*, 2012, **16**, 2671.
330. N. Rahmat, A. Z. Abdullah and A. R. Mohamed, *Renewable Sustainable Energy Rev.*, 2010, **14**, 987.
331. A. K. Sinha, A. Sharma, A. Swaroop and V. Kumar, *Tetrahedron*, 2007, **63**, 1000.
332. S. Rat, D. Mathiron, P. Michaud, J. Kovensky and A. Wadouachi, *Tetrahedron*, 2007, **63**, 12424.
333. S. Rat, M. Bosco, M.-L. Tavernier, P. Michaud, A. Wadouachi and J. Kovensky, *C. R. Chim.*, 2011, **14**, 307.
334. A. El Nemr, S. Ragab and A. El Sikaily, *Carbohydr. Polym.*, 2016, **151**, 1058.
335. D. Kim, J. Choi, G.-J. Kim, S. K. Seol and S. Jung, *Bioresour. Technol.*, 2011, **102**, 7229.
336. D. Kim, J. Choi, G.-J. Kim, S. K. Seol, Y.-C. Ha, M. Vijayan, S. Jung, B. H. Kim, G. D. Lee and S. S. Park, *Bioresour. Technol.*, 2011, **102**, 3639.
337. T. Lieu, S. Yusup and M. Moniruzzaman, *Bioresour. Technol.*, 2016, **211**, 248.
338. R. Priambodo, T.-C. Chen, M.-C. Lu, A. Gedanken, J.-D. Liao and Y.-H. Huang, *Energy Procedia*, 2015, **75**, 84.
339. P. Patil, V. G. Gude, S. Pinappu and S. Deng, *Chem. Eng. J.*, 2011, **168**, 1296.
340. W. Ye, Y. Gao, H. Ding, M. Liu, S. Liu, X. Han and J. Qi, *Fuel*, 2016, **180**, 574.
341. G. Perin, G. Álvaro, E. Westphal, L. H. Viana, R. G. Jacob, E. J. Lenardão and M. G. M. D'Oca, *Fuel*, 2008, **87**, 2838.
342. Y.-C. Lin, S.-C. Chen, T.-Y. Wu, P.-M. Yang, S.-R. Jhang and J.-F. Lin, *J. Taiwan Inst. Chem. Eng.*, 2015, **49**, 72.
343. C.-C. Liao and T.-W. Chung, *Chem. Eng. Res. Des.*, 2013, **91**, 2457.
344. M. Koberg, R. Abu-Much and A. Gedanken, *Bioresour. Technol.*, 2011, **102**, 1073.
345. M.-C. Hsiao, C.-C. Lin and Y.-H. Chang, *Fuel*, 2011, **90**, 1963.
346. J. M. Encinar, J. F. González, G. Martínez, N. Sánchez and A. Pardal, *Fuel*, 2012, **95**, 386.
347. M. Z. Duz, A. Saydut and G. Ozturk, *Fuel Process. Technol.*, 2011, **92**, 308.
348. C. L. Teo and A. Idris, *Bioresour. Technol.*, 2014, **174**, 311.
349. P. D. Patil, V. G. Gude, A. Mannarswamy, P. Cooke, S. Munson-McGee, N. Nirmalakhandan, P. Lammers and S. Deng, *Bioresour. Technol.*, 2011, **102**, 1399.
350. P. D. Patil, V. G. Gude, A. Mannarswamy, P. Cooke, N. Nirmalakhandan, P. Lammers and S. Deng, *Fuel*, 2012, **97**, 822.

351. E. Martinez-Guerra, V. G. Gude, A. Mondala, W. Holmes and R. Hernandez, *Appl. Energy*, 2014, **129**, 354.
352. M. Barekati-Goudarzi, D. Boldor and D. B. Nde, *Bioresour. Technol.*, 2016, **201**, 97.
353. P. D. Patil, H. Reddy, T. Muppaneni, T. Schaub, F. O. Holguin, P. Cooke, P. Lammers, N. Nirmalakhandan, Y. Li, X. Lu and S. Deng, *Bioresour. Technol.*, 2013, **139**, 308.
354. M. Koberg, M. Cohen, A. Ben-Amotz and A. Gedanken, *Bioresour. Technol.*, 2011, **102**, 4265–4269.
355. J. V. Gerpen, *Fuel Process. Technol.*, 2005, **86**, 1097.
356. D. Y. C. Leung, X. Wu and M. K. H. Leung, *Appl. Energy*, 2010, **87**, 1083.
357. A. Abbaszaadeh, B. Ghobadian, M. R. Omidkhah and G. Najafi, *Energy Convers. Manage.*, 2012, **63**, 138.
358. L. Jin, Y. Zhang, J. P. Dombrowski, C.-H. Chen, A. Pravatas, L. Xu, C. Perkins and S. L. Suib, *Appl. Catal., B*, 2011, **103**, 200.
359. M. D. Karkas, B. S. Matsuura, T. M. Monos, G. Magallanes and C. R. J. Stephenson, *Org. Biomol. Chem.*, 2016, **14**, 1853.
360. Z.-J. Yang, Y.-F. Li, Q.-B. Wu, N. Ren, Y.-H. Zhang, Z.-P. Liu and Y. Tang, *J. Catal.*, 2011, **280**, 247.
361. S. Van de Vyver, L. Peng, J. Geboers, H. Schepers, F. de Clippel, C. J. Gommes, B. Goderis, P. A. Jacobs and B. F. Sels, *Green Chem.*, 2010, **12**, 1560.
362. H. Kobayashi, M. Yabushita, T. Komanoya, K. Hara, I. Fujita and A. Fukuoka, *ACS Catal.*, 2013, **3**, 581.
363. D.-m. Lai, L. Deng, Q.-x. Guo and Y. Fu, *Energy Environ. Sci.*, 2011, **4**, 3552.
364. Y. Ogasawara, S. Itagaki, K. Yamaguchi and N. Mizuno, *ChemSusChem*, 2011, **4**, 519.
365. H. Guo, X. Qi, L. Li and R. L. Smith Jr, *Bioresour. Technol.*, 2012, **116**, 355.
366. H. Cai, C. Li, A. Wang, G. Xu and T. Zhang, *Appl. Catal., B*, 2012, **123–124**, 333.
367. Y. Jiang, X. Li, X. Wang, L. Meng, H. Wang, G. Peng, X. Wang and X. Mu, *Green Chem.*, 2012, **14**, 2162.
368. W.-H. Chen, S.-C. Ye and H.-K. Sheen, *Appl. Energy*, 2012, **93**, 237.

CHAPTER 9

Biohydrogen and Biomethane Production

NARCIS HOMS*[a,b], DANIEL MONTANÉ[c] AND PILAR RAMÍREZ DE LA PISCINA[a]

[a]Universitat de Barcelona, Department of Inorganic and Organic Chemistry, Inorganic Chemistry Section, Institute of Nanoscience and Nanotechnology (IN2UB), C/Martí i Franquès 1-11, Barcelona, 08028, Spain; [b]Catalonia Institute for Energy Research (IREC), Materials and Catalysis (MATCAT), Jardins de les Dones de Negre 1, Sant Adrià de Besòs, Barcelona, 08930, Spain; [c]Universitat Rovira i Virgili, Department of Chemical Engineering, Av. Països Catalans, 26, Tarragona, 43007, Spain
*E-mail: narcis.homs@qi.ub.edu

9.1 Introduction and Overview

According to the IEA definition,[1] "Biorefinery is the sustainable processing of biomass into a spectrum of marketable products (food, feed, materials, chemicals) and energy (fuels, power, heat)". Biorefineries are thus facilities that convert biomass – biological materials obtained from living or recently living organisms – into fuel, energy, chemicals or other materials, including feed.[2] So, conceptually, a biorefinery can be envisaged as analogous to a petroleum refinery, where various fuels and chemicals are produced from oil. Even though this analogy is appealing, biorefineries differ from conventional oil refineries in at least one crucial aspect: plant capacity. Conventional oil refineries have plant capacities of a few million tons per year; whereas two fundamental

Green Chemistry Series No. 56
Sustainable Catalysis for Biorefineries
Edited by Francesco Frusteri, Donato Aranda and Giuseppe Bonura

Published by the Royal Society of Chemistry, www.rsc.org

limitations make it impossible to process biomass in plants with such large capacities. Firstly, there is not a unique "biomass"; in fact, biomass includes a wide variety of materials of biological origin that have dissimilar chemical compositions and physical properties. These include lignocellulosic materials (wood and herbaceous plants), sewage sludge, sugary and oleaginous plants, microalgae and seaweeds, and industrial, commercial and post-consumer waste. Because of this diversity, there is no single strategy or technology that is valid for processing all biomasses. Secondly, biomass is a disperse resource at origin, and so its collection and transportation costs have a large impact on plant profitability. The economic advantages of process integration and scaling up rapidly become outweighed by the cost of transporting biomass from increasingly large distances. Consequently, there is a limit to the profitable plant capacity of a biorefinery, which is determined by the availability and distribution of biomass of the necessary type around the processing plant.

Different biorefinery configurations are possible depending on the biomass feedstock to be processed, the primary product streams derived from biomass, the final products of interest, and the technologies used to obtain both the primary and final products.[1] The primary products include C_5/C_6 carbohydrates, lignin, vegetable oils and fats, organic solutions, pyrolysis oil, and syngas; while the technologies used to derive them consist of a wide array of thermal, biological and chemical processes. The final products may similarly be a variety of chemicals, notably including biofuels. The production of refined chemicals other than fuels from biomass, even though less advanced commercially, is a major focus of research and development. Notable current projects, both pilot and commercial, are the transformation of lignin into aromatic derivatives, and the production of novel chemical building blocks from the carbohydrate fraction of biomass. The latter include primary building blocks that can be used to produce specialty chemicals and polymers that could compete with, and eventually replace, existing oil-based products: organic acids (lactic, succinic, acrylic and levulinic), furans (furfural and 5-hydroxymethyl furfural), 1,3-butadiene, alcohols (iso- and *n*-butanol), glycerol and sugars, among others.[3] Most existing commercial and demonstration biorefinery facilities, as well as pilot-plant projects, are oriented towards the production of biofuels. Liquid biofuels such as bioethanol and biodiesel are obtained from carbohydrates and lipids. The principal gaseous biofuels are methane and hydrogen.[4]

There is increasing worldwide interest in the use of CH_4 as a carbon-based fuel for many applications. This is driven by two crucial facts: (i) combustion of CH_4 emits less CO_x per unit of energy produced than combustion of oil or coal; and (ii) CH_4 can be obtained not only from fossil sources – mainly natural gas (NG) reservoirs – but also from local biomass resources. Although not exclusively, the first factor relates to interest in countries where policies to reduce greenhouse gas emissions are being implemented. Meanwhile, the second factor offers a way to increase fuel independence, which can be a serious consideration in regions such as China and Europe. Methane can be produced from dry biomass by thermal gasification and methanation, and from wet biomass by anaerobic digestion or gasification in aqueous media. Hydrogen can also be produced using the same technologies: by thermal gasification of dry biomass with separation from the syngas, by gasification of wet biomass in aqueous media, and biologically (dark fermentation). The existing infrastructure for the distribution and consumption of NG makes the production of CH_4 from biomass more attractive than the production of pure H_2. However, H_2 is also important in the operation of biorefineries, since it is crucial for selective hydrogenations involved in the transformation of the primary products into final products. Recently, the mixture of H_2 and CH_4 (called hythane fuel) has been considered as a convenient way to introduce H_2 as an energy carrier into gaseous fuel systems. Appropriate mixtures of H_2 and CH_4, usually below 10% H_2 v/v, could have the advantage of allowing use of the NG network, thus facilitating its distribution. Moreover, hythane fuel is more energy-efficient and cleaner, regarding CO_2 emissions, than NG. Thus, recent research to tune biomass transformation to yield H_2/CH_4 rich mixtures is not surprising, envisaging the direct production of the corresponding biohythane.

In this chapter, we focus on the latest advances in processes to produce H_2 and CH_4 from biorefinery processing. The review covers the catalytic gasification of biomass in aqueous media, under both sub- and supercritical conditions, for biohydrogen production, and processes for biomethane and biosynthetic NG (Bio-SNG) production.

9.2 Catalytic Gasification of Biomass in Aqueous Media

Thermal gasification with steam/O_2 is a well-known process that is applied to produce syngas and H_2 from biomass. However, this gasification requires a feedstock biomass with a low moisture content for the gasifier to be thermally self-sustaining, so it is ill-suited for treatment of wet biomass such as microalgae, sludge, high-moisture lignocellulose, or aqueous streams loaded with organic matter from the food industry. Hydrothermal processes conducted in liquid phase are an alternative to conventional gasification that allow direct processing of moist biomass and aqueous solutions without the energy penalty incurred by the need to dry the biomass. Hydrothermal gasification may take place at a reaction temperature between *ca.* 200 °C and 600 °C, and pressure from 2 to 40 MPa, generally in the presence of a catalyst that facilitates decomposition of the biomass substrate and the formation of a gaseous product rich in H_2, CO_2 and CH_4. The reaction mechanisms involved are complex and depend on the nature of the organic substrate being converted, the catalyst, and various properties of the aqueous phase such as pH. Overall, the reaction process may be schematized by combining the reversible stoichiometric eqn (9.1)–(9.6) – which comprise substrate dehydration/hydration (C–O scission/formation), dehydrogenation/hydrogenation (C–H scission/formation) and fragmentation/polymerization

(C–C scission/formation) – with reforming of all the organic species to CO and H_2, together with both water gas shift (WGS) and methanation reactions. Some catalysts may also favour C–C forming reactions involving Fisher–Tropsch mechanisms, which represent the inverse of the reforming reaction (9.4).

$$C_xH_yO_z \rightleftarrows C_xH_{y-2}O_{z-1} + H_2O \quad \text{C–O scission/formation} \tag{9.1}$$

$$C_xH_yO_z \rightleftarrows C_xH_{y-2}O_z + H_2 \quad \text{C–H scission/formation} \tag{9.2}$$

$$C_xH_yO_z \rightleftarrows C_{x-a}H_{y-b}O_{z-c} + C_aH_bO_c \quad \text{C–C scission/formation} \tag{9.3}$$

$$C_xH_yO_z + (x-z)H_2O \rightleftarrows (x + y/2 - z)H_2 + xCO \quad \text{reforming/FT synthesis} \tag{9.4}$$

$$CO + H_2O \rightleftarrows H_2 + CO_2 \quad \text{water gas shift} \tag{9.5}$$

$$CO + 3H_2 \rightleftarrows CH_4 + H_2O \quad \text{methanation} \tag{9.6}$$

In practice, it is convenient to distinguish between sub- and supercritical processes, based on the critical point of water (374.2 °C and 22.12 MPa). Gasification in supercritical water (SCWG) takes advantage of the drastic change in properties such as the dielectric constant, ion product and diffusion coefficients with water density near the critical point. Therefore, the properties of the reacting fluid may be tuned to some extent by adjusting temperature and pressure. Processes that take place in the subcritical water range – commonly referred to as aqueous-phase reforming (APR) – use milder reaction conditions. For these, temperature is usually kept below 280 °C and pressure plays a lesser role, since it only needs to be maintained above the boiling point of water at the selected reaction temperature. Even under subcritical conditions, hot pressurized water is an aggressive chemical medium that poses significant challenges for catalyst stability, and much effort has been devoted over recent years to improving the durability of supported metal catalysts. Supercritical conditions are even harsher. Besides research aimed at improving catalyst stability, the development of construction materials that can withstand corrosion and guarantee reactor integrity during SCWG is also a field of intense research.

To assess the efficiency of the catalyst under study with regard to the formation of hydrogen in APR or SCWG, biomass conversion (X_B), carbon conversion to liquid products ($X_{B,L}$), carbon conversion to gas products ($X_{B,G}$) and hydrogen selectivity (S_{H_2}) are used as common process performance indicators, together with the molar ratios of hydrogen to methane (or alkanes) and hydrogen to carbon monoxide.

$$X_B = 100\left(\frac{\text{Amount of biomass converted}}{\text{Amount of feed biomass}}\right) \tag{9.7}$$

$$X_{B,L} = 100\left(\frac{\text{Carbon in liquid products}}{\text{Carbon in feed biomass}}\right) \tag{9.8}$$

$$X_{B,G} = 100\left(\frac{\text{Carbon in gas products}}{\text{Carbon in feed biomass}}\right) \tag{9.9}$$

$$S_{H_2} = 100\left(\frac{\text{Hydrogen gas produced}}{\text{Carbon in gas products}}\right)\left(\frac{1}{R_B}\right) \tag{9.10}$$

where R_B is the stoichiometric reforming ratio, which is the stoichiometric H_2/CO_2 molar ratio at complete conversion of the biomass (or model compound) under study, which is the result of the addition of stoichiometric eqn (9.4) and (9.5).

9.2.1 Catalytic Aqueous-phase Reforming (APR)

APR of biomass-derived polyols (glucose, sorbitol, glycerol, ethylene glycol) and methanol to produce hydrogen was introduced by Cortright and co-workers using a Pt-based catalyst supported on γ-alumina.[5] Since then, the development of catalysts for APR has been an active research field. It is generally accepted that the hydrothermal conversion of polyols in APR proceeds through two competing routes. Figure 9.1 is a scheme of the main reaction paths for glycerol,[6–9] but the same types of mechanisms have been used to explain APR of larger polyols such as xylitol[10] and sorbitol.[11] The gasification route consists of the sequential dehydrogenation and decarbonylation of the polyol chains, and leads to the formation of hydrogen gas and carbon

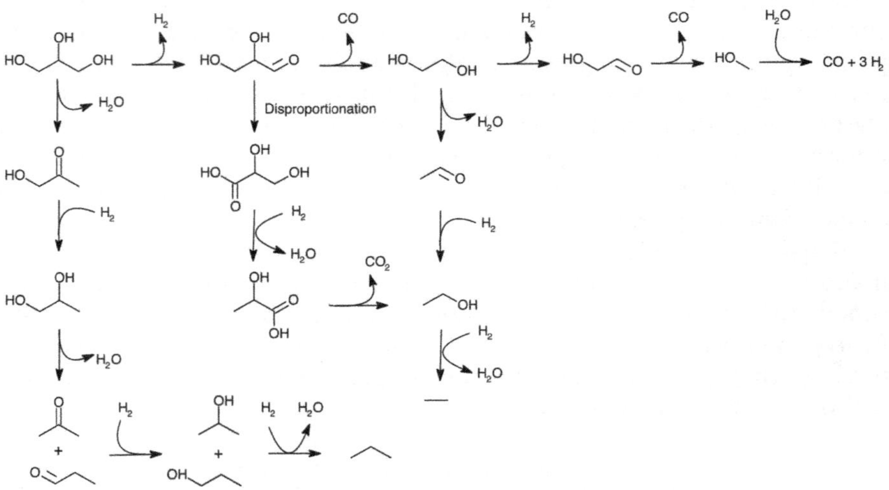

Figure 9.1 Major reaction pathways for the aqueous phase reforming of glycerol.

monoxide. Due to the excess of water, the CO is either further converted to CO_2 and additional H_2 through the WGS reaction, or it may react with H_2 to form CH_4 through the competing methanation reaction. The hydrogenation route leads to the formation of alcohols, and eventually alkanes, through dehydration and hydrogenation. Depending on the catalyst, glyceraldehyde may also react to form carboxylic acids, which are further decomposed by decarboxylation to ethanol and carbon dioxide.[6]

The nature and surface properties of the catalyst determine to a large extent which route will be favoured in APR. Besides catalyst structure, there are additional factors that may influence the selectivity, activity and even long-term stability of the catalyst. Two of these factors are the possible contribution of the aqueous medium to the promotion of parasitic reactions in the liquid phase, and liquid-to-gas mass transfer. Water is not an inert solvent in APR. The ion product of water at APR conditions is higher than that of water at room temperature (for instance, it is 4.19×10^{-12} $mol^2\ L^{-2}$ at 250 °C and 5 MPa). Ionic reactions may develop in the liquid phase,[12] involving both organic products and some catalyst supports such as γ-Al_2O_3, which transforms to boehmite. Furthermore, if carbon dioxide and carboxylic acids are formed during APR, the increased acidity of the aqueous phase may favour the occurrence of homogeneous dehydration reactions. This effect is more important the higher the initial concentration and conversion of the polyol, and it may influence the distribution of liquid products.

Mass transfer is another factor that may influence the selectivity of the overall reaction process, thus making the comparison of results from different types of laboratory reactors more difficult. In APR, hydrogen gas is both a product and a reactant. Because of the temperature and pressure used in APR and the low solubility in water of H_2 and CO, they readily separate from the liquid and form a gaseous phase inside the reactor. Laboratory batch reactors use stirring devices to disperse powdered catalyst into the liquid phase and thereby maximize liquid-to-catalyst mass transfer. Consequently, liquid-to-gas mass transfer is also enhanced. Furthermore, these reactors are usually cold-purged and pressurized with an inert gas before the reaction starts. During the reaction, the excess of inert gas tends to saturate the liquid phase, thus displacing H_2 from the liquid, and it also dilutes the gas phase, resulting in a low partial pressure of H_2. Even if the catalyst is active to form alcohols and alkanes, this route will be hindered because of the low availability of dissolved hydrogen,[13] and the activity and true selectivity of the catalyst towards the formation of gases and liquids will be affected. When continuous packed-bed tubular reactors are used, liquid-to-gas mass transfer is not as efficient as in stirred batch reactors, and the concentration of H_2 in the liquid phase will tend to be higher. In some studies, a sweep current of inert gas is also added to the feed to lower the partial pressure of hydrogen in the gas phase and its concentration in the liquid. This influences the assessment of the real selectivity of the catalyst. Furthermore, using an inert gas to entrain H_2 from the APR reactor and increase the yield makes little sense in industrial practices if high-purity H_2 is the desired product (or a mixture of methane and hydrogen with a significant heating value).

Multiple combinations of supports, active metals and promoters have been tested in efforts to enhance catalyst activity, selectivity to hydrogen, and long-term stability, even though Pt continues to be the metal of choice due to its high activity under APR conditions. Table 9.1 summarizes recent work on the reforming of biomass-derived polyols with the aim of producing hydrogen, grouped according to the main active metal. Table 9.2 shows selected results of key reaction performance indicators. Platinum supported on γ-Al_2O_3 is the APR catalyst that has been studied the most. Wawrzetz[6] conducted a study on the APR of glycerol at high concentrations (7–30 wt%) using a Pt/γ-Al_2O_3 catalyst prepared by impregnation in a tubular flow reactor. It was observed that the γ-Al_2O_3 was not stable and was converted to boehmite during the reaction, even at a time-on-stream of just a few hours. The size of the Pt crystallite influenced selectivity, with lower gas formation the larger the crystallite. Additional experiments with intermediate products (alcohols) revealed that hydrogenolysis of C–C and C–O bonds did not take place under APR conditions. Dehydrogenation and dehydration were the primary glycerol reactions, and the resulting aldehydes and ketones were subsequently converted by decarbonylation and hydrogenation, respectively (see Figure 9.1). Aldehydes could also form organic acids, which decomposed by decarboxylation to yield alcohols. The absence of hydrogenolysis of C–C bonds explained why C_3 and C_2 alkanes were obtained among the liquid products. Around 20% of the hydrogen produced was consumed *in situ* in the generation of liquid products. The effect of the Pt particle size on the selectivity to gaseous products was also reported for the APR of the aqueous fraction of bio-oil in a series of experiments in batch mode at a 25% concentration of bio-oil in water.[12] Studies of the reforming of xylitol[10] and sorbitol[11] confirmed the conversion mechanism based on dehydration/hydrogenation and dehydrogenation/decarbonylation. The Pt load of the catalyst did not influence the selectivity to gas products below 2.77 wt% Pt. However, the catalyst was not stable. Metal dispersion decreased after 5 h of reaction and some carbonaceous deposits formed on the surface of the catalyst, even though catalyst activity was maintained without additional carbon deposition after a time-on-stream of 12 h.

The instability of γ-Al_2O_3 under APR conditions prompted research into alternative supports. Pendem[7] used hydrotalcites (HTs) as the support to prepare Pt catalysts by impregnation. HT with an Al_2O_3 to MgO ratio of 80:20 gave a highly dispersed catalyst, with Pt particles ranging from 2 to 5 nm in diameter. A catalyst with 3 wt% Pt provided 98% conversion of glycerol to liquid products, with a selectivity to 1,3-propanediol of 75% at 250 °C. Most of the hydrogen produced *via* the dehydrogenation route was consumed *in situ* in the hydrogenation of hydroxyacetone to 1,3-propanediol. The gas phase consisted of a mixture of hydrogen and carbon dioxide, and it was proposed that carbonic acid in the aqueous-phase catalysed the dehydration

Table 9.1 Aqueous phase reforming: catalytic systems, methods of preparation and conditions for testing catalytic activity.

Metal	Load (wt%)	Promoters	Load (wt%)	Supports	Preparation	Feedstock	Load (wt%)	T (°C)	P (MPa)	Reactor type	WHSV (h^{-1}) or organic feed to catalyst mass ratio	Time on stream or reaction time (h)	Ref.
Pt	1, 3, 5	—	—	γ-Al_2O_3	Imp.	Glycerol	7–30	225	2.6, 2.9, 4.5	Tubular	0.45–22.7	n.a.	6
Pt	0.3–2.77	—	—	γ-Al_2O_3	Imp.	Xylitol	1–10	225	2.93	Tubular	0.6–2.4	12	10
Pt	0.3–2.77	—	—	γ-Al_2O_3	Imp.	Sorbitol	1–5	225	2.93	Tubular	0.6–2.4	5	11
Pt	2	—	—	γ-Al_2O_3	Imp.	Bio-oil aq.	25	260	n.a.	Batch	20	4	14
Pt	3.0	—	—	HT	Imp.	Glycerol	10	225, 250, 270	3.5, 4.5, 6.0	Batch	10	3	7
Pt	1–10	—	—	CMK-3	Imp.	Ethylene glycol	10	250	4.4	Tubular	2	25	16
Pt	Pt/CeZr, 0.05/0.95	—	—	$Ce_xZr_{1-x}O_2$, $x = 0$–0.95	Cop.	Glycerol	10	250	5.0	Tubular	2.45	50	17
Pt	1–2	—	—	ZnO/Al_2O_3	ALD	1-Propanol	5	250	6.4	Tubular	6	16	18
Pt	0.75	—	—	MgO, Al_2O_3, CeO_2, TiO_2, SiO_2	Imp.	Glycerol	5	225	2.76	Tubular	0.18	40	8
Pt	1–2	—	—	SiO_2, α-Al_2O_3, α-Al_2O_3 with CeO_2 and ZrO_2	Imp.	Glycerol	10	200, 225, 250	1.6, 2.4, 4.4	Batch	2.38	2	15
Pt	3	Ni	11	γ-Al_2O_3	Imp. and Sol–Gel	Glycerol	10	230, 250	30, 50	Tubular	2.6	33	19
Pt	3	Ni	10	γ-Al_2O_3	Imp. and Sol–Gel	Glycerol	10	230, 250	30, 50	Tubular	2.6	32	20

(continued)

Table 9.1 *(continued)*

Metal	Load (wt%)	Promoters	Load (wt%)	Supports	Preparation	Feedstock	Load (wt%)	T (°C)	P (MPa)	Reactor type	WHSV (h^{-1}) or organic feed to catalyst mass ratio	Time on stream or reaction time (h)	Ref.
Pt	3	Re	3	AC	Imp.	Glycerol	10	225	3.03	Tubular	6	n.a.	9
Pt, Rh	2.5, 2	Re	(Rh or Pt) : Re mass ratio 1 : 1.54	AC	Imp.	Glycerol	10	225	n.a.	Batch	80	5.3	21
Pt	3 (total load)	Re	Pt : Re molar ratio 1 : 1	CMK-3, SiO_2, γ-Al_2O_3, AC	Imp.	Ethylene and propylene glycol, glycerol, sorbitol	10	250	4.5	Tubular	2	24	22
Pt	1	Cu	Several ratios	Mg(Al)O mixed oxides with various ratios of Mg/Al	Imp.	Glycerol	10	225	2.9	Batch	3.33	4–5	23
Pt	3	Fe	Pt : Fe molar ratio 1 : 1 to 1 : 5	CMK-9, AC, γ-Al_2O_3	Imp.	Ethylene glycol, glycerol, xylitol and sorbitol	10	230, 260	45	Tubular	1–6	90	24

Ni	5–20	—	—	ZSM-5 with Si/Al molar ratios of 12.5, 25, 100 and 180	Imp.	Phenol	0.1–0.5 M	200, 280	4, 7	Batch	4.7–23.5	3	25
Ni	65	—	—	SiO_2–Al_2O_3	Commercial catalyst	Glycerol	10, 20 and 40	200, 220, 240	1.6, 2.25, 3.35	Batch	4	4	26
Ni	15	Cu, Co, Fe	5, 5, 5	$LaAlO_3$	Cop.	Glycerol	15	250	2	Tubular	0.75	n.a.	27
Co	Co : Zn 1 : 3, 1 : 2, 1 : 1, 1 : 0.5	—	—	ZnO	Cop.	Ethylene glycol	5	225	2.58	Tubular	0.59	8	28
Rh, Ir	5, 5	Re	Re : M molar ratios from 0.2 to 2	TiO_2	Imp.	Ethanol	EtOH : H_2O molar ratio 1 : 10	200	2.00	Batch	0.5 g_{cat} : 1 mol_{EtOH}	10	29
Pd	5	—	—	Fe_3O_4	Cop.	Ethylene glycol	5	225	2.58	Tubular	3.5	130	30

Table 9.2 Aqueous phase reforming: catalyst characteristics, reaction conditions, carbon conversion, yield of H_2 and H_2/alkanes ratio for selected references.

Catalyst	S_{BET} (m^2/g^{-1})	Metal dispersion (%)	Metal particle size (nm)	Feedstock	T (°C)	P (MPa)	WHSV (h^{-1}) or organic feed to catalyst mass ratio	Time on stream or reaction time (h)	Total carbon conversion	Conversion to liquids	Conversion to gases	H_2 yield (%)	H_2/alkanes (mol/mol)	Ref.
Pt(0.79%)/MgO	42	—	—	Glycerol (5%)	225	2.76	0.18	25	—	—	48.0	28.4	9.2 : 1.0	8
Pt(2.77%)/γ-Al_2O_3	209	54	1.6	Xylitol (5%)	225	2.93	1.2	12	95	50	45	20	11.5 : 1.0	10
Pt(2.77%)/γ-Al_2O_3	209	54	1.6	Sorbitol (10%)	225	2.93	0.6	5	90	51	39	31	21.7 : 1.0	11
Pt(1.1%)Ce(4%) Zr(1%)/α-Al_2O_3	—	60	2.1	Glycerol	250	4.4	2.38	2	60	31	29	20	—	15
Pt(7%)/CMK-3	770	55.7	2	EG (10%)	250	4.4	2	25	—	—	75.8	56.7	2.7 : 1.0	16
$Pt_{0.5}Ce_{0.475}Zr_{0.475}O_2$	81	39	2.9	Glycerol (10%)	250	5.0	2.45	50	99.8	4.8	95.0	93	49.0 : 1.0	17
Pt(3%)–Re(3%)/ CMK-3	1035	61.7	1.8	EG (10%)	250	4.4	2	24	89.3	—	—	55	6.6 : 1.0	22
Pt(2.95%)– Fe(2.95%)/ CMK-9	1633	60.9	2.2	EG (10%)	250	4.4	2	48	68.4	—	—	71.1	26.2 : 1.0	24
Co/ZnO (Co : Zn 0.5 : 1.0)	12	—	23	EG (5%)	225	2.58	0.59	8	6.8	1.6	5.2	—	—	28
Pd(5%)/Fe_3O_4	190	—	—	EG (5%)	225	2.58	3.5	6	100.0	0.4	99.6	—	702.0 : 1.0	30

of glycerol to hydroxyacetone, together with the acidic sites on the surface of the catalyst. The influence of the support basicity on the performance of Pt catalysts was studied by Guo[8] using a series of catalysts prepared by impregnation with colloidal dispersions of Pt clusters. It was established that the activity of a 0.75 wt% Pt catalyst decreased when the basicity of the support was lower, in the order: Pt/MgO > Pt/Al_2O_3 > Pt/CeO_2 > Pt/TiO_2 > Pt/SiO_2. This was attributed to the concentration of basic sites on the surface of the material, which facilitated WGS and enhanced hydrogen yield. However, all the catalysts suffered significant deactivation after 20 h on stream. Barbelli[15] used α-Al_2O_3 modified with CeO_2 and ZrO_2 as the support for a Pt catalyst prepared by impregnation. The addition of 4 wt% CeO_2 and 1 wt% ZrO_2 to alumina resulted in high metal dispersion and small Pt particles, and better properties than Pt on unmodified α-Al_2O_3 or SiO_2. The small particle size promoted cleavage of C–C bonds in preference to cleavage of C–O bonds and hydrogenation reactions, which led to better activity and selectivity to H_2 formation. The redox properties of the modified support also favoured the WGS reaction. However, the overall selectivity to liquid products was still higher than that to hydrogen. Fast deactivation was observed by sintering after a few hours of reaction, although particle size, dispersion and activity were maintained when the catalyst was reused in three consecutive reaction batches. The use of ordered mesoporous carbon (CMK-3) as the support for a Pt catalyst prepared by impregnation was tested in the APR of ethylene glycol (EG).[16] The activity of a catalyst with 7 wt% Pt was more than twofold higher than that of a Pt/γ-Al_2O_3 catalyst with the same metal load and dispersion, and it had a lower selectivity to alkanes. The CMK-3 support was stable under APR conditions. The textural properties of the mesoporous CMK-3 were maintained after 25 h on stream. However, even if the rate of hydrogen production and the selectivity to the different products were maintained, metal dispersion decreased and Pt particle size increased significantly, which raises some concern about long-term stability.

Improved catalyst stability has been reported for methods of catalyst preparation other than impregnation. A $Pt_{0.05}Ce_xZr_{0.95-x}O_2$ catalyst, prepared by co-precipitation, was recently studied.[17] The activity of the catalyst was dependent on the Ce : Zr atomic ratio, which influenced the oxidation state of the Pt together with its dispersion, surface area and particle size. The ratio 1 : 1 gave the best results of those tested. The catalyst achieved 95% conversion of carbon to gases in the APR of 10 wt% glycerol, with a hydrogen yield of 93% and a selectivity of 98% at 250 °C, 5.0 MPa and a weight hourly space velocity (WHSV) of 2.45 h^{-1}. Conversion and yields were stable after 50 h on stream and there was no evidence of particle agglomeration in the spent catalyst, which had an average metal particle size and size distribution close to those of the fresh sample. An overlayer of ZnO to prevent sintering has been demonstrated to stabilize Pt particles on Al_2O_3 in catalysts prepared by atomic layer deposition (ALD) for the APR of 1-propanol.[18] A catalyst containing *ca.* 1 wt% Pt prepared by consecutive ALD of Pt and ZnO onto Al_2O_3 nanopowder produced higher selectivity to hydrogen and lower selectivity to alkanes, with propanal being the main liquid product. ZnO suppressed C–O bond cleavage and hydrogen consumption leading to the formation of propane. The size of the Pt nanoparticles was not modified substantially after 15 h on stream at 200 °C and 6.4 MPa.

Addition of several promoters to Pt/γ-Al_2O_3 catalysts has also been studied. El Doukkali *et al.*[19,20] added 10 wt% Ni to 3 wt% Pt on γ-Al_2O_3 catalysts and compared the materials obtained by impregnation with those obtained using a sol–gel procedure. Materials obtained by sol–gel contained well-dispersed active phases and were more resistant against sintering than their impregnation counterparts. As a result, they were more active in the APR of glycerol and had better selectivity to gas products, especially under moderate temperature/pressure conditions (230–250 °C and 3.0–5.0 MPa). However, during the reaction (up to 33 h on stream), a significant part of the γ-Al_2O_3 was structurally transformed into γ-AlOOH in the sol–gel catalysts, changing their surface area, porosity, surface acidity and metal dispersion, while the Ni particles were affected by re-oxidation. In the impregnation catalysts, the metallic Ni particles underwent major agglomeration. No leaching of metals or formation of graphitized carbon deposits was observed in either type of catalyst, but the structural changes the catalysts underwent caused continuous deactivation for the first 24 h. Re is another metal that has been tested as a promoter of Pt APR catalysts. In Pt–Re catalysts prepared by impregnation of activated carbon (AC), Re was partially oxidized under APR conditions, which generated surface acidity, thus changing the dominant reaction pathways with respect to the Pt/AC catalyst.[9] The increased acidity favoured C–O over C–C cleavage, which resulted in higher selectivity to liquid products and alkanes at the expense of hydrogen selectivity. Similar results were reported by Ciftci *et al.*,[21] who investigated Pt–Re/AC and Rh–Re/AC prepared by impregnation in the APR of glycerol. They concluded that alloying with Re added Brønsted acidity to the active surface and increased the formation of products *via* routes involving selective deoxygenation and dehydration reactions. It also increased the WGS activity of the catalyst, which resulted in an overall increase in CO_2 and H_2 yields. In a comprehensive study,[22] Pt–Re catalysts with a total load of 3 wt% metal and a Pt : Re atom ratio of 1 : 1 were supported on CMK-3, SiO_2, γ-Al_2O_3 and AC, and tested in the APR of EC, propylene glycol, glycerol and sorbitol. The reforming activity was dependent on the type of support: it increased in the order γ-Al_2O_3 < SiO_2 < AC < CMK-3 and correlated with the dispersion of the metal. Longer polyol chains tended to produce less conversion, a lower hydrogen yield and a wider variety of liquid products. Carbon-based catalysts (AC and CMK-3) were stable and maintained their textural properties after 25 h on stream. In contrast, the γ-Al_2O_3 and especially the SiO_2 catalyst lost surface area and porosity through sintering and phase transformation from high surface area metastable phases to low surface area phases. The use of Cu as a promoter was tested in Pt catalysts prepared by impregnation of Mg(Al)O mixed oxides with various ratios of Mg/Al (0.54 : 1 to 2.95 : 1), which were produced by calcination of commercial layered double hydroxides (LDHs).[23] The structure of the mixed oxides calcined at 400 °C reverted to the LDH structure under APR conditions, with Brønsted basicity of the catalyst developing and increasing selectivity to hydrogen over methane. LDHs calcined at higher temperatures (600 °C and 900 °C) did not fully revert to the LDH structure after 5 h of reaction, and consequently, metal dispersion and H_2 yield were lower, and the catalyst was less stable. The Mg/Al ratio also influenced H_2 selectivity, which increased when more basic supports were used (higher Mg/Al ratios). The addition of Cu as a promoter at loads from 0.2% to 1% increased H_2 selectivity and suppressed CH_4 formation, which was attributed to the metal–metal interaction between Pt and Cu. Addition of Fe to the Pt in catalysts prepared by impregnation of CMK-9, AC and γ-Al_2O_3 and used in the APR of EG, glycerol, xylitol and sorbitol was reported by Kim and co-workers.[24] CMK-9 resulted in a more active catalyst than AC or commercial γ-Al_2O_3. The catalyst supported on CMK-9 with a Pt : Fe ratio of 1 : 3 yielded the best results for EG conversion. The Fe did not participate in C–C bond breaking but acted as a promoter favouring the WGS reaction and consequently improving the H_2 yield. The rate of H_2 formation remained stable for 90 h of operation at 250 °C and 4.5 MPa using 10 wt% EG at a WHSV of 2.0 h^{-1}. The activity of the catalysts was found to decrease with the molar mass of the polyol, as observed in other catalytic systems.[22] The H_2 selectivity was equivalent to that of alkanes, but no detailed information on their composition or the formation of liquid products was given.

Nickel is also active in APR, although higher metal loads are required to obtain reforming rates close to those of Pt-based catalysts. Because of the capacity of Ni to catalyse C–C breaking reactions, it is probably a more suitable active metal to convert biomass-derived fractions other than carbohydrates, such as lignin, aromatics present in pyrolysis oil, or fatty acids from lipids, which tend to be less reactive because of the scarcity of oxygenated functional groups. Ni/ZSM-5 catalysts prepared by impregnation were tested in the APR of phenol,[25] which is commonly used as a model compound for lignin and the aromatic fraction of bio-oil. An optimization study established that a Si/Al ratio of 25 : 1 yielded optimal support properties. Higher ratios resulted in an excessively acidic surface, while lower ratios hindered Ni dispersion. In fact, selectivity to hydrocarbons was dominant at a Si/Al ratio of 12.5 : 1, and little hydrogen was produced. The optimal Ni load was 16%; 90% phenol conversion and 80% selectivity to hydrogen were obtained with this catalyst at 240 °C and 4.9 MPa. However, no information was provided on the structural changes the catalyst may have suffered during reaction, or on its long-term activity. A commercial high-load Ni catalyst (65 wt%) supported on SiO_2–Al_2O_3 was tested in the APR of concentrated glycerol solutions (10, 20 and 40 wt%) in a batch reactor system.[26] CO_2 was the main gaseous product, accompanied by similar amounts of CH_4 and H_2 at all temperatures (200 °C to 240 °C). The liquid products consisted of alcohols (methanol and ethanol), acetol, and ethylene and propylene glycols, which were formed by partial hydrogenation of glycerol with the H_2 produced by hydrothermal reforming. Propylene glycol was the main liquid product with yields well above those of the gaseous products. No characterization of the spent catalyst or an evaluation of its long-term stability was given. Addition of Cu, Co and Fe as promoters to $Ni/LaAlO_3$ catalysts prepared by co-precipitation was studied in the APR of glycerol at 250 °C, 2 MPa and a WHSV of 0.75 h^{-1}.[27] Cu–Ni gave the best selectivity to H_2, with little CH_4 or CO formed, which correlated with the small metal particles and high Ni dispersion in this catalyst. However, this catalyst also accumulated the largest amount of whisker-type carbon deposits. No information was given on the long-term stability and activity of the catalyst, or on the selectivity to and yields of liquid products.

Other metals have been tested in the preparation of APR catalysts. Co-precipitated Co/ZnO catalysts with Co : Zn ratios of 1 : 3 to 1 : 0.5 were tested in the APR of 5 wt% EC at 225 °C and 2.58 MPa, in a tubular reactor, at a WHSV of 0.59 h^{-1} for 8 h on stream.[28] Activity and selectivity to H_2 increased with the ZnO content, but total conversion was below 10 wt% in the best case. Maximum selectivity to H_2 was obtained with the Co : Zn 1 : 3 catalyst (89%), but conversion to gases was only 5.2 wt%. The main liquid products were methanol, ethanol and acetic acid, but conversion to liquids was below 1.6%. The poor performance of the catalyst in terms of conversion and hydrogen productivity was attributed to the low dispersion of Co. Re-promoted Rh and Ir catalysts supported on TiO_2 by impregnation, with atomic ratios Re/M from 0.2 : 1 to 2 : 1, were tested in the APR of ethanol at 200 °C and 2.0 MPa in a batch reactor system.[29] Addition of Re increased the rates of gas production (H_2, CO_2 and CH_4) by a factor of 3 to 4, but also promoted the formation of acetaldehyde and acetic acid as the main liquid products. The transformation of acetaldehyde to acetic acid proceeded on partially oxidized ReO_x, whereas well-dispersed bimetallic composites acted as active sites for ethanol dehydrogenation and acetaldehyde decomposition to gases. $Ir–Re/TiO_2$ was less active than $Ru–Re/TiO_2$ but produced a gas with higher H_2/CH_4 and H_2/CO_2 ratios and more acetic acid, thus showing that it was less active in acetaldehyde decomposition. A co-precipitated Pd/Fe_3O_4 catalyst was tested in EG APR at 225 °C, 2.58 MPa and a WHSV of 3.5 h^{-1} in a tubular reactor.[30] Complete conversion of EG (5 wt%) was achieved over 50 h, with a selectivity to gases of 99.6%. H_2 selectivity was 94% and the rest of the gas was CO_2 and a very small amount of CH_4 (the $H_2 : CH_4$ molar ratio was 702 : 1). The fraction of liquid products consisted almost entirely of methanol and ethanol. After 130 h on stream, the catalyst activity stabilized at 80% of its initial value while still maintaining selectivity to hydrogen gas.

9.2.2 Catalytic Supercritical Water Gasification (SCWG)

The use of water above its critical point as the reaction medium was developed for the destruction of highly hazardous liquid waste in the so-called supercritical water oxidation process (SCWO).[31,32] The reason why water is an appealing medium is the dramatic change in physical properties that it exhibits at its critical point. For instance, at 500 °C and 25 MPa—common conditions in supercritical water (SCW) reactions—the ion product of water is only 3.16×10^{-25} mol^2 L^{-2} and its static dielectric constant drops from 78.4 at room conditions to only 1.57. The low dielectric constant means that SCW behaves similarly to an organic solvent. Organic compounds and gases become miscible with water above its critical point, thus suppressing the mass transport limitations that are inherent in reacting systems involving aqueous, organic and gaseous phases below the critical temperature of water. The use of SCW in hydrolysis and depolymerization reactions or the reforming of organic molecules became of interest after SCWO. However, the use of SCW poses two important challenges. SCW is a very aggressive medium that attacks most catalyst supports and active phases, thus causing severe deactivation by sintering, phase transformation and in some cases leaching of the active metal. Meanwhile, the low ion product and dielectric constant mean that inorganic salts are insoluble in SCW. This is especially important in the processing of biomass and its fractions since the naturally occurring inorganic species present in biomass will tend to precipitate under supercritical conditions, forming deposits on the walls of the reactor and deactivating catalysts by fouling.[33]

Catalytic SCWG of biomass was first reported by Elliot and co-workers, who demonstrated the capacity of catalysts based on Ni, Ru or Rh to convert wet biomass into a gas product rich in CH_4.[34] Those metals, together with AC, have been widely used in catalytic SCWG studies of different biomass streams.[35,36] Recent work, summarized in Tables 9.3 and 9.4, has focused on different support materials, methods of preparation, and the use of promoters to increase stability in SCW. In a screening study, Zhang and co-workers prepared catalysts by impregnation of γ-Al_2O_3 with Ni, Ru, Cu and Co.[37] The effect of Na, K, Mg and Ru addition to Ni/γ-Al_2O_3 was also tested. It was concluded that Ni and Ru supported on γ-Al_2O_3 were the most active metals in H_2 production, while Cu and Co were less effective at gasifying glucose in SCWG. Ni was only active in a reduced form, whereas Ru was active in both metallic and oxidized (RuO_2) forms. Binary metallic catalysts (NiRu, NiCo, NiCu and RuCu) supported on γ-Al_2O_3 had activities and selectivities that resulted from the additive effects of the component metals. Addition of promoters (K, Ca, Mg or Ru) to the Ni/γ-Al_2O_3 catalyst did not suppress the formation of tar and coke. In fact, tar and coke formation was a major problem for the promoted catalysts since their accumulation tended to block the catalyst bed at a WHSV of 6 h^{-1}, and meant the reactor had to be shut down before the scheduled time-on-stream (8 h). A comparison between γ-Al_2O_3, ZrO_2 and AC as supports for Ni catalyst was briefly assessed, showing that the activity of a catalyst containing 10 wt% Ni decreased in the order: γ-Al_2O_3 > ZrO_2 > AC. Concerning stability, γ-Al_2O_3 was found to be partially transformed to α-Al_2O_3 in the Ru/γ-Al_2O_3 catalyst. This change was not observed in the other catalysts, probably due to the reduced time-on-stream used in all the experiments. The limited performance of Co was also reported by Pairojpiriyakul and co-workers,[38] who impregnated Co into different support materials (ZrO_2, YSZ, La_2O_3, γ-Al_2O_3 and α-Al_2O_3) to prepare catalysts for the SCWG of glycerol. Catalysts containing 10 wt% Co were tested for 4 h in a flow reactor at 450–575 °C,

Table 9.3 Supercritical water gasification: catalytic systems, methods of preparation and conditions for testing catalytic activity.

Metal	Load (wt%)	Promoters	Load (wt%)	Supports	Preparation	Feedstock	Load (wt%)	T (°C)	P (MPa)	Reactor type	WHSV (h^{-1}) or organic feed to catalyst mass ratio	Time on stream or reaction time (h)	Ref.
Ni, Ru, Cu, Co	Depending on metal	Na, K, Mg, Ru	Depending on promoter	γ-Al_2O_3, ZrO_2, AC	Imp.	Glucose	5	600	24	Tubular	3–6	Up to 8	37
Co	10	—	—	ZrO_2, YSZ, L_2O_3, γ-Al_2O_3 and α-Al_2O_3	Imp.	Glycerol	5	450–575	25.0	Tubular	3.15–10.8	4	38
—	—	Ce	1–10	γ-Al_2O_3	Imp.	—	—	500, 600, 700	24.6	Batch	—	3	39
Ni	20	—	—	γ-Al_2O_3, CeO_2–γ-Al_2O_3 (5 wt% Ce)	Imp.	Glucose	9.1	400	24.5	Batch	5	20 min (+40 min heating)	40
Ni	10	—	—	CeO_2–γ-Al_2O_3 (5 wt% Ce), La_2O_3–γ-Al_2O_3 (5 wt% La), MgO–γ-Al_2O_3 (5 wt% Mg), ZrO_2–γ-Al_2O_3 (5 wt% Zr)	Imp.	Glucose	10	400	23.5	Batch	10	20 min (+90 min heating)	41

Ni	25–40	—	—	Mg–Al oxides	Cop.	Glucose	9.1	400	22.5, 25.0	Batch	5	20 min (+40 min heating)	42
Ni	5–25, 20	—	—	ZrO_2, MgO–ZrO_2(Mg : Zr, 1 : 9 to 6 : 9)	Imp.	Glycerol	5–45	600, 700, 800	24.5	Tubular	72–648	60	43
Ni	4.6–25.3	—	—	Zr(Ce,Y)$O_{2-\delta}$	Cop.	Glucose	10	500	23–24	Batch	1	30 min	44
Ni	0–5	Co	0–6	Mg–Al	Imp, Cop.	Lignin	16.7	650	26	Batch	1	n.a.	45
Ni	12	Co	8	AC	Polyol reduction	Phenol	5	475–525	n.a.	Batch	2	1	46
Ni	10–20	Ru	2	γ-Al_2O_3	Imp.	Glucose	4.5	500	27.5	Tubular	3	6 to 20	47
Ni	20	Ru	2	γ-Al_2O_3	Imp.	Aqueous fraction bio-oil	0.7–2.98C	400, 500, 600	27.5	Tubular	3	6	48
Ni	11	Ru	0.5	Al_2O_3	Sol–gel	Glucose	4.5	500	25.0	Batch	4.5	2.5	49

Table 9.4 Supercritical water gasification: catalyst characteristics, reaction conditions, carbon conversion, yield of H_2 and H_2/alkanes ratio for selected references.

Catalyst	S_{BET} (m^2 g^{-1})	Metal dispersion (%)	Metal particle size (nm)	Feedstock	T (°C)	P (MPa)	WHSV (h^{-1}) or organic feed to catalyst mass ratio	Time on stream or reaction time (h)	Total carbon conversion	Conversion to liquids	Conversion to gases	H_2 yield (%)	H_2/alkanes (mol/mol)	Ref.
Ni(10%)/γ-Al_2O_3	41.9	0.7	145	Glucose (5%)	600	24	3	8	99.7	—	—	58.3	5.9 : 1.0	37
Co(10%)/YSZ	9.6	—	—	Glycerol (5%)	500	25	6.45	3.83	94	17	77	53.1	23.3 : 1.0	38
Ni(28.6) Mg(7.1) Al(24.8) O(39.5)	132.8	—	9.39	Glucose (9.09%)	400	22.5	5	0.33	71	—	48.8	25	—	42
Ni(20%)/MgO–ZrO_2 (Mg : Zn atom ratio 0.43 : 1.0)	64	—	—	Glycerol (5%)	700	24.5	72	—	>99	—	—	50.3	6.5 : 1.0	43
$Ni_{0.5}/Zr_{0.4}Ce_{0.4}Y_{0.2}O_{2-\delta}$	15.68	—	—	Glucose (10%)	500	23	1	0.5	—	—	76	35	2.6 : 1.0	44
Ni(20%)–Ru(2%)/γ-Al_2O_3	126.5	3.3	30.4	Glucose (5%)	500	27.5	3	6	100	0	100	—	0.54 : 1.0	47
Ni(20%)–Ru(2%)/γ-Al_2O_3	126.5	—	—	Aq. bio-oil (2.98 wt% C)	700	27.5	3	6	89	—	—	1.18 mol mol_C^{-1}	6.9 : 1.0	48
Ni(10%)–Ru(.5%)/γ-Al_2O_3	316	3.78	9.5	Glucose (4.59%)	500	25	4.5	2.5	—	—	96.3	40	4.1 : 1.0	49

25.0 MPa and a WHSV of 3.15 to 10.8 h^{-1}. The deposition of carbonaceous solids on the catalysts supported on alumina led to bed plugging and the build-up of pressure, even after short reaction times. Even if glycerol conversion was high (especially on Co/γ-Al_2O_3), these catalysts favoured the formation of acetaldehyde and other liquid products, and they exhibited low gasification activity. CO was the main gaseous product. Co/YSZ was the most active and selective to gas products. However, selectivity tended to shift towards the formation of less gas and more liquid products with time-on-stream, which suggests Co-based catalysts are of limited suitability for SCWG.

Most recent work on the SCWG of biomass-derived molecules has chosen Ni as the active metal, either alone or with the addition of Co or Ru as promoters. The stability of the support materials has also been a major concern. Byrd and Gupta[39] studied the stabilization of γ-Al_2O_3 with Ce to obtain catalyst supports that maintain their physical integrity when exposed to SCW. Ceria was chosen because it also acts as a promoter, adding WGS capacity to the catalyst, and it improves the dispersion of supported metals. Addition of 1–10 wt% Ce did slow down the transformation of the γ-phase of alumina to the κ- and α-phases, and it also reduced the loss of porosity and surface area at 24.6 MPa. A Ce load below 2 wt% at 500 °C to 600 °C maintained the highest surface areas after 3 h of exposure to SCW; while at 700 °C, all loads retained around 30–40 m^2 g^{-1}. However, long-term stability of the modified materials under SCW was not tested. The effect of promoting a Ni/γ-Al_2O_3 catalyst (20 wt% Ni) with 5 wt% Ce for the SCWG of glucose at 400 °C and 24.5 MPa in a batch reactor[40] revealed that Ce addition increases the production of H_2 and CO_2. Carbon deposition and coking led to the deactivation of both catalysts, even though Ce in the Ni/CeO_2–γ-Al_2O_3 catalyst inhibited carbon deposition. The decrease in surface area was similar with or without Ce addition. Besides CeO_2–γ-Al_2O_3, other mixed oxides such as La_2O_3–γ-Al_2O_3, MgO–γ-Al_2O_3 and ZrO_2–γ-Al_2O_3 were used as supports for Ni catalysts (10 wt% Ni and 5 wt% Ce, La, Mg or Zr) and tested in the SCWG of glucose at 400 °C and 23.5 MPa.[41] All the catalysts suffered deactivation by sintering of the Ni particles and changes in the support structures when exposed to SCW. Carbon was deposited on the catalysts during reaction, which also contributed to deactivation. The catalyst containing zirconia gave the lowest yield and selectivity to H_2, and it had the largest amount of carbon deposited on the surface, *ca.* 2.55 gC/g_{cat} after one reaction cycle (90 min heating + 20 min at 400 °C), which was graphitic in type. Graphitic carbon deposition on the catalysts containing Mg and La was also important, but these gave better yields of hydrogen gas. The catalyst containing Ce underwent the lowest deposition of carbon (0.86 gC/g_{cat} mainly amorphous) and showed the highest yield and selectivity to H_2. This was attributed to the capacity of CeO_2 to store oxygen that can then gasify carbon deposits. However, surface area and porosity were reduced to half their values for the fresh catalyst, proving that Ce addition did not fully prevent the structural changes to the support in the SCW environment. Ni was supported on Mg–Al oxides with different Mg/Al ratios by co-precipitation and tested in the SCWG of glucose at 400 °C.[42]

Ni/MgO gave the lowest yield of H_2. MgO was not a suitable support due to the difficulty of reducing Ni (which formed $MgNiO_2$) and its transformation into $Mg(OH)_2$ under supercritical conditions. Addition of a small amount of Mg to form Mg–Al enhanced the stability of the Ni catalyst and inhibited the formation of graphitic carbon at the acidic alumina sites, even though some sintering of the Ni particles and partial catalyst deactivation were observed. The catalyst $NiMg_{0.6}Al_{1.9}$ exhibited the best hydrothermal stability and catalytic activity for hydrogen gas formation, attributed to the formation of $MgAl_2O_4$ in the support. This effect of Mg was not reported for MgO–γ-Al_2O_3 catalysts prepared by impregnation,[41] which suggests that co-precipitation provides support materials and catalytic systems with better overall properties than impregnation.

Zirconia has often been used as an alternative support to alumina in the preparation of SCWG catalysts. The addition of MgO was studied to prepare Ni/MgO–ZrO_2 catalysts by impregnation, using Mg : Zr mass ratios of 1 : 9 to 6 : 9.[43] The resulting catalysts were tested in the SCWG of glycerol at 24.5 MPa and temperatures from 600 °C to 800 °C in a tubular reactor (at a WHSV of 72 to 648 h^{-1}), with a time-on-stream of up to 60 h in some experiments. Complete conversion of glycerol and high selectivity to hydrogen were obtained with the Ni/ZrO_2–MgO catalysts over short periods of time, even at a glycerol concentration of 45 wt%. The presence of MgO shifted the composition of the gas products to hydrogen instead of methane. However, at low WHSVs (*i.e.*, higher contact times) less H_2 and more CH_4 were produced, which suggests the reaction initially proceeded through dehydrogenation and reforming, and then the products tended to equilibrium by WGS and methanation (little CO was reported). The catalyst was almost completely deactivated after 60 h on stream, even with the more diluted 5 wt% glycerol solution. Yttrium-stabilized zirconia (YSZ), with and without Ce addition, was used to prepare Ni-based catalysts by impregnation at several Ni/Zr and Zr/Ce ratios.[44] The catalysts were tested in the SCWG of 10 wt% glucose in a batch reactor at 500 °C and 23–24 MPa, using a glucose-to-catalyst mass ratio of 1 : 1 and a reaction time of 30 min. Addition of Ce improved carbon gasification efficiency and selectivity to methane, but consequently decreased selectivity to hydrogen. Overall, the catalysts had good structural stability. No significant changes in the crystalline phases were observed after reaction, even though surface area increased due to SCW weakening the agglomeration of catalyst particles. No carbon deposition was observed on most catalysts, particularly those with a high Ni load and Ce addition. The optimum material was $Ni_{0.5}/Zr_{0.4}Ce_{0.4}Y_{0.2}O_{2-\delta}$, which gave a carbon gasification efficiency of 76% and very similar selectivity to H_2 and CH_4 (*ca.* 35%).

Successful results concerning activity and catalyst stability during the SCWG of resilient biomass have been reported with some bimetallic Ni-based catalysts. The performance of Ni–Co catalysts supported on Mg–Al prepared by impregnation and co-precipitation was studied in the SCWG of lignin at 26 MPa and 650 °C, using a lignin-to-catalyst mass ratio of 1 : 1.[45] Catalysts with Ni–Co were more active and selective than single-metal catalysts with similar loads of active metal. Also, catalysts prepared by co-precipitation were more active than their respective impregnation counterparts. A catalyst containing relative metal mass percentages of Ni/Co/Mg/Al of 2/4/68/26 was the most active in the production of hydrogen and yielded a moderate amount of methane (a H_2 to CH_4 ratio of 1.72 : 1.0). The same catalyst was capable of gasifying cellulose and real biomasses (wheat straw, timothy grass and canola meal), with gas yields of between *ca.* 15 and 20 wt%. However, little information is given on carbon gasification efficiencies and the selectivity to liquid products. Structural changes in the catalysts and their reusability were not reported. Bimetallic Ni–Co catalysts supported on AC prepared by a polyol reduction method were tested in the SCWG of phenol at 475–525 °C.[46] Again, the bimetallic catalyst (12 wt% Ni and 8 wt% Co) was more active in the gasification of phenol than the monometallic Ni or Co catalysts, but it was a little less active than a commercial 5% Ru/AC catalyst. The greater activity of the bimetallic catalyst was attributed to better metal dispersion and smaller metal particles compared to the monometallic catalysts. Carbon gasification efficiencies above 90% were achieved for a 5 wt% phenol solution with no tar formation detected after 1 h of reaction.

Ni–Ru bimetallic catalysts supported on γ-Al_2O_3 were tested in the gasification of glucose in SCW at 500 °C and 27.5 MPa in a tubular reactor at a WHSV of 3 h^{-1}.[47] Catalysts containing Ni and 2 wt% Ru were synthesized by impregnation. When the content of Ni was 10%, the carbon gasification efficiency was only 63%. Increasing the Ni load to 20% achieved 98% gasification efficiency. The catalyst was active for 20 h without any noticeable deactivation, and no formation of char deposits or tar was reported. It was suggested that the main effect of Ru in the supported Ni catalyst was to decrease the reduction temperature of nickel oxide followed by the formation of a Ni_xRu_{1-x} solid solution, which also improved metal dispersion, and also to act as a stabilizing agent preventing leaching of Ni and Al. The catalyst with 20 wt% Ni and 2 wt% Ru was tested in the SCWG of the aqueous fraction of bio-oil (pyrolysis oil) at 500–700 °C, 27.5 MPa, a WHSV of 3 h^{-1}, and feed concentrations of 0.7 and 2.98 wt% total carbon with a time-on-stream of 6 h.[48] The catalyst converted the diluted bio-oil (0.7 wt% C) with a carbon gasification efficiency of over 90% at 600 °C without tar formation; but the tar yield was 3.5 wt% at 500 °C. For concentrated bio-oil (2.98 wt% C), the gasification efficiency was only 19% at 600 °C, although tar was not formed. The temperature of 700 °C was needed to achieve 89% gasification efficiency, showing that lower WHSVs should be used at higher concentrations of the bio-oil feed. Some char was deposited on the catalysts under all conditions. Higher temperatures favoured H_2 over CH_4. The catalyst appeared to be stable for the 6 h of the experiment, since no evidence of active metal leaching was observed; but performance and stability in longer experiments was not evaluated. In an attempt to increase activity, the sol–gel synthesis of Ni–Ru/Al_2O_3 catalysts using two different templates (Pluronic P-123 and CTAB) was recently reported by Hossain and co-workers.[49] Catalysts with 11 wt% Ni and 0.5 wt% Ru were synthesized and compared to a Ni–Cu/γ-Al_2O_3 catalyst with the same metal load prepared by the conventional incipient wetness impregnation method. The templated catalysts had larger surface areas and greater metal dispersion, as well as smaller active metal particle diameters. The metal particles in the sol–gel catalysts exhibited stronger interactions with alumina than in the impregnated catalyst, which prevented the aggregation of metal particles during reaction. The catalysts were tested in the SCWG of glucose at 500 °C and 25.0 MPa in a batch reactor. The CTAB-derived catalyst was slightly more active than the one prepared with P-123, but both achieved high hydrogen yields and produced less CO than the impregnated catalysts. Carbon gasification efficiencies were over 95% after 2.5 h of reaction, with total organic carbon (TOC) destruction close to 100%. Detailed analysis of the liquid phase only showed the presence of minor amounts of intermediate compounds (furfural, phenol and toluene) after 0.5 h of reaction. Minor amounts of amorphous carbon were detected on the spent catalysts, but no graphitic carbon was present.

Besides hydrogen production, attempts have been made to adjust SCWG conditions for the production of methane-rich CH_4/H_2 mixtures that could be used to obtain biohythane. The gasification of microalgae at 430 °C using Pt/C, Ru/C, Pd/C and Rh/C catalysts has been reported to increase the H_2 and CH_4 yields by 2–3 and 3–9 times respectively, compared with the non-catalytic reaction.[50] Catalytic SCWG of a fuel derived from municipal solid waste using a RuO_2/γ-Al_2O_3 catalyst in the presence of NaOH at 500 °C, resulted in high yields of H_2 and CH_4.[51] Enhanced CH_4 and H_2 production has been reported for the catalytic SCWG of wood sawdust using a pre-processing step in subcritical water in the presence of alkaline and solid acid compounds at 8 MPa and 280 °C. Catalytic SCWG of the pre-treated solid with a Ru/Al_2O_3 catalyst at 500 °C and 30 MPa increased H_2 and CH_4 yields by 51% and 61%, respectively, when compared with the combined yield of the two gases from the direct gasification of the original sawdust under identical reaction conditions.[52] Several Ni_xRu_y/γ-Al_2O_3 catalysts were tested in the SCWG of glucose, in attempts to achieve complete conversion of all the carbon in glucose at 500 °C and 27.5 MPa. As a function of the composition of the catalyst, increased CH_4 and/or H_2 production was observed; with a Ni(30%wt)Ru(25%wt)/γ-Al_2O_3 catalyst, production of a CH_4/H_2 mixture with a molar ratio of *ca.* 1 : 1 was reported with a 0.9 mol/mol carbon gasification efficiency.[47]

9.3 Biomethane and Biosynthetic Natural Gas (Bio-SNG)

In this section of the chapter, we will consider the production of CH_4 from different biochemical and thermochemical techniques to process biomass (Figure 9.2). Besides the CH_4 obtained from biogas production, CH_4 is associated with different biorefinery processes. We will refer to the CH_4 directly obtained *via* these routes as biomethane. Moreover, other carbon-containing compounds, namely carbon oxides, CO and CO_2 (CO_x), either from the routes

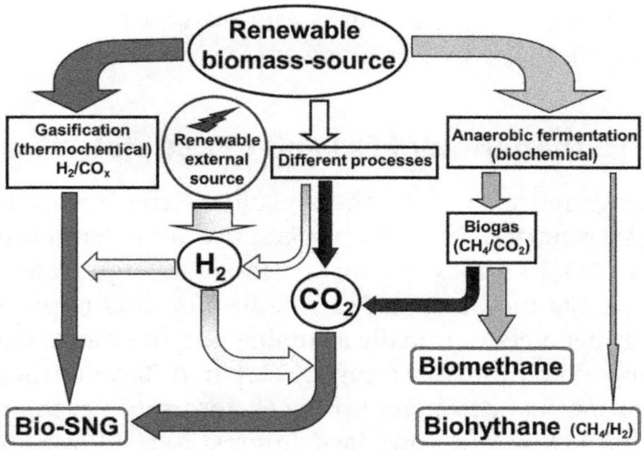

Figure 9.2 Main routes to biomethane and biosynthetic natural gas within biorefinery activities.

that produce biomethane or from other sources that are present in the activities of a biorefinery, and in general terms from the processing of biomass, can be considered as further sources of CH_4 *via* the methanation reaction. In this case, the major route to CH_4 is *via* its synthesis by catalytic hydrogenation of the CO_x feedstock. We will refer to this synthesized CH_4 as Bio-SNG, which is distinguished from the general term of SNG, for which the CO_x source can also have a fossil origin, such as coal. To obtain Bio-SNG we must consider the necessary H_2 source, which can ideally be considered to result from the activities of the biorefinery, as already indicated in the first part of this chapter. In fact, biomass gasification leads to the simultaneous production of H_2/CO_x mixtures; and some specific transformation processes applied to specific biomass, *e.g.*, algae, yield the simultaneous production of CH_4 and H_2 (biohythane) (Figure 9.2). However, we should also consider the potential application of external H_2 derived from water electrolysis using renewable electricity. More specifically, the occasional excess energy derived from inconstant sources, *e.g.*, wind-derived electricity. In fact, this is associated with the Power-to-Gas concept, according to which, from a broad perspective, electrical energy is converted to gas: either H_2 or after hydrogenation of CO_x, CH_4 (Figure 9.2).

Via this route, either Bio-SNG or SNG could be produced, which can easily be introduced into the NG network and used as fuel in combustion engines. At this point, it is necessary to recall that despite the fact that CO_x hydrogenation incurs an additional cost, besides the advantages of the volumetric energy density of CH_4 (40.0 MJ m^{-3}) compared to that of H_2 (12.7 MJ m^{-3}), storage of the synthesized CH_4 is simpler and cheaper than that of the corresponding H_2 within existing CH_4 network facilities.

Analysis of the strategies and applications of the Power-to-Gas concept is beyond the scope of this chapter. However, we direct the reader to recently published reports and reviews that include technological and economic analysis, environmental implications and, of course, are related to SNG, Bio-SNG and biomethane in different ways.[53–59]

9.3.1 The Methanation of CO and CO_2: Catalysts and Reactors

Catalytic hydrogenation of CO_2, the so-called Sabatier reaction, is a well-known process reported in 1902 by Sabatier and Senderens.[60] However, methanation of CO_2 has not yet been developed at large scale, and only a few plants based on CO methanation can be found. This is probably because, to date, NG has been economically available and the methanation reactions have been considered more of a clean-up step in different processes, as in the synthesis of ammonia or to clean-up the H_2 for proton exchange membrane (PEM) fuel cells.[61] However, nowadays, interest in the Power-to-Gas concept for energy-efficient storage of renewable electricity and to facilitate direct use of the SNG produced has increasingly driven efforts to develop efficient production of CH_4 from H_2 and CO_2. Commercial catalysts for methanation (regardless of whether CO or CO_2 is used as a reagent) are available, and at the laboratory level, great efforts are being made to optimize methanation catalysts, as we discuss below. However, as we will also see later, reactor design faces a great challenge in the form of the necessary control of the temperature because of the exothermic reactions involved, meaning that efficient heat removal is highly desirable throughout the process. Moreover, considering that within the Power-to-Gas context it is necessary to contemplate the possibility of working under a non-constant production regime, the methanation reactor should preferably be designed for fast start-up and shutdown.

In what follows, we present a brief description of two main points: advances in catalysts and in reactors, based on the characteristics of the process.

The CO_2 hydrogenation reaction requires 4 moles of H_2 per mole of CO_2, a higher ratio than for the conversion of CO_2 to formic acid, methanol, or Fischer–Tropsch products. The Sabatier reaction is an exothermic reaction (9.11):

$$CO_{2(g)} + 4H_{2(g)} \rightarrow CH_{4(g)} + 2H_2O_{(g)},\ \Delta H^0{}_{298K} = -164.9\ \text{kJ mol}^{-1} \quad (9.11)$$

which can also be regarded as the reverse of the industrially well-developed methane steam reforming reaction (MSR). In fact, the Sabatier process can be regarded as a combination of an endothermic reaction, the reverse WGS (RWGS) (9.12), and the highly exothermic CO methanation reaction (9.13); the combination results in an exothermic process. That means that during the process, a part of the calorific value of the reagents is released as heat. Thus, if the heat released during the process is not used, the process is energetically very inefficient.

$$CO_{2(g)} + H_{2(g)} \leftrightarrow CO_{(g)} + H_2O_{(g)},\ \Delta H^0{}_{298K} = +41.2\ \text{kJ mol}^{-1} \quad (9.12)$$

$$CO_{(g)} + 3H_{2(g)} \leftrightarrow CH_{4(g)} + H_2O_{(g)},\ \Delta H^0{}_{298K} = -206.4\ kJ\ mol^{-1} \qquad (9.13)$$

From a thermodynamic point of view, the methanation reaction of CO_2 (9.11) is a highly favourable process ($\Delta G^0{}_{298K} = -113.6$ kJ mol^{-1}). This indicates that the chemical equilibrium can be displaced far towards the products and nearly complete conversion of CO_2 to CH_4 is theoretically possible. The equilibrium shift towards the products is favoured at low temperature and by an increase in pressure. At 0.1 MPa, the process is thermodynamically favoured only up to *ca.* 500 °C. When the temperature exceeds about 550 °C, the RWGS reaction (9.12) dominates. Higher temperatures, *e.g.*, 800 °C, can be used at higher pressures, *e.g.*, 3 MPa to favour CO_2 methanation. Moreover, in order to obtain a high CH_4 yield and avoid carbon deposition, the H_2/CO_2 ratio should not be lower than 4, even at 3 MPa.[62]

Meanwhile, the reduction of CO_2 to CH_4 is a process involving 8 electrons that has important kinetic restrictions and from a practical point of view, it requires a good balance between conversion, reaction rate and energy consumption. Therefore, an industrially acceptable catalyst that can provide the necessary reaction rate and selectivity at as low a temperature as possible is required. Moreover, for such purposes, small-volume reactors and appropriate heat exchange design are also necessary.

Catalysts for the hydrogenation of CO_x have been the subject of many studies in the past.[63,64] Catalysts that are active in the methanation of CO are in general also capable of hydrogenating CO_2. The most suitable catalysts for the methanation of CO_2 were also much studied in the last century[65–70] and they have been extensively reviewed in recent years.[71–73] After the report of using Ni by Sabatier and Senderens,[60] Fischer, Tropsch and Dilthey[74] established the decreasing methanation activity of different metals tested in the methanation of carbon oxides, in the order: Ru > Ir > Rh > Ni > Co > Os > Pt > Fe > Mo > Pd > Ag. Among different group 8–10 metals, extensive work by Vannice[75] on the basis of their specific activity resulted in the CO methanation order: Ru > Fe > Ni > Co > Rh > Pd > Pt > Ir. The temperature-conversion performance of noble metal-based catalysts for CO hydrogenation can be summarized as: Ru ≫ Rh ≫ Pt > Pd. However, the CH_4 selectivity order does not follow the same trend; Pd chiefly yields CH_4 while Ru produces the highest C_{2+} fraction. Nowadays, supported catalysts based on Ru, Rh, Pt, Ni, Fe and Co are the most studied for CO_2 hydrogenation. At low temperatures, noble metal-based catalysts show better activity than Ni;[76,77] a Rh/γ-Al_2O_3 catalyst has been reported to be active in the methanation of CO_2 at atmospheric pressure and temperatures of 25–150 °C.[78] However, the Ni-based systems are by far the more commonly used; and this is undoubtedly due to their relatively high activity, CH_4 selectivity, low price and availability.

The activity and selectivity of supported Ni catalysts are strongly influenced by the Ni loading, the size of the Ni metal particles and metal–support interactions.[72] The nature of the support and the method of preparation of the catalyst also play important roles in the activity and selectivity in the hydrogenation of CO_2. Various supports including TiO_2, SiO_2, Al_2O_3, CeO_2, ZrO_2 and zeolites, as well as the use of different mixed compositions, precursors, promoters and preparation methods, are the subject of continuing research.[79–87] For highly effective CO_2 methanation, the design of new catalysts is still necessary; a high space velocity is recommended to test the activity of catalysts far from thermodynamic equilibrium.[72]

While the methanation of CO_2 is a relatively simple reaction, the reaction mechanism has been studied in depth but remains difficult to determine because it appears to be strongly dependent on the characteristics of the catalyst.[65,69–73,86–88] Reviews of recent studies of the hydrogenation of CO_2 to methane have been published.[89–91] Mechanistic studies have led to two main proposals. The first mechanism involves conversion of CO_2 to CO before the methanation and then the reaction progresses to the methanation of CO through different key intermediates. The second proposal involves the methanation of CO_2 to CH_4 without the formation of CO as a reaction intermediate; in this case, CO_2 methanation occurs *via* surface carbonates and formate species, which are further hydrogenated to CH_4. Recently reported mechanistic studies of CO_2 methanation on supported Ni catalysts suggest that basic characteristics of the support control the catalytic activity.[81] In Ni/ceria–zirconia, H_2 would be dissociated at Ni^0 sites while CO_2 would be activated at the weak basic sites of the ceria–zirconia support to form carbonates.[92] This differs from Ni–silica catalysts on which both CO_2 and H_2 are activated on the Ni^0 particles. The effect of the precursor used in the preparation of Ni/CeO_2 for CO_2 methanation was recently elucidated.[93] On Ni/CeO_2, CO_2 hydrogenation proceeds *via* the adsorption of CO_2 onto the ceria surface and stepwise hydrogenation to methane through hydrogen carbonate and formate intermediates by the hydrogen that spills over from the Ni particles. With the catalyst prepared using a chloride precursor, this reaction pathway is not available due to the ceria surface being blocked by chlorine, thus inhibiting CO_2 methanation.[93] Recent work has focused on the mechanism of CO_2 methanation over noble metal-based catalysts. For Rh/γ-Al_2O_3,[94] it was found that CO_2 is not dissociatively adsorbed at the support and mainly forms hydrogen carbonate; the dissociative adsorption of H_2 proceeds on Rh. Then, the activated H_2 reacts mainly with the hydrogen carbonate species forming CO adsorbed on Rh and formate species stably adsorbed on the support. For Ru/CeO_2,[95] CO_2 methanation undergoes the formate route, and the formate dissociation to methanol catalysed by the oxygen vacancy of the support is the rate-determining step. In contrast, in Ru/α-Al_2O_3 catalysts with the absence of an oxygen vacancy, CO_2 methanation undergoes the CO route over the Ru surface.[95]

Focusing on Ni-based catalysts for Bio-SNG applications, we must take into account the stability of the catalytic systems to ensure good operation of the methanation process. Unfortunately, under usual reaction conditions, Ni-based catalysts present deactivation due to several effects: carbon deposition, particle sintering, carbonyl formation and poisoning by contaminants in the feedstock. The last factor will be considered deeply in the following sections when analyzing biomethane and Bio-SNG production from biomass processing in more detail (through biogas and gasification). Regarding the other factors, it is well known that the deactivation of supported metal catalysts by carbon formation is a serious problem in methanation processes. The carbon formation reactions:

$$2CO \leftrightarrow CO_2 + C, \Delta G = -172 \text{ kJ mol}^{-1} \quad (9.14)$$

$$CO + H_2 \leftrightarrow C + H_2O, \Delta G = -131.3 \text{ kJ mol}^{-1} \quad (9.15)$$

$$CO_2 + 2H_2 \leftrightarrow C + 2H_2O, \Delta G = -90.1 \text{ kJ mol}^{-1} \quad (9.16)$$

$$CH_4 \leftrightarrow 2H_2 + C, \Delta G = 74.9 \text{ kJ mol}^{-1} \quad (9.17)$$

indicate that carbon formation is less favoured in CO_2 hydrogenation than in that of CO. However, at high temperatures, carbon can form from methane (9.17). Catalyst deactivation caused by carbon deposits could be prevented through optimization of reaction parameters such as the temperature/pressure conditions and the H_2/CO_2 ratio, as well as by the addition of steam to the reactant mixture. Of course, another way to avoid the unwanted deactivation is by modifying the catalyst composition, for example by adding different promoters[96,97] that facilitate carbon gasification and thus provide more resistance against deposits.

Ni particle sintering during the highly exothermic CO_2 methanation process is another challenge for Ni catalysts. Moreover, water is formed in the methanation reaction, which could further accelerate the sintering process. Particle migration and coalescence rather than Ostwald ripening was found to govern the sintering of Ni particles.[98,99] Particle sintering can be controlled by increasing the metal–support interaction through new preparation routes, and by adding promoters, among other means.[96,96,100,101] Finally, depending on the reaction conditions, the volatile carbonyl $Ni(CO)_4$ can form from small Ni particles, leading to loss of the active phase.

When considering the catalytic reactor necessary to carry out CO_2 methanation, several factors must be taken into account. For concentrated CO_2 feedstocks, that is, with no-diluted CO_2 streams, the stoichiometric feed implies a mixture of 20% CO_2 v/v and 80% H_2 v/v. As stated above, the reaction (9.11) is highly exothermic and under adiabatic conditions, a high temperature could be reached. Assuming a gas hourly space velocity (GHSV) of 5000 h^{-1} and total CO_2 conversion, about 2 MW of heat per m^3 of catalyst bed needs to be removed.[53] So, the greatest challenge in the overall process is controlling the temperature of the reaction, which could also affect the stability of the catalyst; efficient heat removal is essential, which is directly related to the reactor design. Different types of reactor have been used: fixed-bed, fluidized-bed, three-phase and more recently, structured reactors.[53,102–105] Each of these systems presents a different approach to the management of the heat released. There are basically three options to eliminate the heat generated by the reaction: elimination between reactors in series, *in situ* heat removal in the reactor or a combination of the two.

In the case of eliminating heat between fixed-bed reactors, the reactors are adiabatic: they are thermally insulated and do not exchange heat with the environment. In the first reactor, the concentration of reagents is the highest, which generates a high temperature in the reactor. The concentration of CO_2 and H_2 decreases at each step, while the CH_4 content increases. The gas is cooled prior to supplying it to the next reactor. This cooling of the gas should fulfil two objectives: to reduce the operating temperature of the next reactor and to separate out the water produced. Both objectives help move the reaction thermodynamically towards the products; the first *via* temperature reduction and the second through the elimination of a reaction product. This process is repeated until a gas of sufficient quality to be injected into the NG network is obtained, which usually requires a series of 2 to 5 reactors.

In the case of *in situ* heat removal, a single isothermal fixed-bed reactor is used. An isothermal reactor maintains a relatively constant temperature throughout the reactor *via* efficient cooling. There are several types of isothermal reactors: a refrigerant can pass through helical or straight tubes within the reactor, plate heat exchangers can be used or the reactor can be jacketed. An isothermal reactor operates at a temperature lower than that of the first adiabatic reactor considered previously and the process set-up is simpler than for adiabatic reactors. However, the reactor is more expensive and the steam obtained is of poor quality. The use of fluidized-bed reactors leads to almost isothermal operating conditions inside the reactor with highly effective heat removal. The use of a single reactor with a simplified design is economical. However, in this case, complete CO_2 conversion may not be achieved.[53] Moreover, under fluidized conditions, serious attrition and deactivation of the catalyst can occur. In three-phase reactors, a liquid phase with a high heat capacity is introduced, which allows more effective temperature control; the reactor can operate under almost isothermal conditions. However, these slurry reactors suffer from the instability of the suspension liquid.[106] Structured reactors, such as monolith reactors, have been developed to overcome the drawbacks of fixed-bed reactors. Two main aspects can be improved: the considerable drops in pressure and the existence of hot spots. Depending on the material, the radial heat transport can be enhanced by two or three orders of magnitude.[107] New developments in structured reactors are related to the sorption-enhanced methanation concept. In this set-up, the water produced is removed from the gas phase by a water-adsorbent catalyst carrier; this water removal must be combined with a regeneration step.[108]

9.3.2 Processes to Biomethane

The main source of biomethane is CH_4 derived from the direct biochemical transformation of biomass. The appropriate anaerobic digestion of different substrates produces the so-called biogas. Biogas plants are numerous in Europe, China and the USA.[53,109] Raw biogas from anaerobic digestion processes of different types of biomass and landfill is a gaseous mixture that mainly contains CH_4 and CO_2 in variable amounts (CH_4, 50–70%; CO_2, 30–50%) and other minor compounds. The substrate used, the technology and the collection method can all affect not only the productivity but also the composition of raw biogas. The use of biowaste to produce biogas by anaerobic digestion can be envisaged,[110] as can biogas from the abundant dairy fibre in manure.[111] The simultaneous anaerobic digestion of two or more substrates for improvement of conventional anaerobic digestion of pure substrates in the biomethane production has been reviewed.[112]

Besides CH_4 and CO_2, raw biogas also contains small amounts, usually below 2%, of NH_3, H_2S, H_2, O_2, N_2, siloxanes and halogenated-hydrocarbons, as well as up to 10% water. Despite the possibility of using biogas directly for some purposes, the presence of some amounts of harmful components usually makes cleaning raw biogas to remove these unwanted components essential. Moreover, the calorific value of the raw gas is quite low compared to that of NG. So, the cleaning and upgrading biogas to biomethane is one of the most attractive technological processes within the biorefinery concept. The separation of H_2O, CO_2 and other contaminants requires the application of several processes. Extensive reviews of appropriate biogas cleaning and upgrading technologies have recently been published.[113–115] Besides classical water scrubbing and chemical absorption using amines, novel technologies such as cryogenic and biological cleaning methods are being considered. Furthermore, membrane-based processes for upgrading biogas to biomethane have recently been assessed.[116,117] There is also a comprehensive review of the state of the art of biogas upgrading technologies with special emphasis on biotechnologies for the removal of CO_2, H_2S, siloxanes and halogenated-hydrocarbons.[118] Energetic analysis of various scenarios for the production of biomethane from biogas taking these aspects into account has also recently been undertaken.[119]

Since biogas quality strongly affects overall system performance in terms of cost and efficiency, integrated optimization of both biogas upgrading and biomethane utilization is needed. To further lower the cost of biogas upgrading, the separated CO_2 can also be used to financial advantage. As the resulting CO_2 can be used for further methanation to Bio-SNG (see below), additional considerations concerning the purity of the separated CO_2 must be taken into account. In this respect, the CO_2 from biogas plants using residues as the substrate for anaerobic digestion has high sulphur loads.[53] The tolerance of Ni-based catalysts to H_2S is quite low, and even if most of the other impurities have been eliminated *via* different cleaning/upgrading processes, further cleaning of sulphur upstream of the methanation reactor would be necessary.

Thus, although the main interest in biogas plants is related to biomethane production, biogas can also be regarded as a CO_2 source. After separation from the biogas in the process of upgrading it to biomethane, the resulting CO_2 can be used as the feedstock for methanation using H_2 from water electrolysis performed using renewable electricity, *e.g.*, wind-based electricity.

This would bring us close to the Power-to-Gas concept, and new strategies need to be developed not only for the catalyst but also for the reactor. Specifically, the methanation reactor should be designed to operate under dynamic conditions. If it did not, high-capacity H_2 storage would be required to maintain a constant flow to the methanation reactor and thereby permit operation under steady-state conditions; obviously, this would increase the overall costs. The reactor temperature may be difficult to maintain under repeated shut down operations, leading to considerable variation of temperature in the catalyst bed. This could cause physical damage to or sintering of the active phase, thus seriously shortening the catalyst lifetime. Moreover, in fluctuating operation mode, the system could remain in standby, with the catalyst exposed to an atmosphere enriched in CO_2, and bulk-like oxidation of the Ni-based catalyst has been observed under fluctuating CO_2 methanation conditions.[120]

9.3.3 Processes to Produce Bio-SNG

Within the activities of the biorefinery, besides the biogas production considered above, different processes can result in CO_2 as a by-product. It is produced as a result of the biomass fermentation processes used for bioethanol production and from different reforming processes, as already explained in the first part of this chapter. However, to address the production of Bio-SNG *via* the methanation of the CO_2 that could be collected from these sources, several factors must first be taken into account: the costs of CO_2 capture from the effluent, and its purification and necessary concentration. These considerations make these sources not very attractive for further methanation at present.

However, the Bio-SNG process has been proposed as a way to valorize the thermal gasification of biomass.[121–124] Moreover, we should recall here that the gasification of bioslurry feedstocks obtained from remote biomass pyrolysis units has been proposed as an appropriate syngas source for Fischer–Tropsch synthesis of liquid hydrocarbons.[125] The overall Bio-SNG process starting from biomass gasification can be regarded as being composed of three main stages: the gasification of the biomass into raw syngas, whose main dry components are H_2 and CO_x; appropriate cleaning before the methanation step; and upgrading of the crude Bio-SNG to the NG network specifications by removing water and excess CO_2.

In the gasification of biomass, the syngas obtained varies in its composition, depending on the gasifier and conditions used. Only when operating at low temperatures (<500 °C) under low air ratios, can CH_4 be present. Under the usual operating conditions, *i.e.* above 800 °C and using oxygen-rich atmospheres, the syngas produced mainly contains H_2 and CO_x.[126] This syngas can then be used to obtain Bio-SNG through methanation. In this context, approaches involving sorption-enhanced steam hydrogasification have been reported for the control of the H_2/C and steam/C ratios. Together with the use of a WGS reactor prior to the methanation stage, these approaches can contribute to an increase in CH_4 productivity.[127]

However, besides the H_2/CO_x mixture, the raw syngas from biomass gasification contains undesired particles, H_2S, ammonia, HCl and alkali metals, as well as minor amounts of hydrocarbons other than CH_4. Most of the impurities can be separated from the gas by filters and cyclones, and then with further absorptive processes such as alkaline washing or the use of active carbon and ZnO beds.[121] Among the hydrocarbons, tars and lower hydrocarbons, such as ethylene, must be eliminated to avoid further deactivation of the methanation catalyst, and serious operational and maintenance problems due to polymerization processes or corrosion and blockages of the methanation reactor. Tar formation is probably intrinsic to biomass gasification and still represents a technical barrier for this technique. Much research has focused on the use of catalytic technology to assist the clean-up of the syngas produced. Indeed, catalytic strategies offer the possibility of transforming the tars into useful gas compounds and thereby increasing the energy content of the syngas product. Cheap materials such as dolomite and olivine[128] as well as different Ni- and Rh-based catalysts have been used.[129] A few percent of ethylene content, specifically when a methane-rich producer gas from the biomass gasifier is used, is a challenge for long-term catalyst stability when using an adiabatic fixed-bed reactor. Ethylene is a recognized carbon precursor at the high temperatures reached in the methanation reactor. Therefore, ethylene must be eliminated prior to the methanation step, through conversion to less reactive species, *e.g.*, by hydrogenation. In contrast, ethylene appears not to be a concern when using an isothermal fluidized-bed reactor, and no upstream ethylene removal or conversion is needed.[102]

After methanation, the raw Bio-SNG has to be dried and, if necessary, CO_2 must be removed (*e.g.*, by amine absorption), as well as H_2, using membrane technology. The standard minimum CH_4 content allowed for NG pipeline injection is 95% with maxima of CO_2 and H_2 of 2.5% and 5%, respectively.

Various approaches have already considered the integration into biomass gasification units of H_2 from water electrolysis using surplus renewable electricity.[130] Several benefits can be expected from such a coupling of the Power-to-Gas and Bio-SNG concepts: the increased H_2 availability would allow the use of increased H_2/C ratios, thus also avoiding an additional WGS reactor, and the O_2 from electrolysis could be used to increase the efficiency of the biomass gasification process. Moreover, in this case, the heat from the methanation reactor could be used not only for gasification and the recovery of the amine solution from the CO_2 recovery unit, but also in the electrolysis itself, if a solid oxide electrolytic cell (SOEC) is used. All these parameters still allow an increase of the overall efficiency of a future Bio-SNG process.

9.3.4 Processes for *In situ* Upgrading Biogas

In order to improve the overall production of biomethane/Bio-SNG, different strategies have been developed for the *in situ* conversion of CO_2 to CH_4. Interest in the concept of Power-to-Gas and biogas upgrading has increased efforts to develop new ideas for the direct hydrogenation of CO_2 in the mixture: the methanation of biogas, avoiding the separation of CO_2. Obviously, the interest resides in the fact that by converting the CO_2 contained in the biogas, the production of biomethane/Bio-SNG from the existing biogas plants could be almost doubled. This could be attractive for biogas installations with a small capacity, for which upgrading biogas may not be economically viable. Two approaches can be contemplated: firstly, using biochemical methods; and secondly, catalytic hydrogenation of the biogas.

In the first approach, chemoautotrophic microbial conversion of CO_2 to CH_4 is based on the action of hydrogenotrophic methanogens of the *Archae* domain as biocatalysts capable of using CO_2 as the carbon source (electron acceptor) and H_2 as the energy source (electron donor). Thus, the bioconversion of CO_2 to CH_4 using an external H_2 injection to improve the upgrading of the biogas to biomethane has been extensively reviewed[118,131,132] since the reported upgrading from 60% to 96% of CH_4 with no H_2S detected in the off-gases using *Methanobacterium thermoautotrophicum*.[133] The syngas from biomass gasification containing H_2 and CO_x could also be upgraded to CH_4 *via* the biocatalytic route.[118] However, due to the poor solubility of H_2 in the solution under the operating conditions, H_2 mass transfer limitation is a major engineering challenge; high gas residence times are necessary and the use of continuously stirred tanks has been proposed.[53,118] Biochemical routes have also been reported to tune anaerobic fermentation processes towards the production of CH_4/H_2 mixtures, biohythane. Two-stage fermentation could provide biohythane with a range of H_2 content; the H_2 content depends on the feedstock composition and operational parameters.[134] Recent studies have reported the improvement of CH_4 and H_2 production by using different operational factors and bacteria in the anaerobic fermentation of different substrates, such as algae, organic waste including food waste, sewage sludge, waste of cellulose or sugar materials and contaminated waste-water.[135–139]

Adopting the second approach, several companies have demonstrated that the catalytic methanation of CO_2 using a biogas feedstock can be successfully achieved to upgrade the carbon to a methane content of at least 96%, which allows for its direct use as fuel.[53] In this respect, an upgraded model has been recently reported for evaluating an integrated power-to-methane system based on fluctuating energy sources, renewable electrolysis, anaerobic fermentation to biogas and *in situ* methanation.[140] The largest plant operating is probably that of ETOGAS for the Audi company in Wertle (Germany). There, H_2 is produced from three 2 MW units of alkaline electrolyzers. However, in that plant, in contrast to previous plants with smaller capacities, the source of CO_2 is from upgraded biogas. The fixed-bed methanation reactor operates at low GHSV to produce less mechanical wear of the Ni-based catalyst. Haldor Topsoe has a patented process for methanation of syngas called TREMP (Topsoe Recycle Methanation Process), which consists of three adiabatic methanation reactors.[53] The TREMP process uses steam recovery to heat the reactant gases. The catalysts are home-developed by Haldor Topsoe; in the first reactor, the catalyst must be more thermally stable than in the next two reactors. The proposal for *in situ* biomethane/Bio-SNG production uses H_2 from SOEC technology and non-upgraded biogas as the carbon source. In this case, the use of a direct biogas source containing about 65% CH_4 allows the use of only one adiabatic catalytic reactor to reach a CH_4 content of 96% to 98%. This is an additional advantage compared to the use of a pure separated CO_2 source for methanation; only a small heat flow needs to be added to the electrolysis process, in addition to the steam obtained from the methanation reactor.

9.4 Conclusion

Biorefineries offer multiple side streams and coproducts of little direct value, in addition to the primary chemical and biofuel products. These coproducts should be valorized to improve the economic viability and profitability of the overall biorefining process and make efficient use of biogenic carbon sources. Among the possibilities available for this, conversion of these secondary sources into gaseous biofuels may be the option of choice, especially when dealing with moist biomass and aqueous solutions that have a high content of organic matter. CH_4 (biomethane/Bio-SNG) and H_2-enriched CH_4 (bio-hythane) are the target gaseous biofuel products, due to the lack of existing infrastructure for handling and storing large amounts of hydrogen gas. However, H_2 production may be of interest in biorefineries where selective hydrogenation of platform intermediate molecules is performed to manufacture fine chemicals. As discussed in this review, several technologies are under development with the aim of producing gaseous biofuels; each of them has its specific advantages, but at the same time, all currently suffer from unsolved limitations. We have analyzed the catalytic aqueous-phase gasification of biomass (APR and SCWG). The aggressive nature of highly pressurized water used at high temperatures limits the lifetime of the catalysts required for the process, and this is a key factor that hinders the industrialization of these technologies. Although significant advances have been made, we think that the design and synthesis of catalytic materials based on combinations of metals and carriers that are more stable under APR and SCWG conditions using novel techniques is paramount and should guide further research in this field. Catalytic methanation of carbon oxides (CO_x) – either CO_2 in biogas, captured CO_2, or CO/CO_2 in syngas – is currently the technique that is closest to full industrialization, and in fact, some solutions are already available commercially. However, the possible effects on catalyst stability and activity caused by the broad range of trace contaminants in CO_x-rich biorefinery streams are still of concern. Finally, another key factor that contributes to maximizing profitability of the biorefinery is full energy integration between processes to take advantage of sources of high-quality waste heat, such as that available from methanation reactors. Clearly, much research into the multiple possibilities offered by biorefining for CH_4 and H_2 production is still needed, but the benefits are already tangible and we cannot afford to waste the further opportunities it offers.

References

1. E. de Jong, A. Higson, P. Walsh and M. Wellisch, *Bio-based Chemicals: Value Added Products from Biorefineries. IEA Bioenergy Task 42*, 2012.
2. A. Hagan and A. Wong, *The Future of Industrial Biorefineries*, World Economic Forum, 2010, p. 210610.
3. E4tech, RE-CORD and WUR. *From the Sugar Platform to Biofuels and Biochemicals. Final Report for the European Commission, Contract No. ENER/C2/423-2012/SI2.673791*, 2015.
4. *Biorefining. Sustainable and Synergetic Processing of Biomass into Marketable Food & Feed Ingredients, Chemicals, Materials and Energy (Fuels, Power, Heat). IEA Bioenergy Task 42*, ed. R. van Ree and A. van Zeeland, 2014.
5. R. D. Cortrigth, R. R. Davda and J. A. Dumesic, *Nature*, 2002, **418**, 964.
6. A. Wawrzetz, B. Peng, A. Hrabar, A. Jentys, A. A. Lemonidou and J. A. Lercher, *J. Catal.*, 2010, **269**, 411.
7. C. Pendem, P. Gupta, N. Chaudhary, S. Singh, J. Kumar, T. Sasaki, A. Datta and R. Bal, *Green Chem.*, 2012, **14**, 3107.
8. Y. Guo, M. U. Azmat, X. Liu, Y. Wang and G. Lu, *Appl. Energy*, 2012, **92**, 218.
9. L. Zhang, A. M. Karim, M. H. Engelhard, Z. Wei, D. L. King and Y. Wang, *J. Catal.*, 2012, **287**, 37.
10. H. A. Duarte, M. E. Sad and C. R. Apesteguía, *Int. J. Hydrogen Energy*, 2017, **42**, 4051.
11. H. A. Duarte, M. E. Sad and C. R. Apesteguía, *Int. J. Hydrogen Energy*, 2016, **41**, 17290.
12. E. Furimsky, *Ind. Eng. Chem. Res.*, 2013, **52**, 17695.
13. I. Gandarias, P. L. Arias, J. Requies, M. B. Güemez and J. L. G. Fierro, *Appl. Catal., B*, 2010, **97**, 248.
14. A. Chen, P. Chen, D. Cao and H. Lou, *Int. J. Hydrogen Energy*, 2015, **40**, 14798.
15. M. L. Barbelli, F. Pompeo, G. F. Santori and N. N. Nichio, *Catal. Today*, 2013, **213**, 58.
16. T.-W. Kim, H.-D. Kim, K.-E. Jeong, H.-J. Chae, S.-Y. Jeong, C.-H. Lee and C.-H. Kim, *Green Chem.*, 2011, **13**, 1718.
17. A. S. Larimi, M. Kazemeini and F. Khorasheh, *Appl. Catal., A*, 2016, **523**, 230.
18. Y. Lei, S. Lee, K.-B. Low, C. L. Marshall and J. W. Elam, *ACS Catal.*, 2016, **6**, 3457.
19. M. El Doukkali, A. Iriondo, P. L. Arias, J. Requies, I. Gandarías, L. Jalowiecki-Duhamel and F. Dumeignil, *Appl. Catal., B*, 2012, **125**, 516.
20. M. El Doukkali, A. Iriondo, J. F. Cambra, I. Gandarías, L. Jalowiecki-Duhamel, F. Dumeignil and P. L. Arias, *Appl. Catal., A*, 2014, **472**, 80.
21. A. Ciftci, D. A. J. M. Ligthrt and E. J. Hensen, *Green Chem.*, 2014, **16**, 853.
22. H.-D. Kim, H. J. Park, T.-W. Kim, K.-E. Jeong, H.-J. Chae, S.-Y. Jeong, C.-H. Lee and C.-U. Kim, *Catal. Today*, 2012, **185**, 73.

23. D. A. Boga, R. Oord, A. M. Beale, Y.-M. Chung, P. C. A. Bruijnincx and B. M. Weckhuysen, *ChemCatChem*, 2013, **5**, 529.
24. M.-C. Kim, T.-W. Kim, H.-J. Kim, C.-U. Kim and J. W. Bae, *Renewable Energy*, 2016, **95**, 396.
25. B. Yan, W. Li, J. Tao, N. Xu, X. Li and G. Chen, *Int. J. Hydrogen Energy*, 2017, **42**, 6674.
26. A. Seretis and P. Tsiakaras, *Renewable Energy*, 2016, **97**, 373.
27. Y. H. Park, J. Y. Kim, D. J. Moon, N. C. Park and Y. C. Kim, *Res. Chem. Intermed.*, 2015, **4**, 9603.
28. X. Chu, J. Liu, B. Sun, R. Dai, Y. Pei, M. Qiao and K. Fan, *J. Mol. Catal. A: Chem.*, 2011, **335**, 129.
29. T. Nozawa, A. Yoshida, S. Hikichi and S. Naito, *Int. J. Hydrogen Energy*, 2015, **40**, 4129.
30. J. Liu, B. Sun, J. Hu, Y. Pei, H. Li and M. Qiao, *J. Catal.*, 2010, **274**, 287.
31. M. Modell, U.S. Patent 4338199, 1982.
32. M. Modell, in *Standard Handbook of Hazardous Waste Treatment and Disposal*, ed. H. M. Freeman, McGraw-Hill, New York, 1989, p. 8.11.
33. S. I. Kawasaki, T. Oe, S. Itoh, A. Suzuki, K. Sue and K. Arai, *J. Supercrit. Fluids*, 2007, **42**, 241.
34. D. C. Elliot, J. Sealock Jr and E. G. Baker, *Ind. Eng. Chem. Res.*, 1993, **32**, 1542.
35. Y. Guo, S. Z. Wang, D. H. Xu, Y. M. Gong, H. H. Ma and X. Y. Tang, *Renewable Sustainable Energy Rev.*, 2010, **14**, 334.
36. P. Azadi and R. Farnood, *Int. J. Hydrogen Energy*, 2011, **36**, 9529.
37. L. Zhang, P. Champagne and C. Xu, *Int. J. Hydrogen Energy*, 2011, **36**, 9591.
38. T. Pairojpiriyakul, E. Croiset, W. Kiatkittipong, K. Kiatkittipong, A. Arpornwichanop and S. Assabumrungrat, *Int. J. Hydrogen Energy*, 2013, **38**, 4368.
39. A. J. Byrd and R. B. Gupta, *Appl. Catal., A*, 2010, **381**, 177.
40. Y. Lu, S. Li, L. Guo and X. Zhang, *Int. J. Hydrogen Energy*, 2010, **35**, 7161.
41. Y. Lu, Y. Zhu, S. Li, X. Zhang and L. Guo, *Int. J. Hydrogen Energy*, 2014, **67**, 125.
42. S. Li, L. Guo, C. Zhu and Y. Lu, *Int. J. Hydrogen Energy*, 2013, **38**, 9688.
43. Q. Liu, L. Liao, Z. Liu and X. Dong, *J. Energy Chem.*, 2013, **22**, 665.
44. J. Huang, X. Lian, L. Wang, C. Zhu, H. Jin and R. Wang, *Int. J. Hydrogen Energy*, 2017, **42**, 4613.
45. K. Kang, M. Shakouri, R. Azargohar, A. K. Dalai and H. Wang, *Catal. Lett.*, 2016, **146**, 2596.
46. Q. Guan, S. Chen, Y. Chen, J. Gu, B. Li, R. Miao, Q. Chen and P. Ning, *Int. J. Hydrogen Energy*, 2017, **42**, 6511.
47. I. Behnia, Z. Yuan, P. Charpentier and C. Xu, *Fuel Process. Technol.*, 2016, **143**, 27.
48. I. Behnia, Z. Yuan, P. Charpentier and C. Xu, *AIChE J.*, 2016, **62**, 2786.
49. M. Z. Hossain, M. B. I. Chowdhury, Q. Alsharari, A. K. Jhawar and P. Charpentier, *Fuel Process. Technol.*, 2017, **159**, 55.

50. J.-L. Jiao, P. Duan and F. Wang, *Prepr. - Am. Chem. Soc., Div. Energy Fuels*, 2016, **61**, 20.
51. E. Yildirir, J. A. Onwudili and P. T. Williams, *Waste Biomass Valorization*, 2017, **8**, 359.
52. J. A. Onwudili and P. T. Williams, *RSC Adv.*, 2013, **3**, 12432.
53. M. Götz, J. Lefebvre, F. Mörs, A. McDaniel Koch, F. Graf, S. Bajohr, R. Reimert and T. Kolb, *Renewable Energy*, 2016, **85**, 1371.
54. M. Bailera, P. Lisbona, L. M. Romeo and S. Espatolero, *Renewable Sustainable Energy Rev.*, 2017, **69**, 292.
55. S. Pellegrino, A. Lanzini and P. Leone, *Renewable Sustainable Energy Rev.*, 2017, **70**, 266.
56. P. Collet, E. Flottes, A. Favre, L. Raynal, H. Pierre, S. Capela and C. Peregrina, *Appl. Energy*, 2017, **192**, 282.
57. X. Zhang, C. Bauer, C. L. Mutel and K. Volkart, *Appl. Energy*, 2017, **190**, 326.
58. F. D. Meylan, V. Moreau and S. Erkman, *Energy Policy*, 2016, **94**, 366.
59. E. P. Ahern, P. Deane, T. Persson, B. O. Gallachóir and J. D. Murphy, *Renewable Energy*, 2015, **78**, 648.
60. P. Sabatier and J. B. Senderens, *C. R. Acad. Sci.*, 1902, **134**, 514.
61. J. Rostrup-Nielsen and L. J. Christiansen, Concepts in Syngas manufacture, in *Catalytic Science Series*, ed. G. J. Hutchings, Imperial College Press, London, 2011, ch. 2, vol. 10, pp. 134–140.
62. J. Gao, Y. Wang, Y. Ping, D. Hu, G. Xu, F. Gu and F. Su, *RSC Adv.*, 2012, **2**, 2358.
63. M. A. A. Aziz, A. A. Jalil, S. Triwahyono and A. Ahmad, *Green Chem.*, 2015, **17**, 2647.
64. G. A. Mills and F. W. Steffgen, *Catal. Rev.: Sci. Eng.*, 1974, **8**, 159.
65. G. D. Weatherbee and C. H. Bartholomew, *J. Catal.*, 1982, **77**, 460.
66. C. K. Vance and C. H. Bartholomew, *Appl. Catal.*, 1983, **7**, 169.
67. J. H. Chiang and J. R. Hopper, *Ind. Eng. Chem. Prod. Res. Dev.*, 1983, **22**, 225.
68. G. D. Weatherbee and C. H. Bartholomew, *J. Catal.*, 1984, **87**, 352.
69. S. Fujita, H. Terunuma, M. Nakamura and N. Takezawa, *Ind. Eng. Chem. Res.*, 1991, **30**, 1146.
70. I. A. Fisher and A. T. Bell, *J. Catal.*, 1996, **162**, 54.
71. P. Frontera, A. Macario, M. Ferraro and P. L. Antonucci, *Catalysts*, 2017, **7**, 59.
72. J. Gao, Q. Liu, F. Gu, B. Liu, Z. Zhong and F. Su, *RSC Adv.*, 2015, **5**, 22759.
73. W. Wang, S. Wang, X. Ma and J. Gong, *Chem. Soc. Rev.*, 2011, **40**, 3703.
74. F. Fischer, H. Tropsch and P. Dilthey, *Brennst.-Chem.*, 1925, **6**, 265.
75. M. A. Vannice, *J. Catal.*, 1975, **31**, 449.
76. A. L. Kustov, M. Frey, K. E. Larsen, T. Johannessen, J. K. Norskov and C. H. Christensen, *Appl. Catal., A*, 2007, **320**, 98.
77. T. Abe, M. Tanizawa, K. Watanabe and A. Taguchi, *Energy Environ. Sci.*, 2009, **2**, 315.
78. M. Jacquemin, A. Beuls and P. Ruiz, *Catal. Today*, 2010, **157**, 462.

79. S. Abelló, C. Berrueco, F. Gispert-Guirado and D. Montané, *Catal. Sci. Technol.*, 2016, **6**, 2305–2317.
80. S. Abate, C. Mebrahtu, E. Giglio, F. Deorsola, S. Bensaid, S. Perathoner, R. Pirone and G. Centi, *Ind. Eng. Chem. Res.*, 2016, **55**, 4451.
81. H. Muroyama, Y. Tsuda, T. Asakoshi, H. Masita, T. Okanishi, T. Matsui and K. Eguchi, *J. Catal.*, 2016, **343**, 178.
82. H. Takano, Y. Kirihata, K. Izumiya, N. Kumagai, H. Habazaki and K. Hashimoto, *Appl. Surf. Sci.*, 2016, **388**, 653.
83. K. Zhao, W. Wang and Z. Li, *J. CO2 Util.*, 2016, **16**, 236.
84. R. Zhou, N. Rui, Z. Fan and C. Liu, *Int. J. Hydrogen Energy*, 2016, **41**, 22017.
85. N. Bette, J. Thielemann, M. Schreiner and F. Mertens, *ChemCatChem*, 2016, **8**, 2903.
86. A. Westermann, B. Azambre, M. C. Bacariza, I. Grac, M. F. Ribeiro, J. M. Lopes and C. Henriques, *Catal. Today*, 2017, **283**, 74.
87. J. Ashok, M. L. Ang and S. Kawi, *Catal. Today*, 2017, **281**, 304.
88. M. Marwood, R. Doepper and A. Renken, *Appl. Catal., A*, 1997, **151**, 223.
89. S. Rönsch, J. Schneider, S. Matthischke, M. Schlüter, M. Götz, J. Lefebvre, P. Prabhakaran and S. Bajohr, *Fuel*, 2016, **166**, 276.
90. X. Su, J. Xu, B. Liang, H. Duan, B. Hou and Y. Huang, *J. Energy Chem.*, 2016, **25**, 553.
91. N. Homs, J. Toyir and P. Ramírez de la Piscina, in *New and Future Developments in Catalysis: Activation of Carbon Dioxide*, ed. S. L. Suib, Elsevier, Amsterdam, 2013, ch. 1, pp. 1–26.
92. P. A. Ussa Aldana, F. Ocampo, K. Kobl, B. Louis, F. Thibault-Starzyk, M. Daturi, P. Bazin, S. Thomas and A. C. Roger, *Catal. Today*, 2013, **215**, 201.
93. M. V. Konishcheva, D. I. Potemkin, S. D. Badmaev, P. V. Snytnikov, E. A. Paukshtis, V. A. Sobyanin and V. N. Parmon, *Top. Catal.*, 2016, **59**, 1424.
94. D. Heyl, U. Rodemerck and U. Bentrup, *ACS Catal.*, 2016, **6**, 6275.
95. F. Wang, S. He, H. Chen, B. Wang, L. Zheng, M. Wei, D. G. Evans and X. Duan, *J. Am. Chem. Soc.*, 2016, **138**, 6298.
96. D. Hu, J. Gao, Y. Ping, L. Jia, P. Gunawan, Z. Zhong, G. Xu, F. Gu and F. Su, *Ind. Eng. Chem. Res.*, 2012, **51**, 4875.
97. Q. Liu, J. Gao, M. Zhang, H. Li, F. Gu, G. Xu, Z. Zhong and F. Su, *RSC Adv.*, 2014, **4**, 16094.
98. J. Sehested, J. A. P. Gelten and S. Helveg, *Appl. Catal., A*, 2006, **309**, 237.
99. J. R. Rostrup-Nielsen, K. Pedersen and J. Sehested, *Appl. Catal., A*, 2007, **330**, 134.
100. G. Du, S. Lim, Y. Yang, C. Wang, L. Pfefferle and G. Haller, *J. Catal.*, 2007, **249**, 370.
101. S. He, C. Li, H. Chen, D. Su, B. Zhang, X. Cao, B. Wang, M. Wei, D. G. Evans and X. Duan, *Chem. Mater.*, 2013, **25**, 1040.
102. J. Kopyscinski, T. J. Schildhauer and S. M. A. Biollaz, *Fuel*, 2010, **89**, 1763.
103. F. Feng, G. H. Song, L. Hong Shen and J. Xiao, *Clean Technol. Environ. Policy*, 2016, **18**, 965.

104. T. J. Schildhauer and S. M. A. Biollaz, in *Synthetic Natural Gas from Coal, Dry Biomass, and Power-to-Gas Application*, ed. T. J. Schildhauer and S. M. A. Biollaz, John Wiley & Sons, Inc., Hoboken, NJ, USA, 2016, ch. 8, pp. 221–230.
105. B. Swain, S.-H. Kang, J.-H. Kim, K.-J. Jung, Y.-D. Yoo, K.-J. Kim, D. J. Koh and J.-H. Ryu, *Int. J. Energy Res.*, 2017, **41**, 353.
106. J. Lefebvre, M. Götz, S. Bajohr, R. Reimert and T. Kolb, *Fuel Process. Technol.*, 2015, **132**, 83.
107. C. Janke, M. S. Duyar, M. Hoskins and R. Farrauto, *Appl. Catal., B*, 2014, **152–153**, 184.
108. S. Walspurger, G. D. Elzinga, J. W. Dijkstra, M. Sarić and W. G. Haije, *Chem. Eng. J.*, 2014, **242**, 379.
109. C. Liu, J. Wang, X. Ji, H. Qian, L. Huang and X. Lu, *Chin. J. Chem. Eng.*, 2016, **24**, 920.
110. A.-N. Zeng and M. Kaltschmitt, *Eng. Life Sci.*, 2016, **16**, 211.
111. M. R. Pelaez-Samaniego, R. L. Hummel, W. Liao, J. Ma, J. Jensen, C. Kruger and C. Frear, *Renewable Sustainable Energy Rev.*, 2017, **72**, 254.
112. G. Lovato, S. M. Ratusznei and J. A. D. Rodrigues, *Curr. Biochem. Eng.*, 2016, **3**, 200.
113. Q. Sun, H. Li, J. Yan, L. Liu, Z. Yu and X. Yu, *Renewable Sustainable Energy Rev.*, 2015, **51**, 521.
114. O. W. Awe, Y. Zhao, A. Nzihou, D. P. Minh and N. Lyczko, *Waste Biomass Valorization*, 2017, **8**, 267.
115. F. Bauer, T. Persson, C. Hulteberg and D. Tamm, *Biofuels, Bioprod. Biorefin.*, 2013, **7**, 499.
116. G. Valenti, A. Arcidiacono and J. A. Nieto Ruiz, *Biomass Bioenergy*, 2016, **85**, 35.
117. C. Micale, *Energy Procedia*, 2015, **82**, 971.
118. R. Muñoz, L. Meier, I. Diaz and D. Jeison, *Rev. Environ. Sci. Bio/Technol.*, 2015, **14**, 727.
119. B. Wu, X. Zhang, D. Bao, Y. Xu, S. Zhang and L. Deng, *Bioresour. Technol.*, 2016, **206**, 155.
120. B. Mutz, H. W. P. Carvalho, S. Mangold, W. Kleist and J.-D. Grunwaldt, *J. Catal.*, 2015, **327**, 48.
121. S. Rönsch and M. Kaltschmitt, *Biomass Convers. Biorefin.*, 2012, **2**, 285.
122. S. Heyne, H. Thunman and S. Harvey, *Biomass Convers. Biorefin.*, 2013, **3**, 337.
123. H. Wagner, C. Wulf and M. Kaltschmitt, *Biomass Convers. Biorefin.*, 2015, **5**, 103.
124. A. Molino and G. Braccio, *Fuel*, 2015, **139**, 425.
125. G. Jacobs and B. H. Davis, in *Thermochemical Conversion of Biomass to Liquid Fuels and Chemicals*, ed. M. Croker, RSC Energy and Environment Series, Cambridge, 2010, ch. 5, pp. 95–121.
126. J. K. Neathery, in *Thermochemical Conversion of Biomass to Liquid Fuels and Chemicals*, ed. M. Croker, RSC Energy and Environment Series, Cambridge, 2010, ch. 4, pp. 67–93.

127. Z. Liu, J. M. Norbeck, A. S. K. Raju, S. Kim and C. S. Park, *Energy Convers. Manage.*, 2016, **126**, 256.
128. S. Rapagnà, N. Jand, A. Kiennemann and P. U. Foscolo, *Biomass Bioenergy*, 2000, **19**, 187.
129. A. M. Steele, S. Poulston, K. Magrini-Bair and W. Jablonski, *Catal. Today*, 2013, **214**, 74.
130. M. Gassner and F. Maréchal, *Energy*, 2008, **33**, 189.
131. A. Bensmann, R. Hanke-Rauschenbach, R. Heyer, F. Kohrs, D. Benndorf, U. Reichl and K. Sundmacher, *Appl. Energy*, 2014, **134**, 413.
132. G. Leonzio, *Chem. Eng. J.*, 2016, **290**, 490.
133. K. A. Strevett, R. F. Vieth and D. Grasso, *Biochem. Eng. J.*, 1995, **58**, 71.
134. A. Xia, J. Cheng and J. D. Murphy, *Biotechnol. Adv.*, 2016, **34**, 451.
135. L. Ding, J. Cheng, A. Xia, A. Jacob, M. Voelklein and J. D. Murphy, *Bioresour. Technol.*, 2016, **218**, 224.
136. B. Si, Z. Liu, Y. Zhang, J. Li, R. Shen, Z. Zhu and X. Xing, *Int. J. Hydrogen Energy*, 2016, **41**, 4429.
137. A. Elreedy, M. Fujii and A. Tawfik, *Bioresour. Technol.*, 2017, **223**, 10.
138. C.-Y. Lin, C.-Y. Chu, C.-H. Lay and H.-J. Leu, *Curr. Biochem. Eng.*, 2016, **3**, 16.
139. J. Cheng, L. Ding, R. Lin, L. Yue, J. Liu, J. Zhou and K. Cen, *Appl. Energy*, 2016, **184**, 1.
140. T. Zoss, E. Dace and D. Blumberg, *Appl. Energy*, 2016, **170**, 278.

CHAPTER 10

Biochar Production, Activation and Application as a Promising Catalyst

SHURONG WANG*[a], LINGJUN ZHU[a] AND ZHONGQING MA[a,b]

[a]Zhejiang University, State Key Laboratory of Clean Energy Utilization, Zheda Road 38, Hangzhou, 310027, China; [b]Zhejiang Agriculture & Forestry University, School of Engineering, Huancheng North Road 88, Lin'an, 311300, China
*E-mail: srwang@zju.edu.cn

10.1 Introduction

Biochar is a porous carbon-rich material, arising from the biomass thermal degradation process in inert or very low stoichiometric oxygen atmospheres.[1,2] Currently, biochar has been notably applied in some areas because of its versatile physicochemical characteristics (Figure 10.1). These promising potential applications include: energy production for a high heating value, acid soil remediation because of its alkaline pH value and being rich in plant nutrients, carbon sequestration due to the high content of carbon, an excellent interim form for activated carbon production, or use as a catalyst precursor and fuel cell material due to its strong pore structure and low electrical resistivity.[3–7]

Green Chemistry Series No. 56
Sustainable Catalysis for Biorefineries
Edited by Francesco Frusteri, Donato Aranda and Giuseppe Bonura

Published by the Royal Society of Chemistry, www.rsc.org

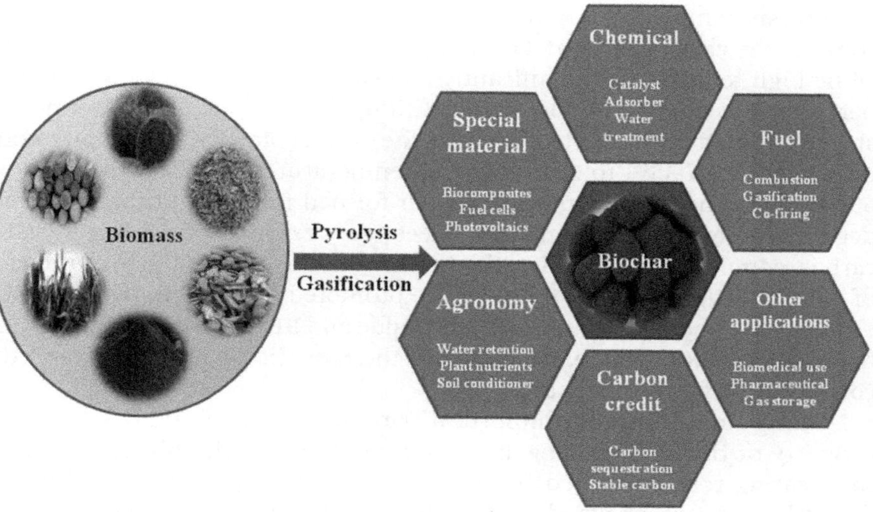

Figure 10.1 The production and application of biochar.

These applications are highly related to the basic properties of biochar, which are significantly affected by the production methods, feedstock properties, operating conditions, and reactor types.[8,9] The conventional production of biochar is pyrolysis and gasification technology, and now microwave assisted pyrolysis (MWP) and hydrothermal carbonization (HTC) technology are another two new-fashioned approaches. This chapter will introduce the biochar production process, biochar activation and upgrading approach, and its application as a promising catalyst.

10.2 Biochar Production

10.2.1 Conventional Pyrolysis (CP)

In the recent past, pyrolysis has been developed as a promising thermochemical conversion technology to convert biomass material into three products, namely biochar, bio-oil and bio-gas.[10–14] The typical pyrolysis equipment is a carbonizing kiln, fluidized bed, and rotating cone reactor coupled with a downstream particle separation and bio-oil condensation device. Depending on the different heating rates and solid residence times, pyrolysis is divided into three main types: slow, fast and flash, resulting in significant differences in the yields and basic characteristics of the solid, liquid and gaseous products. For slow pyrolysis, the dominant product is biochar, while bio-oil is dominant for fast and flash pyrolysis.[15,16] The yield and properties of biochar are significantly affected by operating conditions, such as temperature, heating rate, solid residence time, particle size, *etc.*

Pyrolysis temperature is the most important parameter, which mainly controls the yield of biochar. Low temperature favors a higher biochar yield, while high temperature significantly reduces the biochar yield. As the temperature increased from 300 to 700 °C, the yield of biochar from pine nut shell decreased from 66% to 26%,[1] and the yield of biochar from poplar wood decreased from 35% to 28.5% as the temperature increased from 300 to 600 °C.[2] At a high temperature, biochar formed from the primary thermal degradation reaction will undergo a secondary reaction of the heavy hydrocarbon component, increasing the bio-oil and bio-gas products at the cost of biochar.[9] High pyrolysis temperature promoted the content of the carbon element, high heating value (HHV), pH value and BET surface area (S_{BET}), and the graphitization degree, which led to the reduction of H/C, O/C, and the content of surface functional groups.[17]

Heating rate is another important parameter controlling the yield and property of the biochar. Low heating rate increases the biochar yield. As the heating rate decreased from 50 to 10 °C min^{-1}, the yield of biochar derived from poplar wood and safflower seed cake increased by 3–6%, while the carbon content, HHV, and pH of the biochar decreased.[2,18] At low heating rate, the possibility of secondary pyrolysis reactions was reduced, resulting in a higher biochar yield. Higher heating rate involved in the fast pyrolysis will avoid the second cracking between volatiles and solid charcoal, which favors the bio-oil production and hinders the biochar formation.[19–21]

Residence time also affects the biochar yield. High vapor residence time will supply sufficient time for repolymerization reactions to maximize the biochar yield.[22] Particle size of biomass material mainly affects the heat and mass transfer between the particle surface and the core. Increasing the thickness of particles results in a significant temperature gradient and favors the char yield.[23]

The characteristics of the feedstock affect the production of biochar as well. Generally speaking, the high lignin content in wood will result in a high biochar yield compared with agricultural residues as a result of the stability of lignin to thermal degradation, and biochar from agricultural residues is finer and less robust with low mechanical strength.[24] Besides, biochar from woody feedstock contains a low content of ash, whereas biomass with high mineral content such as grass and straw residues produces biochar with high ash contents.

10.2.2 Gasification

Gasification is another thermochemical process that converts biomass into gaseous fuels through its partial oxidation with air, oxygen, steam or their mixture.[25,26] The gasification process can be divided into four stages: namely drying, pyrolysis, oxidation/combustion, and gasification.[27,28] The gas is used as a clean fuel in the boiler or engine for heat and electricity supply. However, biomass could not be completely converted into gaseous products, and part of the biomass energy is still kept in the biochar as a gasification by-product. Comparing the downdraft fixed bed with a fluidized bed gasifier, the downdraft fixed bed gasifier produces the maximum biochar with a yield of 15–25%. Table 10.1 shows the properties of biochar derived from the downdraft fixed bed. The HHV of woody biochar (31.76 MJ kg^{-1}) is much higher than those for rice husk and rice straw biochar, while the content of silicon in the rice husk char is the highest, reaching the value of 22.1%.

Biochars from conventional pyrolysis (CP) and gasification are physically and chemically different. Compared with the biochar from CP, the biochar from gasification has a smaller particle size and more macropores due to the rapid devolatilization.[29] The particle density and BET surface area of biochar from corn stove gasification are 1.92 g cm^{-3} and 23.9 m^2 g^{-1}, respectively, while those from slow pyrolysis are 1.54 g cm^{-3} and 20.9 m^2 g^{-1}, respectively.[29] Gasification biochar is much more condensed with around 17 rings per cluster than pyrolysis biochar with around 7–8 rings per cluster.[29]

10.2.3 Microwave Assisted Pyrolysis (MWP)

MWP has been demonstrated as a potential alternative technology to CP. The advantage of MWP is its energy transfer rather than heat transfer, non-contact with material, high selectivity and volumetric heating, and quick startup and shutdown.[30] Microwave irradiation is a form of electromagnetic radiation. The wavelength is ranging from 1 mm to 1 m with the frequency of 300 MHz to 300 GHz. And the frequencies of 915 and 2450 MHz are the most widely used in the domestic microwave ovens.[31,32]

However, the biomass itself is a poor microwave adsorber, and sometimes it is difficult to reach the desired pyrolysis temperature and heating rate. Therefore, solid microwave absorbers (MWAs) are always added in the feedstock to improve the energy transfer. The MWAs are mainly composed of carbon-based (biochar, activated carbon, graphite, and silica carbide) and metal-based materials (metal, metallic oxide and hydroxide).

Table 10.2 shows the biochar yields from different feedstocks at different MWP operating conditions with a fixed frequency of 2.45 GHz. The biochar yield is affected by the species of feedstock and MWAs, MWA adding proportion, and the pyrolysis temperature. High temperature and high ratio of absorber/feedstock lead to the remarkable reduction of biochar yield.

Compared with the CP, the biochar from MWP has been reported to have higher quality, especially the pore structure. The surface area and volume of MWP biochar is much higher than that of CP biochar, meanwhile the micropores of MWP biochar are more uniform and cleaner due to the uniform volatile release without the appearance of a temperature gradient in

Table 10.1 Properties of biochar from downdraft fixed bed gasifier.

	Proximate analysis (%)				Ultimate analysis (%)						
Species	Volatiles	Fixed carbon	Ash	HHV (MJ kg^{-1})	C	H	O	N	S	Si	Reference
Wood char	11.1	85.4	3.5	31.76	81.6	2.1	14.6	0.4	0.13	/	27
Rice husk char	5.2	52.6	42.2	20.46	60.9	1.8	13.5	0.8	0.1	22.1	28
Rice straw char	12.5	65.8	21.7	22.15	68.6	1.5	8.4	0.4	0.076	2.3	28

Table 10.2 Biochar yields from different feedstocks at different MWP operating conditions.

Feedstock	Power (kw)	MW absorbers	Proportion (FS : MWAs)	Pyrolysis T (°C)	Biochar yield (%)	Reference
Pine sawdust	2	Activated carbon	8 : 2	300–600	57.1–19.91	33
Pine sawdust	2	/	/	400–800	26–13	33
Oil palm fiber	0.4	Biochar	1 : 0.25, 1 : 0.5, 1 : 1	237	88, 52, 66	34
Oil palm shell	0.4	Biochar	1 : 0.25, 1 : 0.5, 1 : 1	237	67, 42, 44	34
Switch grass	0.75	K_3PO_4	9 : 1, 9 : 2, 9 : 3	400	31, 33, 44	35
Coffee husk	/	Biochar	/	500–1000	30–15	36
Corn stover	0.7	/	/	550–650	25.7–25.1	37
Sewage sludge	0.7	Activated carbon	9 : 1	400	12.5	38

the feedstock. Therefore, biochar from MWP with a strong pore structure can be used as the contaminant adsorber, the precursor for activated carbon production, or even the catalyst support.

10.2.4 Hydrothermal Carbonization (HTC)

In recent years, hydrothermal carbonization, also called wet pyrolysis, has been described, which is a new-fashioned thermochemical conversion process where biomass is treated in the presence of subcritical liquid water under a high pressure (1–5 MPa) and low reaction temperature (150–375 °C).[39–42] Compared to the CP, the advantage of HTC is that it can convert wet biomass to biochar with relative high yields without drying before the pyrolysis process.[43,44] The experiment is mainly carried out in a lab-scale air-tight container that withstands certain pressure and temperature.[45] Recently, the majority of research on biomass HTC has been focused on the property evolution of the solid product hydrochar, while it has been less focused on the gaseous and liquid product because of hydrochar's potential value-added application in bio-energy, chemical engineering and contaminant treatment. Till now, no pilot-scale or commercial large-scale HTC plants have been reported in operation.

Compared with the CP, the significant difference in the operation parameters is lower temperature, higher reaction pressure and the different environmental medium using water.[46] The water medium plays an important role in the HTC process and acts as an acid solvent. The critical point of liquid water is at 374 °C and 22.1 MPa, below this point is subcritical water, while above refers to supercritical water.[43] Subcritical water is always

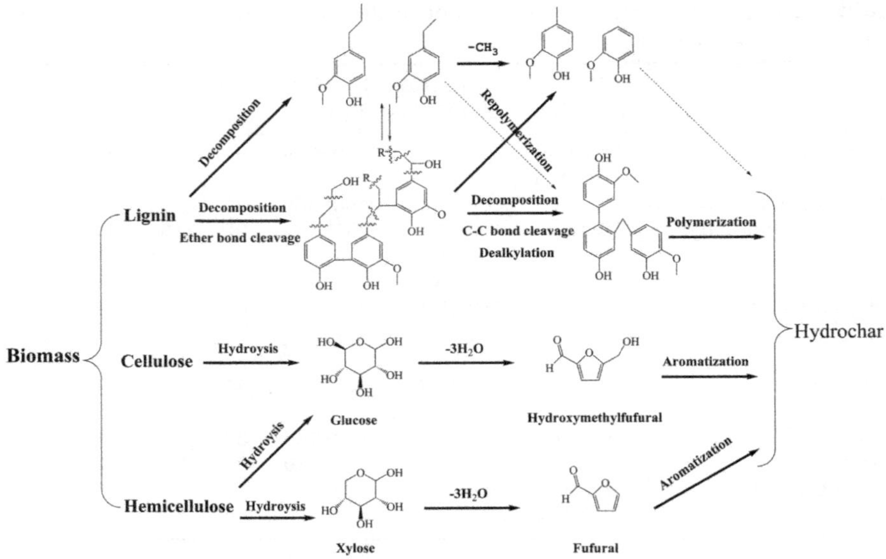

Figure 10.2 Proposed pathways for the biomass components involved in the formation of hydrochar. Reproduced from ref. 41 with permission from Progress in Chemistry.

employed in biomass HTC. With the subcritical condition, a high degree of ionization occurs. The water will dissociate into acidic hydronium ions (H_3O^+) and hydroxide ions (OH^-), and enhance the solubility of the biomass component.[47]

The reaction mechanism of biomass HTC is considered as a comprehensive decomposition of its three components (cellulose, hemicellulose and lignin, shown in Figure 10.2). The reactions are composed of hydrolysis, dehydration, decarboxylation, aromatization and recondensation. Hydrolysis is the first step where H^+ from water media reacts with biomass components and breaks the ester and ether bonds, *etc.*, resulting in the formation of intermediates, such as soluble oligomers like saccharides (glucose and xylose), and phenol and phenolic derivatives. Then, the dehydration and decarboxylation reactions occur on the intermediate to produce saccharides and phenolic derivatives with lower molecular weight. Finally, these products continue to undergo the polymerization, aromatization, and condensation reactions, and are converted into hydrochar.[40,41]

Pyrolysis temperature and time are undoubtedly the main operating parameters controlling the yield and properties of hydrochar.[48] Table 10.3 shows the hydrochar yields and properties from different feedstocks at different operating conditions. In general, as the temperature increases, the content of fixed carbon, carbon element, ash, and the value of HHV rise,

Table 10.3 Hydrochar yields and properties from different feedstocks and operating conditions.

Feedstock	T (°C)	Time (h)	Yield (%)	Proximate analysis (%)			Ultimate analysis(%)			HHV (MJ kg^{-1})	References
				Vol.	F.C.	Ash	C	H	O		
Pine wood	225	20	/	51.31	47.38	1.31	67.55	5.6	24.94	27.12	46
Pine wood	245	20	/	48.89	49.71	1.4	69.86	5.41	22.69	28.39	46
Pine wood	265	20	/	47.15	51.12	1.73	74.22	5.54	17.91	29.57	46
Coconut fiber	220	0.5	/	69.8	24.0	6.2	62.47	5.28	31.09	24.7	46
Coconut fiber	250	0.5	/	67.9	27.1	5.0	67.10	5.20	26.43	26.7	46
Coconut fiber	300	0.5	/	53.6	42.1	4.3	73.22	5.09	20.21	29.4	46
Coconut fiber	350	0.5	/	56.6	38.5	4.9	73.37	4.52	28.70	28.7	46
Coconut fiber	375	0.5	/	42.6	48.8	8.6	78.27	4.31	30.60	30.6	46
Eucalyptus leaves	200	0.5	/	72.5	20.2	7.3	61.11	6.13	60.74	25.3	46
Eucalyptus leaves	250	0.5	/	70.1	23.0	6.9	62.30	5.47	30.35	25.0	46
Eucalyptus leaves	300	0.5	/	61.2	31.7	7.1	68.87	6.00	22.79	28.7	46
Eucalyptus leaves	350	0.5	/	56.2	33.9	9.9	70.50	5.93	20.45	29.4	46
Eucalyptus leaves	375	0.5	/	43.2	42.6	14.2	72.19	4.81	19.85	28.7	46
Palm fruit bunches	150	0.3	76	74.48	17.41	4.01	49.58	5.91	43.37	20.01	51
Palm fruit bunches	250	0.3	62	63.11	27.41	4.16	54.30	4.17	38.29	22.07	51
Palm fruit bunches	350	0.3	49	47.52	42.41	4.32	66.02	4.15	28.43	27.2	51
Water hyacinth	240	1	/	/	/	28.94	45.99	4.22	17.96	18.41	45
Water hyacinth	240	4	/	/	/	32.84	46.51	4.09	13.76	19.15	45
Water hyacinth	240	8	/	/	/	29.16	48.97	4.63	14.47	20.63	45
Water hyacinth	240	12	/	/	/	32.45	47.2	4.3	13.15	19.8	45
Water hyacinth	240	20	/	/	/	34.37	46.39	3.6	12.71	19.06	45

while the yield of hydrochar and the content of volatiles, hydrogen and oxygen elements decrease. A similar variation tendency of the basic properties is observed by increasing the pyrolysis time. The species of biomass also obviously affects the formation of hydrochar. Generally, high content of lignin enhances the yield of hydrochar because lignin is difficult to degrade due to its complex aromatic matrix. In contrast, cellulose and hemicellulose are easy to decompose to form bio-oil.[49]

A significant difference of the properties between hydrochar and CP-char was obtained. Garlapalli *et al.*[47] reported the difference of digestate biochar from HTC (180–260 °C) and CP (400–800 °C). The yield of biochar from HTC (46.3%) at 260 °C was almost double that from CP (26%) at 800 °C, while the HHV from CP (33 MJ kg^{-1}) was slightly higher than that from HTC (29.9 MJ kg^{-1}). The CP had a higher pyrolysis temperature than HTC, which would lead to more exhaustive thermal degradation of the biomass component, and result in a lower yield of biochar. During the thermal degradation process of CP, the result of ^{13}C-NMR analysis indicated that more oxygen-containing and alkyl-C-containing components would convert into aryl-C containing polycyclic aromatic components with higher values of C/H and C/O.[17] Therefore, the carbon content and HHV of CP-char was higher than hydrochar. Liu *et al.*[50] compared the chars from coconut fibers and eucalyptus leaves by HTC and low temperature pyrolysis (LTP) at 250 °C, and found that hydrochar had higher energy density while pyrolytic biochar had higher energy yield. They also observed that the pyrolytic biochar retained almost all major ash-forming metals while the hydrochar preserved less than 40%, and the pyrolytic biochars showed higher reactivities than the hydrochars.

However, the biochar obtained directly from HTC generally had pore microporosity and low surface area (less than 100 m^2 g^{-1}). The surface area of hydrochar is much lower than CP biochar produced at high temperature (100–400 m^2 g^{-1}) and activated carbons (>1000 m^2 g^{-1}).[43] This is far from meeting the high value-added application of hydrochar as a special materia for catalysis, adsorption or energy storage. Therefore, a variety of activation techniques is necessary to improve the surface area and create more controlled porosity.

10.3 Biochar Activation and Upgrading

The biochar produced from the above-mentioned thermochemical conversion technologies has to be activated to enhance the pore volume, control the pore diameter, and improve the surface area. The activation process is mainly performed by three methods, namely physical activation, chemical activation, and the physicochemical activation (a combination of physical and chemical activation).[52] Activation of biochar is an important quality upgrading approach to enlarge the application range of the biochar.

10.3.1 Physical Activation

Physical activation, also called gas activation, usually uses steam and CO_2 to activate biochar at high reaction temperatures (>700 °C). The reaction mechanism of biochar steam activation is presented as eqn (10.1). The dominant reaction in steam activation of biochar is the water gas reaction, part of the carbon in biochar is converted into CO in this endothermic reaction. Carbon dioxide (CO_2) is another physical activating agent to improve the pore structure of biochar. The Boudouard reaction (eqn (10.2)) between a carbon atom and CO_2 is also an endothermic reaction. During the activation process, the carbon element in the biochar is gradually depleted by the activation reaction, and a new pore structure is formed with an increased BET surface area.

$$C + H_2O \rightarrow CO + H_2 \; (\Delta H = -130 \text{ kJ mol}^{-1}) \tag{10.1}$$

$$C + CO_2 \rightarrow 2CO \; (\Delta H = -170 \text{ kJ mol}^{-1}) \tag{10.2}$$

The characteristics of activated carbon derived from physical activation of biochar are mainly affected by activating agents, activating temperature, activating time and the species of biochar, *etc.* Table 10.4 shows the characteristics of activated carbon derived from physical activation of pyrolyzed biochar. Using the physical activation method, steam is more effective in producing a relatively higher surface area than CO_2.[53] The smaller molecular size of water is responsible for facilitating diffusion within the biochar's porous structure. However, in the commercial process of activated carbon, the combined usage of steam and CO_2 is performed to activate the biochar in steps, in order to obtain the highest adsorbability.

The activating temperature is the dominant parameter controlling the reaction rate and yield of activated carbon. The temperature for physical activation of biochar is usually between 800 and 1000 °C. In order to promote the production capacity and obtain the higher surface area, higher temperature is always applied. However, higher temperature will lead to higher burn-off rate, and thus largely reduce the yield of activated carbon.[1,54] Chen *et al.*[1] and Gu *et al.*[54] investigated the effect of activation temperature on the properties of activated carbon derived from pine nut shell and camellia shell, respectively. The optimal temperature was 850 °C with the maximum S_{BET} values of 1017 and 659 $m^2 g^{-1}$, respectively.

Fixing other operating conditions, as the activating time increases, the yield of activated carbon decreases, while the specific pore volume and average pore size increase.[57,58] The developing of pore structure could be divided into two stages as the reaction time increased. The primary stage is the developing of micropores. In the second stage, the micropore is gradually

Table 10.4 Characteristics of activated carbon derived from physical activation of pyrolyzed biochar.

Biochar feedstock	Activating agent	Temp. (°C)	Time (min)	Yield (%)	S_{BET} ($m^2 g^{-1}$)	$V_{Tot.}$ ($cm^3 g^{-1}$)	$V_{Mic.}$ ($cm^3 g^{-1}$)	$V_{Mes.}$ ($cm^3 g^{-1}$)	References
Rice husk	Steam	700	60	/	363	0.164	/	/	3
Pine nut shell	Steam	750	60	56	776	/	/	/	1
Pine nut shell	Steam	850	60	36	1017	/	/	/	1
Pine nut shell	Steam	950	60	23	761	/	/	/	1
Camellia shell	Steam	800	120	32.5	621	/	/	/	54
Camellia shell	Steam	850	120	33.1	659	/	/	/	54
Camellia shell	Steam	900	120	19.9	613	/	/	/	54
Almond shell	Steam	850	30	47.3	601	/	0.341	0.034	55
Walnut shell	Steam	850	30	62.3	792	/	0.442	0.082	55
Almond tree	Steam	850	30	41.5	1080	/	0.568	0.382	55
Olive stone	Steam	850	30	50.6	813	/	0.455	0.1	55
P. O. fibers	Steam	600	20	85.3	375	0.056	0.016	0.036	56
P. O. fibers	Steam	600	120	76.6	496	0.086	0.018	0.061	56
P. O. fibers	Steam	600	300	68.3	615	0.16	0.021	0.118	56
P. O. fibers	Steam	600	720	21.7	313	0.707	0.016	0.61	56
Olive stones	CO_2	850	30	80.7	446	/	0.243	0.021	57
Olive stones	CO_2	850	60	63.5	572	/	0.319	0.036	57
Olive stones	CO_2	850	90	54.1	761	/	0.406	0.059	57
Olive stones	CO_2	850	120	34.7	778	/	0.406	0.068	57
Palm kernel shell	CO_2	900	60	80.9	600	0.2845	0.2532	0.0313	58
Palm kernel shell	CO_2	900	120	71.8	695	0.3391	0.2948	0.0443	58
Palm kernel shell	CO_2	900	180	59.7	807	0.3833	0.3390	0.0443	58

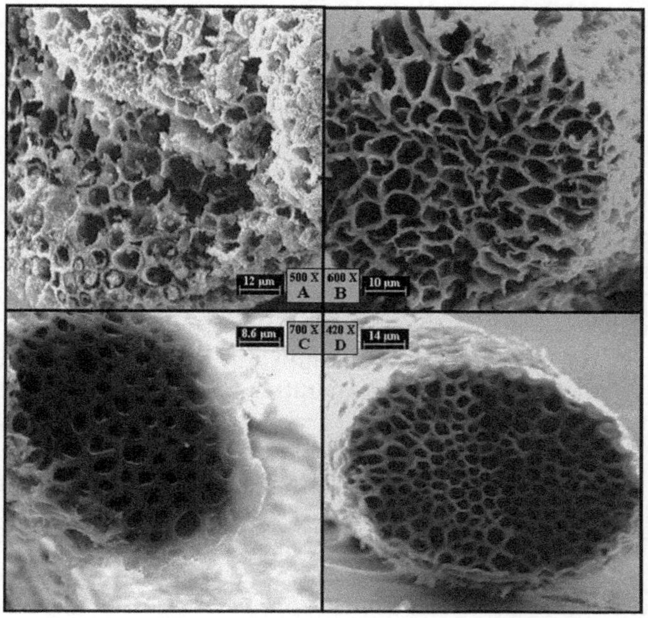

Figure 10.3 SEM images of *Posidonia oceanica* fiber activated carbon derived from pyrolyzed biochar and activated by steam at various times: (A) 20 min, (B) 120 min, (C) 300 min and (D) 720 min. Reprinted from *Journal of Hazardous Materials*, **165**, M. C. Ncibi, V. Jeanne-Rose, B. Mahjoub, C. Jean-Marius, J. Lambert, J. J. Ehrhardt, Y. Bercion, M. Seffen and S. Gaspard, Preparation and characterisation of raw chars and physically activated carbons derived from marine *Posidonia oceanica* (L.) fibers, 240–249, Copyright 2009 with permission from Elsevier.[56]

enlarged into a mesopore by the activation reaction, and the maximum value of S_{BET} is reached when the burn-off rate is close to 40–60%. Figure 10.3 shows the SEM images of *Posidonia oceanica* (L.) fiber activated carbon derived from steam activation at various times.[56] As the char is activated for longer periods, the pores begin to be more individualized and "clean". More tar component originally adhered on the pore surface will be eliminated by the activation reaction.

10.3.2 Chemical Activation

In chemical activation, biochar is impregnated with a chemical activating agent followed by a heating process under an inert atmosphere. Then, the pore structure is improved by subsequent dehydration and oxidation. Compared with physical activation, chemical activation has the advantages of lower treatment temperature and time, higher activated carbon yield and surface area, and better controlled microporosity. However, the drawbacks of chemical activation are the corrosion of the apparatus by chemicals, high cost of chemicals, and difficult recovery of chemicals.[52]

The acids (H_3PO_4,[62] H_2SO_4[63] and HCl[64]), alkali hydroxide (KOH[59] and NaOH[65]), and other chemicals (K_2CO_3[60] and $ZnCl_2$[61]) are the representative chemical activating agents to produce the biochar-based activated carbon. For chemical activation of biomass, a one step activation method is always performed wherein chemicals are impregnated into a biomass precursor using different impregnation ratios. The chemicals normally act as dehydrating agents and oxidants, and prevent the formation of tar or ash to increase the yield of activated carbon. The final step of the chemical activation is a washing step to remove the extra chemicals using acid, alkali or water. The properties of activated carbon are mainly affected by the type and dose of activating agent, and reaction temperature. Table 10.5 shows the characteristics of activated carbon derived from chemical activation of pyrolyzed biochar.

Among the various chemical activating agents, H_3PO_4 and $ZnCl_2$ are the most commonly used agents for the activation of lignocellulosic biomass materials in the practical factory because of their low cost.[66] Activation with $ZnCl_2$ can produce a higher surface area than that with H_3PO_4 because with $ZnCl_2$, biochar primarily forms a microporous structure, while H_3PO_4 is more effective in producing mesopores and resulting in a higher total pore volume and pore diameter.[67] However, H_3PO_4 is preferred in the commercial production process of activated carbon because it is more environmentally friendly than $ZnCl_2$.

KOH is reported to be more effective in developing the porosity with a narrow pore size distribution than H_3PO_4 and $ZnCl_2$.[68] Eqn (10.3) presents the activated mechanism of the biochar using the activation agent KOH. The alkali metal intercalates in the carbon lattice during the activation process and plays the role of electron donor to promote the activation reaction. The oxygen element in the alkali hydroxide has the ability to remove the carbon atom crossing linkage in the crystallites.[69]

$$6KOH + 2C \leftrightarrow 2K + 3H_2 + 2K_2CO_3 \tag{10.3}$$

Impregnation ratio, defined as the weight ratio of the activating agent to the biochar, is one of the most important parameters to control the quality of the activated carbon.[70] In general, the impregnation ratio is set in the range of 1–4. High impregnation ratio leads to high surface area and total pore volume, while the yield will decrease. The increased porosity is mainly attributed to the release of the tar existing in the cross-linked framework of the feedstock. As the impregnation ratio increased from 1 to 4 using $ZnCl_2$, the surface area of activated carbon derived from safflower seed press cake increased from 619 $m^2\ g^{-1}$ to 801 $m^2\ g^{-1}$, and the total pore volume increased from 0.33 to 0.393 $cm^3\ g^{-1}$.[61] Using the activation agent of NaOH and KOH, as the impregnation ratio increased from 1

Table 10.5 Characteristics of activated carbon derived from chemical activation of pyrolyzed biochar.

Biochar feedstock	Act. agent	Imp. ratio[a]	T (°C)	Time (min)	Yield (%)	S_{BET} ($m^2\ g^{-1}$)	$V_{Tot.}$ ($cm^3\ g^{-1}$)	$V_{Mic.}$ ($cm^3\ g^{-1}$)	$V_{Mes.}$ ($cm^3\ g^{-1}$)	Ave. pore size (nm)	References
Rice straw	KOH	1	700	/	/	772	0.422	/	/	2.18	59
Sewage sludge	KOH	1	700	/	/	782	0.606	/	/	3.01	59
Soybean oil cake	KOH	1	600	60	6.25	600	0.291	0.143	0.148	9.41	60
Soybean oil cake	KOH	1	800	60	4.11	618	0.299	0.213	0.086	9.98	60
Soybean oil cake	K_2CO_3	1	600	60	16.22	643	0.336	0.272	0.064	10.43	60
Soybean oil cake	K_2CO_3	1	800	60	11.56	1352	0.680	0.400	0.280	10.05	60
Safflower seed press cake	$ZnCl_2$	1	900	120	19.9	619	0.33	0.223	0.107	2.13	61
Safflower seed press cake	$ZnCl_2$	2	900	30	47.3	703	0.344	0.263	0.081	1.94	61
Safflower seed press cake	$ZnCl_2$	3	900	30	62.3	719	0.371	0.269	0.102	2.23	61
Safflower seed press cake	$ZnCl_2$	4	900	30	41.5	801	0.393	0.284	0.109	1.96	61

[a]Impregnation ratio is the ratio of the weight of the activating agent to biochar.

to 3, the surface area of activated carbon derived from sewage sludge char increased from 689 $m^2 g^{-1}$ and 853 $m^2 g^{-1}$ to 853 $m^2 g^{-1}$ and 1686 $m^2 g^{-1}$, respectively.[65]

Activating temperature is another important operating parameter that can significantly affect the properties of activated carbon. In general, the increase of the activating temperature will lead to the lower yield of activated carbon, and usually an increase in the surface area.[71] For the soybean oil cake derived biochar using the activating agent of K_2CO_3, as the activation temperature increased from 600 °C to 800 °C, the yield of activated carbon decreased from 16.22% to 11.56%, the surface area increased from 643 $m^2 g^{-1}$ to 1352 $m^2 g^{-1}$, and the average pore diameter decreased from 10.43 nm to 10.05 nm.[60] Angin *et al.*[61] investigated the effect of activating temperature (600 °C to 900 °C) on the pore structure of the activated carbon derived from safflower seed press cake using the activating agent of $ZnCl_2$. The yield of activated carbon decreased from 73.24% to 29.06% at the impregnation ratio of 4:1. The surface area increased from 249.3 to 801.5 $m^2 g^{-1}$, and the total pore volume, micropore volume and mesopore volume rapidly increased. The average pore diameter decreased from 2.42 nm to 1.96 nm.

10.4 Biochar as Catalyst and Catalyst Support

It is known that biomass contains trace amounts of some metal elements, including K, Ca, Na, Mg, Al, Fe, Cu, *etc.* The most abundant inorganic salts in biomass are alkali salts and alkali-earth metal salts, especially potassium and sodium salts.[72] After pyrolysis, the contents of these metals are concentrated in the biochar (pyrolysis residue), and as a result, the primary biochar has high contents of alkali and alkali-earth metals (AAEMs) especially in the herbaceous plant derived biochar. Additionally, the activated biochar generally has a relatively high surface area and pore volume possessing a multi-level porous structure with both micropores and mesopores. And the functional groups on the surface of biochar would facilitate its surface chemistry modification. Furthermore, the active components can be easily separated from the biochar-supported catalyst using a simple oxidation method.[73] Because of all of these properties, biochar has great potential for application as a catalyst and catalyst support.

10.4.1 Biochar Used as a Catalyst Directly

Due to the presence of AAEMs and transition elements as well as the functional groups, biochar can be used as a catalyst directly. Biochar from different kinds of biomass has different contents of metal elements and surface properties, and accordingly, it is often applied as a catalyst in the thermochemical conversion of biomass and tar removal directly.

10.4.1.1 The Application in Thermochemical Conversion of Biomass

Hydrogen production from biomass steam gasification with biochar as catalyst has been investigated in a two-stage fixed bed reactor by Yao *et al.*[74] The two stages are: (1) pyrolysis stage, where biomass is decomposed into char, volatiles and gases; and (2) reforming stage where volatiles/gases were reformed to produce gases like CO and H_2 in the presence of steam and catalyst. The commercial activated carbon and three kinds of biochar, namely wheat stalk char (WC), cotton stalk char (CC) and rice stalk char (RC), were loaded in the second stage to study their effects on the biomass steam gasification by comparing the biomass pyrolysis and steam reforming results. It was found that biochar has great influence on both the H_2 yield and H_2/CO_x ratio in the biomass pyrolysis. Compared to biomass pyrolysis without catalyst, 3–7 times higher H_2 yields were obtained for the biochar steam gasification. The H_2/CO_x ratio increased from the primary value of 0.28 to a high value of 2.94 for the wheat biochar. This gas composition of biomass gas is very beneficial for its application in alcohols and hydrocarbons synthesis because all the required H_2/CO ratios for these synthesis processes are above 2. The main reason for the promotion of gas production, especially H_2 yield, can be ascribed to the presence of AAEMs with high concentration in the biochars. There is about 19.0 wt% K in the WC sample and around 15.85 wt% Ca in the CC sample.

Ren *et al.*[75] studied the hydrocarbon and hydrogen-rich syngas production by microwave assisted biomass catalytic pyrolysis and bio-oil upgrading over biochar catalysts. Corn stover biochar produced from MWP at 650 °C and activated biochars were used as the catalysts. The biochar catalysts were found to remarkably enhance the yield of syngas enriched in H_2, CO, and CH_4. However, the bio-oil yield significantly decreased with the increase of the ratio of biochar to biomass. Surprisingly, the bio-oil quality was improved by biochar catalysis. The biochar is a good cost-competitive catalyst in biomass conversion and bio-oil upgrading.

The reaction mechanism of biomass pyrolysis catalyzed by biochar is shown in Figure 10.4.[75] The surface of the biochar catalyst is enriched in functional groups such as hydroxyl and carboxylic, thus a large amount of acid sites forms on the surface of biochar, which can catalyze the reactions such as dehydration and cracking. As for the production of hydrocarbons in the bio-oil from biomass pyrolysis, there are two main routes. One is from cellulose and hemicellulose and the other is from lignin. With the presence of biochar, the intermediate chemicals can be catalyzed to produce more aromatic hydrocarbons with further reactions. In addition, during the further reactions of intermediates such as dehydration, decarboxylation and decarbonylation, CO_x and H_2O are also formed, which contributes to the change of gas production. Simultaneously, the cracking of guaiacols leads to the promotion of CH_4 yield. Furthermore, the inter-reactions of light gaseous

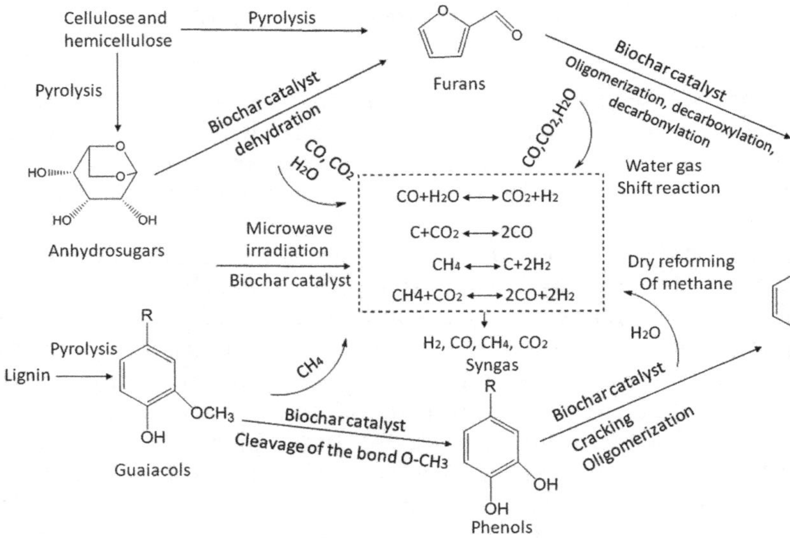

Figure 10.4 Reaction pathway of biomass catalytic pyrolysis and bio-oil upgrading for hydrocarbon and syngas production using biochar catalyst under microwave heating. Reproduced from ref. 75 with permission from The Royal Society of Chemistry.

molecules are another important reason for the promotion of syngas. Through the water gas shift and Boudouard reactions catalyzed by biochar, the yield of syngas increased obviously. The inherent AAEMs in biomass have also been suggested to enhance the water gas shift reaction and the decomposition of hydrocarbons during biomass pyrolysis.[76] In addition, transition metals are the other important elements for the promotion of the water gas shift reaction in the biochar.[77,78]

10.4.1.2 The Application in Tar Removal for Clean Syngas

Another direct utilization of biochar as catalyst is to removal the tar in the syngas from biomass pyrolysis or gasification. Syngas produced from gasification of biomass during the pyrolysis step contains a considerable amount of tar. Being a mixture of oxygenated hydrocarbons, phenolics, aromatic hydrocarbons and poly aromatic hydrocarbons, tar exhibits toxicity and is detrimental to downstream processes.[79,80] Therefore, the decomposition of tar is of great importance in gasification.

Bhandari *et al.*[81] compared the performance of original biochar and activated carbon derived from biochar for the removal of tar using toluene as a model compound in a fixed bed tubular catalytic reactor at temperatures of 700 and 800 °C. Biochar was generated using switch grass as a biomass feedstock. A KOH activation method was used to obtain activated carbon.

Both the original biochar and activated carbon were effective in tar removal. And the activated carbon catalysts resulted in higher (92%) toluene removal efficiency than biochar (79%). After activation, the residual metal element contents in activated carbon are much lower than that of the original biochar. However, activated carbon had a higher surface area of about 900 m^2 g^{-1} and pore volume of about 0.4 cm^3 g^{-1} while the surface area of original biochar is only about 60 m^2 g^{-1}. This indicates that the pore structure and surface properties of biochar have a great effect on the catalytic performance of biochar in the removal of tar. Pine bark biochar generated by slow pyrolysis (950 °C) was used as a low-cost catalyst to decompose toluene by Mani *et al.*[82] Fractional conversion of toluene with biochar as a catalyst was achieved up to 94% at 900 °C. Arrhenius analysis indicated that the activation energy (91 kJ mol^{-1}) of tar conversion over pinebark biochar was comparable to that of synthetic catalysts (*e.g.*, 80.24 kJ mol^{-1} for Ni/Mayenite and 196 kJ mol^{-1} for olivine) and only about 25% of that for thermal cracking. The catalytic performance of biochar can be attributed to the high BET surface area and mineral contents. Generally, biochar generated at high temperature has a significantly higher surface area compared to that generated by lower temperature pyrolysis. El-Rub *et al.*[83] compared biochar to other catalysts such as calcinated dolomite, olivine, commercial nickel catalyst, FCC catalyst, and biomass ash using the model tar compounds phenol and naphthalene in a fixed bed reactor. More than 98% phenol can be converted at 900 °C even without a catalyst. Nevertheless, only biochar catalyst has comparable performance for naphthalene conversion with commercial Ni catalyst. Therefore, biochar could be used as a catalyst with good performance in tar removal.

10.4.2 Biochar-supported Heterogeneous Catalysts

It is known that a well-developed pore structure and high surface area, which are likely to favor the uniform dispersion of active components and to stabilize these against sintering, are essential for a good catalyst support. Due to the high surface area and multi-level pore structure, activated biochar was used to prepare many kinds of catalysts and was applied in reactions such as methanation, Fischer–Tropsch synthesis (FTS) and hydrotreating of bio-oil.

Using methanation as a probe reaction, biochar was studied by Zhu *et al.* as a new promising catalyst support.[84] Biochar from rice husk pyrolysis was activated by a chemical and physical process. The activated biochar (ABC) showed graphite-like morphology and had a BET surface area of more than 1000 m^2 g^{-1}, which is about 25 times higher than that of the original BC. Unlike the commercial activated carbon with the BET surface area mostly contributed by the micropores, the microporous and mesoporous structures coexisted in ABC and its supported catalysts. This multi-level pore structure was considered to facilitate the mass transfer for catalytic reaction.[85,86]

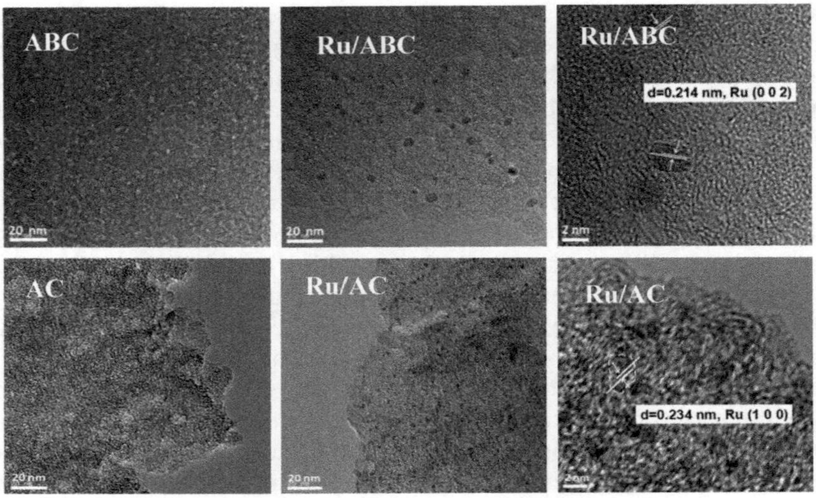

Figure 10.5 TEM and HRTEM images for ABC, Ru/ABC, AC and Ru/AC. Reproduced from ref. 84, http://dx.doi.org/10.1002/ese3.58[84] Copyright 2015 The Authors. Published under the terms of the CC BY 4.0 licence, https://creativecommons.org/licenses/by/4.0/.

Figure 10.5 compares the TEM images of bare and Ru loaded ABC and AC samples.[84] Large numbers of random pores were observed on the ABC, and Ru particles were uniformly distributed on the Ru/ABC catalysts with the particle size slightly larger than that of Ru/AC. As for the catalytic performance of the Ru/ABC catalyst in methanation, a high CH_4 selectivity of 98% and a CO conversion of 100% were obtained under the proper reaction conditions, which is superior or comparable to the conventional AC supported Ru catalyst. In addition, there was no deactivation observed over the ABC supported Ru catalyst for syngas methanation after as long as 100 h on stream. This indicated that the Ru/ABC catalyst showed an excellent catalytic activity and stability in the syngas methanation reaction.

It has been reported that the catalytic activity of the CO–CO_2 mixture in methanation is poor due to the presence of CO_2.[87] Wang *et al.* prepared Ru/ABC catalysts with lauan pyrolysis biochar as the support and applied them in the methanation of bio-syngas.[88] The conversion of CO and CO_2, as well as the CH_4 selectivity and yield, was low when using original bio-syngas as the feedstock because the H_2/C ratio was only 0.6 while the corresponding theoretical H_2/C is 3.3 for bio-syngas according to its CO and CO_2 concentrations. Subsequently, additional H_2 was supplemented to increase the H_2/C ratio in bio-syngas. As a result, using the 0.5 Ru/ABC catalyst at 693 K with bio-syngas (H_2/C = 4) gave a CO conversion of 97%, CO_2 conversion of 55% and CH_4 selectivity of 92%.

FTS is a well-known way to produce synthetic fuels from syngas. Because the product distribution follows the Anderson–Schulz–Flory (ASF) model in conventional FTS, using an efficient catalyst is key for the production of a specific range of products with high selectivity.[89] Yan *et al.*[90] prepared a carbon-encapsulated iron nanoparticles catalyst with biochar as the support material for the FTS of bio-syngas to produce liquid hydrocarbons. The biochar was obtained from a typical fast pyrolysis of pinewood. The catalyst was synthesized by an impregnation method followed by drying at 110 °C for a week to obtain iron-impregnated pine biochar and then by thermal treatment at 1000 °C for 1 h to obtain a core–shell structure catalyst. The FTS results indicate that the biochar-based iron nanoparticles catalyst has high activity in the conversion of bio-syngas and good selectivity towards liquid hydrocarbons. CO conversion of 95% and liquid hydrocarbons selectivity up to 70% were obtained by using this biochar-based catalyst and olefins are the dominant products in liquid hydrocarbons. In addition, over a 1500 h testing period, the catalysts showed very good stability against deactivation. They pointed out that the strong metal–carbon support interaction was the main reason for the good performance of the catalyst. After the thermal-treating at 1000 °C, the Fe_xC catalyst formed by the strong interaction between metal and carbon. Thus, the metal iron active sites were fully packed and prevented from sintering even under severe FTS conditions.

Hong *et al.*[91] prepared a new catalyst by introducing Ni onto the surface of biochar obtained from microalgae biomass pyrolysis and applied it in the hydrotreating of bio-oil for the oxygen and nitrogen removal. The results showed that the Ni/biochar catalyst could be very effective in the upgrading process to produce liquid products containing a large amount of hydrocarbons, especially *n*-heptadecane, which is a very important component of diesel fuel.

10.4.3 Biochar-based Solid Acid Catalyst

Biodiesel is a mixture of alkyl ester compounds produced from the reaction (transesterification or esterification) of alcohol with vegetable oil or animal fat composed of triglycerides and free fatty acids. Transesterification is a reaction in which the main component of fat and oil, triglyceride, reacts with alcohol over a catalyst producing fatty acid methyl ester and glycerol.[92] In esterification, free fatty acids (FFAs) and alcohol react to produce biodiesel and water.[93] Biochar material has high contents of oxygen (27–34 wt%) mostly in the form of acidic groups such as phenolic and carboxylic. These groups can improve the catalytic activity by increasing the acidity and improving the adsorption of reactant molecules on the catalyst surface. Recently, solid carbon catalysts derived directly from biomass *via* fast pyrolysis biochars were explored as biodiesel catalysts.

Dehkhoda *et al.*[94] prepared two carbon-based solid acid catalysts by sulfonating pyrolysis char with concentrated or fuming sulfuric acids. Both of the catalysts were activated chemically with KOH to increase the specific surface area and were applied to the transesterification of vegetable oil and the esterification of FFA. Concentrated sulfonated biochar exhibited high activity only for the esterification of free fatty acid, whereas fuming sulfonated char showed a higher transesterification activity than concentrated sulfonated biochar. Results showed the catalyst with the highest surface area and acid density had the highest catalytic activity for the production of biodiesel from canola oil in the presence of methanol as the reagent. They also found that the catalyst with a higher surface area showed higher transesterification activity among the catalysts with similar acid densities.

Kastner *et al.*[95] used the biochar generated from the pyrolysis of pelletized peanut hulls, pine logging residues, and wood chips at 400–500 °C as catalyst support. They sulfonated those biochars and commercial activated carbon using H_2SO_4 at 100, 150, and 200 °C for 12 h and using SO_3 gas at 23 °C. These sulfonated carbon materials were applied for the esterification of FFAs in vegetable oil and animal fat (5–15 wt% FFA) with methanol. The esterification of fatty acids for 30–60 min resulted in 90–100% conversion at 55–60 °C. When the catalysts were reused, the char sulfonated at 100 °C exhibited the highest reaction rate and the lowest conversion reduction. Therefore, biochar has great potential in developing reusable, porous carbon supported acid catalysts for biodiesel generation.

10.4.4 Biochar-based Electrochemical Catalyst

Carbonaceous materials with high electrical conductivity, porosity and stability can also be used as an electrode material in fuel cells or super-capacitors.[96] Biochars may effectively take the place of the commercial carbon-based catalysts in electrocatalytic applications. Major electrocatalysis applications of these biomass-derived carbons include methanol electrooxidation[97] and oxygen reduction reaction (ORR).[98]

Paulo *et al.*[99] prepared a modified carbon paste electrode (MCPE) using mercury ions (Hg^{2+}) supported on a biochar surface obtained from castor oil cake and used for electrode preparation. The proposed electrode was evaluated for preconcentration and determination of zinc(II) ions present in commercial samples (collyrium and ointment) by differential pulse anodic stripping voltammetry (DPASV). They found high analytical performance of the biochar/mercury based electrode applied for construction of the electroanalytical sensor. Using the best set of experimental conditions, a linear response for zinc(II) ions was observed over a concentration range of 5.0×10^{-7} to 3.0×10^{-5} mol L^{-1}.

Nieva Lobos *et al.*[100] evaluated the biochar produced by fast pyrolysis at 350 °C of untreated and acid-treated cellulose as a support for Cu–Ru@Pt core–shell nanoparticles (2.9–3.5 nm) for the electrooxidation of methanol in acid media. The nano-catalyst supported onto the biochar from acid-treated cellulose exhibited a high electroactive surface area (38 m^2 g^{-1}) and high performance in the methanol electrooxidation reaction, reaching a turnover number (the molecule number of methanol converted per Pt active site per second) value of 0.151 molec. (sites s)$^{-1}$ at 0.5 V.

Biochar can also be used as a low-cost anode material in a microbial fuel cell (MFC).[101] Tyler M. *et al.*[102] obtained graphitic biochar (BC) using high temperature gasification and alkaline post-treatment of wood-based biomass. The BC was evaluated as a manganese oxide electrocatalytic support (MnO_x/BC) by cyclic voltammetry. The results showed that MnO_x/BC had electrochemical features typical of MnO_2 with a current density of 0.9 mA cm^{-2}. BC showed satisfactory maximum power densities of 146.7 mW m^{-2} (BC) and 187.8 mW m^{-2} (MnO_x/BC), compared with Vulcan Carbon (VC) (156.8 mW m^{-2}) and manganese oxide VC composites (MnO_x/VC) (606.1 mW m^{-2}).

10.4.5 Biochar-based Photocatalyst

Current photocatalytic oxidation using commercial TiO_2 exhibited critical shortcomings such as quick recombination of electrons and electron holes, difficult separation of the nano to micro-scale TiO_2 powders after treatment, and low reactivity with low concentrations of contaminants in water. The quick recombination of electrons and electron holes lowers photocatalytic efficiency while the separation of fine TiO_2 powder after treatment increases energy consumption.[103]

To overcome these limitations, the present study suggested a biochar-supported TiO_2 (TiO_2/biochar) as an effective photocatalyst to eliminate sulfamethoxazole (SMX) in water. The biochar had excellent conductivity, which would lower the recombination rate of electrons and electron holes during the photocatalysis for enhancing photocatalytic oxidation. Efficient attachment of TiO_2 onto the biochar was allowed due to the large specific surface area, rich functional groups and easy surface modification of the biochar. The biochar-supported TiO_2 was also easily separated and reused after treatment of SMX in water. In addition to the support for TiO_2, the biochar demonstrated high adsorption capacity of SMX, which allowed SMX to undergo phase transfer from a dilute solution onto the biochar surface where the SMX was immobilized and concentrated. This led to much more efficient photocatalytic oxidation of SMX rather than in the dilute aqueous phase. The enhanced photocatalytic performance of TiO_2/carbon composites has been mainly attributed to the porosity of the carbon support, and/or strong interfacial electronic effects.[104]

Matos[105] found that the photocatalytic activity for hydrogen photoproduction under visible light irradiation increased by a factor of up to 3 times on TiO_2/biochars compared to that obtained on neat commercial TiO_2. The results reinforce the suggestion that micropores in carbon materials play the role of microreactors and therefore the free-semiconductor photocatalytic activity by carbon materials under solar irradiation is possible and therefore has important potential applications.

Many studies have shown that an introduction of oxygen groups to carbon surfaces not only plays a photo-assisting role and enhances TiO_2

photoactivity in the degradations of various organic contaminants[106] but also enhances hydrogen photoproduction by water splitting under artificial solar conditions.

Acknowledgements

The authors are grateful for the financial support from the National Natural Science Foundation of China (51661145011,51476142), the National Science and Technology Supporting Plan Through Contract (2015BAD15B06), the Foundation of the State Key Laboratory of Clean Energy Utilization (ZJU-CEU2016001). The authors also sincerely appreciate Shi Yin, Xiaoliu Wang, Haizhou Lin and Gongxin Dai for their great help on literature collection, illustration arrangement and text proofreading.

References

1. D. Chen, X. Chen, J. Sun, Z. Zheng and K. Fu, *Bioresour. Technol.*, 2016, **216**, 629.
2. D. Chen, Y. Li, K. Cen, M. Luo, H. Li and B. Lu, *Bioresour. Technol.*, 2016, **218**, 780.
3. S. Nanda, A. K. Dalai, F. Berruti and J. A. Kozinski, *Waste Biomass Valorization*, 2016, **7**, 201.
4. K. Z. Qian, A. Kumar, H. L. Zhang, D. Bellmer and R. Huhnke, *Renewable Sustainable Energy Rev.*, 2015, **42**, 1055.
5. J. S. Cha, S. H. Park, S. C. Jung, C. Ryu, J. K. Jeon, M. C. Shin and Y. K. Park, *J. Ind. Eng. Chem.*, 2016, **40**, 1.
6. D. Liu, S. Yu, Y. Shen, H. Chen, Z. Shen, S. Zhao, S. Fu, Y. Yu and B. Bao, *Ind. Eng. Chem. Res.*, 2015, **54**, 1697.
7. M. Ahmad, A. U. Rajapaksha, J. E. Lim, M. Zhang, N. Bolan, D. Mohan, M. Vithanage, S. S. Lee and Y. S. Ok, *Chemosphere*, 2014, **99**, 19.
8. T. Kan, V. Strezov and T. J. Evans, *Renewable Sustainable Energy Rev.*, 2016, **57**, 1126.
9. M. Tripathi, J. N. Sahu and P. Ganesan, *Renewable Sustainable Energy Rev.*, 2016, **55**, 467.
10. H. Chen, G. Lin, Y. Chen, W. Chen and H. Yang, *Energy Fuels*, 2016, **30**, 1579.
11. Y. Chen, H. Yang, X. Wang, S. Zhang and H. Chen, *Bioresour. Technol.*, 2012, **107**, 411.
12. Y. Q. Chen, H. P. Yang, X. H. Wang, W. Chen and H. P. Chen, *Energy Fuels*, 2016, **30**, 414.
13. D. Chen, J. Mei, H. Li, Y. Li, M. Lu, T. Ma and Z. Ma, *Bioresour. Technol.*, 2017, **228**, 62.
14. D. Chen, K. Cen, X. Jing, J. Gao, C. Li and Z. Ma, *Bioresour. Technol.*, 2017, **233**, 150.

15. Z. Q. Ma, D. Y. Chen, J. Gu, B. F. Bao and Q. S. Zhang, *Energy Convers. Manage.*, 2015, **89**, 251.
16. L. Frost, E. Elangovan and J. Hartvigsen, *Can. J. Chem. Eng.*, 2016, **94**, 636.
17. Y. Zhang, Z. Ma, Q. Zhang, J. Wang, Q. Ma, Y. Yang, X. Luo and W. Zhang, *BioResources*, 2017, **12**, 4652.
18. D. Angin, *Bioresour. Technol.*, 2013, **128**, 593.
19. J. Zhang, J. Liu and R. L. Liu, *Bioresour. Technol.*, 2015, **176**, 288.
20. D. Chen, J. Zhou and Q. Zhang, *Bioresour. Technol.*, 2014, **169**, 313.
21. G. Cimo, J. Kucerik, A. E. Berns, G. E. Schaumann, G. Alonzo and P. Conte, *J. Agric. Food Chem.*, 2014, **62**, 1912.
22. H. J. Park, Y.-K. Park and J. S. Kim, *Fuel Process. Technol.*, 2008, **89**, 797.
23. J. M. Encinar, F. J. Beltrán, A. Bernalte, A. Ramiro and J. F. González, *Biomass Bioenergy*, 1996, **11**, 397.
24. A. Demirbas, *J. Anal. Appl. Pyrolysis*, 2004, **72**, 243.
25. J. A. Ruiz, M. C. Juárez, M. P. Morales, P. Muñoz and M. A. Mendívil, *Renewable Sustainable Energy Rev.*, 2013, **18**, 174.
26. M. Asadullah, *Renewable Sustainable Energy Rev.*, 2014, **29**, 201.
27. Z. Q. Ma, Y. M. Zhang, Q. S. Zhang, Y. B. Qu, J. B. Zhou and H. F. Qin, *Energy*, 2012, **46**, 140.
28. J. R. Kastner, S. Mani and A. Juneja, *Fuel Process. Technol.*, 2015, **130**, 31.
29. C. E. Brewer, K. Schmidt-Rohr, J. A. Satrio and R. C. Brown, *Environ. Prog. Sustainable Energy*, 2009, **28**, 386.
30. J. Li, J. J. Dai, G. Q. Liu, H. D. Zhang, Z. P. Gao, J. Fu, Y. F. He and Y. Huang, *Biomass Bioenergy*, 2016, **94**, 228.
31. D. J. Macquarrie, J. H. Clark and E. Fitzpatrick, *Biofuels, Bioprod. Biorefin.*, 2012, **6**, 549.
32. R. Luque, J. A. Menendez, A. Arenillas and J. Cot, *Energy Environ. Sci.*, 2012, **5**, 5481.
33. A. Mamaeva, A. Tahmasebi, L. Tian and J. Yu, *Bioresour. Technol.*, 2016, **211**, 382.
34. A. A. Salema and F. N. Ani, *Bioresour. Technol.*, 2011, **102**, 3388.
35. B. A. Mohamed, C. S. Kim, N. Ellis and X. Bi, *Bioresour. Technol.*, 2016, **201**, 121.
36. Y. Fernández and J. A. Menéndez, *J. Anal. Appl. Pyrolysis*, 2011, **91**, 316.
37. L. Zhu, H. Lei, L. Wang, G. Yadavalli, X. Zhang, Y. Wei, Y. Liu, D. Yan, S. Chen and B. Ahring, *J. Anal. Appl. Pyrolysis*, 2015, **115**, 149.
38. Y. Yu, J. Yu, B. Sun and Z. Yan, *J. Anal. Appl. Pyrolysis*, 2014, **106**, 86.
39. S. K. Hoekman, A. Broch and C. Robbins, *Energy Fuels*, 2011, **25**, 1802.
40. A. Funke and F. Ziegler, *Biofuels, Bioprod. Biorefin.*, 2010, **4**, 160.
41. W. Jiaoyan, L. Wei, W. Qiong and L. Shouxin, *Prog. Chem.*, 2015, **28**, 121.
42. J. A. Libra, K. S. Ro, C. Kammann, A. Funke, N. D. Berge, Y. Neubauer, M.-M. Titirici, C. Fühner, O. Bens and J. Kern, *Biofuels*, 2011, **2**, 71.
43. H. S. Kambo and A. Dutta, *Renewable Sustainable Energy Rev.*, 2015, **45**, 359.

44. M. T. Reza, J. Andert, B. Wirth, D. Busch, J. Pielert, J. G. Lynam and J. Mumme, *Appl. Bioenergy*, 2014, **1**, 71.
45. Y. Gao, X. Wang, J. Wang, X. Li, J. Cheng, H. Yang and H. Chen, *Energy*, 2013, **58**, 376.
46. Z. Liu, A. Quek, S. K. Hoekman and R. Balasubramanian, *Fuel*, 2013, **103**, 943.
47. R. K. Garlapalli, B. Wirth and M. T. Reza, *Bioresour. Technol.*, 2016, **220**, 168.
48. B. Hu, K. Wang, L. Wu, S.-H. Yu, M. Antonietti and M.-M. Titirici, *Adv. Mater.*, 2010, **22**, 813.
49. S. Nizamuddin, H. A. Baloch, G. J. Griffin, N. M. Mubarak, A. W. Bhutto, R. Abro, S. A. Mazari and B. S. Ali, *Renewable Sustainable Energy Rev.*, 2017, **73**, 1289.
50. Z. G. Liu and R. Balasubramanian, *Appl. Energy*, 2014, **114**, 857.
51. G. K. Parshetti, S. Kent Hoekman and R. Balasubramanian, *Bioresour. Technol.*, 2013, **135**, 683.
52. M. A. Yahya, Z. Al-Qodah and C. Z. Ngah, *Renewable Sustainable Energy Rev.*, 2015, **46**, 218.
53. F. Rodriguez-Reinoso, M. Molina-Sabio and M. Gonzalez, *Carbon*, 1995, **33**, 15.
54. J. Gu, J. Zhou, H. Ma, M. Ma and M. Xing, *Trans. Chin. Soc. Agric. Eng.*, 2015, **31**, 233.
55. J. F. González, S. Román, J. M. Encinar and G. Martínez, *J. Anal. Appl. Pyrolysis*, 2009, **85**, 134.
56. M. Ncibi, V. Jeanne-Rose, B. Mahjoub, C. Jean-Marius, J. Lambert, J. Ehrhardt, Y. Bercion, M. Seffen and S. Gaspard, *J. Hazard. Mater.*, 2009, **165**, 240.
57. S. Román, J. González, C. González-García and F. Zamora, *Fuel Process. Technol.*, 2008, **89**, 715.
58. G.-G. Choi, S.-J. Oh, S.-J. Lee and J.-S. Kim, *Bioresour. Technol.*, 2015, **178**, 99.
59. J. S. Cha, J.-C. Choi, J. H. Ko, Y.-K. Park, S. H. Park, K.-E. Jeong, S.-S. Kim and J.-K. Jeon, *Chem. Eng. J.*, 2010, **156**, 321.
60. T. Tay, S. Ucar and S. Karagöz, *J. Hazard. Mater.*, 2009, **165**, 481.
61. D. Angin, E. Altintig and T. E. Kose, *Bioresour. Technol.*, 2013, **148**, 542.
62. B. Shen, J. Chen, S. Yue and G. Li, *Fuel*, 2015, **156**, 47.
63. F.-S. Zhang, J. O. Nriagu and H. Itoh, *Water Res.*, 2005, **39**, 389.
64. J. H. Ko, Y.-H. Kwak, K.-S. Yoo, J.-K. Jeon, S. H. Park and Y.-K. Park, *J. Mater. Cycles Waste Manage.*, 2011, **13**, 173.
65. A. Ros, M. A. Lillo-Ródenas, E. Fuente, M. A. Montes-Morán, M. J. Martín and A. Linares-Solano, *Chemosphere*, 2006, **65**, 132.
66. A. J. Romero-Anaya, M. A. Lillo-Ródenas, C. Salinas-Martínez de Lecea and A. Linares-Solano, *Carbon*, 2012, **50**, 3158.
67. L. J. Konwar, J. Boro and D. Deka, *Renewable Sustainable Energy Rev.*, 2014, **29**, 546.
68. A. E. Creamer, B. Gao and M. Zhang, *Chem. Eng. J.*, 2014, **249**, 174.

69. S. Niaz, T. Manzoor and A. H. Pandith, *Renewable Sustainable Energy Rev.*, 2015, **50**, 457.
70. Y. A. Alhamed, *J. King Abdulaziz Univ., Sci.*, 2006, **17**, 75.
71. O. Ioannidou and A. Zabaniotou, *Renewable Sustainable Energy Rev.*, 2007, **11**, 1966.
72. S. Wang and Z. Luo, *Pyrolysis Biomass*, De Gruyter, 2017.
73. H. Xiong, M. A. M. Motchelaho, M. Moyo, L. L. Jewell and N. J. Coville, *J. Catal.*, 2011, **278**, 26.
74. D. Yao, H. Qiang, D. Wang, H. Yang, C. Wu, X. Wang and H. Chen, *Bioresour. Technol.*, 2016, **216**, 159.
75. S. Ren, H. Lei, L. Wang, Q. Bu, S. Chen and J. Wu, *RSC Adv.*, 2014, **4**, 10731.
76. S. Hu, L. Jiang, Y. Wang, S. Su, L. S. Sun, B. Y. Xu, L. M. He and J. Xiang, *Bioresour. Technol.*, 2015, **192**, 23.
77. H. Lei, S. Ren and J. Julson, *Energy Fuels*, 2009, **23**, 3254.
78. W. H. Chen, J. G. Jheng and A. B. Yu, *Int. J. Hydrogen Energy*, 2008, **33**, 4789.
79. A. R. Mudinoor, D. Bellmer, L. Marin, A. Kumar and R. Huhnke, *Trans. ASABE*, 2011, **54**, 1819.
80. S. Anis and Z. A. Zainal, *Renewable Sustainable Energy Rev.*, 2011, **15**, 2355.
81. P. N. Bhandari, A. Kumar, D. D. Bellmer and R. L. Huhnke, *Renewable Energy*, 2014, **66**, 346.
82. S. Mani, J. R. Kastner and A. Juneja, *Fuel Process. Technol.*, 2013, **114**, 118.
83. Z. A. El-Rub, E. A. Bramer and G. Brem, *Fuel*, 2008, **87**, 2243.
84. L. Zhu, S. Yin, Q. Yin, H. Wang and S. Wang, *Energy Sci. Eng.*, 2015, **3**, 126.
85. D. Özçi̇men and A. Ersoy-Meri̇çboyu, *Renewable Energy*, 2010, **35**, 1319.
86. V. R. Surisetty, A. K. Dalai and J. Kozinski, *Appl. Catal., A*, 2011, **393**, 50.
87. R. Razzaq, H. Zhu, J. Li, U. Muhammad, C. Li and S. Zhang, *Ind. Eng. Chem. Res.*, 2013, **52**, 2247.
88. S. Wang, H. Wang, Q. Yin, L. Zhu and S. Yin, *New J. Chem.*, 2014, **38**, 4471.
89. S. Wang, Q. Yin, J. Guo, B. Ru and L. Zhu, *Fuel*, 2013, **108**, 597.
90. Q. Yan, C. Wan, J. Liu, J. Gao, F. Yu, J. Zhang and Z. Cai, *Green Chem.*, 2013, **15**, 1631.
91. K. D. N. Hong, V. V. Pham and T. D. Hai, *Catal. Lett.*, 2016, **146**, 1.
92. Y. L. Edgar Lotero, D. E. Lopez, K. Suwannakarn, D. A. Bruce and J. G. Goodwin Jr, *Ind. Eng. Chem. Res.*, 2005, **44**, 5353.
93. A. M. Dehkhoda and N. Ellis, *Catal. Today*, 2013, **207**, 86.
94. A. M. Dehkhoda, A. H. West and N. Ellis, *Appl. Catal., A*, 2010, **382**, 197.
95. J. R. Kastner, J. Miller, D. P. Geller, J. Locklin, L. H. Keith and T. Johnson, *Catal. Today*, 2012, **190**, 122.
96. K. Qian, A. Kumar, H. Zhang, D. Bellmer and R. Huhnke, *Renewable Sustainable Energy Rev.*, 2015, **42**, 1055.

97. X. Zhao, J. Zhu, L. Liang, J. Liao, C. Liu and W. Xing, *J. Mater. Chem.*, 2012, **22**, 19718.
98. C. Han, S. Wang, J. Wang, M. Li, J. Deng, H. Li and Y. Wang, *Nano Res.*, 2014, **7**, 1809.
99. P. R. d. Oliveira, A. C. Lamy-Mendes, J. L. Gogola, A. S. Mangrich, L. H. Marcolino Jr and M. F. Bergamini, *Electrochim. Acta*, 2015, **151**, 525.
100. M. L. Nieva Lobos, J. M. Sieben, V. Comignani, M. Duarte, M. A. Volpe and E. L. Moyano, *Int. J. Hydrogen Energy*, 2016, **41**, 10695.
101. K. Ganesh and J. R. Jambeck, *Bioresour. Technol.*, 2013, **139**, 383–387.
102. T. M. Huggins, J. J. Pietron, H. Wang, Z. J. Ren and J. C. Biffinger, *Bioresour. Technol.*, 2015, **195**, 147.
103. J. R. Kim and E. Kan, *J. Environ. Manage.*, 2016, **180**, 94.
104. C. Ma, D. Wu, X. Yao, X. Liu, M. Wei, Y. Liu, Z. Ma, P. Huo and Y. Yan, *J. Alloys Compd.*, 2017, **712**, 486.
105. J. Matos, *Top. Catal.*, 2015, **59**, 394.
106. F. J. López-Tenllado, A. Marinas, F. J. Urbano, J. C. Colmenares, M. C. Hidalgo, J. M. Marinas and J. M. Moreno, *Appl. Catal., B*, 2012, **128**, 150.

Subject Index